T0206112

Bayesian Psychometric Modeling

Chapman & Hall/CRC
Statistics in the Social and Behavioral Sciences Series

Aims and scope

Large and complex datasets are becoming prevalent in the social and behavioral sciences and statistical methods are crucial for the analysis and interpretation of such data. This series aims to capture new developments in statistical methodology with particular relevance to applications in the social and behavioral sciences. It seeks to promote appropriate use of statistical, econometric and psychometric methods in these applied sciences by publishing a broad range of reference works, textbooks and handbooks.

The scope of the series is wide, including applications of statistical methodology in sociology, psychology, economics, education, marketing research, political science, criminology, public policy, demography, survey methodology and official statistics. The titles included in the series are designed to appeal to applied statisticians, as well as students, researchers and practitioners from the above disciplines. The inclusion of real examples and case studies is therefore essential.

Published Titles

Analyzing Spatial Models of Choice and Judgment with R
David A. Armstrong II, Ryan Bakker, Royce Carroll, Christopher Hare, Keith T. Poole, and Howard Rosenthal

Analysis of Multivariate Social Science Data, Second Edition
David J. Bartholomew, Fiona Steele, Irini Moustaki, and Jane I. Galbraith

Latent Markov Models for Longitudinal Data
Francesco Bartolucci, Alessio Farcomeni, and Fulvia Pennoni

Statistical Test Theory for the Behavioral Sciences
Dato N. M. de Gruijter and Leo J. Th. van der Kamp

Multivariable Modeling and Multivariate Analysis for the Behavioral Sciences
Brian S. Everitt

Multilevel Modeling Using R
W. Holmes Finch, Jocelyn E. Bolin, and Ken Kelley

Ordered Regression Models: Parallel, Partial, and Non-Parallel Alternatives
Andrew S. Fullerton and Jun Xu

Bayesian Methods: A Social and Behavioral Sciences Approach, Third Edition
Jeff Gill

Multiple Correspondence Analysis and Related Methods
Michael Greenacre and Jorg Blasius

Applied Survey Data Analysis
Steven G. Heeringa, Brady T. West, and Patricia A. Berglund

Informative Hypotheses: Theory and Practice for Behavioral and Social Scientists
Herbert Hoijtink

Generalized Structured Component Analysis: A Component-Based Approach to Structural Equation Modeling
Heungsun Hwang and Yoshio Takane

Bayesian Psychometric Modeling
Roy Levy and Robert J. Mislevy

Statistical Studies of Income, Poverty and Inequality in Europe: Computing and Graphics in R Using EU-SILC
Nicholas T. Longford

Foundations of Factor Analysis, Second Edition
Stanley A. Mulaik

Linear Causal Modeling with Structural Equations
Stanley A. Mulaik

Age–Period–Cohort Models: Approaches and Analyses with Aggregate Data
Robert M. O'Brien

Handbook of International Large-Scale Assessment: Background, Technical Issues, and Methods of Data Analysis
Leslie Rutkowski, Matthias von Davier, and David Rutkowski

Generalized Linear Models for Categorical and Continuous Limited Dependent Variables
Michael Smithson and Edgar C. Merkle

Incomplete Categorical Data Design: Non-Randomized Response Techniques for Sensitive Questions in Surveys
Guo-Liang Tian and Man-Lai Tang

Handbook of Item Response Theory, Volume 1: Models
Wim J. van der Linden

Handbook of Item Response Theory, Volume 2: Statistical Tools
Wim J. van der Linden

Handbook of Item Response Theory, Volume 3: Applications
Wim J. van der Linden

Computerized Multistage Testing: Theory and Applications
Duanli Yan, Alina A. von Davier, and Charles Lewis

Chapman & Hall/CRC
Statistics in the Social and Behavioral Sciences Series

Bayesian Psychometric Modeling

Roy Levy
Arizona State University
Tempe, Arizona, USA

Robert J. Mislevy
Educational Testing Service
Princeton New Jersey, USA

CRC Press is an imprint of the
Taylor & Francis Group, an **informa** business
A CHAPMAN & HALL BOOK

CRC Press
Taylor & Francis Group
6000 Broken Sound Parkway NW, Suite 300
Boca Raton, FL 33487-2742

First issued in paperback 2020

ISBN-13: 978-1-4398-8467-6 (hbk)
ISBN-13: 978-0-367-73709-2 (pbk)

Library of Congress Cataloging-in-Publication Data

Names: Levy, Roy (Statistics professor)
Title: Bayesian psychometric modeling / Roy Levy and Robert J. Mislevy.
Description: Boca Raton : Taylor & Francis Group, 2016. | Series: Chapman & Hall/CRC statistics in the social and behavioral sciences | Includes bibliographical references and index.
Identifiers: LCCN 2015045985 | ISBN 9781439884676 (alk. paper)
Subjects: LCSH: Psychometrics--Mathematical models. | Bayesian statistical decision theory.
Classification: LCC BF39.2.B39 L49 2016 | DDC 150.1/519542--dc23
LC record available at http://lccn.loc.gov/2015045985

Visit the Taylor & Francis Web site at
http://www.taylorandfrancis.com

and the CRC Press Web site at
http://www.crcpress.com

To Paige, who continues to make all that I do possible, and to Ella, Adrian, and Skylar, who have joined Paige in making it all worthwhile.—Roy

To Robbie, still my best friend after years of this kind of thing, and to Jessica and Meredith, who must not think this kind of thing is so bad because they do it too.—Bob

Contents

Section II Psychometrics

Preface

Context and Aims

Psychometrics concerns the principles and practices of reasoning in assessment. As a discipline, it came of age at a time marked by the dominance of frequentist approaches to probability, statistical modeling, and inference. Despite this backdrop, Bayesian methods have enjoyed widespread acceptance as solutions to certain problems in psychometrics. Nevertheless, as of this writing, *conventional* psychometric practice—fitting models, estimating parameters, evaluating model-data fit, using results to make or equate tests, evaluate fairness, and so on—employs mainly frequentist approaches. The vast majority of psychometric texts couch most if not all of their presentations in a frequentist paradigm (Bartholomew, Knott, & Moustaki, 2011; Bollen, 1989; Collins & Lanza, 2010; Dayton, 1999; De Ayala, 2009; Embretson & Reise, 2000; Hambleton & Swaminathan, 1985; Kline, 2010; McDonald, 1999; Mulaik, 2009; Rupp, Templin, & Henson, 2010; Skrondal & Rabe-Hesketh, 2004; van der Linden & Hambleton, 1997). Similarly, much of quantitative training in psychometrics and related methods are dominated by frequentist approaches to inference, with training in Bayesian methods being rather rare (Aiken, West, & Millsap, 2008).

Bayesian approaches to psychometrics and related latent variable modeling scenarios have attracted more attention recently, aided in part by shifts in the larger statistical landscape regarding the viability of conventional approaches to inference (Clark, 2005; Goodman, 2008; Kruschke, Aguinis, & Joo, 2012; Rodgers, 2010; Wagenmakers, 2007) and remarkable advances in statistical computing (Brooks, Gelman, Jones, & Meng, 2011; Gilks, Richardson, & Spiegelhalter, 1996b) that have opened up new possibilities for Bayesian psychometrics (Levy, 2009). McGrayne (2011) provided a nontechnical account of the history of Bayesian inference, and the current crest in interest.

The maturation of Bayesian psychometrics is in part reflected by the presence of chapters on such models in general Bayesian statistical texts (Congdon, 2006; Jackman, 2009; Kaplan, 2014), the inclusion of Bayesian methods as part of conventional practice (e.g., Bartholomew et al., 2011; Skrondal & Rabe-Hesketh, 2004), and even more strongly in recent textbook length treatments of Bayesian approaches to psychometric and related latent variable modeling paradigms, such as Lee's (2007) *Structural equation modeling: A Bayesian approach*; Fox's (2010) *Bayesian item response modeling: Theory and applications*; and Almond, Mislevy, Steinberg, Williamson, and Yan's (2015) *Bayesian networks in educational assessment*.

The present effort operates in this lineage, advancing a Bayesian perspective on psychometrics as an alternative to conventional approaches that is not only viable, but in many respects preferable. More specifically, we believe that Bayesian approaches offer distinct and profound advantages in achieving many of the goals of psychometrics. There are lots of things we do in psychometrics. Some involve comparing individuals or groups to standards or to each other, or deciding what a student should work on next, either inside or outside of the formal boundaries of the assessment. Other activities include designing tasks, investigating their psychometric properties for populations of examinees, creating tests out of many tasks, and equating or linking different tests. We also examine threats to inference, develop and connect constructs, and deal issues such as those with missing data (planned and unplanned), all in the service of characterizing an existing assessment or informing future assessment efforts. In addition, psychometrics can aid in checking and revising beliefs about the nature of the domain, how people acquire and synthesize proficiency in the domain, and how their behavior reflects that. We touch on some of these, in particular many that have traditionally been associated with the use of statistical models. Our position is that Bayesian approaches to reasoning, including but not limited to the use of full probability models, offer considerable advantages in many if not all of these sorts of applications. A central purpose of this book is to document those approaches and describe their advantages.

In advocating for Bayesian psychometrics, the *why* is just as important as the *how*. We aim to show both how to do psychometrics using Bayesian methods and why many of the activities in psychometrics align with if not call for Bayesian thinking. To be sure, we aim for the reader to come away knowing how to use Bayesian methods to answer key questions such as "Now that we have conducted an assessment, what do we think about this examinee?" But we also aim for the reader to come away seeing why that question quite naturally lends itself to Bayesian inference. Broadening this out, we argue that a Bayesian perspective on inference seamlessly aligns with the goals of psychometrics and related activities in assessment. As a result, adopting a Bayesian approach can aid in unifying seemingly disparate—and sometimes conflicting—ideas and activities in psychometrics. In short, a Bayesian approach allows for one to adopt a single coherent perspective that provides conceptually and practically a broader set of tools and procedures to get work done.

This is a statistics book, but one where the focus is on concepts and ideas surrounding psychometrics, assessment, and Bayesian inference. Mathematical terminology and expressions abound, and although we do some mathematics of probability and statistics, our aim is to do so at a level that illuminates concepts. Along the way we do some history, etymology, and philosophy. Our forays into the first two are decidedly of the armchair variety. For the last, we believe we are on more rigorous footing, although we do not propose that this is a comprehensive account of the issues involved.

It is not our aim, then, to present a balanced approach of all Bayesian and frequentist approaches to modeling and estimation in psychometrics; that would take a book several times as long. We are trying to present a particular Bayesian way of thinking, for a reader who already knows something about psychometrics and more familiar inferential approaches from frequentist and perhaps some Bayesian perspectives. Sometimes particular methods from one of those approaches are more efficient in particular applied problems, and in fact we use them ourselves at times in our own work. We do note some of these perspectives and methods along the way, when they add insights, bring out connections, or are important in the historical development of popular psychometric practices.

In the chapter on the normal model, for example, the reader already knows how to estimate a mean and standard deviation, understands the procedures in terms of least squares and probably maximum likelihood, and maybe even conjugate Bayesian analysis. Our rather long discussion on this basic model, including details of Markov chain Monte Carlo (MCMC) estimation that are not at all necessary if one only wants to know about the first two moments of a given dataset, is meant to help the reader see this familiar problem from a less familiar perspective: in terms of what is known, what is unknown, how they are related in terms of distributions, and how this knowledge can be visualized directly in terms of distributions and samples from them. Seeing inference in the normal model from this perspective adds a considerable insight and a unified way of thinking that will reveal itself in the subsequent presentations of factor analysis, item response theory, Bayes nets, and other psychometric models that are historically seen as more different than we think they ought to be.

Organization of the Book

This book is organized into two parts, followed by two appendices. Each of the two parts is prefaced by a short introduction that lays out the topics of the coming chapters. Briefly, Chapters 1 through 6 comprise the first part of the book, where foundational principles and statistical models are introduced. This is not intended to serve the purposes of a full introductory textbook on Bayesian procedures. It is a selected treatment of selected topics, oriented toward how they will be called into service in Chapters 7 through 15 that comprise the second part of the book where we focus more directly on psychometrics. We have endeavored to cover many of the popular psychometric models, and in doing so we have traded depth for breadth. Nevertheless, we aim to provide a launching point for readers who would like more depth. To this end, we list supporting references throughout, and in many of the chapters the concluding section contains a bibliographic note where readers may find additional references on these topics, including references to extensions beyond what is covered in the chapter. The first appendix provides details on the full conditional distributions presented in Chapters 6, 8, 9, 11, and 13. The second appendix offers a brief summary of the probability distributions used in the book.

Terminology and Notation

Most of the language and examples used, and applications covered, come from educational assessment. There is, however, a wider world of psychometrics, as the ubiquity of measurement error and the generality of probabilistic models have led to the wide application of psychometric models throughout education, the social sciences, and the life sciences. Importantly, the principles of assessment, psychometrics, and Bayesian modeling cut across disciplines. What we have to say about the features of the models primarily in educational assessment applies with little difference in other disciplines.

We use the notation $p(\cdot)$ to refer to (a) the probability of an event, (b) the probability mass function for a discrete random variable, or (c) the probability density function for a continuous random variable. In the latter case, we use the terms "distribution" and "density" interchangeably. We hope that the usage is clear from context, and that the use of a simple notation for these variations on the probability concept allows us to better focus on conceptual variations that serve our purposes. Chief among them is conditioning notation. We write $p(B \mid A)$ to refer to the conditional probability for B given A. When necessary, we use $P(\cdot)$ to denote the probability that a discrete variable takes on a particular value as a function of other entities.

We primarily use x to denote observable data. We use the subscript i to refer to examinees (persons, subjects, and cases) and j to refer to observables. Observables are the dependent variables in psychometrics, the variables we are trying to model in terms of variables for properties of examinees and observational situations. (For familiar tests, e.g., the observable variables in classical test theory and often in factor analysis are test scores; in item response theory and latent class analysis, the observables are item scores, which usually correspond 1-1 to test items.) We use bold typeface to represent collections, vectors, or matrices, with subscripts indicating what the collection is specific to. For example, x_{ij} refers to the data from examinee i for observable j, $\mathbf{x}_i = (x_{i1},\ldots, x_{iJ})$ is the collection of J such observables that comprise the data for examinee i, and $\mathbf{x} = (\mathbf{x}_1,\ldots, \mathbf{x}_n)$ is the full collection of data from n examinees.

When using a standard distribution, we also use notation based on its name to communicate as much. For example, if x has a normal distribution with mean μ and variance σ^2, we write $x \sim N(\mu,\sigma^2)$ or $p(x) = N(\mu,\sigma^2)$. In some cases, we write the arguments of the distribution with subscripts to indicate what the distribution refers to. We routinely highlight the dependence structures of random variables, for example, writing the preceding as $x \mid \mu,\sigma^2 \sim N(\mu,\sigma^2)$ and $p(x \mid \mu,\sigma^2) = N(\mu,\sigma^2)$, respectively, to highlight that what is on the left side of the conditioning bar depends on what is on the right side of the conditioning bar. The entities conditioned on may be other random variables or they may be fixed quantities, such as values known or chosen by the analyst.

We have attempted to use notational schemes that are conventional in the different areas of statistics and psychometrics. Rather than advance a general notation that cuts across these areas, we have attempted to use notational schemes that are commonly found in each tradition, that is, we use typical regression notation in Chapter 6, factor analysis notation in Chapter 9, item response theory notation in Chapter 11, and so on. Table 7.2 summarizes the notation used in the psychometric models that comprise the second part of the book.

As a consequence, several notational elements are reused. For example, α is used as a parameter in the beta and Dirichlet distributions, as the acceptance probability in Metropolis and Metropolis samplers, and as a transformed discrimination parameter in item response theory. We hope that the intended interpretation is clear from the context.

A particularly acute challenge arises with the use of θ, which is commonly used to refer to a generic parameter in statistical texts, and a latent variable (or person parameter) in certain psychometric traditions. We attempt to participate in both traditions. We use θ to refer to a generic parameter as we develop core Bayesian ideas in the first part of the book. When we pivot to psychometrics beginning with Chapter 7, we use θ to refer to a latent variable, and then use it as such in several of the subsequent chapters. Again, we hope that the intended interpretation is clear from the context.

Examples and Software

Throughout the book, we illustrate the procedures using examples, primarily from educational assessments. When fitting Bayesian models, we primarily use WinBUGS (Spiegelhalter, Thomas, Best, & Lunn, 2007), which has several advantages that recommend it for our purposes. First, it is flexible enough to accommodate all of the models we cover in the book. Second, the WinBUGS code for the model closely corresponds to the mathematical and graphical expressions of the models we use in this book, a point we illustrate and elaborate on in Chapter 2. Third, it can be called from other software packages (see http://www.mrc-bsu.cam.ac.uk/software/bugs/calling-winbugs-1-4-from-other-programs/), so that users with varied preferences for statistical software may integrate WinBUGS into their toolkit of statistical software.

In addition, we use the Netica software package (Norsys, 1999–2012) to illustrate the use of Bayesian networks in Chapter 14. In preparing the examples, we have also made use of the R statistical programming environment (R Core Team, 2014) for certain analyses, including several packages: BaM (Gill, 2012), coda (Plummer, Best, Cowles, & Vines, 2006), R2WinBUGS (Sturtz, Ligges, & Gelman, 2005), MCMCpack (Martin, Quinn, & Park, 2011), mcmcplots (Curtis, 2015), poLCA (Linzer & Lewis, 2011), pscl (Jackman, 2014), and RNetica (Almond, 2013), as well as the code given by Nadarajah and Kotz (2006), and code of our own writing.

Throughout the book, we present `WinBUGS code in this font`, commenting on features of the code pertinent to the discussion. We used WinBUGS for our analyses, but the code presented here should work with minimal modification in other versions of BUGS (Lunn, Jackson, Best, Thomas, & Spiegelhalter, 2012) and the closely related JAGS software (Plummer, 2003).

Our use of these software packages should not be taken as an endorsement of them over others. Happily, there are many programs that can be used for fitting Bayesian psychometric and statistical models, including some such as WinBUGS that are intended to be general and so require the user to express the model, and others that have predefined models built in.

Online Resources

Additional materials including the datasets, WinBUGS code, R code, and Netica files used in the examples, and any errors found after the book goes to press, are available on the book's website, www.bayespsychometrics.com. Included there is our current contact information, where you can send us any comments.

Roy Levy
Arizona State University, Tempe, Arizona

Robert J. Mislevy
Educational Testing Service, Princeton, New Jersey

Acknowledgments

Our thinking on matters discussed in the book has been influenced by a number of scholars, and these influences manifest themselves in different ways in the book. Some of those have been direct collaborations, and several of our examples have appeared in articles and books written with these colleagues. When introducing such examples, we cite our prior articles and books with these coauthors. Some of Roy's work on Bayesian psychometrics emerged from projects funded by Cisco, the National Center for Research on Evaluation, Standards, & Student Testing (CRESST) at the University of California at Los Angeles, the Institute for Education Sciences (IES), and Pearson. In addition to Cisco and CRESST, Bob's work was also supported by grants from the Office of Naval Research and the Spencer Foundation. We are grateful to John Behrens and Kristen DiCerbo for their support during their time leading research efforts at Cisco and now at Pearson. We thank Dennis Frezzo, Barbara Termaat, and Telethia Willis for their support while leading research efforts at Cisco. We also thank Eva Baker and Greg Chung of CRESST for their support, and Allen Ruby and Phill Gagné, Program Officers for IES's Statistical and Research Methodology in Education program. The findings and opinions expressed in this book are those of the authors and do not represent views of the IES or the US Department of Education.

Other interactions, projects, and conversations have shaped our thinking and influenced this book in more subtle but pervasive ways. We are also grateful to a number of colleagues who have supported us in more general ways. We thank Russell Almond, John Behrens, Darrell Bock, Jaehwa Choi, Kristen DiCerbo, Joanna Gorin, Sam Green, Greg Hancock, Geneva Haertel, Charlie Lewis, Andre Rupp, Sandip Sinharay, Linda Steinberg, Marilyn Thompson, David Williamson, and Duanli Yan for their collaboration, insight, and collegiality.

We also thank the students, participants, and institutions that have hosted our courses and workshops on Bayesian methods at Arizona State University, the University of Maryland, the University of Miami, and meetings of the National Council on Measurement in Education. To the students and participants, this book was written with you in mind. We hope you didn't mind us piloting some of these materials on you.

We thank Arizona State University for granting Roy a sabbatical to work on this book, and ETS for supporting Bob's work as the Frederic M. Lord Chair in Measurement and Statistics.

We are indebted to a number of colleagues who read and provided critiques of drafts of the chapters. The feedback provided by Jaehwa Choi, Katherine Castellano, Jean-Paul Fox, Shelby Haberman, and Dubravka Svetina was invaluable, and we wish to thank them all for doing us this great service. Of course, they are not responsible for any shortcomings or problems with the book; that resides with the authors. We thank Kim Fryer for managing the reviewing and editing process at ETS. We are indebted to Rob Calver, our editor at Chapman & Hall/CRC Press, for his support, encouragement, and patience. We also thank Kari Budyk, Alex Edwards, Sarah Gelson, Rachel Holt, and Saf Khan of Chapman & Hall/CRC Press for their assistance.

Roy's interests in Bayesian psychometrics began when he was Bob's student at the University of Maryland, and it was there over a decade ago that the first inklings of what would become this book took shape. The collaboration on this book is but the latest in professional and intellectual debts that Roy owes to Bob. For his mentorship, support, friendship, and influence over all these years, in too many ways to recount here, Roy cannot thank Bob enough. Bob, in turn, cannot express the deep satisfaction in playing some role in the experiences that have shaped a leading voice in a new generation of psychometrics—not to mention a treasured friend and well-matched collaborator.

We close with thanks to those closest to us, who may be the happiest to see this book completed. Roy wishes to thank Paige, whose support is only outdone by her patience. Bob thanks Robbie, also noting patience and support, spiced with tenacity and humor.

Section I

Foundations

The first part of the book lays out foundational material that will be leveraged in the second part of the book. In Chapter 1, we set out background material on assessment, psychometrics, probability, and model-based reasoning, setting the stage for developments to come. Chapters 2 and 3 provide an introduction to Bayesian inference and treat Bernoulli and binomial models. Whereas Chapter 2 focuses on the machinery of Bayesian inference, Chapter 3 delves into more conceptual issues. Chapter 4 reviews models for normal distributions. Chapter 5 discusses strategies for estimating posterior distributions, with an emphasis on Markov chain Monte Carlo methods used in the remainder of the book. Chapter 6 treats basic regression models. The second part of the book will draw from and build off Chapters 2 through 6. Our treatment of the material in Chapters 2 through 6 is somewhat cursory, as our goals here are to cover things only at a depth necessary for us to exploit them in our treatment of psychometric models later in the book. Importantly, each of these topics could be treated in more depth, and most introductory Bayesian texts do so; excellent accounts may be found in Bernardo and Smith (2000), Congdon (2006), Gelman et al. (2013), Gill (2007), Jackman (2009), Kaplan (2014), Kruschke (2011), Lynch (2007), and Marin and Robert (2007).

1

Overview of Assessment and Psychometric Modeling

Assessment is an integral aspect of many institutions and disciplines including education, psychology, medicine, jurisprudence, public policy, and business. Most of us are likely quite familiar with assessment in the context of educational or professional settings. We suspect that through the course of their education or occupation every reader of this text has taken more tests than they can recall—some of which may be remembered as excellent assessments, no doubt quite a few that engendered skepticism, and still many more that have been forgotten.

Assessment dates back millennia, transcending geographical and cultural boundaries (Clauser, Margolis, & Case, 2006; Dubois, 1970; Wainer, 2000; Zwick, 2006), with forms of educational testing that we would easily recognize as such today—complete with associated ramifications for students, teachers, schools, and policymakers—emerging in the mid-nineteenth century (Reese, 2013). However, it was not until early in the twentieth century that psychometrics crystallized as a distinct discipline focusing on issues such as scoring and the characterization of tasks, with strong connections to aspects of assessment design, administration, and validation.* This book describes approaches to psychometrics from the perspective of Bayesian inference and statistical modeling.

This chapter lays the grounding for this effort, providing an overview of assessment and the central challenges that we will subsequently argue may be advantageously addressed by adopting a Bayesian perspective.

1.1 Assessment as Evidentiary Reasoning

On the surface, tests are just that—tests. We take a test, we get a score, and it is natural to think that is all there is to it. Despite appearances, assessment is not really about assigning scores—it is about reasoning from what we observe people say, do, or produce to broader conceptions of them that we have not—and often cannot—observe.† In education, we seek to reason from what students say or do to what they know or can do more broadly. In medical diagnosis, we seek to reason from a patient's physical characteristics, statements, and environment to their affliction. In personnel decisions, we seek to reason from an applicant's response to interview questions to how they would perform in the job. In jurisprudence, we seek to reason from testimony to determine guilt or innocence.

More specifically, we view assessment as an instance of *evidentiary reasoning*. This view, and the use of probability-based modeling to carry out this reasoning, draws heavily from

* See Jones and Thissen (2007) for a historical account of psychometrics and alternative perspectives on its origins.

† Apologies to Glenn Shafer, quoted in Section 1.2.2.

Schum's (1987, 1994) account of evidentiary reasoning as reasoning from what is known to what is unknown in the form of explanations, predictions, or conclusions. In the remainder of this chapter, we provide an overview of this perspective and how it pertains to psychometrics; see Mislevy (1994) for a more thorough account. Much of the rest of this book may be seen as elaborating on this overview, providing the mechanics of how to accomplish these activities, and discussing the concepts and implications of this perspective.

Observed data become evidence when they are deemed relevant for a desired inference through establishing relations between the data and the inferential target. We often employ data from multiple sources of information to serve as evidence. These may be of similar type (e.g., test questions with the same format targeting the same proficiency, testimony from multiple witnesses) or of quite different type (e.g., an applicant's resume in addition to her interview, a patient's family medical history and a survey of hazardous materials in her home, crime scene photos in addition to witness testimony). Evidence may be contradictory (e.g., conflicting testimony from witnesses, a student succeeds at a hard task but fails at an easy one), and almost always falls short of being perfectly conclusive.

These features have two implications. First, properly marshaling and understanding the implications of the evidence is difficult. Inference is a messy business. Second, owing to the inconclusive nature of the evidence, we are necessarily uncertain about our inferences. To begin to address these, the remainder of this section describes tools that aid in representing the act of inference and uncertainty.

In evidentiary reasoning, an argument is constructed to ground inferences. Toulmin (1958) diagrams offer a visual representation of the structure of evidentiary arguments, and we present versions of them that align with inference in assessment (Mislevy, Steinberg, & Almond, 2003). Figure 1.1 depicts a generic diagram; Figure 1.2 depicts a simplified example from educational assessment (a one-item test, with a highly informative item!),* where

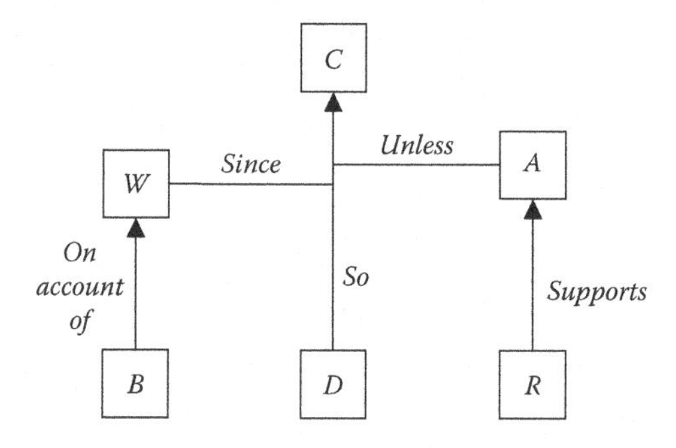

FIGURE 1.1
Toulmin diagram for the structure of an argument. The flow of reasoning is from Data (*D*) to a Claim (*C*), through a Warrant (*W*). The Warrant is a generalization that flows the other way: If a Claim holds, certain kinds of Data usually follow. Backing (*B*) is empirical and theoretical support for the Warrant. It may not hold in a particular case, for reasons expressed as alternative explanations (*A*), which may be supported or weakened by Rebuttal (*R*) data. (Mislevy, R. J., Steinberg, L. S., & Almond, R. G. (2003). On the structure of educational assessments. *Measurement: Interdisciplinary Research and Perspectives*, 1, 3–62, figure 2. With permission of CRESST.)

* One-item tests are not unheard of historically, even in high-stakes circumstances (Wainer, 2000). However, we typically do not have any one item that is so informative as to be conclusive about an examinee. As a result of these and other considerations, assessments are typically constructed using multiple items, as developed in Section 1.3.

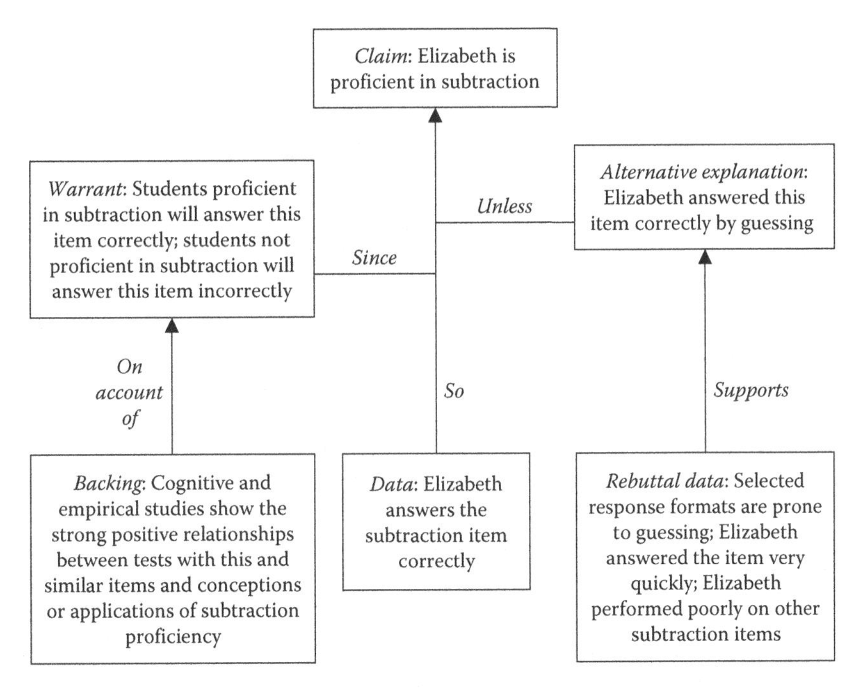

FIGURE 1.2
Example Toulmin diagram for the structure of the assessment argument. This example shows reasoning from Elizabeth's correct response to a claim about her proficiency.

we seek to reason about a student's subtraction proficiency based on her performance on a selected-response item designed to measure subtraction proficiency.

Inference in a particular instance flows along the central arrow from the *data* to the *claim*; in educational assessment, from the observations of what students say, do, or produce to hypotheses about their capabilities more broadly. In the example, we aim to proceed from the observation that Elizabeth selected the option corresponding to the correct value on the item to support the claim that she is proficient in subtraction.

This reasoning is justified by the *warrant*, a generalization that runs in the other direction, and summarizes the main thrust of the evidentiary relation between the data and the claim. Warrants require justification, which comes in the form of *backing*. In assessment, this comes from substantive theory and/or prior empirical research.

In assessment arguments, the warrant typically expresses the relation by stating that when some constellation of conditions expressed in the claim holds, certain observations will occur. In the example, the warrant expresses that students proficient in subtraction will correctly answer the item, and students who are not proficient in subtraction will incorrectly answer the item. Importantly, note that the direction of the warrant is from the claim to the data which may be depicted as

$$C \xrightarrow{W} D.$$

The flow from the claim to data along the direction of the warrant is *deductive*, in that it proceeds from the general to the particular. In this example, it is from a claim about what people like Elizabeth, both those who are proficient and those who are not, typically do, to Elizabeth's behavior on a particular item. When particular data are observed we must reason from these data to a particular claim, back up through the warrant in the opposite direction.

Warrants have varying degrees of strength, reflecting the strength of the relationship between the claims and the data. We begin with an extreme case of formal syllogisms, which are powerful warrants (Jaynes, 2003; Mislevy et al., 2003). We will bring uncertainty and multiple items into the picture shortly, but an initial error-free example illustrates the basic logic and structure. In this special case, the first component of the warrant is the major premise (P1), the claim is the minor premise (P2), and the data follow inevitably.

P1	W_1:	Students proficient in subtraction will answer this item correctly.
P2	C:	Elizabeth is proficient in subtraction.
∴	D_p:	Elizabeth will correctly answer this item.

The notation of D_p indicates that the conclusion of this argument is framed as a *predictive* or *prospective* account of the data, rather than the observed value. In general, deductive arguments are well equipped to reach prospective conclusions. Given the general statement about Elizabeth's proficiency and the warrant, we can make predictions or inferences about what will happen in particular instances.

Importantly, the directional flow of the inference as it actually plays out is in the *opposite* direction from how the warrant is laid out here. We do not start with a premise about student capabilities and end up with a prospective statement about what the student will do. Rather, we start with observations of the student's behaviors and seek conclusions about their capabilities. That is, we would like to reverse the directional flow of the warrant and reason *inductively* from the particular observations of Elizabeth's behaviors to reach conclusions about her proficiency broadly construed. That is, we would like to reason as follows:

P1	W_1:	Students proficient in subtraction will answer this item correctly.
P2	D:	Elizabeth correctly answered this item.
∴	C:	Elizabeth is proficient in subtraction.

However, such an argument is invalid; D and W_1 are insufficient to conclude C, as it could be that students not proficient in subtraction will correctly answer subtraction items. What is needed is the second component of the warrant. Using W_2 as the major premise, we have the valid argument.

P1	W_2:	Students not proficient in subtraction will answer this item incorrectly.
P2	D:	Elizabeth correctly answered this item.
∴	C:	It is not the case that Elizabeth is not proficient in subtraction (i.e., Elizabeth is proficient in subtraction).

Figure 1.3 depicts another example, where we reason from observing an incorrect response from another student, Frank, to a claim about his proficiency. This example is similar to that in Figure 1.2, but has different claims and explanations as it is based on different data, namely, Frank providing an incorrect response. Note that the warrant and the backing are the same regardless of whether the student provides a correct or incorrect answer. The warrant is constructed to provide an inferential path regardless of the statuses of the particular claim and data. The warrant spells out expected behavior given all possible values of proficiency. In the simple example here, a student is either proficient or is not proficient, and the warrant specifies what is expected under both these possibilities. Note that the warrant explicitly links different behaviors with different states of proficiency. This differential association embodies the psychometric concept of discrimination, and is crucial to conducting inference.

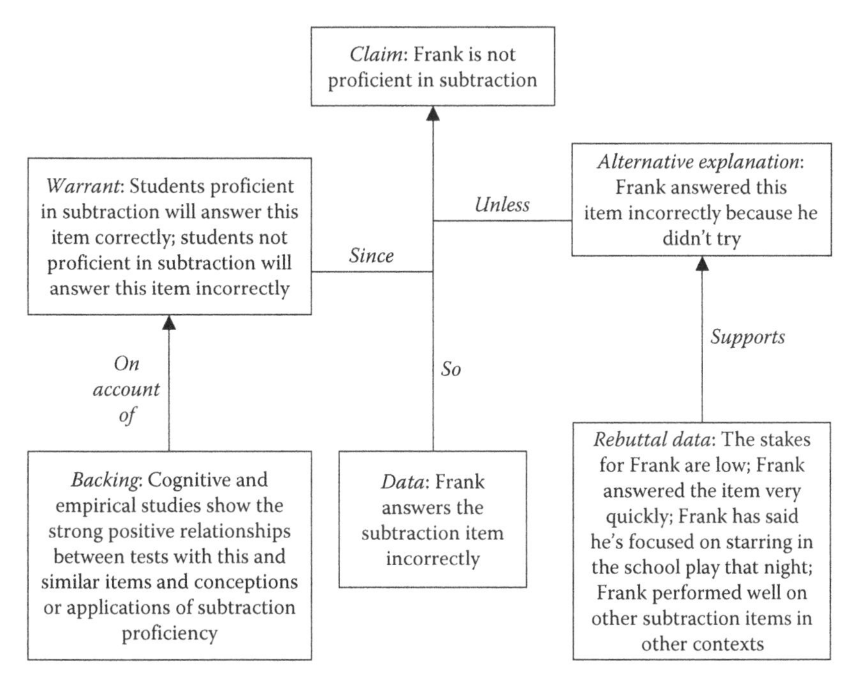

FIGURE 1.3
Example Toulmin diagram for the structure of the assessment argument. This example shows reasoning from Frank's incorrect response to a claim about his lack of proficiency.

TABLE 1.1
A Deterministic Warrant Connecting a Student's Proficiency Status with Her Response to an Item

	Response	
Proficiency	**Correct**	**Incorrect**
Proficient	Yes	No
Not Proficient	No	Yes

Note: Rows correspond to possible proficiency states; columns correspond to possible outcomes. Reading across a row gives the *Response Outcome* conditional on that proficiency status.

The warrant is summarized in Table 1.1, which is akin to a truth table for the propositions that the student is proficient and the response is correct. The deductive flow of the argument proceeds from the student's status in terms of proficiency to her response, contained in the rows of the table. The inductive flow proceeds from the student's response to her status in terms of proficiency, contained in the columns of the table.

In real-world situations, however, inferences are not so simple. There are almost always *alternative explanations* for the data, which in turn require their own support in the form of *rebuttal* observations or theory. In educational assessment, alternative explanations are ever-present, a point has been articulated in a variety of ways over the years. Rasch (1960, p. 73) warned that

> Even if we know a person to be very capable, we cannot be sure that he will solve a difficult problem, nor even a much easier one. There is always the possibility that he fails—he may be tired or his attention led astray, or some other excuse may be given.

Holland (1990, p. 579) cautioned us that

> It is important to remember that performance on a test can depend on many factors in addition to the actual test questions—for example, the conditions under which the test was given, the conditions of motivation or pressure impinging on the examinee at the time of testing, and his or her previous experiences with tests...

In the context of discussing college entrance exams, Lord and Novick (1968, p. 30) stated that

> Most students taking...examinations are convinced that how they do on a particular day depends to some extent on "how they feel that day." A student who receives scores which he considers surprisingly low often attributes this unfortunate circumstance to a physical or psychological indisposition or to some more serious *temporary* state of affairs not related to the fact that he is taking the test that day.

Returning to the example in Figure 1.2, one alternative explanation is that Elizabeth correctly answered the item because she guessed, which is supported by a theoretical understanding of the response process on selected-response (multiple-choice) items, perhaps as well as other empirical evidence, say, how quickly she answered, or her performance on other related tasks. Still other possible explanations exist (e.g., that she cheated), which require rebuttal observations of their own (e.g., Elizabeth gave the same answer as another examinee positioned nearby, and answered much easier subtraction items incorrectly). In the second example in Figure 1.3, one alternative explanation is that Frank did not try to answer the item correctly, which is supported by recognizing that the stakes are low for Frank and that his attention is focused on another upcoming activity.

As these examples illustrate, each alternative explanation undercuts the application of the warrant in some way. The explanations that Elizabeth guessed or cheated stand in opposition to the warrant's assertion that students not proficient in subtraction will answer the item incorrectly. The explanation that Frank did not try contradicts the warrant's assertion that students proficient in subtraction will answer the item correctly. Generally speaking, the possibility of alternative explanations weakens the relationship between the data and claim. Our evidence is almost always inconclusive, yielding inferences and conclusions that are qualified, contingent, and uncertain (Schum, 1994).

In assessment, we are usually not interested in an examinee's performance on a test in its own right, as it is in some sense contaminated with the idiosyncrasies of performance at that instance (i.e., on that day, at that time, under those conditions, with these particular tasks, etc.). Rather, we are interested in broader conception of the examinee, in terms of what they can do in a larger class of situations, or in terms of their capabilities more broadly construed. In an ideal case, variation in performance is due solely to variation in these more broadly construed capabilities. Rarely, if ever, do we have such a circumstance. Performance is driven not just by the broader capabilities, but also by a variety of other factors. The implication is that we must recognize that a test, no matter how well constructed, is an imperfect measure of what is ultimately of interest and what we would really like to know. A century ago, this disconnect between the observed performance and the more broadly construed capabilities that are of interest began to be formalized into the encompassing term *measurement error.*

Reasoning in assessment relies on the relationship between the examinee's capabilities and what we observe them say, do, or produce, which is weakened by the presence of measurement error. Table 1.1 simply will not do, as it ignores the possibility of measurement error and its resulting disconnect between the proficiency of inferential interest and the observed data used as evidence. In his review of the field of Measurement of Learning and Mental Abilities at the 25th anniversary of the Psychometric Society in 1961, Gulliksen described "the central problem of test theory" as "the relation between the ability of the individual and his [or her] observed score on the test" (Gulliksen, 1961, p. 101). Such a characterization holds up well today, with a suitably broad definition of "ability" and "test score" to encapsulate the wide variety of assessment settings. Understanding and representing the disconnect between observed performance and the broader conception of capability is central to reasoning in assessment.

As a result of measurement error, our reasoning in assessment—in educational contexts, reasoning from what we observe students say, do, or produce to what they know or can do in other situations or as more broadly construed—is an instance of reasoning under uncertainty. Because of the imperfect nature of our measurement, we are uncertain about our resulting inference. Reasoning from the limited (what we see a student say, do, or produce) to the more general (student capabilities broadly construed) is necessarily uncertain, and our inference or conclusion is necessarily possibly incorrect.

1.2 The Role of Probability

1.2.1 Enter Probability

What tools should we employ to represent this disconnect, and our uncertainty about the relationship between observed performances and broader conceptions of capability? At the 50th anniversary of the Psychometric Society in 1986, Lewis built off Gulliksen's remarks to observe that much progress in test theory had been made "by treating the study of the relationship between responses to a set of test items and a hypothesized trait (or traits) of an individual as a problem of statistical inference" (Lewis, 1986, p. 11). The centrality if not necessity of framing the problem in this way is evident in statements such as the following remark from Samejima (1983, p. 159), who drew a direct line between the notion of measurement error and a probabilistic approach.

> There may be an enormous number of factors eliciting [a student's] specific overt reactions to a stimulus, and, therefore, it is suitable, even necessary, to handle the situation in terms of the probabilistic relationship between the two.

This view has come to dominate modern psychometrics, and reflects the utility of employing the language and calculus of probability theory to communicate the imperfection of evidence, and the resulting uncertainty in inferences regardless of discipline (Schum, 1994). Much of this book is devoted to illuminating how adopting a probabilistic approach to psychometric modeling aids not only in representing the disconnect between proficiency and performance, but also in bridging that divide by employing a framework within which we can reason from one to the other, all the while properly acknowledging the distinction between them.

As discussed in Section 1.1, in deductive reasoning we proceed from the general to the particular(s), seeing the latter as arising from well-understood relationships between the two. Deterministic logic is a paradigmatic example of deductive reasoning, but probabilistic reasoning can be deductive as well. In assessment, deductive reasoning flows from the unknown or latent variable thought of as the general characterization in the claim (say, of examinees' capabilities) to the observable variables (OVs) that represent the particular (say, their behavior on tasks). The possibility of alternative explanations undermines the strength of the evidentiary linkage between the claim and the data. In these cases, the warrant no longer assures that a certain outcome follows necessarily from the claim. Rather, deductive reasoning starts with examinee capability as the given and, for each of the possible values of that capability, proceeds to arrive at the *likely* behavior on tasks.

Framing the warrant probabilistically quantifies how likely the behaviors are for an examinee with particular capabilities. In educational assessment, we employ a probability model to quantify how likely certain behaviors on tasks are, given the student's proficiencies. A bit more formally, letting x denote the variable corresponding to the student's behavior on a task (e.g., response to an item) and θ denote the variable corresponding to the student's proficiency status, the relationship is formulated as $p(x \mid \theta)$—that is, the probability distribution of x conditional on the value of θ. Returning to the examples, Table 1.2 illustrates these concepts. The rows in the table represent the probabilities of the different possible responses for the different possibilities of the student's proficiency status. If we begin with the premise that Elizabeth is proficient at subtraction, we arrive at the probability of .70 that she correctly answers the item, and (its complement) the probability of .30 that she incorrectly answers the item. Likewise for Frank, we begin with the premise that he is not proficient at subtraction and arrive at the probability of .20 that he correctly answers the item, and (its complement) the probability of .80 that he incorrectly answers the item. Setting up the probability model relating the variables as $p(x \mid \theta)$, that is, where variables corresponding to observed behavior are modeled as stochastically dependent on variables corresponding to unobservable variables representing broadly conceived capabilities, is ubiquitous throughout different psychometric modeling traditions. In Chapter 7, we explore connections between the construction of archetypical psychometric models and Bayesian approaches to modeling. For the moment, it is sufficient to recognize that the flow of the probability model directly supports the deductive reasoning; for example, the rows of Table 1.2 are probability distributions.

TABLE 1.2

A Probabilistic Warrant Connecting a Student's Proficiency Status with Her Response to an Item

	Response	
Proficiency	**Correct**	**Incorrect**
Proficient	.70	.30
Not Proficient	.20	.80

Note: Rows correspond to possible proficiency states; columns correspond to possible outcomes. Reading across a row gves the *Probabilities for a Response Outcome* conditional on that proficiency status.

In inductive reasoning, we reverse this flow of reasoning and proceed from the particular(s) to the general. In assessment, inductive reasoning starts with examinee behavior on tasks and proceeds to arrive at likely values of examinee proficiency. How to do so is not as straightforward as in deductive reasoning; in Table 1.2, we set up the *rows* as probability distributions, but we cannot interpret the *columns* as probability distributions. The warrant is framed as a probability model with the flow from examinee proficiency to likely behaviors. To reverse this flow, we employ probability calculus. In particular, we will employ Bayes' theorem to facilitate this reversal, supporting probability-based reasoning from particulars to general (i.e., from performances on tasks to proficiency) as well as from general to particulars (i.e., from proficiency to performances on tasks). But we are getting a bit ahead of ourselves. Before turning to the mechanics of Bayesian inference in Chapter 2, we lay out more of the necessary groundwork.

1.2.2 Model-Based Reasoning

A model is a simplified version of a real-world scenario, wherein salient features relevant to the problem at hand are represented, while others are suppressed, as constructed by the analyst to facilitate the desired inferences. We adopt a philosophical position that asserts that our models are necessarily wrong, though hopefully useful. This sentiment, most closely associated with George Box (1976, 1979, Box & Draper, 1987), has been widely reinforced throughout the statistical literature (e.g., Freedman, 1987; Gelman & Shalizi, 2013), as well as the literatures associated with different psychometric modeling paradigms (e.g., Edwards, 2013; MacCallum, 2003; McDonald, 2010; Preacher, 2006; Thissen, 2001). Instead of being correct, our hope for a model is that it (a) represents the relationships among the relevant entities, (b) provides the machinery for making inferences about what is unknown based on what is known, and (c) offers mechanisms for effectively communicating results to take actions in the real world.

The beliefs encoded in a model are a subset of a richer, fuller understanding of the world, simplified in ways that are suited to the inferential task at hand. Recognizing that a model is not a perfect representation of the world, the fidelity of the model to the real-world situation at hand is nevertheless critical. Matching the simplifications made and complexities preserved in specifying a model with the purpose of modeling is crucial as models that ignore important aspects of the problem are less likely to yield inferences that are useful for taking action in the world.

Figure 1.4 lays out a schematic adapted from Mislevy (2010) that communicates the process of modeling (for a related perspective for modeling in physics, see Hughes, 1997; see also Morey, Romeijn, & Rouder, 2013, for connections to statistical modeling). The space at the bottom left next to the picture of Earth refers to the real-world situation of inferential interest. Following the upward-pointing arrow, the plane in the middle refers to the entities and relationships in the *model-space*. As the silhouette of Rodin's The Thinker suggests, this model-space is defined by our thinking about the world, as opposed to the world itself. Note that not all the elements of the real-world situation are included in the model-space. Those that are included are depicted as polygons, indicating that the fuzzy, rough, or jagged edges of the real-world problem are approximated in the model-space by simplified or abstracted versions with smoothed lines and sharp corners. Still other elements may be introduced in the model-space, depicted in Figure 1.4 with the diagonally filled polygons, without referring to entities in the real world.

In the assessment example of Figure 1.2, the bottom left represents the real-world phenomena of Elizabeth struggling with the problems, writing answers, erasing, rewriting,

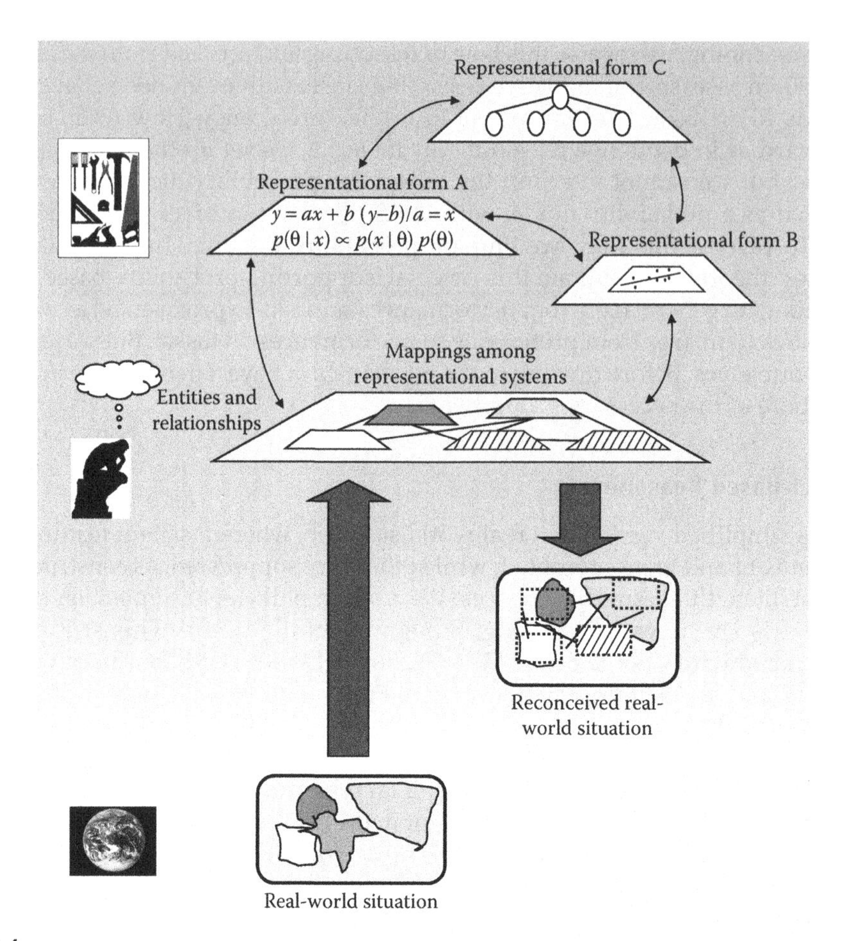

FIGURE 1.4
Schematic of model-based reasoning. A real-world situation is depicted at the lower left. It is mapped into the entities and relationships of the model in the narrative space in the middle plane. The lower right is a reconception of the real-world situation, blending salient features of the situation with the structures the model supplies. The upper boxes represent symbol systems that models may have to aid reasoning in the model plane. Mathematical formulations, diagrams, and graphs are examples of symbol systems used in psychometrics. (With kind permission from Taylor & Francis: *Research Papers in Education*, Some implications of expertise research for educational assessment. *25*, 2010, 253–270, Mislevy, R. J. Figure 8.)

some quickly, some slowly. Only the correctness of her final responses is represented in the model-space. The model-space also contains a variable for Elizabeth's proficiency, and relationships that indicate that it determines probability distributions for all the responses—relationships that are not apparent in the world.

Narrowing our focus to that of *probability models*, choices made in specifying such models rely on a mixture of considerations including beliefs about the problem, theoretical conjectures, conventions, computational tractability, ease of interpretation, and communicative power (Gelman & Shalizi, 2013). In the model expressed in Table 1.2, we have a dichotomous variable for subtraction proficiency that can take on one of two values, Proficient or Not Proficient. This amounts to an intentionally simplified framing of the more complicated real-world situation in light of the desired inferences and available evidence. If we had other purposes, or other evidence that informed on the examinee's proficiency in a more fine-grained manner, we may have specified a variable with more than two categories, reflecting multiple stages to proficiency, or continuously, reflecting that one can have more or less proficiency.

As another example, it is doubtful that any distribution has been called into service in statistical modeling of data more than the normal distribution. However, it is rarely the case in which variables are actually normally distributed. We have yet to come across one variable in real-world phenomena whose distribution exactly follows a normal distribution. Why then are normal distributions so popular? Several reasons come to the forefront. First, there are (many!) situations in which a normal distribution is a good approximation to the actual distribution in that it captures the salient patterns. Second, normal distributions can be expressed in terms of a few, interpretable parameters; in the bivariate case we have two means, two standard deviations, and one correlation that capture central tendency, variability, and association, respectively. Third, we have well-established mathematical tools for working with normal distributions (e.g., fitting it to data, and computing probabilities associated with it). Thus, when we call a normal distribution into service to represent the distribution of a variable, we are employing it to *model* the variable and the data—that is to (a) approximate the real-world situation and (b) provide mechanisms for effective and efficient summaries for reasoning and communication.

Returning to the middle plane of Figure 1.4, once we have built the model, we then conduct analyses on the entities in the model-space using the machinery of the model. This typically involves employing knowledge representations (Markman, 1999) that externalize information in forms conducive to reasoning. These are depicted above the model-space in Figure 1.4 next to an icon indicating that they are tools for conducting our work in the model-space. For the example in Figure 1.2, the symbol system at the left is the mathematical formulation of the measurement model, the one at the center is a diagrammatic representation of the model, and the one at the right could be a graph of the relationships. Table 1.2 is an example of a knowledge representation that facilitates deductive reasoning from an examinee's proficiency status to their likely response to an item. Of course, we will also wish to conduct inductive reasoning from the examinee's response to their proficiency status. As we will see, we will accomplish this via applications of Bayes' theorem. In Chapter 2, Bayes' theorem itself is expressed via multiple knowledge representations, including equation forms that guide computations (in Equation 2.5) and tree diagrams (Figure 2.1). The latter is a more efficacious representation for humans reasoning through frequency problems of this sort (Gigerenzer, 2002); the former may be more efficacious when embedded in larger syntactical expressions of probability structures, reasoning with continuous variables, and for use in programming computer algorithms. In psychometrics, we employ knowledge representations found in quantitative methods, including Bayes' theorem and other mathematical equations, graphical models, and graphical representations of data (Mislevy et al., 2010; see also Mislevy, 2010). Frequently, working with the machinery of the model involves gaining an understanding of the model's parameters (e.g., by statistical estimation), which are summaries of features of the representation that reside in the model-space.

Based on our use of such tools in the model-space, we construct a reconceived notion of the real-world situation, as represented by the downward-pointing arrow in Figure 1.4. This model-based conception of the situation does not perfectly match the real-world situation, but allows us to make inferences and decisions. The background shading in Figure 1.4 reflects the extent to which the entities reside in our accounting of the real world as opposed to the real world itself: the real-world situation is unshaded, reflecting its status in the real world; the model plane is completely shaded, reflecting that it resides squarely in our accounting of the world; the reconceived situation is in the partial shade, indicating that it is a blend of the actual real-world situation and the structures supplied by the model. In the example from Figure 1.2, the reconceived situation still addresses only the correctness of Elizabeth's responses, but sees them as outcomes of a random process that depends on her proficiency parameter. This parameter is introduced in the model-space, and may be retained in our reconceived notion of the real-world situation. The model provides us an estimate, as well as an indication of its accuracy, and predictions for Elizabeth's likely responses on other items she was not even presented.

In summary, in *probability-based reasoning* we use a formal probability framework to capture imperfect relationships, represent our uncertainty, and reason through the inferential situation at hand. We begin with a real-world situation (bottom left of Figure 1.4) and through our understanding of the salient features of this situation we construct a simplified version in terms of a probability model (the middle layer of Figure 1.4). We then apply tools, including the calculus of probabilities, to the entities in the modeled version. The results represent, in the probability framework, the implications and conclusions that constitute the reconceived real-world situation (bottom right of Figure 1.4). In our development of psychometric modeling, we will routinely employ probability theory as the machinery of inference, reflecting a sentiment expressed by Shafer (as quoted in Pearl, 1988, p. 77): "Probability is not really about numbers; it is about the structure of reasoning."

1.2.3 Epistemic Probability

Having introduced the notion of using probability to aid in representing and bridging the divide between what is of inferential interest and the observations that constitute evidence for the desired inference, we take this opportunity to clarify how we conceive of probability and the interpretations we ascribe to the term. We subscribe to a perspective that views probability as a language of expressing uncertain beliefs, commonly referred to as a *subjective, degree-of-belief,* or *epistemic* interpretation (de Finetti, 1974; see also Bernardo & Smith, 2000; Jackman, 2009; Kadane, 2011; Kyburg & Smokler, 1964). For expository reasons, it is useful to contrast our view with that of frequentist interpretations of probability that ground much of conventional approaches to statistical modeling. Informally, a frequentist position conceives of probability as a limiting value of a relative frequency of a repeatable process. For example, the probability that a flipped coin will land on heads is the long-run frequency that it lands on heads in repeated flips. From this perspective, probability is property of the world (e.g., it is a property of the repeated flips of the coin), and strongly relies on a hypothetical if not actual collective (e.g., the repeated flips of the coin).

From our perspective, the statement that a flipped coin will land on heads with a particular probability is an expression of a particular belief. Probabilities are represented as numbers between 0 and 1; these endpoints mean that I am certain the coin will not land on heads and certain the coin will land on heads, respectively. Importantly, people may have different beliefs, and therefore may have different probabilities. Suppose we are standing in a casino, waiting for an employee of the casino to flip a coin, who will then pay out money to those who have correctly wagered on the outcome of the coin flip. I think that the probability of the coin landing heads is .5. You, on the other hand, had witnessed a casino manager discretely conversing with and accepting a thick envelope from a known mobster, who has now bet heavily on the coin coming up heads. If you think that these events are connected and the mobster has influenced things such that he is likely (perhaps very likely) to win, your probability might then be considerably larger than .5. Moreover, the same person might have different probabilities at different points in time. If you whisper to me what you observed, I might then change my beliefs and assert that the probability is also larger than .5.

That different people may assert different probabilities highlights a key aspect of our perspective, which is that probability is not a property of the world, but rather a property of the analyst or reasoning agent. This is summarized by this view being referred to as one of *subjective* or *personal* probability. Our interpretation is that probability is an *epistemic* rather than an ontological concept, in that it is a property of the analyst's thinking about the world. For me to initially say that the probability of heads is .5 and then to say the probability is larger than .5 after you recount what you observed reflects a change in *my thinking about the world*, not a change in the world itself.

To readers apprehensive about this interpretation, we might emphasize several related points. First, this conception is by no means new, and dates to the origins of modern mathematical treatments of probability (Hacking, 1975). Second, this perspective is commensurate with the usual mathematical expressions and restrictions: probabilities must be greater than or equal to 0, the probability of an event that is certain is 1, and the probability of the union of disjoint events is equal to the sum of the probabilities associated with each event individually. See Kadane (2011) and Bernardo and Smith (2000) for recent accounts on the alignment between an epistemic probability interpretation and the familiar mathematical machinery of probability.

Third, this is the way that probability is often used in scientific investigations. For example, The International Panel on Climate Change (IPCC, 2014) reported that it was extremely likely, meaning the probability was greater than .95, that humans caused more than half of the observed increase in global average surface temperature between the years 1951 and 2010.

Fourth, this is indeed the usual way that probability is used in everyday language. For example, consider the following statement from an account of the United States Government's reasoning on whether to pursue an assault on a compound that indeed turned out to be the location of Osama bin Laden (Woodward, 2011).

> Several assessments concluded there was a 60 to 80 percent chance that bin Laden was in the compound. Michael Leiter, the head of the National Counterterrorism Center, was much more conservative. During one White House meeting, he put the probability at about 40 percent. When a participant suggested that was a low chance of success, Leiter said, "Yes, but what we've got is 38 percent better than we have ever had before."

Here, we have a statement expressing different values for the probability held by different reasoning agents, and at different times. This reflects how a probability is an expression of uncertainty that is situated in terms of what is believed by a particular person at a particular time.

These examples additionally highlight how, from the epistemic perspective, probability can be used to discuss a one-time event. Indeed, it is difficult to conceive of statements about changes in global temperatures and Osama Bin Laden's whereabouts as having a justification from a frequentist perspective. If probability is, by definition, a long-run frequency, we would have to object to statements such as these, not on the grounds that they are wrong, but on the grounds that they nonsensical. Yet we suspect that these sorts of statements are quite natural to most readers.

Similarly, probability can be used to refer to characterize beliefs about (one-time) events in the past for which we are uncertain. These include things we may never know with certainty, such as the causes of changes in global temperatures or who wrote the disputed

Federalist Papers (see Mosteller & Wallace, 1964, for a Bayesian analysis), and even things we can know with certainty, such as whether Martin Van Buren was the ninth President of the United States.*

Importantly, the notion of epistemic probability is not antithetical to long-run relative frequencies. Relative frequencies obtained from situations deemed to approximate the repeated trials that ground the frequentist interpretation may be excellent bases for belief. Suppose I am interested in the probability that a given person will be a driver in a car accident in the next year and consult recent data that suggest that the proportion of drivers of this age that get in an accident in a year is .12. I might then assert that the probability that the person will be in an accident is .12. Of course, other possibilities exist. I might believe this person to be a worse driver, or more reckless, than others of the same age, and might assert that the probability is a bit higher than .12. In either case, I am making a judgment and ultimately the probability is an expression of my beliefs (Winkler, 1972).

Long-run relative frequencies can also serve as an important metaphor for thinking through what is ultimately an expression of belief, even if the subject is a one-time event that is not amenable to repeated trials. Sports fans, journalists, prognosticators, commentators, and talk show hosts routinely talk about the probability that one team wins an upcoming game—such talk is all but inescapable in sports media before championship games. One mechanism for doing so is by referring to hypothetical long-run frequencies; for example, I might explain my probability that a team wins the game—which will only be played once—by saying "if they played the game 100 times, this team would win 85 of them." The notion of relative frequency over some repetition is a useful metaphor, but ultimately the probability statements are expressions of our uncertain beliefs.

Our perspective will manifest itself throughout the book in the language we use, such as referring to probability distributions as representing our beliefs. But our perspective is just that—our perspective. Many others exist; see Barnett (1999) for a thorough review and critique of the dominant perspectives on probability and the statistical inferential frameworks that trade on these perspectives. More acutely, one need not subscribe to the same perspective as we do to employ Bayesian methods (Jaynes, 2003; Novick, 1964; Senn, 2011; Williamson, 2010).

1.3 The Role of Context in the Assessment Argument

The presentation of the assessment argument to this point has focused on the examinee. We have stated that the inferential task at hand is to reason from what we observe an examinee say, do, or produce to their capabilities or attributes more broadly conceived. Correspondingly, the claim in the assessment argument is formulated as an expression about the examinee. Likewise, the data in the assessment argument on the surface are simply statements about what the examinee has said, done, or produced. However, examinees behave in context. What we observe as data is really an examinee's behavior in a particular situation (Mislevy, 2006).† The data are properly thought of as emerging from the amalgamation of two contributing sources, the examinee and the context. Importantly, the

* He was not. Van Buren was the eighth President; William Henry Harrison was the ninth.

† More properly, it is a particular evaluation of a behavior. We will say more about this later, but different aspects of the same assessment performance could be identified and evaluated for different purposes.

warrant also refers to both the examinee and the context, expressing the likely behavior for an examinee in a particular situation. In simple situations such as an examinee taking a single item, the role of the context is implicit, as it is in the quantitative expression of the warrant in Table 1.2 and even the notation $p(x \mid \theta)$, where x and θ are viewed as examinee variables.

However, once we move to more complicated situations, it becomes apparent that the simple framing we have developed this far will not suffice. Perhaps the most basic extension of our example is the use of *multiple* items to assess proficiency. If the responses to each item have exactly the same evidentiary bearing on proficiency, it may suffice to assume that the warrant expressed in Table 1.2 holds for every item response for an examinee. More generally, we would like to allow them to differ, representing the possibilities that the evidentiary bearing of our observations might vary with the situation. This is expressed in a narrative form casually by saying that some items are harder, some items are easier, and we should take that into account in our inference. Accordingly, we may specify each item to have a possibly unique conditional probability table associated with it. When we turn to these more complicated statistical models as a means to quantitatively express more complicated warrants, we will have parameters summarizing the inferential relevance of the situation in addition to the parameters summarizing the examinee. Accordingly, our notation for denoting the conditional probability for variables standing for observations will have to expand, for example, to $p(x \mid \theta, \omega)$, where here ω stands for a parameter summarizing the features of the context.

It is almost always the case that there remain potentially relevant features of the context that lurk unstated in the warrant. Returning to the example of Frank (Figure 1.3), an alternative explanation for his incorrect response is that he did not try on account of the stakes being low and his attention being focused elsewhere. In addition to undermining the inferential force of the data on the inference, this calls attention to a presumed but unstated aspect of the warrant, namely, that examinees are motivated to answer the question correctly.

In principle, the list of possibly relevant features of the context is limitless ("Are the examinees motivated?" "Is the assessment administered in a language foreign to them?" "Is there enough light in the room for them to read?") and choices need to be made about what to formally include in the warrant. What may be ignorable in one case may not be in another. Issues of motivation play no role in certain assessment arguments, such as those based on blood tests in medicine, but a crucial role in others, such as those based on witness testimony in jurisprudence and intelligence analysis. Considerations of the language of the test may not be important when a teacher gives a test to the students she has had in class for months and knows well, but may be critical when she gives the test to the student who just transferred to the school from a country whose people predominantly speak a different language.

If previously unarticulated aspects of the context are deemed important, one option is to expand the warrant and the data we collect. Returning to the example of Frank incorrectly answering an item, we may expand the warrant by framing it as conditional on whether the examinee is motivated to perform well. For example, we may say that if the examinee is motivated to perform well, the linkage between the data and the claim as articulated in Figure 1.3 holds; if the examinee is not motivated to perform well, another linkage is needed. Expanding the complexity of the warrant may call for additional data to be considered, in this case data that have bearing on whether the examinee is motivated.

An expansion of the evidentiary narrative of the warrant manifests itself in a corresponding expansion of the statistical model that is the quantitative expression of the warrant (Mislevy, Levy, Kroopnick, & Rutstein, 2008). Returning to the example, we may specify a mixture model, where the conditional probability structure in Table 1.2 holds for examinees who are motivated and a different conditional probability structure holds for examinees who are not motivated, along with a probability that an examinee is motivated and possibly a probability structure relating other data that has evidentiary bearing on whether the examinee is motivated.

1.4 Evidence-Centered Design

The ingredients involved in assessment argumentation and inference can be fleshed out by adopting the framework of evidence-centered design (ECD), which lays out the fundamental entities, their connections, and actions that take place in assessment (Mislevy et al., 2003). ECD provides a terminology and sets of representations (Mislevy et al., 2010) for use in designing, deploying, and reasoning from assessments. It is *descriptive* in the sense that all forms of assessment may be framed in terms of ECD, regardless of their varying surface features. ECD is also *prescriptive* in the sense that it provides a set of guiding principles for the design of assessment, the aim of which is to articulate how the assessment can be used to make warranted inferences. Importantly, it is *not* prescriptive in the sense of dictating what particular forms, models, and representations to use, nor particular operational processes to build assessment arguments and the machinery that instantiates them. Rather, ECD (a) helps us understand the argumentation behind the use of particular psychometric models like those covered in this book, and (b) helps us through the assessment development process that might lead to the use of such models, but (c) does not *require* the use of such models.

A quotation from Messick (1994, p. 16) is useful to understand the idea of an assessment argument and the perspective of ECD:

> [We] would begin by asking what complex of knowledge, skills, or other attributes should be assessed, presumably because they are tied to explicit or implicit objectives of instruction or are otherwise valued by society. Next, what behaviors or performances should reveal those constructs, and what tasks or situations should elicit those behaviors? Thus, the nature of the construct guides the selection or construction of relevant tasks as well as the rational development of construct-based scoring criteria and rubrics.

ECD provides a framework for working through the many details, relationships, and pieces of machinery that make an operational assessment, including the psychometric models. It is the grounding in the assessment argument that gives meaning to the variables and functions in a psychometric model; modeling questions that seem to be strictly technical on the surface can often only be answered in the context of the application at hand.

To ground such discussions when we need them, this section provides a brief overview of ECD in terms of the layers depicted in Figure 1.5. The layering in this figure indicates a layering in the kind of work an assessment designer is doing, from initial background research in Domain Modeling through actual administering, scoring, and reporting the results of an operational assessment in Assessment Delivery. The ideas and elements constructed in a given layer are input for work at the next layer. (The actual design process is generally iterative, such as moving from design to implementation of a prototype and going back to revise scoring procedures or psychometric models in the Conceptual Assessment Framework [CAF], or even finding more foundational research is needed.) Figure 1.5 is laid

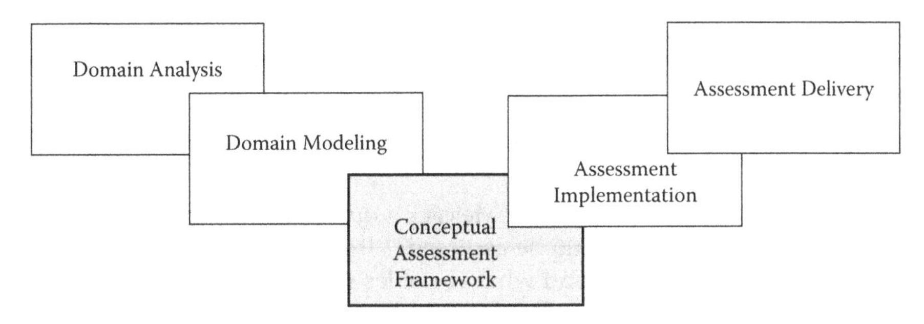

FIGURE 1.5
Layers of ECD. The layers represent successively refined and specific kinds of work test developers do, from studying the domain in Domain Analysis, through developing an assessment argument in Domain Modeling, to specifying the models and forms of the elements and processes that will constitute the assessment in the CAF, to constructing those elements in Assessment Implementation, to administering, scoring, and reporting in an operational assessment as specified in the Assessment Delivery layer. Actual design usually requires iterative movements up and down the layers.

out in such a way to highlight the role of the CAF. Conceptually this is a transition point, which facilitates our pivoting from our understandings about the domain (the layers to the left) to the actual assessment the examinees experience (the layers to the right). Most of the material in this book concerns this layer, though as we will see in various examples, aspects of the other layers will have bearing on the specifications here. Our focus will be on how ECD allows us to see the role played by the psychometric models that are the focus of this book, and we will give short shrift to many features of ECD. Complementary treatments that offer greater depth on different aspects or in different contexts may be found in Mislevy et al. (2003); Mislevy, Almond, and Lukas (2004); Mislevy and Riconscente (2006); Mislevy (2013); Behrens, Mislevy, DiCerbo, and Levy (2012); and Almond et al. (2015).

ECD was developed in the context of educational assessment, and in our treatment here we preserve the language of ECD that reflects this origin, in particular where the examinees are *students* in educational environments. In Section 1.4.4, we discuss the wider application of the ideas of ECD to other assessment environments.

1.4.1 Domain Analysis and Domain Modeling

In Domain Analysis, we define the content of the domain(s) to be assessed, namely the subject matter, the way people use it, the situations they use it in, and the way they represent it. Resources here include subject matter expertise, task analyses, and surveys of artifacts (e.g., textbooks). In Domain Modeling, the information culled in Domain Analysis is organized in terms of relationships between entities, including observable behaviors we might see people do that constitute evidence of proficiency, in what situations those actions could occur or be evoked. It is during this stage that we lay out the assessment argument that will be instantiated when tasks are built, the assessment is delivered, student performances are scored, and inferences are made. This often involves articulating what claims we would like to be able to make, Toulmin diagrams for assessment arguments, and design patterns for tasks.

1.4.2 Conceptual Assessment Framework

In the CAF, we take the results from Domain Modeling and, in light of the purposes and constraints, we devise the blueprint for the assessment. Our focus will be on the three central models that address the questions articulated by Messick (1994, quoted in Section 1.4).

They are the Student, Task, and Evidence models depicted in Figure 1.6. They address Messick's questions in a way that becomes a blueprint for jointly developing the tasks and psychometric models.

1.4.2.1 Student Model

The Student Model addresses the first of Messick's questions, "what complex of knowledge, skills, or other attributes should be assessed?" It is here that we articulate what the desired claims and inferences are, and what variables will be used to represent beliefs to support those claims and inferences. The Student Model lays out the relevant configuration of the aspects of proficiency to be assessed and therefore represents choices regarding what aspects of the proficiency identified in the Domain Analysis and Domain Modeling will be included in the assessment as variables in the psychometric model. We build the Student Model by specifying what have variously been termed student model variables (SMVs), proficiency model variables, or competency model variables. Modern psychometric models are characterized by their use of *latent* variables as SMVs, which reflects that ultimately what is of inferential interest about students cannot be observed directly. These latent SMVs are shown as circles in Figure 1.6, consistent with a convention in structural equation modeling. Several SMVs are shown in the figure, as occurs in factor analysis and multivariate item response theory. Unidimensional item response theory, latent class models, and most applications of classical test theory only have one SMV.

1.4.2.2 Task Models

A Task Model answers the last of Messick's questions, "what behaviors or performances should reveal those constructs, and what tasks or situations should elicit those behaviors?" It specifies the situations in which relevant evidence can be collected to inform upon the values of the variables in the Student Model and thereby yield inferences about students. Task Models specify two main aspects: first, what are the features of the tasks, activities, or situations presented to the student? In the Task Models box in Figure 1.6, the two icons on the right represent the stimulus features and conditions of the task the student interacts with. Second, what are the work products—the things students say, create, or do in these situations—that will be collected? The jumble of shapes in the icon on the left of the Task Models box represents the work products a student creates, which could be simply marks on an answer sheet, or a log of actions in a simulation task, or a video recording of a dance performance.

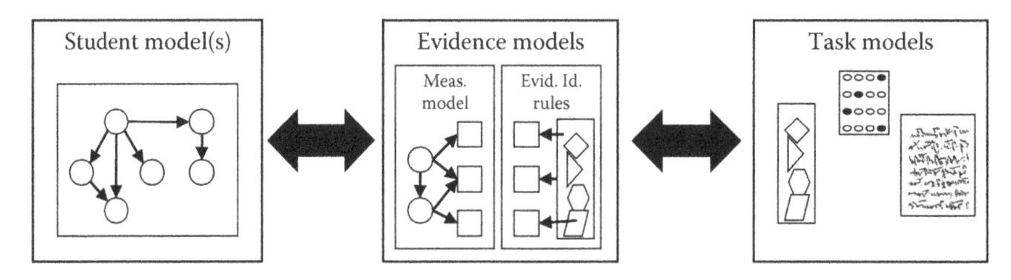

FIGURE 1.6
Three central models of the CAF. Circular nodes in the Student Model represent SMVs, with arrows indicating dependence relationships. Task Models consist of materials presented to the examinee, and work products collected from the examinee. These work products are subjected to evidence identification rules in the Evidence Models yielding OVs represented as boxes, which are modeled as dependent on the SMVs in the psychometric or measurement model. (With kind permission from Taylor & Francis: *Educational Assessment*, Psychometric and evidentiary advances, opportunities, and challenges for simulation-based assessment. *18*, 2013, 182–207, Levy, R.)

1.4.2.3 Evidence Models

An Evidence Model connects the work products collected as specified in the Task Model to the variables in the Student Model. This is achieved via the two ingredients of *evidence identification rules* and the *psychometric* or *measurement model*. Evidence identification rules declare how the work products will be evaluated. The Evidence Identification Rules box on the right in the Evidence Model shows arrows from the work product(s) created by a student (the same jumble of shapes depicted in the Task Models) to values of OVs, which are shown as squares to contrast with the latent SMVs. The arrows represent the evidence identification rules themselves, which might take the form of a scoring key to determine whether the selected responses on multiple-choice items are correct, automated rules to evaluate the efficiency of an investigation that is represented in a log file, or a rubric that human raters apply to evaluate videos of a dance. There can be multiple stages of evidence identification rules, as when an automated essay-scoring procedure first identifies linguistic features of a text and then uses logistic regression models to produce a score. Furthermore, when hierarchies such as finer-grained features of a complex performance are first identified and then summarized, an analyst might choose to fit a psychometric model at either level, depending on the purposes of the exercise (e.g., fitting a classical test theory model to test scores, or item response theory models to the item scores they comprise).

Turing to the left side of the Measurement Models box, a psychometric or measurement model specifies the relationships between the OVs (the same squares identified via evidence rules) and the SMVs (the same circles in the Student Model). Modern psychometric modeling is characterized as specifying these relationships by modeling the OVs as stochastically dependent on latent SMVs.* The arrows in the Measurement Model box in the Evidence Model represent conditional probability distributions that lie at the heart of psychometric models. Table 1.3 classifies a number of such models in terms of their specifications regarding the nature of the SMVs and OVs. Readers familiar with classical test theory will note that it need not be framed as a latent variable model with specified distributional forms, though there are advantages to conceiving of it as such (Bollen, 1989; McDonald, 1999).

TABLE 1.3
Taxonomy of Popular Psychometric Models

		Latent SMVs	
		Continuous	**Discrete**
OVs	**Continuous**	Classical test theory Factor analysis	Latent profile analysis
	Discrete	Item response theory	Latent class analysis Bayesian networks Diagnostic classification models

* Note that the observable variables are identified by their role as the dependent variables in the psychometric model, not by the nature of the tasks or the evaluation rules per se. We avoid ambiguity by using the terms *task* for the situations in which persons act, *work products* for what they produce, and *observables* for the pieces of evidence identified from them by whatever means as the lowest level of the psychometric model. For example, in multiple-choice tests, the word "items" is used to refer to tasks, responses, evaluated responses, and the dependent variables in item response models. This does not usually cause problems because these distinct entities are in one-to-one correspondence. Such correspondences need not hold when the performances and the evidentiary relationships are more complex, and more precise language is needed.

1.4.3 Assessment Implementation and Assessment Delivery

Returning to the layers of ECD in Figure 1.5, in Assessment Implementation we manufacture the assessment, including authoring the tasks, building automated extraction, parsing, and scoring processes to move from work products to OVs, and forming the statistical models.

Assessment Delivery can be described as a four-process architecture (Almond, Steinberg, & Mislevy, 2002), depicted in Figure 1.7. Assessment Delivery cycles through the steps of (a) Task Selection (i.e., what should the student work on next?); (b) task presentation (i.e., delivery of the chosen task); (c) evidence identification through the application of evidence identification rules, in which the student work products are evaluated to produce values for OVs; and (d) evidence accumulation, which occurs when values for the OVs are entered into the measurement model to produce estimates or updated values for the SMVs. Assessments may vary in terms of what decisions are made (e.g., linear assessments that fix the selection and presentation order of tasks a priori vs. adaptive assessments that choose a task based on previous examinee performance, Wainer et al., 2000, vs. assessments that randomly choose the next task, Warner, 1965), and who makes these decisions (e.g., examiner- vs. examinee-governed task and SMV selection in adaptive testing, Levy, Behrens, & Mislevy, 2006; Wise, Plake, Johnson, & Roos, 1992), such that they appear quite different on the surface. Nevertheless, the cycle depicted in Figure 1.7 is a general account of the ongoings of an assessment. Likewise, though assessments vary considerably in how they instantiate the

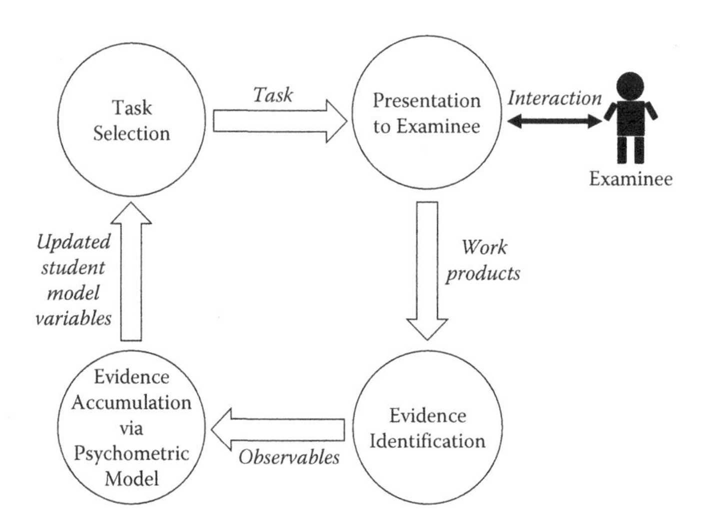

FIGURE 1.7
A four-process architecture of Assessment Delivery. Activities are determined in the Task Selection process. The selected task(s) is/are presented to an examinee (e.g., given a test booklet and an answer sheet to fill out, presented a simulation environment to take troubleshooting and repair actions to repair a hydraulics system in an airplane). The work products(s) are passed on to the Evidence Identification process, where values of OVs are evaluated from them. Values of the OVs are sent to the Evidence Accumulation process, where the information in the OVs is used as evidence to update beliefs about the latent variables in the Student Model, as can be done with a psychometric model. The updated beliefs about SMVs may be used in selecting further tasks, as in adaptive testing. (This is a modification of figure 1 of CSE Technical Report 543, *A Sample Assessment Using the Four Process Framework*, by Russell G. Almond, Linda S. Steinberg, and Robert J. Mislevy, published in 2001 by the National Center for Research on Evaluation, Standards and Student Testing (CRESST). It is used here with permission of CRESST.)

layers in ECD, the framework provides a general account of the work that needs to be done in developing an assessment to enact an assessment argument that facilitates our reasoning from the observed examinee behavior to the inferential targets.

1.4.4 Summary

ECD may be seen as an effort to define what inferences about examinees are sought, what data will be used to make those inferences, how that data will be obtained, how the data will be used to facilitate the inferences, and why the use of that particular data in those particular ways is warranted for the desired inferences. Working clockwise from the top left of Figure 1.7, Assessment Delivery may be seen as the underlying architecture behind seeking the data, collecting it, processing it, and then synthesizing it to facilitate inferences.

The psychometric or measurement model plays a central role in the last of these activities, evidence accumulation. As the junction point where the examinee's behaviors captured by OVs are used to inform on the SMVs that represent our beliefs about the examinee's proficiency, the measurement model is the distillation of the assessment argument. More abstractly, it dictates how features of examinee performances, which in and of themselves are innocent any claim to relevance for our desired inferences, are to be used to conduct those inferences.

The terminology of ECD reflects its origins in educational assessment, but the notions may be interpreted more broadly. For example, "student model variables" may be viewed as aspects of inferential interest for those being assessed, such as the political persuasion of legislators. In this example, bills play the role of tasks or items, evidence identification amounts to characterizing the legislators' votes on the bills (e.g., in favor or opposed), and a psychometric model would connect observables capturing these characterizations to the "student" model variable, namely the political persuasion of the legislator (the "student").

1.5 Summary and Looking Ahead

Assessment refers to the process of reasoning from what we observe people say, do, or produce to broader conceptions of them that we do not, and typically cannot, observe. Measurement error is a broad term that refers to the disconnect between what we can observe and what is ultimately of inferential interest. The presence of measurement error weakens the evidentiary linkage between what we know and what we would like to infer.

As in other disciplines, reasoning in assessment can be a messy business. Measurement error implies that our inferences are based on data that may be fallible, or only sensible in light of understanding contextual factors, or jointly contradictory. As a result the evidence is necessarily inconclusive—it always admits multiple possibilities. Psychometric models are motivated by the fundamental tenet that in education, the social sciences, and many other disciplines, what is ultimately of most interest cannot be directly observed. This principle cuts across disciplines where the presence of measurement error dictates that an observed variable may be *related to* or an *indicator* of what is really of interest, but no matter how well the assessment is devised, it is *not the same thing* as what is of interest. Recognizing this, modern psychometric models are characterized by the use of latent variables to represent what is of inferential interest.

To meet the challenges of representing and synthesizing the evidence, we turn to probability as a language for expressing uncertainty, and in particular probability model-based reasoning for conducting inference under uncertainty. A probability model is constructed for the entities deemed relevant to the desired inference, encoding our uncertain beliefs about those entities and their relationships. An ECD perspective suggests that this model is a distillation of a larger assessment argument, in that it is the junction point that relates the observed data that will serve as evidence to the inferential targets.

Though surface features vary, modern psychometric modeling paradigms typically structure observed variables as stochastically dependent on the latent variables. As we will see, building such a model allows for the use of powerful mathematical rules from the calculus of probabilities—with Bayes' theorem being a workhorse—to synthesize the evidence thereby facilitating the desired reasoning. Adopting a probability framework to model these relationships allows us to characterize both what we believe and why we believe it, in terms of probability statements that quantify our uncertainty. We develop these ideas by discussing Bayesian approaches to most of the psychometric modeling families listed in Table 1.3. Along the way we will see instances where Bayesian approaches represent novel, attractive ways to tackle thorny measurement problems. We will also see instances where Bayesian approaches are entrenched and amount to doing business as usual, even when these approaches are not framed or referred to as Bayesian.

2

Introduction to Bayesian Inference

This chapter describes principles of Bayesian approaches to inference, with a focus on the *mechanics* of Bayesian inference. Along the way, several conceptual issues will be introduced or alluded to; we expand on these and a number of other conceptual issues in Chapter 3. To develop the mechanics of Bayesian inference, we first review frequentist approaches to inference, in particular maximum likelihood (ML) approaches, which then serves as a launching point for the description of Bayesian inference. Frequentist inference and estimation strategies have been the dominant approaches to psychometric modeling, mirroring their dominance in statistics generally in the past century. Provocatively, as discussed throughout the book, Bayesian approaches have become well accepted in certain areas of psychometrics, but not others.

We then turn to Bayesian inference, beginning with a review of foundational principles. We illustrate the basic computations using two running examples. The first involves two discrete variables each with two values. This serves to illustrate the computations involved in working through Bayes' theorem in the simplest situation possible. The second example involves modeling the parameter of Bernoulli and binomial random variables. Such a context is typically discussed in introductory probability and statistics courses and texts, and as such should be familiar to most readers. This latter example illustrates a number of key features of Bayesian modeling and will be a basis for a number of the discussions in Chapter 3, as well as a basis for item response models discussed in Chapter 11.

2.1 Review of Frequentist Inference via ML

Let $\mathbf{x}$ and θ denote the full collections of observed data and model parameters, respectively. In frequentist approaches, a model is typically constructed by specifying the conditional distribution of the data given the model parameters, denoted as $p(\mathbf{x} \mid \theta)$. In this paradigm, the data are treated as random, modeled by the conditional distribution $p(\mathbf{x} \mid \theta)$ and most model parameters are treated as fixed (i.e., constant) and unknown. Model fitting and parameter estimation then comes to finding point estimates for the parameters. Popular estimation routines include those in ML and least-squares traditions. In many situations, these estimation routines can provide consistent, efficient, and unbiased parameter estimates and asymptotic standard errors, which can be used to construct confidence intervals and tests of model-data fit.

Conceptually, ML estimation seeks to answer the question, "What are the values of the parameters that yield the highest probability of observing the values of the data that were in fact observed?" ML estimation seeks the single best set of values for the parameters, where the notion of "best" is operationalized as maximizing the likelihood function, defined as the conditional probability expression $p(\mathbf{x} \mid \theta)$ when viewed as a function of the parameters θ. In practice, once values for $\mathbf{x}$ are observed, they can be entered into the conditional probability expression to induce a likelihood function, denoted as $L(\theta \mid \mathbf{x})$. This likelihood function $L(\theta \mid \mathbf{x})$ is the same expression as the conditional probability expression $p(\mathbf{x} \mid \theta)$. The difference in notation reflects that when viewed as a likelihood function, the values of the data are known and the expression is viewed as varying over different possible values of θ. ML estimation then comes to finding the values of θ that maximize $L(\theta \mid \mathbf{x})$.

In frequentist approaches, the (ML) estimator is a function of the data, which have a distribution. This renders the estimator to be a random variable, and a realization of this random variable, the ML estimate (MLE), is obtained given a sample of data from the population. Standard errors of these estimates capture the uncertainty in the estimates, and can be employed to construct confidence intervals. Importantly, these estimation routines for many psychometric models typically rely on asymptotic arguments to justify the calculation of parameter estimates, standard errors, or the assumed sampling distributions of the parameter estimates and associated test statistics.

Moreover, the interpretations of point estimates, standard errors, and confidence intervals derive from the frequentist perspective in which the parameters are treated as *fixed* (constant). In this perspective, it is inappropriate to discuss fixed parameters probabilistically. Distributional notions of uncertainty and variability therefore concern the parameter estimates, and are rooted in the treatment of the data as random. The standard error is a measure of the variability of the parameter estimator, which is the variability of the parameter *estimates* on repeated sampling of the data from the population. Likewise, the probabilistic interpretation of a confidence interval rests on the sampling distribution of the interval on repeated sampling of data, and applies to the process of interval estimator construction. Importantly, these notions refer to the variability and likely values *of a parameter estimator* (be it a point or interval estimator), that is, the distribution of *parameter estimates* on repeated sampling. This approach supports probabilistic deductive inferences in which reasoning flows from the general to the particulars, in this case from the parameters θ assumed constant over repeated samples to the data $\mathbf{x}$. A frequentist perspective does not support probabilistic inductive inferences in which reasoning flows from the particulars, here, the data $\mathbf{x}$, to the general, here, the parameters θ. In frequentist inference, probabilistic statements refer to the variability and likely values of the parameter *estimator*, not the parameter *itself*.

2.2 Bayesian Inference

Bayesian approaches to inference and statistical modeling share several features with frequentist approaches. In particular, both approaches treat the data as random and accordingly assign the data a distribution, usually structured as conditional on model parameters, $p(\mathbf{x} \mid \theta)$. It expresses how, through the lens of the model, the analyst views the chances of possible values of data, given possible values of the parameters. As discussed in Section 2.1, once data are observed and this is taken as a function of the model parameters, it is a likelihood function. As such, the likelihood function plays a key role in Bayesian inference just as in ML estimation.

However, a Bayesian approach differs from frequentist approaches in several key ways. Frequentist approaches typically treat model parameters as fixed. In contrast, a fully

Bayesian approach treats the model parameters as *random*,* and uses distributions to model our beliefs about them.† Importantly, this conceptual distinction from how frequentist inference frames parameters has implications for the use of probability-based reasoning for inductive inference, parameter estimation, quantifying uncertainty in estimation, and interpreting the results from fitting models to data. Illuminating this difference and how it plays out in psychometric modeling is a central theme of this book.

In a Bayesian analysis, the model parameters are assigned a *prior distribution*, a distribution specified by the analyst to possibly reflect substantive, *a priori* knowledge, beliefs, or assumptions about the parameters, denoted as $p(\theta)$. Bayesian inference then comes to synthesizing the prior distribution and the likelihood to yield the *posterior distribution*, denoted as $p(\theta \mid \mathbf{x})$. This synthesis is given by Bayes' theorem. For discrete parameters θ and datum $\mathbf{x}$ such that $p(\mathbf{x}) \neq 0$, Bayes' theorem states that the posterior distribution is

$$\begin{aligned} p(\theta \mid \mathbf{x}) &= \frac{p(\mathbf{x},\theta)}{p(\mathbf{x})} \\ &= \frac{p(\mathbf{x} \mid \theta)p(\theta)}{p(\mathbf{x})} \\ &= \frac{p(\mathbf{x} \mid \theta)p(\theta)}{\sum_{\theta^*} p(\mathbf{x} \mid \theta^*)p(\theta^*)}. \\ &\propto p(\mathbf{x} \mid \theta)p(\theta). \end{aligned} \tag{2.1}$$

To explicate, we work through an example in medical assessment and diagnosis drawn from Gigerenzer (2002). Consider women aged 40–50 with no family history and symptoms of cancer. For this population, the proportion of women that have breast cancer is .008. Suppose we conduct a mammography screening aimed at the detection of breast cancer. If the patient has breast cancer, the probability of a positive result on the mammography screening is .90. If the patient does not have breast cancer, the probability of a positive mammography screening result is .07. A patient undergoes the mammography screening, and the result is positive. What then should we infer about the patient? Let us frame the situation in a fairly simple way, namely where a patient either has or does not have breast cancer. Note that this reflects a modeling choice—there are stages to cancer, or more continuously, one can have more or less cancer. The simplification here reflects a particular way of framing the more complicated real-world situation in light of the desired inferences.

Gigerenzer (2002) argued for the use of a natural frequency representation in reasoning. The situation can be framed in natural frequency terms, with slight rounding, as follows. Eight of every 1,000 women have breast cancer. Of these eight women with breast cancer, seven will have a positive mammogram. Of the remaining 992 women who do not have breast cancer, 69 will have a positive mammogram. Imagine a sample of women who have positive mammograms in screenings. What proportion of these women actually has breast cancer? Figure 2.1 contains a graphic adapted from Gigerenzer (2002) that further

* Specifically, they are random in the sense that the analyst has uncertain knowledge about them.

† Though this characterization applies to Bayesian inference broadly and will underscore the treatment in this book, there are exceptions. For example, in some approaches to Bayesian modeling certain unknown entities, such as those at the highest level of a hierarchical model specification, are not treated as random and modeled via distributions, and instead are modeled as fixed and estimated using frequentist strategies (Carlin & Louis, 2008).

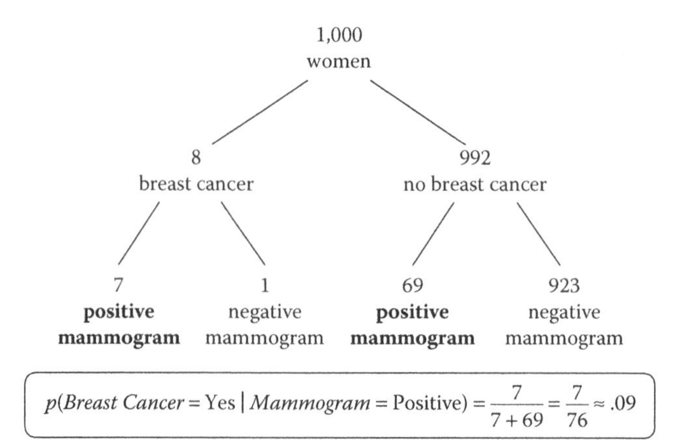

FIGURE 2.1
Natural-frequency representation and computation for Breast Cancer diagnosis example. (From Gigerenzer, G. (2002). *Calculated risks: How to know when numbers deceive you*. New York: Simon & Schuster, figure 4.2. With permission.)

illuminates the situation and the answer. On this description of 1,000 women, we see that 76 will have positive mammograms. Of these 76, only 7 actually have breast cancer. The proportion of women with positive mammograms who have breast cancer is $7/76 \approx .09$.

This is an instance of Bayesian reasoning. The same result is obtained by instantiating (2.1). Let θ be a variable capturing whether the patient has breast cancer, which can take on the values of Yes or No. Let x be a variable capturing whether the result of the mammography, which can take on the values of Positive and Negative. To work through this problem, the first step that is commensurate with frequentist perspective is to define the conditional probability of the data given the model parameter, $p(x \mid \theta)$. This is given in Table 2.1, where the rows represent the probability of a particular result of the mammography given the presence or absence of breast cancer.

We note that ML strictly works off of the information in Table 2.1 corresponding to the observed data. In this case, the column associated with the positive mammography result is the likelihood function. ML views the resulting values of .90 for breast cancer being present and .07 for breast cancer being absent as a likelihood function. Maximizing it then comes to recognizing that the value associated with having breast cancer (.90) is larger than the value for not having breast cancer (.07). That is, $p(x = \text{Positive} \mid \theta = \text{Yes})$ is larger than $p(x = \text{Positive} \mid \theta = \text{No})$. The ML estimate is therefore that θ (breast cancer) is "Yes," and this is indeed what the data (mammography result) has to say about the situation.

In a Bayesian analysis, we combine the information in the data, expressed in the likelihood, with the prior information about the parameter. To specify a prior distribution for

TABLE 2.1
Conditional Probability of the Mammography Result Given Breast Cancer

	Mammography Result (x)	
Breast Cancer (θ)	**Positive**	**Negative**
Yes	.90	.10
No	.07	.93

TABLE 2.2

Prior Probability of Breast Cancer

Breast Cancer (θ)	Probability
Yes	.008
No	.992

the model parameter, θ, we draw from the statement that the patient has no family history of cancer nor any symptoms, and the proportion of women in this population that have breast cancer is .008. Table 2.2 presents the prior distribution, $p(\theta)$.

With these ingredients, we can proceed with the computations in Bayes' theorem. The elements in Tables 2.1 and 2.2 constitute the terms in the numerator on the right-hand side of the second through fourth lines of (2.1). Equation (2.2) illustrates the computations for the numerator of Bayes' theorem, with x positive. The bold elements in the conditional probability of the data are multiplied by their corresponding elements in the prior distribution. Note that the column for *Mammography Result* taking on the value of Positive is involved in the computation, as that was the actual observed value; the column for *Mammography Result* taking on the value of Negative is not involved.

$$p(x=\text{Positive}\mid\theta)p(\theta) \quad = \quad p(x=\text{Positive}\mid\theta) \quad \times \quad p(\theta)$$

Breast Cancer	*Mammography Result*: Positive	*Mammography Result*: Negative		*Breast Cancer*			*Breast Cancer*		
Yes	**.90**	.10	×	Yes	**.008**	=	Yes	**.00720**	(2.2)
No	**.07**	.93		No	**.992**		No	**.06944**	

The denominator in Bayes' theorem is the *marginal probability* of the observed data under the model. For discrete parameters this is given by

$$p(\mathbf{x}) = \sum_{\theta} p(\mathbf{x}\mid\theta)p(\theta). \tag{2.3}$$

For the current example, we have one data point (x) that has a value of Positive, and

$$\begin{aligned} p(x=\text{Positive}) &= p(x=\text{Positive}\mid\theta=\text{Yes})p(\theta=\text{Yes}) + p(x=\text{Positive}\mid\theta=\text{No})p(\theta=\text{No}) \\ &= (.90)(.008)+(.07)(.992) \\ &= .07664. \end{aligned} \tag{2.4}$$

Continuing with Bayes' theorem, we take the results on the right-hand side of (2.2) and divide those numbers by the marginal probability of .07664 from (2.4). Thus we find that the posterior distribution is, rounding to two decimal places, $p(\theta=\text{Yes}\mid x=\text{Positive})\approx .09$, $p(\theta=\text{No}\mid x=\text{Positive})\approx .91$, which captures our belief about the patient after observing the test result is positive. Equation (2.5) illustrates the entire computation in a single representation.

$$p(\theta \mid x = \text{Positive}) = \frac{p(x = \text{Positive} \mid \theta)p(\theta)}{p(x = \text{Positive})}$$

$p(x = \text{Positive} \mid \theta)$

Breast Cancer	*Mammography Result*	
	Positive	Negative
Yes	**.90**	.10
No	**.07**	.93

$\times$ $p(\theta)$

Breast Cancer	
Yes	**.008**
No	**.992**

$=$

Breast Cancer	
Yes	**.00720**
No	**.06944**
	.07664

$= \dfrac{\cdots}{p(x = \text{Positive})}$ (denominator: .07664)

$\approx$ $p(\theta \mid x = \text{Positive})$

Breast Cancer	
Yes	**.09**
No	**.91**

(2.5)

The marginal probability of the observed data $p(\mathbf{x})$ in the denominator of (2.1) serves as a normalizing constant to ensure the resulting mass function sums to one. Importantly, as the notation and example shows, $p(\mathbf{x})$ does not vary with the value of θ. As the last line of (2.1) shows, dropping this term in the denominator reveals that the posterior distribution is proportional to the product of the likelihood and the prior. In the computation in (2.5), all of the information in the posterior for *Breast Cancer* is contained in the values of .00720 for Yes and .06944 for No. However, these are not in an interpretable probability metric. Dividing each by their sum of .07664 yields the values that are in a probability metric—the posterior distribution. This last step is primarily computational. The conceptual implication of the last line of (2.1) is that the key terms relevant for defining the posterior distribution are $p(\mathbf{x} \mid \theta)$ and $p(\theta)$. In addition, as discussed in Chapter 5, the proportionality relationship has important implications for estimation of posterior distributions using simulation-based methods.

Equation (2.1) instantiates Bayes' theorem for discrete parameters. For continuous parameters θ,

$$\begin{aligned} p(\theta \mid \mathbf{x}) &= \frac{p(\mathbf{x},\theta)}{p(\mathbf{x})} \\ &= \frac{p(\mathbf{x} \mid \theta)p(\theta)}{p(\mathbf{x})} \\ &= \frac{p(\mathbf{x} \mid \theta)p(\theta)}{\int_{\theta} p(\mathbf{x} \mid \theta)p(\theta)d\theta} \\ &\propto p(\mathbf{x} \mid \theta)p(\theta). \end{aligned} \tag{2.6}$$

Similarly, for models with both discrete and continuous parameters, the marginal distribution is obtained by summing and integrating over the parameter space, as needed.

2.3 Bernoulli and Binomial Models

To facilitate the exposition of a number of features of Bayesian inference, we develop Bayesian approaches to modeling data from Bernoulli processes. Let x be an observable Bernoulli random variable taking on one of two mutually exclusive and exhaustive values, coded as 1 and 0. Examples of such variables may be the result of a coin flip, whether or not a product is faulty, or in assessment contexts a dichotomously scored response to a task or stimulus. As is common in educational assessment, let the values of 1 and 0 represent a correct and an incorrect response, respectively. The probability that x takes on a value of 1 is an unknown parameter, θ, and the probability that x takes on a value of 0 is $1 - \theta$.

The conditional probability of the variable may be written as

$$p(x \mid \theta) = \text{Bernoulli}(x \mid \theta) = \theta^{x}(1-\theta)^{1-x}. \tag{2.7}$$

Suppose now we have a collection of J independent and identically distributed (i.i.d.) Bernoulli random variables, $\mathbf{x} = (x_1, \ldots, x_J)$, each with parameter θ. (Actually, instead of assuming the variables are i.i.d., what we require for our development is that the variables are considered *exchangeable* [de Finetti, 1974], a notion we discuss in more detail in Chapter 3. For the present purposes, it is sufficient to conceive of them as i.i.d. variables.) The variables in $\mathbf{x}$ may represent the results from a collection of J trials, where each trial is deemed a success or not. For example, we may have dichotomously scored responses to J tasks, where it is assumed the probability of correct (i.e., success) on any task is θ. The conditional probability of the data may be written as

$$p(\mathbf{x} \mid \theta) = \prod_{j=1}^{J} p(x_j \mid \theta) = \prod_{j=1}^{J} \theta^{x_j}(1-\theta)^{1-x_j}. \tag{2.8}$$

Let $y = \sum_{j=1}^{J} x_j$ denote the total number of 1s in $\mathbf{x}$. Continuing with the educational assessment example, y is the total number of correct responses in the set of responses to the J tasks. The conditional probability of y, given θ and J, then follows a binomial mass function

$$p(y \mid \theta, J) = \text{Binomial}(y \mid \theta, J) = \binom{J}{y} \theta^{y}(1-\theta)^{J-y}. \tag{2.9}$$

Suppose then that we observe $y = 7$ successes in $J = 10$ tasks. In the following two sections, ML and Bayesian approaches to inference are developed for this example.

2.3.1 Frequentist Modeling for Binomial Distributions via ML Estimation

In a frequentist approach, the data are treated as random, following a binomial model as in (2.9). The parameter θ is viewed as fixed and unknown, in need of estimation. Figure 2.2 plots the likelihood function for observing $y = 7$ with $J = 10$. The ML estimator is

$$\hat{\theta} = \frac{y}{J}. \tag{2.10}$$

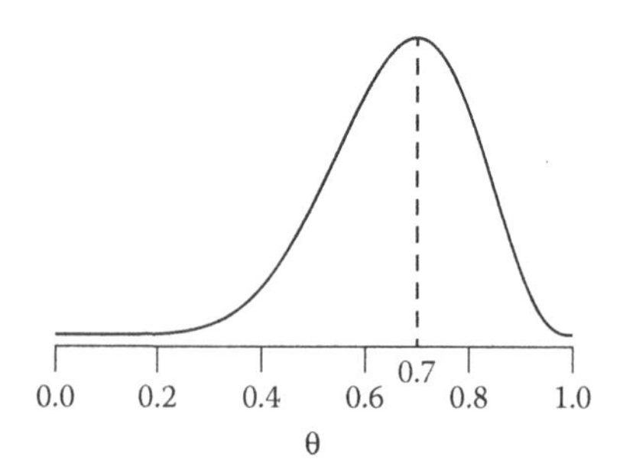

FIGURE 2.2
Likelihood function for a binomial model with $y = 7$ with $J = 10$. The maximum likelihood estimate (MLE) of θ is .7, and is indicated via a vertical line.

In the current example, the MLE is .7, the point where the highest value of the likelihood function occurs, as readily seen in Figure 2.2. The standard error of the ML estimator is

$$\sigma_{\hat{\theta}} = \sqrt{\frac{\theta(1-\theta)}{J}}. \tag{2.11}$$

As θ is unknown, substituting in the MLE yields the estimate of the standard error

$$\hat{\sigma}_{\hat{\theta}} = \sqrt{\frac{\hat{\theta}(1-\hat{\theta})}{J}}. \tag{2.12}$$

In the current example, $\hat{\sigma}_{\hat{\theta}} \approx .14$. When J is large, a normal approximation gives the usual 95% confidence interval of $\hat{\theta} \pm 1.96\hat{\sigma}_{\hat{\theta}}$, or (.42, .98). In the present example, calculating binomial probabilities for the possible values of y when θ actually is .7 tells us that 96% of the MLEs would be included in the score confidence interval [.4, .9]. This interval takes into account the finite possible values for the MLE, the bounded nature of the parameter, and the asymmetric likelihood.

2.3.2 Bayesian Modeling for Binomial Distributed Data: The Beta-Binomial Model

Rewriting (2.6) for the current case, Bayes' theorem is given by

$$p(\theta \mid y, J) = \frac{p(y \mid \theta, J)p(\theta)}{p(y)} = \frac{p(y \mid \theta, J)p(\theta)}{\int_{\theta} p(y \mid \theta, J)p(\theta)d\theta} \propto p(y \mid \theta, J)p(\theta). \tag{2.13}$$

To specify the posterior distribution, we must specify the terms on the right-hand side of (2.13). We now treat each in turn.

The first term, $p(y \mid \theta, J)$, is the conditional probability for the data given the parameter. Once data are observed and it is viewed as a likelihood function of the parameter, the leading term

$$\binom{J}{y}$$

can be dropped as it does not vary with the parameter. It is repeated here with this one additional expression

$$p(y \mid \theta, J) = \text{Binomial}(y \mid \theta, J) = \binom{J}{y} \theta^{y}(1-\theta)^{J-y} \propto \theta^{y}(1-\theta)^{J-y}. \tag{2.14}$$

The second term on the right-hand side of (2.13), $p(\theta)$, is the prior probability distribution for the parameter θ. From the epistemic probability perspective, the prior distribution encodes the analyst's beliefs about the parameter *before* having observed the data. A more spread out, less informative, *diffuse* prior distribution may be employed to represent vague or highly uncertain prior beliefs. More focused beliefs call for a prior distribution that is heavily concentrated at the corresponding points or regions of the support of the distribution. Once data are observed, we employ Bayes' theorem to arrive at the posterior distribution, which represents our updated beliefs arrived at by incorporating the data with our prior beliefs. Section 3.2 discusses principles of specifying prior distributions in greater detail; suffice it to say that we often seek to employ distributional forms that allow us to flexibly reflect a variety of beliefs, are reasonably interpretable in terms of the problem at hand, as well as those that ease the computational burden.

In the current context of a binomial, a popular choice that meets all of these goals is a beta distribution,

$$p(\theta) = p(\theta \mid \alpha, \beta) = \text{Beta}(\theta \mid \alpha, \beta) \propto \theta^{\alpha-1}(1-\theta)^{\beta-1}, \tag{2.15}$$

where the first equality indicates that the probability distribution depends on parameters α and β. The beta distribution is a particularly handy kind of prior distribution for the binomial, a conjugate prior. We will say more about this in Chapter 3 too, but for now the key ideas are that the posterior will have the same functional form as the prior and there are convenient interpretations of its parameters.

A beta distribution's so-called shape parameters α and β tell us about its shape and the focus. The sum $\alpha + \beta$ indicates how informative it is. When $\alpha = \beta$ it is symmetric; when $\alpha < \beta$ it is positively skewed, and when $\alpha > \beta$ it is negatively skewed. When $\alpha < 1$ there is an asymptote at 0 and when $\beta < 1$ there is an asymptote at 1. Beta(1,1) is the uniform distribution. When both $\alpha > 1$ and $\beta > 1$, it has a single mode on (0,1). Figure 2.3 contains density plots for several beta distributions. We see that when used as priors, Beta(5,17) is more focused or informative than Beta(6,6), which is more diffuse or uninformative. Beta(6,6) is centered around .5, while Beta(5,17) is positively skewed, with most of its density around .2. The formulas for its mean, mode, and variance are as follows (see also Appendix B):

$$\text{Mean:} \quad \frac{\alpha}{(\alpha+\beta)}$$

$$\text{Mode:} \quad \frac{\alpha-1}{(\alpha+\beta-2)}$$

$$\text{Variance:} \quad \frac{\alpha\beta}{(\alpha+\beta)^2(\alpha+\beta+1)}.$$

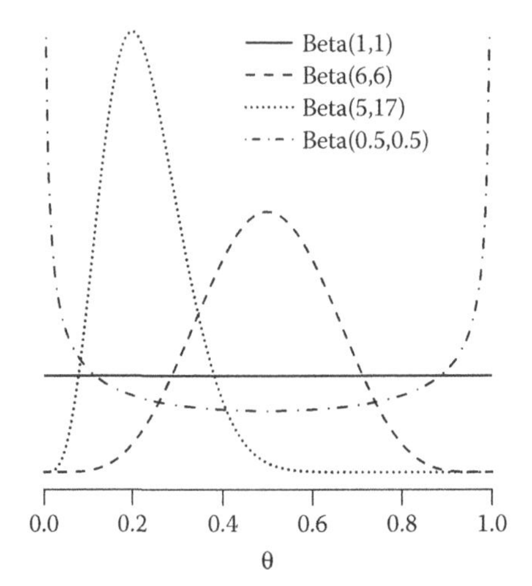

FIGURE 2.3
Density plots for beta distributions, including Beta(1,1) (solid line); Beta(6,6) (dashed line); Beta(5,17) (dotted line); Beta(.5,.5) (dash–dotted line).

The form of the beta distribution expressed in the rightmost side of (2.15) suggests a useful interpretation of the parameters of the beta distribution, α and β. The rightmost sides of the likelihood expressed in (2.14) and the beta prior in (2.15) have similar forms with two components: θ raised to a certain exponent, and $(1-\theta)$ raised to a certain exponent. For the likelihood, these exponents are the number of successes (y) and number of failures ($J-y$), respectively. The similarity in the forms suggests an interpretation of the exponents and hence the parameters of the beta prior distribution, namely, that the information in the prior is akin to $\alpha - 1$ successes and $\beta - 1$ failures.

It can be shown that, with the chosen prior and likelihood, the denominator for (2.13) follows a beta-binomial distribution. For the current purposes, it is useful to ignore this term and work with the proportionality relationship in Bayes' theorem. Putting the pieces together, the posterior distribution is

$$\begin{aligned} p(\theta \mid y, J) &\propto p(y \mid \theta, J) p(\theta \mid \alpha, \beta) \\ &= \text{Binomial}(y \mid \theta, J)\text{Beta}(\theta \mid \alpha, \beta) \\ &\propto \theta^{y}(1-\theta)^{J-y}\theta^{\alpha-1}(1-\theta)^{\beta-1} \\ &\propto \theta^{y+\alpha-1}(1-\theta)^{J-y+\beta-1}. \end{aligned} \tag{2.16}$$

The form of the posterior distribution can be recognized as a beta distribution. That is,

$$p(\theta \mid y, J) = \text{Beta}(\theta \mid y + \alpha, J - y + \beta). \tag{2.17}$$

Turning to the example, recall that $y = 7$ and $J = 10$. In this analysis, we employ a Beta(6,6) prior, depicted in Figure 2.4. Following the suggested interpretation, the use of the Beta(6,6) prior distribution is akin to modeling prior beliefs akin to having seen five successes and five failures. Accordingly, the distribution is symmetric around its mode of .5, embodying the beliefs that our best guess is that the value is most likely near .5, but we are not very confident, and it is just as likely to be closer to 0 as it is to be closer to 1. Figure 2.4 also plots the likelihood and the posterior distribution. Following (2.17), the posterior is a

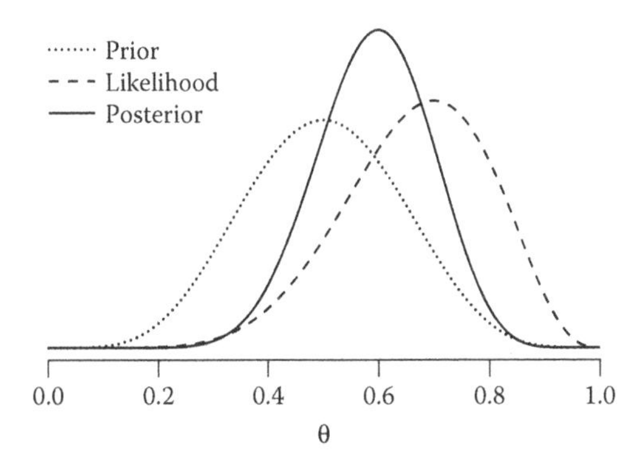

FIGURE 2.4
Prior, likelihood, and posterior for the beta-binomial model with a Beta(6,6) prior for θ, $y = 7$, and $J = 10$.

Beta distribution, with parameters $y + \alpha = 7 + 6$ and $J - y + \beta = 10 - 7 + 6$, or more simply, Beta(13,9). Its mean (.59) and mode (.60) are similar because the parameters of the posterior (13 and 9) are similar and so it is not far from being symmetric.

2.4 Summarizing Posterior Distributions

The posterior distribution constitutes the "solution" to the analysis. This probability distribution summarizes the entirety of our belief about θ from both the prior and the information in the data. We will see that in some problems using this entire posterior is exactly what we need to do. But for communicating the results more concisely, researchers often prefer summaries in the form of a few key features of the posterior. The posterior can in fact be summarized using familiar strategies for characterizing distributions. Common point summaries of central tendency include the mean, median, or mode. Common summaries of variability include the posterior standard deviation, variance, and interquartile range. Percentiles of the posterior may be used to construct interval summaries. Generally, a $100(1-\alpha)\%$ *central credibility interval* extends from the $100(\alpha/2)\%$ile to the $100(1-\alpha/2)\%$ile.* For example, with $\alpha = .05$, the interval defined by the 2.5th and 97.5 th percentiles of the posterior constitute the 95% central credibility interval. Alternatively, the $100(1-\alpha)\%$ *highest posterior density (HPD) interval* is the interval that contains $(1-\alpha)\%$ of the space of θ where the posterior density of θ is maximized. Informally, it is the interval that covers $(1-\alpha)\%$ of the posterior where the relative probability of any value of θ inside the interval is greater than that of any value outside the interval. The central posterior and credibility intervals will often be similar, and will be equal when the posterior is unimodal and symmetric. In some cases, the highest posterior region is discontinuous, rather than an interval (e.g., in bimodal distributions, see Gill, 2007 p. 49). For the example of the beta-binomial model, the posterior is a Beta(13,9) distribution. The posterior mean, mode, and standard deviation are .59, .60, and .10, respectively. The 95% central credibility interval is (.38, .78) and the 95% HPD interval is (.39, .79). Figure 2.5 plots the Beta(13,9) posterior, shading the 95% HPD interval, and locating the mode and the mean.

* The use of α here refers to a probability, and is not to be confused with the use of α as a parameter of a beta distribution.

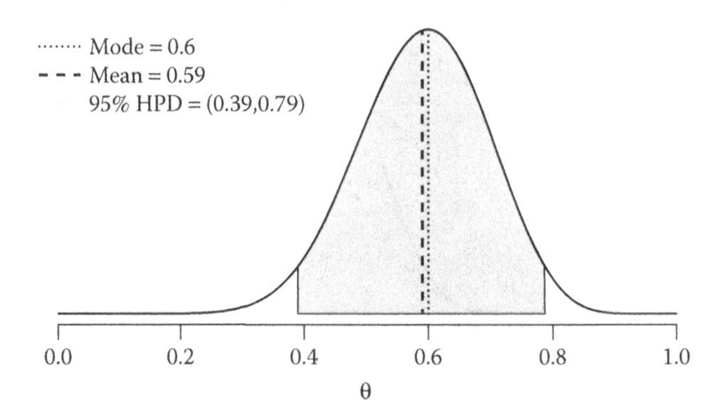

FIGURE 2.5
Beta(13,9) posterior, with shaded 95% highest posterior density (HPD) interval, mode, and mean.

Importantly, the posterior distribution, and these summaries, refers to the *parameter*. The posterior standard deviation characterizes our uncertainty about the parameter in terms of variability in the distribution that reflects our belief, after incorporating evidence from data with our initial beliefs. Similarly, the posterior intervals are interpreted as direct probabilistic statements of our belief about the unknown parameter; that is, we express our degree of belief that the parameter falls in this interval in terms of the probability .95. Similarly, probabilistic statements can be asserted for ranges of parameters or among multiple parameters (e.g., the probability that one parameter exceeds a selected value, or the probability that one parameter exceeds another).

2.5 Graphical Model Representation

Bayesian models may be advantageously represented as a particular type of graphical model (Lunn, Spiegelhalter, Thomas, & Best, 2009). We briefly describe a graphical modeling approach to representing and constructing models for Bayesian analyses.

An important class of graphical models with connections to Bayesian modeling is acyclic, directed graphs (frequently referred to as *directed acyclic graphs*, DAGs), which consist of the following elements:

- A set of entities/variables represented by ellipses (circles) or boxes (rectangles) and referred to as *nodes*.
- A set of directed edges, represented by one-headed arrows, indicating dependence between the entities/variables/nodes. A node at the destination of an edge is referred to as a *child* of the node at the source of the edge, its *parent*.
- For each variable with one or more parents, there is a conditional probability distribution for the variable given its parent(s).
- For each variable without parents, there is a probability distribution for the variable.

The graphs are directed in that all the edges are directed, represented by one-headed arrows, so that there is a "flow" of dependence. Further, the graphs are acyclic in that, when moving along paths from any node in the direction of the arrows, it is impossible to return to that node. The graph also contains a number of *plates* associated with indexes, used to efficiently represent many nodes. Following conventions from path diagrams in structural equation modeling (Ho, Stark, & Chernyshenko, 2012) we employ rectangles to represent observable entities/variables and circles to represent latent or unknown variables/entities. We further discuss and compare path diagrams to DAGs in Chapter 9.

The structure of the graph conveys how the model structures the joint distribution. Let $\mathbf{v}$ denote the full collection of entities under consideration. The joint distribution $p(\mathbf{v})$ may be factored according to the structure of the graph as

$$p(\mathbf{v}) = \prod_{v \in \mathbf{v}} p\big(v \mid pa(v)\big), \tag{2.18}$$

where $pa(v)$ stands for the parents of v; if v has no parents, $p\big(v \mid pa(v)\big)$ is taken as the unconditional (marginal) distribution of v. Equation (2.18) indicates that we can represent the structure of the joint distribution in a model in terms of the graph. For each variable, we examine whether the node in the graph has parents. If it does not, there is an unconditional distribution for the variable. If it does have parents, the distribution for the variable is specified as conditional on those parents. Thus, the graph reflects key dependence and (conditional) independence relationships in the model* (see Pearl, 2009).

Figure 2.6 presents two versions of a DAG for the beta-binomial model. The first version depicts the graph where the node for y is modeled as child of the node for θ, which has no parents. This reflects that the joint distribution of the random variables is constructed as $p(y,\theta) = p(y \mid \theta)p(\theta)$. The second version is an expanded graph that also contains the entities in the model that are known or fixed, communicating all the entities that are involved in specifying the distributions. More specifically, the graph reflects that

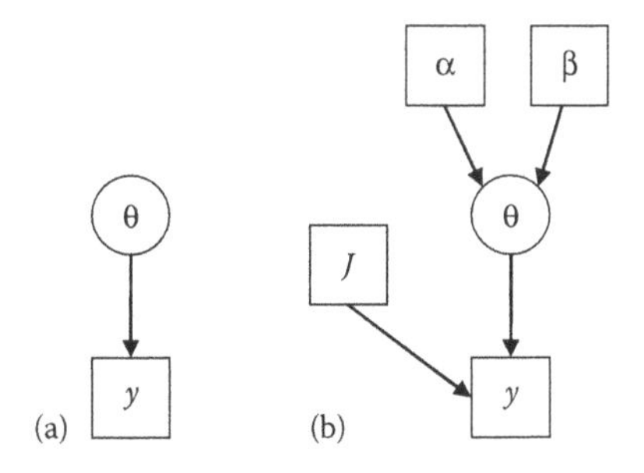

FIGURE 2.6
Two versions of a directed acyclic graph for the beta-binomial model: (a) depicting entities in the joint distribution; (b) additionally depicting known or fixed entities.

* We note that the conditional independence relationships expressed in the graph are those explicitly stated by the model, visible in an ordering of the variables. There may be additional conditional independence relationships among the variables that are not illustrated in a given ordering of the variables, but would show up under different orderings of variables.

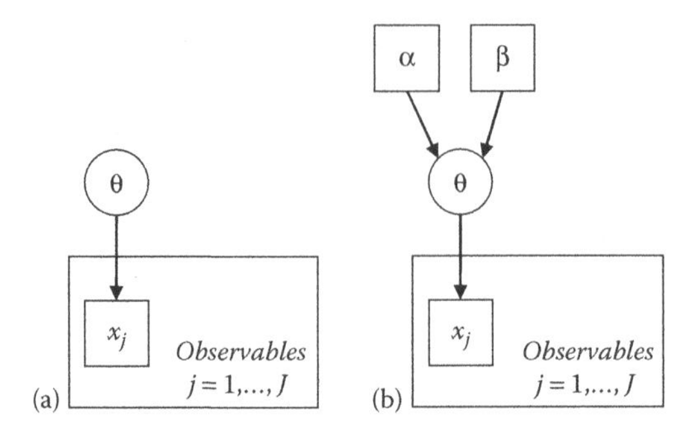

FIGURE 2.7
Two versions of a directed acyclic graph for the beta-Bernoulli model: (a) depicting entities in the joint distribution; (b) additionally depicting known or fixed entities.

- The distribution of y involves θ and J; $p(y \mid \theta, J) = \text{Binomial}(\theta, J)$.
- The distribution of θ involves α and β; $p(\theta \mid \alpha, \beta) = \text{Beta}(\alpha, \beta)$.
- The remaining nodes, J, α, β are known values.*

This version of the graph is potentially misleading in that it depicts θ as having parents, whereas in the model, θ is substantively exogenous in the sense that it does not depend on any other variables. However, it is particularly helpful in three respects. First, it lays out all the entities in the model, including values that are fixed by the design or are chosen by the analyst. Second, it aids in the writing of the WinBUGS code for the model; see Section 2.6 for an example. Third, it is useful in building models. Note that the nodes with no parents are all rectangles, which reflects that all such nodes are known entities. In this graph, if there is an unknown entity in the model (i.e., is a circle) it is assigned a distribution that invokes parent entities. If these entities are known or chosen values, they are rectangles. If however these entities are themselves unknown (i.e., they are circles), they must have parents. This process of adding layers of parents continues until the only remaining entities are known values (rectangles).

Figure 2.7 presents two versions of a DAG for the beta-Bernoulli model, where again the second version is an expanded version of the first, illustrating all the entities that are involved in specifying the distributions. In the beta-Bernoulli model, the individual observables (i.e., the xs), rather than their sum (y), are modeled as dependent on θ. The presence of multiple observables is indicated by the *plate*, depicted as a rectangle that surrounds the node for x_j; the presence of the subscript indicates that there are multiple xs, namely, one for each value of j. The text in the lower-right corner of the plate indicates that the index j takes on values from $1,\ldots,J$. Note that θ is *not* inside the plate. This communicates that it does *not* vary over j, rather, there is a single θ that serves as the parent variable for all the xs.

2.6 Analyses Using WinBUGS

The WinBUGS code for the model and a data statement for the beta-binomial example are given below. In WinBUGS, text following a "#" on any line is ignored, and is used below to insert comments. The first section of the code contains the syntax for the model, and

* J is known because that is the size of the sample in the experiment. α and β are known because we have chosen them to represent our belief about θ before we see the data.

is contained in the braces—the left-hand brace "{" following the "model" statement at the top, and the right-hand brace "}" toward the bottom. Within the braces, the two (noncommented) lines of code give the prior distribution for θ and the conditional distribution of y given θ. The (noncommented) line after the right-hand brace is a data statement that contains known values that are passed to WinBUGS.

```
-----------------------------------------------------------------------
#########################################################################
# Model Syntax
#########################################################################
model{

#########################################################################
# Prior distribution
#########################################################################
theta ~ dbeta(alpha,beta)

#########################################################################
# Conditional distribution of the data
#########################################################################
y ~ dbin(theta, J)

} # closes the model syntax

#########################################################################
# Data statement
#########################################################################
list(J=10, y=7, alpha=6, beta=6)
-----------------------------------------------------------------------
```

We conducted an analysis in WinBUGS for this example, running a chain for 50,000 iterations. We will have more to say about the underlying mechanisms of the estimation strategies implemented in WinBUGS in Chapter 5. For the moment, it is sufficient to state that the analysis produces an empirical approximation to the posterior distribution in that the iterations constitute draws from the posterior distribution. Table 2.3 presents summary statistics based on these draws. WinBUGS does not produce the HPD interval; these values were obtained using the coda package in R. Figure 2.8 plots a smoothed density of the 50,000 draws as produced by the coda package. A similar representation is available from WinBUGS. The empirical approximation to the posterior from WinBUGS is quite good. Note the similarity in shape of the density in Figure 2.8 to the Beta(13,9) density in Figures 2.4 and 2.5. The empirical summaries of the distribution in Table 2.3 are equal to their analytical values, up to two decimal places.

We emphasize how the mathematical expressions of the model, the DAG, and the WinBUGS code cohere. For the beta-binomial, for example, Figure 2.9 depicts the coherence

TABLE 2.3

Summary Statistics for the Posterior for the Beta-Binomial Model with a Beta(6,6) Prior for θ, $y = 7$, and $J = 10$ from the WinBUGS Analysis

Parameter	Mean	Median	Standard Deviation	95% Central Credibility Interval	95% Highest Posterior Density Interval
θ	0.59	0.59	0.10	(0.38, 0.78)	(0.39, 0.79)

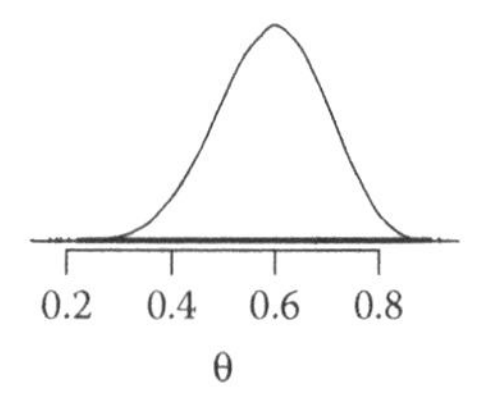

FIGURE 2.8
Smoothed density of the draws from WinBUGS for the posterior for the beta-binomial model with a Beta(6,6) prior for $y = 7$, and $J = 10$.

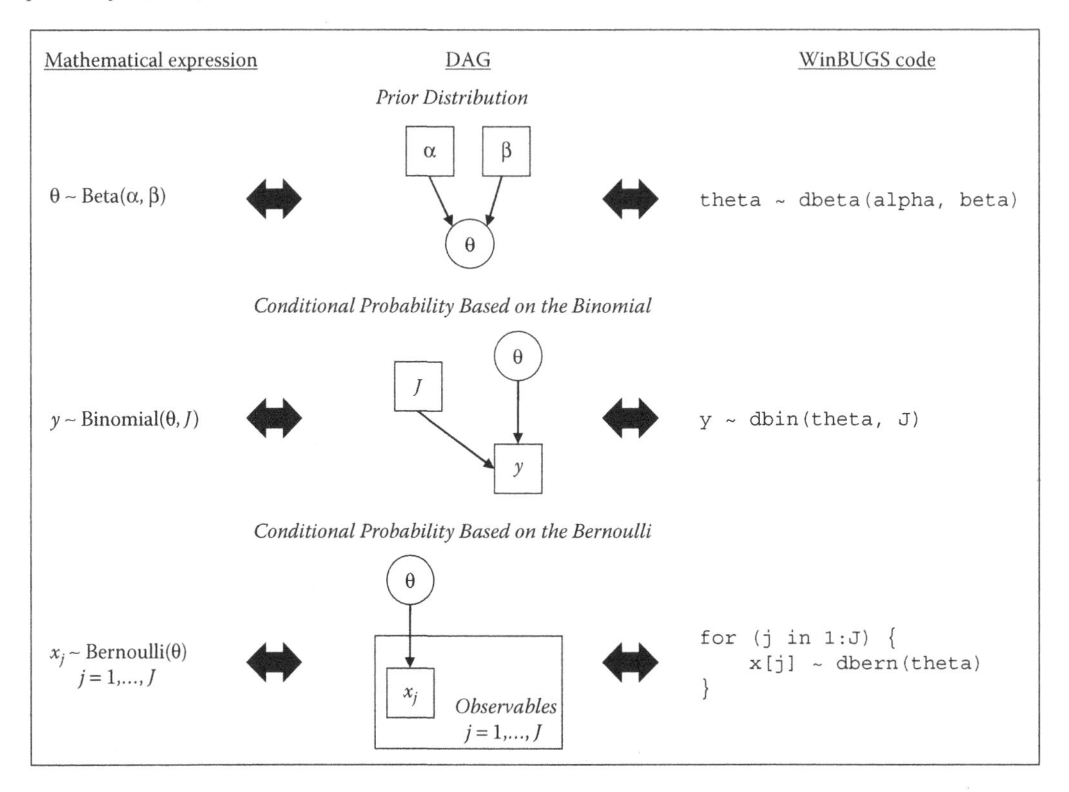

FIGURE 2.9
Correspondence among the WinBUGS code and data statement for the beta-binomial and beta-Bernoulli models. The top row communicates the prior distribution. The second row communicates the conditional probability of the data based on the binomial model. The third row communicates the conditional probability of the data based on the Bernoulli model.

for the prior distribution and the conditional distribution of the data in the top and middle rows, respectively. (The bottom row shows the DAG and the WinBUGS code when we instead use the Bernoulli specifications for the J replications individually.) Importantly, the correspondence between the mathematical model and the DAG is a general point of Bayesian modeling, which simplifies model construction and calculations of posterior distributions (Lauritzen & Spiegelhalter, 1988; Pearl, 1988). The correspondence between these features and the WinBUGS code is fortuitous for understanding and writing the WinBUGS code. In this case, the correspondence is exact. In other cases, the correspondence between the code and the mathematical expressions/DAG will be less than exact. Nevertheless, we find it useful in our work to conceive of models using these various representations: clarity can be gained—and troubleshooting time can be reduced—by laying out and checking the correspondence between the mathematical expressions, DAG, and code. For example, it is seen that any child in the graph appears on the left-hand side of a mathematical expression and the WinBUGS code. For any such node in the graph, its parents are the terms that appear on the right-hand side of the mathematical expression and the WinBUGS code.

The WinBUGS code for the model and a data statement for the beta-Bernoulli example are given below.

```
-------------------------------------------------------------------------
#########################################################################
# Model Syntax
#########################################################################
model{

#########################################################################
# Prior distribution
#########################################################################
theta ~ dbeta(alpha,beta)

#########################################################################
# Conditional distribution of the data
#########################################################################
for(j in 1:J){
  x[j] ~ dbern(theta)
}

} # closes the model syntax

#########################################################################
# Data statement
#########################################################################
list(J=10, x=c(1,0,1,0,1,1,1,1,0,1), alpha=6, beta=6)
-------------------------------------------------------------------------
```

There are two key differences from the beta-Binomial code. First, the code for the individual variables are modeled with a "for" statement, which essentially serves as a loop over the index j. Inside the "for" statement (i.e., the loop), each individual x_j is specified as following a Bernoulli distribution with parameter θ. The second key difference is that the data being supplied are the values of the individual xs for the 10 variables, rather than their sum.

Again, we emphasize the coherence among the different representations of the model. The last row in Figure 2.9 contains the conditional probability of the data, expressed, mathematically, as a DAG, and as the WinBUGS code. Note that repeated structure over j is communicated in each representation: with a list in the mathematical expression, as a plate in the DAG, and as a "for" statement in WinBUGS.

2.7 Summary and Bibliographic Note

We have introduced the mechanics of Bayesian inference using two examples: (a) two dichotomous variables in the context of the medical diagnosis example, and (b) Bernoulli and binomial models using beta prior distributions. Comprehensive introductory treatments of

Bayesian data analysis, modeling, and computation generally and in the contexts covered in this chapter may be found in a number of texts, including those cited in the introduction to Part I. To this, we add that Lunn et al. (2013) provided an account rooted in graphical models introduced in this chapter, as well as details on the use of WinBUGS and its variants.

To summarize the principles of the current chapter on the mechanics of Bayesian analyses, we can hardly do better than Rubin, who described a Bayesian analysis as one that (1984, p. 1152)

> treats known values [$\mathbf{x}$] as observed values of random variables [via $p(\mathbf{x} \mid \theta)$], treats unknown values [θ] as unobserved random variables [via $p(\theta)$], and calculates the conditional distribution of unknowns given knowns and model specifications [$p(\theta \mid \mathbf{x})$] using Bayes's theorem.

Following Gelman et al. (2013), this is accomplished in several steps:

1. Set up the full probability model. This is the joint distribution of all the entities, including observable entities (i.e., data $\mathbf{x}$) and unobservable entities (i.e., parameters θ) in accordance with all that is known about the situation. However, specifying the joint distribution $p(\mathbf{x},\theta)$ may be very difficult for even fairly simple situations. Returning to the medical diagnosis example, it may be difficult to articulate the bivariate distribution for *Breast Cancer* and *Mammography Result* directly. Instead, we structure the joint distribution by specifying the marginal distribution for *Breast Cancer*, that is, without appeal to *Mammography Result*, and then specify a conditional distribution for *Mammography Result* given *Breast Cancer*. This is an instance of the general strategy, in which we factor the joint distribution as $p(\mathbf{x},\theta) = p(\mathbf{x} \mid \theta)p(\theta)$.
2. Condition on the observed data ($\mathbf{x}$) and calculate the conditional probability distribution for the unobservable entities (θ) of interest given the observed data. That is, we obtain the posterior distribution $p(\theta \mid \mathbf{x})$ via Bayes' theorem

$$p(\theta \mid \mathbf{x}) = \frac{p(\mathbf{x},\theta)}{p(\mathbf{x})} = \frac{p(\mathbf{x} \mid \theta)p(\theta)}{p(\mathbf{x})} \propto p(\mathbf{x} \mid \theta)p(\theta).$$

3. Do all the things we are used to doing with fitted statistical models; examples include but are not limited to: examining fit, assessing tenability or sensitivity to assumptions, evaluating the reasonableness of conclusions, respecifying the model if warranted, and summarizing results. (Several of these activities are treated in Chapter 10.)

Turning to the specifics of the examples, Bayesian approaches to medical diagnosis of the sort illustrated here date to Warner, Toronto, Veasey, and Stephenson (1961). The medical diagnosis model may be seen as a latent class (Chapter 13) or Bayesian network (Chapter 14) model, and principles and extensions of this model are further discussed in those chapters. Lindley and Phillips (1976) provided an account of Bayesian approaches to inference with Bernoulli and binomial models using beta prior distributions, strongly contrasting it with conventional frequentist approaches. Novick, Lewis, and Jackson (1973) discussed the extension to the case of multiple binomial processes, arguing that a Bayesian approach assuming exchangeability offers advantages in criterion-referenced assessment, particularly for short tests.

Finally, we have developed the mechanics of Bayesian inference in part by contrasting it with frequentist inference. In doing so, we have alluded to a number of reasons why adopting a Bayesian approach is advantageous. We take up this topic in greater detail in the next chapter. Before doing so, it is important to note that our account of frequentist inference here is far from comprehensive. For our purposes, the key idea is the reversal of the inferential direction that the Bayesian approach affords. Aside from a theoretical view of *probability* per se, there are a variety of methods of *statistical inference,* that focus on the distribution of results (statistics, likelihood functions, etc.) that arise from repeated samples of the data given the fixed but unknown parameters. It is this directionality that we are referring to in the way we use "frequentist" as a contrast to the Bayesian paradigm that requires the full probability model (including priors), where we then base inference on the posterior distribution of parameters and make probability statements about them conditional on data.

Exercises

2.1 Reconsider the breast cancer diagnosis example introduced in Section 2.2. Suppose another patient reports being symptomatic and a family history such that the proportion of women with her symptoms and family history that have breast cancer is .20. She undergoes a mammography screening, and the result is positive. What is the probability that she has breast cancer? Contrast this with the result for the original example developed in Section 2.2.

2.2 Reconsider the breast cancer diagnosis example introduced in Section 2.2, for an asymptomatic woman with no family history; the proportion of such woman that have breast cancer is .008. Suppose that instead of undergoing a mammography screening, the woman is administered a different screening instrument where the probability of a positive result given the presence of breast cancer is .99, and the probability of a positive result given the absence of breast cancer is .03. Suppose the result is positive. What is the probability that she has breast cancer? Contrast this with the result for the original example developed in Section 2.2.

2.3 Revisit the beta-binomial example for a student who has $y = 7$ successes on $J = 10$ attempts.

a. Suppose you wanted to encode minimal prior information, representing beliefs akin to having seen 0 successes and 0 failures. What prior distribution would you use?

b. Using the prior distribution chosen in part (a), obtain the posterior distribution for θ analytically as well as the solution from WinBUGS. How does the posterior distribution compare to the prior distribution and the MLE, for example in terms of shapes, and the posterior means and modes for the prior and posterior versus the MLE?

c. Suppose you wanted to encode prior information that reflects beliefs that the student is very capable and would likely correctly complete 90% of all such tasks, but to express your uncertainty you want to assign the prior a weight akin to 10 observations. That is, your prior belief is akin to having seen nine successes and one failure. What prior distribution would you use?

d. Using the prior distribution chosen in part (c), obtain the posterior distribution for θ analytically as well as the solution from WinBUGS. How does the posterior distribution compare to the prior distribution and the MLE?

e. Suppose you wanted to encode prior information that reflects beliefs that the student is not very capable and would likely correctly complete 10% of all such tasks, but to express your uncertainty you want to assign the prior a weight akin to 10 observations. That is, your prior belief is akin to having seen one success and nine failures. What prior distribution would you use?

f. Using the prior distribution chosen in part (e), obtain the posterior distribution for θ analytically as well as the solution from WinBUGS. How does the posterior distribution compare to the prior distribution and the MLE?

2.4 Exercise 2.3(f) asked you to consider the situation where your prior beliefs were akin to having seen one success and nine failures, and then you went on to observe $y = 7$ successes on $J = 10$ attempts.

a. Keeping the same prior distribution you specified in Exercise 2.3(f), suppose you had observed 14 successes on 20 tasks. What is the posterior distribution for θ? Obtain an analytical solution for the posterior as well as the solution from WinBUGS.

b. Keeping the same prior distribution, suppose you had observed 70 successes on 100 tasks. What is the posterior distribution for θ? Obtain an analytical solution for the posterior as well as the solution from WinBUGS.

c. What do your results from parts (a)–(b) and Exercise 2.3(f) indicate about the influence of the prior and the data (likelihood) on the posterior?

2.5 Suppose you have a student who successfully completes all 10 tasks she is presented.

a. What is the MLE of θ?

b. Using a prior distribution that encodes minimal prior information representing beliefs to having seen 0 successes and 0 failures, obtain the posterior distribution for θ. Obtain an analytical solution as well as the solution from WinBUGS. How does the posterior compare to the prior distribution, as well as the MLE?

c. Suppose before observing the student's performance, you had heard from a colleague that she was a very good student, and the student would likely correctly complete 80% of all such tasks. Suppose you wanted to encode this information into a prior distribution, but only assigning it a weight akin to 10 observations. What prior distribution would you use?

d. Using the prior distribution chosen in part (c), obtain the posterior distribution for θ. Obtain an analytical solution as well as the solution from WinBUGS. How does the posterior compare to the prior distribution, as well as the MLE?

3

Conceptual Issues in Bayesian Inference

Chapter 2 described the mechanics of Bayesian inference, including aspects of model construction and representation, Bayes' theorem, and summarizing posterior distributions. These are foundational *procedural* aspects of Bayesian inference, which will be instantiated in a variety of settings throughout the book. This chapter treats foundational *conceptual* aspects of Bayesian inference.

The chief aims of this chapter are to introduce concepts of Bayesian inference that will be drawn upon repeatedly throughout the rest of the book, and to advance the argument concerning the alignment of Bayesian approaches to statistical inference. In addition, this chapter serves to further characterize features of Bayesian approaches that are distinct from frequentist approaches, and may assuage concerns of those new to Bayesian inference. We suspect that readers steeped in frequentist traditions might not feel immediately comfortable treating parameters as random, specifying prior distributions, and using the language of probability in reference to parameters. For those readers, this chapter may serve as an introduction, albeit far from a comprehensive one, to the arguments and principles that motivate adopting a Bayesian approach to inference.

In Section 3.1, we cover how the prior and data combine to yield the posterior, highlighting the role of the relative amounts of information in each. We then discuss some principles used in specifying prior distributions in Section 3.2. In Section 3.3, we provide a high-level contrast between the Bayesian and frequentist approaches to inference. In Section 3.4, we introduce and connect the key ideas of exchangeability and conditional independence to each other, and to Bayesian modeling. In Section 3.5, we lay out reasons for adopting a Bayesian approach to conducting analyses. In Section 3.6, we describe four different perspectives on just what is going on in Bayesian modeling. We conclude the chapter in Section 3.7 with a summary and pointers to additional readings on these topics.

3.1 Relative Influence of the Prior Distribution and the Data

In Section 2.3, we introduced the Beta-binomial model, illustrating the case of $y = 7$ successes in $J = 10$ attempts and a Beta(6,6) prior distribution which, following Bayes' theorem, yields a Beta(13,9) posterior distribution. Figure 2.4 depicted the situation graphically. We pick up on this example to illustrate the relative influences and effects of the information in the prior and the data.

3.1.1 Effect of the Information in the Prior

To illustrate the effects of the information in the prior, we reanalyzed the Beta-binomial model using the same data ($y = 7$, $J = 10$) and different priors. Figure 3.1 summarizes the reanalyses of the binomial data using Beta(1,1), Beta(10,2), and Beta(2,10) priors (see Exercise 2.3).

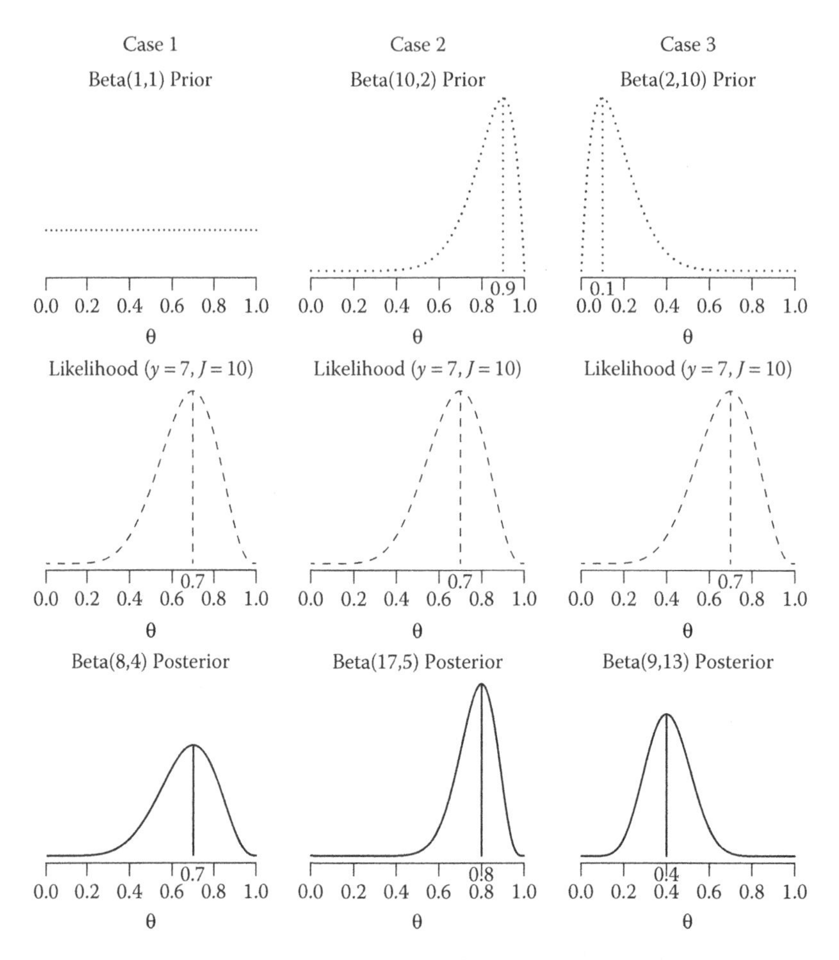

FIGURE 3.1
Prior, likelihood function, and posterior for the Beta-binomial model with $y = 7$ with $J = 10$ and three different cases defined by different prior distributions. Reading down the columns presents the prior, likelihood, and posterior for each case. Where applicable, the mode of the distribution and the MLE is indicated via a vertical line.

Summaries of the resulting posteriors are given in Table 3.1, which also lists the results from the preceding ML analysis and the Bayesian analysis using a Beta(6,6) prior.

As depicted in the first column of Figure 3.1, the Beta(1,1) distribution is uniform over the unit line, which is aligned with the interpretation that it encodes 0 prior successes and 0 prior failures in the information in the prior. The posterior distribution is the Beta(8,4) distribution. This posterior has exactly the same shape as the likelihood. This is because the posterior density function (i.e., the height of the curve for the posterior) is just the normalized product of the height of the curve for the likelihood and the height of the curve for the prior. Accordingly, the posterior mode is identical to the maximum likelihood estimate (MLE) (.70). The second column in Figure 3.1 displays the analysis using a Beta(10,2) prior that encodes nine prior successes and one prior failure. The prior here is concentrated toward larger values, and in conjunction with the likelihood that is peaked at .7, the posterior density is concentrated around values near .8. The third column in Figure 3.1 displays

TABLE 3.1

Summaries of Results for Frequentist and Bayesian Analyses of Binomial Models for Dichotomous Variables

					Point		Variability	95% Interval	
Approach	y	J			**MLE**[a]		**Estimate of SE**[b]	**Confidence Interval**	
Frequentist	7	10			0.70		0.14	(0.39, 0.90)	
	y	J	**Prior**	**Posterior**	**Posterior Mean**	**Posterior Mode**	**Posterior SD**[c]	**Central Credibility Interval**	**Highest Posterior Density Interval**
Bayesian	7	10	Beta(6,6)	Beta(13,9)	0.59	0.60	0.10	(0.38, 0.78)	(0.39, 0.79)
	7	10	Beta(1,1)	Beta(8,4)	0.67	0.70	0.13	(0.39, 0.89)	(0.41, 0.91)
	7	10	Beta(10,2)	Beta(17,5)	0.77	0.80	0.09	(0.58, 0.92)	(0.60, 0.93)
	7	10	Beta(2,10)	Beta(9,13)	0.41	0.40	0.10	(0.22, 0.62)	(0.21, 0.61)
	14	20	Beta(2,10)	Beta(16,16)	0.50	0.50	0.09	(0.33, 0.67)	(0.33, 0.67)
	70	100	Beta(2,10)	Beta(72,40)	0.64	0.65	0.05	(0.55, 0.73)	(0.55, 0.73)

[a] MLE = Maximum Likelihood Estimate.
[b] SE = Standard Error.
[c] SD = Standard Deviation.

the analysis using a Beta(2,10) prior that encodes one prior success and nine prior failures. Here, the prior is concentrated closer to 0, and the resulting posterior is concentrated more toward moderate values of θ.

Figure 3.1 illustrates the effect of the information in the prior, holding the information in the data (i.e., the likelihood) constant. In all cases, the posterior distribution is a synthesis of the prior and the likelihood. When there is little or no information in the prior (left column of Figure 3.1), the posterior is essentially a normalized version of the likelihood. To the extent that the prior carries information, it will influence the posterior such that the posterior will be located "in between" the prior and the likelihood. In the middle and right columns of Figure 3.1, this can be seen by noting that the posterior mode is in between the prior mode and the maximum of the likelihood.

Viewed from a lens that focuses on the influence of the prior on the posterior, we may say that the posterior gets *shrunk* from the data (likelihood) toward the prior distribution. To further demonstrate this point, consider a situation in which we observe a student correctly complete all 10 tasks she attempts (Exercise 2.5). To model our prior beliefs that she is very capable, but with uncertainty akin to this being based on 10 observations, we employ the Beta(9,3) distribution, depicted as a dotted line in Figure 3.2. The dashed line in Figure 3.2 depicts the likelihood for $y = 10$ successes on $J = 10$ attempts. Following Bayes' theorem, the posterior is the Beta(19,3) distribution, depicted as a solid line in Figure 3.2. The likelihood monotonically increases with θ, yielding a maximum when $\theta = 1$. Substantively, a value of 1 for θ means that the probability that the student will correctly complete any such task is 1.0; there is no chance that she will not correctly complete any such task. Although 1 is in fact the MLE, we do not think concluding that $\theta = 1$ is reasonable.* In the Bayesian analysis, the Beta(19,3) posterior distribution falls in between the prior distribution and the likelihood. Relative to the point that maximizes the likelihood (1.0), the posterior mode of .90 is shrunk toward the prior mode of .80. Based on using the mode (or similarly the mean of .86) as a point summary, we would conclude that the student has a high probability of completing such tasks. Such a statement could be buttressed by an expression of uncertainty through, say, the posterior standard deviation of

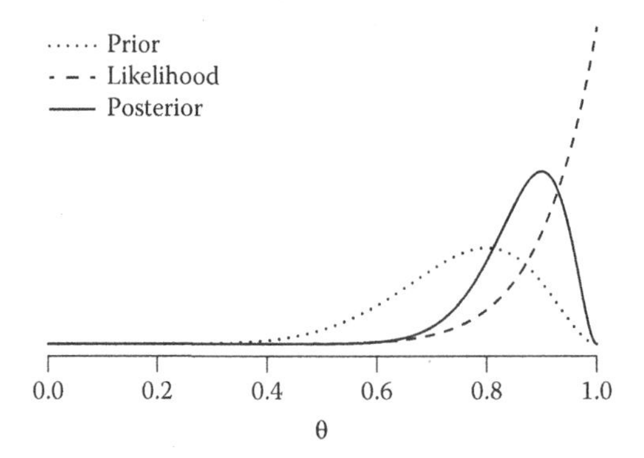

FIGURE 3.2
Beta(9,3) prior, likelihood function, and Beta(19,3) posterior for the Beta-binomial example with $y = 10$ and $J = 10$.

* If you disagree, note that the same conclusion would be reached by seeing nine successes in nine attempts, similarly eight successes in eight attempts, and so on down to one success in one attempt. If you think that is reasonable, the first author would like to report that he just came back from the basketball court where he took and successfully made a free throw. Please feel free to conclude that he will never miss a free throw. Unfortunately, everyone who has ever played basketball with him knows better.

.07 or a 95% HPD interval of (.72, .98), meaning that the probability that θ is in the interval (.72, .98) is .95. We may view this Bayesian analysis as one in which prior distribution *reins in* what is suggested by the data, which is subject to substantial sampling variability in small samples. In this case, the prior distribution guards against the somewhat nonsensical conclusion suggested by the data alone.

This tempering of what is suggested strictly by the data in light of what is already believed is common in statistical inference, as in practices of denoting and handling outliers. It is likewise common in inference more broadly: prior beliefs prevent us from abandoning the laws of physics when we see David Copperfield make something "disappear," much as they prevented the scientific community from doing so when it was reported that a particle was measured as traveling faster than the speed of light (Matson, 2011), much as they prevented the scientific community from concluding that precognition exists when certain analyses suggested they do (Bem, 2011; Wagenmakers, Wetzels, Borsboom, & van der Maas, 2011). The point is not that such things are wrong—perhaps precognition does occur or particles do move faster than the speed of light—the point is that it is appropriate to reserve concluding as much when presented with finite data situated against a backdrop of alternative beliefs.

3.1.2 Effect of the Information in the Data

It is instructive to consider the effects of the information in the data. Continuing with the Beta-binomial model, (2.17) is repeated here,

$$p(\theta \mid y, J) = \text{Beta}(\theta \mid y + \alpha, J - y + \beta),$$

and expresses how the posterior distribution is influenced by the prior and the data. Consider now what occurs as the amount of data increases. As J increases, the posterior distribution becomes increasingly dominated by the data, and the influence of the prior diminishes. Figure 3.3 illustrates the results for J = 10, 20, and 100, holding constant the prior distribution and the proportion of successes y/J = .7 (Exercise 2.4). Numerical summaries for these cases are contained in Table 3.1. Specifically, the prior distribution is the Beta(2,10), which concentrates most of the density toward lower values of θ, which does not align with the information in the data encoded in the likelihood. The precision of the data increases with J, represented by the likelihood becoming more peaked around .7. As a result, the posterior increasingly resembles the likelihood. This represents a general principle in Bayesian modeling: as the amount of data increases the posterior increasingly resembles the likelihood, which represents the contribution of the data, rather than the prior.* That is, the data "swamp" the prior, and increasingly drives the solution, reflecting Savage's principle of stable estimation (Edwards, Lindman, & Savage, 1963). The implication is that, for a large class of models, analysts that specify even wildly different prior distributions but allow for the possibility that the data may suggest otherwise will find that their resulting posterior distributions will become increasingly similar with more data (Blackwell & Dubins, 1962; c.f. Diaconis & Freedman, 1986).

* This is the usually case. Exceptions include situations where the support of the prior does not include the values suggested by the data. As an extreme case, if the prior states the probability of a discrete parameter taking on a particular value θ_0 is $p(\theta = \theta_0) = 0$, no amount of data is going to yield a posterior where $p(\theta = \theta_0 \mid \mathbf{x}) > 0$.

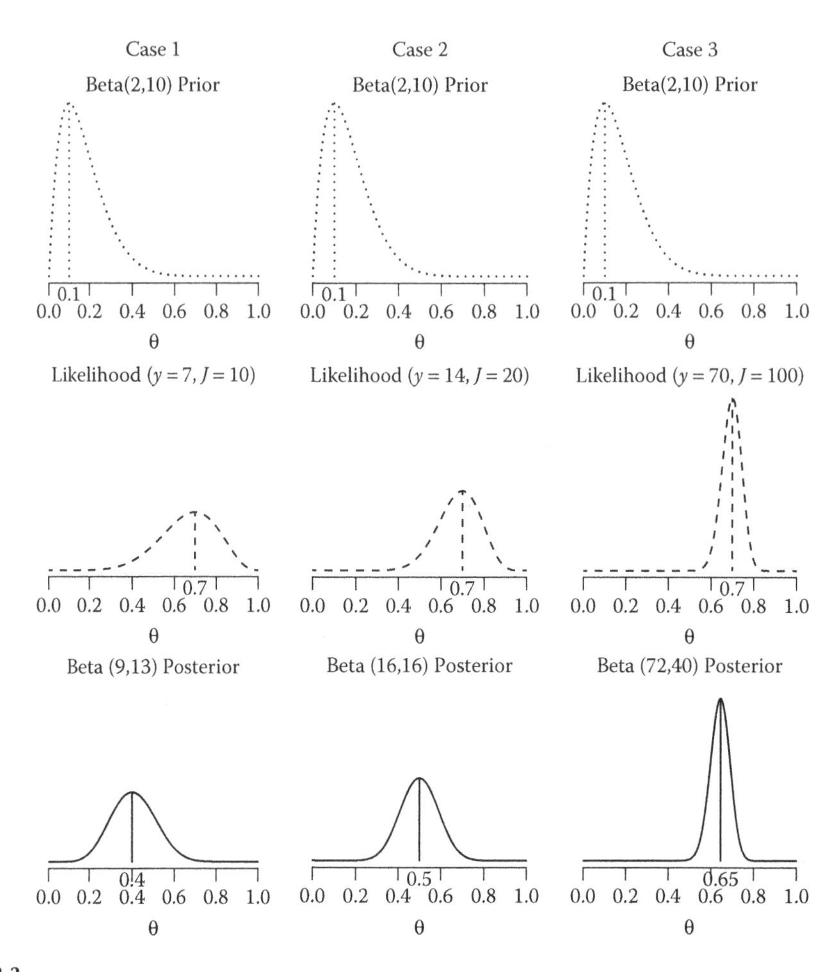

FIGURE 3.3
Prior, likelihood function, and posterior for the Beta-binomial model with a Beta(2, 10) prior and three different cases defined by different sample sizes. Reading down the columns presents the prior, likelihood, and posterior for each case. Where applicable, the mode of the distribution and the MLE is indicated via a vertical line.

3.1.3 Summary of the Effects of Information in the Prior and the Data

As the preceding examples in Figures 3.1 and 3.3 depict, the posterior distribution is a synthesis of the information in the prior and the data (expressed as the likelihood). Generally speaking, the influence of the data may be increased by (a) increasing sample size and/or (b) using a more diffuse prior, reflecting less prior information. Symbolically,*

$$p(\theta \mid x) \propto p(x \mid \theta)p(\theta) \xrightarrow[\text{prior information} \to 0]{\text{sample size} \to \infty\text{,or}} p(\theta \mid x) \propto p(x \mid \theta). \tag{3.1}$$

For the case of data modeled with a binomial distribution, the beta prior distribution makes for a convenient way to express the relative amounts of information in each of the

* Exactly what "prior information → 0" means needs further specification in any given case. How to specify minimally informative priors is a topic of much discussion in the literature (e.g., Bernardo & Smith's [2000] "reference" priors, see also Gelman et al. [2013] on alternative approaches on specifying minimally informative priors).

likelihood and the data. Specifically, a Beta(α,β) distribution carries the information akin to $\alpha + \beta - 2$ data points, where there are $\alpha - 1$ successes and $\beta - 1$ failures. The ease of interpretability of the parameters is one reason for using a beta as a prior in this context, a topic we turn to next.

3.2 Specifying Prior Distributions

There are a number of considerations in specifying the prior distributions including (Winkler, 1972): consistency with substantive beliefs, richness, ease of interpretation, and computational complexity. First, prior distributions should be consistent with, and embody, substantive beliefs about the parameters. This may be based on the meaning of the parameter and the role that it plays. In the models for binomial data, θ is the probability of success, and is theoretically bounded by 0 and 1. A distribution with support from 0 to 1 (e.g., a beta) is therefore more warranted than one that has support outside that range (e.g., a normal). In addition to these theoretical considerations, an important source for choosing prior distributions can come from prior research. For example, if prior research (say, from analyses of an already existing dataset, or from published research) has yielded or informed beliefs for the parameters, the prior distribution can be based on such beliefs. Indeed, the posterior distribution from one study can serve as a prior distribution for another suitably similar study. In this way, the use of a prior distribution allows for the formal inclusion of past research into the current analysis.

A second consideration is the richness, or flexibility of the distribution. The beta distribution is more flexible than the uniform distribution in the sense that the beta includes the uniform (the Beta(1,1) is uniform) but also allows for other shapes over the unit interval, including unimodal and symmetric, skewed, and U-shaped (see Figure 2.3). Similarly, using t distributions rather than normal distributions allows for thicker-tailed distributions. This flexibility is appealing; the more flexible or rich the family of distributions, the more potential that family has, as rich families can be used to model more and differing prior knowledge and belief structures. Thus using flexible prior distributions allows for more nuanced specifications and sensitivity analyses.

A third consideration is interpretability; to the extent that the prior and its parameters can be easily interpreted and understood, the more useful the prior is. The beta distribution is a fairly interpretable prior for the binomial parameter θ, in that the parameters of the beta may be interpreted in terms similar to that of the data, namely the number of successes and failures. If the prior is easily interpretable, it is easier to embody alternative prior beliefs or hypotheses. This also supports sensitivity analyses, which involve assessing the extent to which the posterior is driven by the prior. A basic sensitivity analysis involves rerunning the model with different reasonable prior distributions. To the extent that the prior is easily interpretable, it is easier to translate the set of reasonable belief structures into their corresponding distributions.

A fourth consideration in the specification of the prior distributions is the ease with which the posterior distribution can be obtained. The choice of prior distributions impacts the computation needed to obtain or estimate posterior distributions. In particular, *conjugate* prior distributions are those that, when combined with the likelihood, yield a posterior distribution that is a member of the same family as the prior. This keeps the computational burden at a minimum. For example, a beta prior distribution for the unknown parameter of a binomial, when combined with the binomial likelihood induced by observing the data, yields a posterior that is also a beta distribution. Hence, the beta distribution is a conjugate prior for this parameter. The concept of conjugacy is closely linked with the tractability of the likelihood, expressed in part by the presence of sufficient statistics; these are summaries of the data that capture all the relevant information about the parameter contained in the full data. We can identify a conjugate prior by expressing the conditional distribution for the data in terms of their sufficient statistics, and then view them as defining constants that govern the distribution of the parameter (Novick & Jackson, 1974). In the case of i.i.d. Bernoulli variables considered in Chapter 2, the sufficient statistics were the number of successes y and the total number of attempts J. The conditional probability for the data can be expressed in terms of these two summaries, as in (2.14). Viewing these constants and the parameter θ as a random variable, we can recognize the form of the beta distribution in (2.15), that is, the conjugate prior. See Novick and Jackson (1974) for a detailed step-by-step treatment of this approach for this and other examples.

Conjugate priors are available when the conditional probability distribution for the data (i.e., the likelihood) is a member of the general exponential family (Bernardo & Smith, 2000). However, conjugate priors are often not available in complex multivariate models. This may be the case even when the model is constructed out of pieces that afford conjugate priors if taken individually, but when assembled with the other pieces do not. In other cases, the use of a conjugate multivariate prior implies modeling a dependence between parameters that is not consistent with substantive beliefs or is overly difficult to work with conceptually. One strategy is then to specify priors for parameter separately, and use priors that would be conjugate if all other parameters were known. Though complete conjugacy is lost, the computation of the posterior distribution is still somewhat simplified. These are sometimes referred to as conditionally conjugate, semiconjugate, and generalized conjugate priors (Gelman et al., 2013; Jackman, 2009). Importantly, the use of such priors can ease the computational burden and greatly speed up the necessary computing time of modern simulation-based approaches to estimate posterior distributions, a topic we treat in Chapter 5. As a consequence, many applications of Bayesian psychometric modeling employ such priors. For some software packages they are the default, and, in some cases, the only options for prior distributions.

Historically, the use of conjugate priors was important to facilitate analytical computation of posterior distributions. In light of modern simulation-based estimation methods (Chapter 5), this issue is less critical than before, though certain choices are relatively less computationally demanding. Nevertheless, the modern estimation techniques have, in a sense, freed Bayesian analysts from a dependence on conjugate priors. Accordingly, we recommend consideration of substantive knowledge as the primary criterion. In many situations, we should gladly pay the price of computational complexity for accuracy of modeling one's belief structure. If the prior is flexible, computationally efficient, and easily interpretable, all the better. The development of the models in Chapters 4, 6, 8, 9, 11, 13, and 14 will involve priors specified in light of these considerations, and will further communicate what has typically been done in practice.

Specifying a prior comes to specifying not just a family of distributions, but a particular distribution (i.e., a member of the family). Two broad strategies are described here. The first strategy involves modeling existing research. Returning to the Beta-binomial model for student performance on tasks, if prior research indicates that the probability of a student correctly completing these tasks is around .9, then the Beta(10,2) prior distribution would be a sensible choice. To the extent that existing research is substantive, this should be conveyed in the prior. For example, a Beta(α,β) can be thought of as akin to encoding the results of $\alpha + \beta - 2$ trials. The quantity $\alpha + \beta - 2$ may be thought of as the pseudo-sample size expressed in the prior, as the information in the prior is akin to seeing that number of observations. Thus, one can select the parameters for the prior in such a way that they correspond to the size of the existing research. As a number of examples in the book illustrate, often the parameters of prior distributions can be framed in terms of sample sizes, which more easily allows for this approach. To date, most approaches to incorporating prior research have followed these lines, where prior research is essentially translated into a prior distribution. A related approach to basing prior distributions on existing work involves formally modeling the results of prior research with the current data. As an early effort in this vein, De Leeuw and Klugkist (2012) described several alternatives in the context of regression modeling.

A second strategy for constructing priors involves modeling beliefs of experts. It may be the case that though experts can formulate their prior beliefs, there is difficulty in translating them into values of parameters of distributions, particularly if experts are unfamiliar with the language of probability. Families that have natural interpretations for parameters (e.g., in terms of relative strength, or sample size) may make this easier. For example, the interpretation of the parameters of the beta distribution in terms of the number of successes and failures makes it easier to express prior beliefs. When working with those who have expertise in the subject matter at hand, but not probabilistic reasoning, a useful strategy is to elicit prior beliefs in a form that the experts are comfortable with and then transform them as needed. For example, Almond (2010) suggested asking experts to express their beliefs about the difficulty of tasks by placing them into broad categories of "easy," "medium," and "hard," which were than translated into probability distributions for parameters capturing difficulty in the psychometric model.

3.3 Comparing Bayesian and Frequentist Inferences and Interpretations

In frequentist inference, a point estimate is obtained for the parameter. In ML, for example, the likelihood is maximized to yield the point estimate. In a Bayesian analysis, the likelihood and prior distribution are synthesized to yield the posterior distribution. Visually, the posterior distribution "falls in between" the prior distribution and the likelihood, as seen in Figures 3.1 and 3.3. The notion that the posterior distribution is a synthesis of prior and the likelihood, and therefore falls "in between" the two will manifest itself repeatedly in our tour of Bayesian analysis.

To highlight key differences between frequentist and Bayesian approaches to modeling and estimation, we will draw from a metaphor sketched by Levy and Choi (2013). They characterized the likelihood as a mountain range, where the goal of ML is to find the highest peak among all the mountains in the mountain range. The middle row of Figure 3.1 depicts this mountain range for the binomial likelihood for the original example. Most ML estimation routines are akin to starting at some point and climbing to the top of a mountain blindfolded, preventing the climber from identifying the highest peak, or even broadly surveying the terrain. Instead, the climber must incrementally move in ways that seem optimal in light of the limited, localized knowledge obtained by the feel of mountain's slope at her feet. This is fairly straightforward for simple mountain ranges (e.g., the binomial likelihood in the middle row of Figure 3.1). However, a considerable challenge is that there might be multiple peaks in the mountain range (i.e., multiple *local maxima* exist), and that the peak that one ends up climbing depends on where one starts the climb, as some peaks are hidden from one's view at a particular location (i.e., the *global maximum* might not be reached). Unfortunately, there is no guarantee that one will obtain the global maximum as the stopping point of the estimation routine will depend on the actual starting value. Difficulties with local maxima and other threats to model convergence are usually exacerbated in complex models. A related difficulty is that many ML (and other frequentist) estimation routines typically involve derivatives of the likelihood function, which may be difficult to obtain in complex models. Software built for statistical modeling families, such as generalized linear latent and mixed models (Rabe-Hesketh, Skrondal, & Pickles, 2004) or structured covariance matrices (Muthén & Muthén, 1998–2013), obviates this concern for models that can be framed as a member of that family. However, for models that cannot be defined in those terms and cannot be fitted by the software, analysts are left to their own devices to obtain the derivatives and implement the estimation routines. This is likely a deterrent from using such models.

In a Bayesian analysis, the terrain of interest is a *different* mountain range, namely the posterior distribution. As a synthesis of the prior distribution and the likelihood, the posterior distribution "falls in between" its two ingredients, as seen in Figures 3.1 through 3.3. And instead of seeking the highest peak, a fully Bayesian analysis seeks to map the entire terrain of peaks, valleys, and plateaus in this mountain range. That is, in a Bayesian analysis the solution is not a *point*, but rather a *distribution*.* This difference is relevant to the conceptual foundations of Bayesian analyses of interest here, and has implications for estimation discussed in Chapter 5.

As illustrated in Figure 3.1, the mountain ranges may be fairly similar (column 2) or quite different (column 3), as they depend on the relative strength of the information in the prior and the likelihood, and the degree of agreement of that information. With large sample sizes and/or diffuse priors, the posterior distribution usually strongly resembles the likelihood. If the prior is quite diffuse, the posterior is essentially a normalized likelihood.† In this case, the mode of the posterior distribution is theoretically equivalent to the MLE. Similarly, as sample size increases, the data tend to swamp the prior so that the likelihood comes to dominate the posterior (Figure 3.3); the posterior mode is then asymptotically equivalent to the MLE. More generally, it can be shown that, under fairly general conditions, the posterior distribution asymptotically converges to a normal distribution, the central tendency and variability of which can be approximated by the maximum and curvature of the likelihood (Gelman et al., 2013). As a result, the posterior

* The goal of obtaining the posterior distribution for unknowns lies at the heart of Bayesian analysis. This goal is fully and immediately achievable with conjugate priors. For models that do not enjoy the benefit of having conjugate priors at the ready, this goal has historically been aspirational rather than achievable, and analysts have relied on tools to approximate the full posterior distribution. For example, in maximum a posteriori estimation in item response theory, we apply the same computational algorithms as used in ML, only now applied to the posterior. The resulting point identified as the maximum and the analytical expression of the spread of the surface are then interpreted as features of the full posterior and used to flesh out a normal approximation of the full posterior. As we discuss in more detail in Chapter 5, breakthroughs in statistical computing and the recognition of how they can be leveraged in Bayesian modeling have opened up the possibilities of obtaining increasingly better estimates of the posterior distribution.

† In many cases a uniform prior at least conceptually represents a maximally diffuse prior, and we employ such a specification to convey the ideas about a maximally diffuse prior in our context of binomial models. However, the application of uniform priors to enact a diffuse prior is not without issue, as a uniform prior on one scale need not yield a uniform prior when translated to another scale. These issues and alternative approaches to specifying diffuse or noninformative priors are discussed by Gelman et al. (2013).

mode (and mean and median) is asymptotically equivalent to the MLE, and the posterior standard deviation is asymptotically equivalent to the ML standard error. The asymptotic similarity between posterior summaries and ML estimates provides a connection between Bayesian and frequentist modeling. Frequentist analyses can often be thought of as limiting cases of Bayesian analyses, as sample size increases to infinity and/or the prior distribution is made increasingly diffuse. Bayesian estimation approaches, which map the entire mountain range, may therefore be useful for exploring, maximizing, and finding multimodality in the likelihood.

In finite samples, when the prior is fairly diffuse and/or the sample size is large enough, the posterior summaries are often approximately equal to the estimates from ML. As the results in Table 3.1 illustrate, it is often the case that values emerging from frequentist and Bayesian analyses are similar: the MLE is similar to the posterior mode and sometimes the mean (or any other measure of central tendency), the standard error is similar to the posterior standard deviation, and the confidence interval is similar to a posterior credibility interval. Importantly, however, even when posterior point summaries, standard deviations, and credibility intervals are *numerically similar* to their frequentist counterparts, they carry *conceptually different* interpretations. Those arising from a Bayesian analysis afford probabilistic statements and reasoning about the *parameters*; those arising from a frequentist analysis do not. In a frequentist analysis, distributional notions are invoked by appealing to notions of repeated sampling and apply to parameter estimators, that is, the behavior of parameter estimates on repeated sampling.

To illustrate this point, consider the situation in which we observe a student correctly complete all 10 tasks she attempts (see Exercise 2.5). However, rather than model prior beliefs that the student is very capable, let us model maximal uncertainty by using the Beta(1,1) prior distribution. The analysis is depicted in Figure 3.4. As the prior is uniform, the Beta(11,1) posterior is simply a normalization of the likelihood, and takes on exactly the same shape as the likelihood, such that in Figure 3.4 the curve for the posterior obscures that of the likelihood. Accordingly, the posterior mode (i.e., the value that maximizes the posterior density) of 1 is the same value that maximizes the likelihood. However, their interpretations are different. The posterior mode is a summary of the (posterior) distribution; the likelihood is not a distribution. In this analysis, we have encoded no prior information and found that the posterior is completely driven by the likelihood and that a summary of the posterior is numerically identical to a corresponding feature of the likelihood. But as a Bayesian analysis,

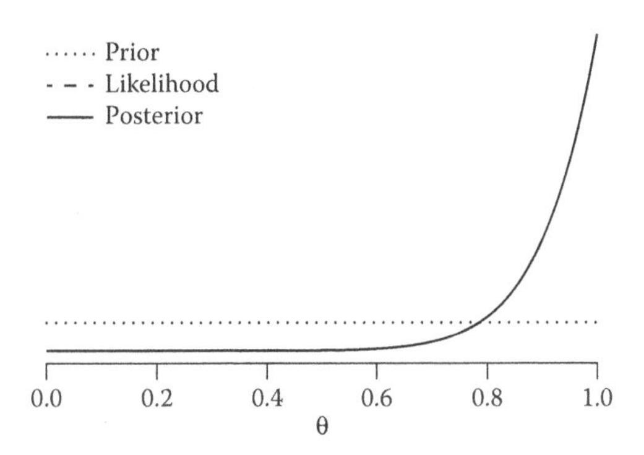

FIGURE 3.4
Beta(1,1) prior, likelihood function, and Beta(11,1) posterior for the Beta-binomial example with $y = 10$ and $J = 10$. As the prior is uniform, the shape of the posterior is the same as, and visually obscures, the likelihood.

the posterior affords a probabilistic interpretation of the parameter; a frequentist analysis based on the likelihood does not.*

To sharpen this point, consider notions of uncertainty in both frameworks in terms of the standard error and the posterior standard deviation. The value of $\theta = 1$ is a boundary of the parameter space, which poses problems for ML theory, and the standard error of the MLE in this case is 0. Conceptually, this may be understood by recognizing that the standard error reflects the sampling variability of the data, given the value of the parameter. If $\theta = 1$, then each attempt will be successful. There is no variability to speak of, so the standard error is 0.† This characterization of uncertainty is by no means nonsensical. This characterization of uncertainty is associated with the deductive inference, in which we reason from a given value the parameter (here $\theta = 1$) to the particulars in the data. But this is not the inference at hand, and so this is not the summary of uncertainty we want. Instead, we would like to characterize the uncertainty associated with the inductive inference, in which we reason from the known particulars (the data) to the parameter, θ. This is what the Bayesian analysis provides. For the Beta(11, 1) posterior distribution, we may summarize the uncertainty in terms the posterior standard deviation, .08, which communicates the uncertainty of the parameter.

Similarly, we can express our uncertainty about parameter via posterior probability intervals, such as the 95% HPD interval of (.76, 1) for the Beta(11, 1), which is a probabilistic statement about the parameter. A Bayesian approach allows us to use the language of probability to *directly* discuss what is of inferential interest—parameters, hypotheses, models, and so on—rather than indirectly as in frequentist approaches. This allows analysts to "say what they mean and mean what they say" (Jackman, 2009, p. xxviii), and overcome the limitations and avoid the pitfalls of the probabilistic machinery of frequentist inference (e.g., Goodman, 2008; Wagenmakers, 2007). In assessment, the desire to directly discuss examinees' student model variables and the confusions brought on by doing so using conventional approaches are longstanding traditions (Dudek, 1979). By adopting a Bayesian approach, we can more clearly, coherently, and correctly frame and achieve the desired inference.

3.4 Exchangeability, Conditional Independence, and Bayesian Inference

3.4.1 Exchangeability

In this section, we develop ideas and practices associated with exchangeability (de Finetti, 1974), one of the foundational elements in Bayesian model conceptualization and construction. The current treatment focuses on the conceptual implications for modeling.

* Although Bayesian and likelihood solutions are numerically similar in straightforward problems, this need not be the case with more complicated models such as hierarchical models, models with covariate structures, multiple sources of evidence, or complex patterns of missingness. The Bayesian approach provides a principled way to think about them. Historically it was not the case, but now, the Bayesian machinery is more flexible than the frequentist machinery for tackling such problems, and it is always the same way of thinking about them (Clark, 2005).

† In the case of a binomial, an exact confidence interval for the probability of success can be obtained even when the MLE is 1 (e.g., Agresti & Coull, 1998). In complex models where approximate confidence intervals are used based on standard errors, boundaries pose problems for the construction of those intervals.

Let us return to the context where $\mathbf{x} = (x_1, \ldots, x_J)$ is a collection of J Bernoulli variables (e.g., coin flips). We can think about exchangeability in a few ways, with varying levels of statistical formality. At a narrative level, absent any statistical aspects, exchangeability amounts to saying that we have the same beliefs about the variables in question (prior to observing their values, or course). Do we think anything different about the first variable (coin flip) than the second? Or the third? Do we think anything different about *any* of the variables? If the answers to these questions are no, we may treat the variables as exchangeable.

A bit more formally, the collection of variables are *exchangeable* if the joint distribution $p(x_1, \ldots, x_J)$ is invariant to any re-ordering of the subscripts. To illustrate, for the case of 1 success in 5 variables, asserting exchangeability amounts to asserting that

$$p(1, 0, 0, 0, 0) = p(0, 1, 0, 0, 0) = p(0, 0, 1, 0, 0) = p(0, 0, 0, 1, 0) = p(0, 0, 0, 0, 1).$$

This amounts to asserting that only the total number of successes is relevant; it is irrelevant *where* in the sequence the successes occur (Diaconis & Freedman, 1980a). More generally, a collection of random variables are exchangeable if the joint distribution is invariant to any permutation (reordering, relabeling) of the random variables.

A remarkable theorem proved by de Finetti (1931, 1937/1964) and later generalized by Hewitt and Savage (1955), states that for the current case of dichotomous variables, in the limit as $J \to \infty$,

$$p(x_1, \ldots, x_J) = \int_0^1 \prod_{j=1}^{J} \theta^{x_j} (1-\theta)^{1-x_j} dF(\theta), \tag{3.2}$$

where $F(\theta)$ is a distribution function for θ. The left-hand side of (3.2) is the joint distribution of the variables. The first term inside the integral on the right-hand side, $\prod_{j=1}^{J} \theta^{x_j} (1-\theta)^{1-x_j}$, is the joint probability for a collection of i.i.d. Bernoulli variables conditional on the parameter θ (see Equation 2.8). The second term inside the integral, $F(\theta)$, is a distribution function for θ. Thus, de Finetti's so-called representation theorem reveals that the joint distribution of an infinite sequence of exchangeable dichotomous variables may be expressed as a mixture of conditionally independent distributions. Joint distributions for finite sets of exchangeable variables can be approximated by conditional i.i.d. representations, with decreasing errors of approximation as J increases (Diaconis & Freedman, 1980b).

De Finetti's theorem has been extended to more general forms, with real-valued variables and mixtures over the space of distributions (Bernardo & Smith, 2000). In our development in the rest of this chapter, we will focus on the case where $F(\theta)$ is absolutely continuous, in which case we obtain the probability density function for θ, $p(\theta) = F(\theta)/d\theta$ and the theorem may be expressed as

$$p(x_1, \ldots, x_J) = \int \prod_{j=1}^{J} p(x_j \mid \theta) dF(\theta), \tag{3.3}$$

or

$$p(x_1, \ldots, x_J) = \int_{\theta} \prod_{j=1}^{J} p(x_j \mid \theta) p(\theta) d\theta. \tag{3.4}$$

We discuss the implications of making assumptions of continuity and positing various parametric forms in Section 3.4.7.

The right-hand side of (3.4) is expressed in as a graph in Figure 3.5, here using the ellipses to condense the presentation of all the *x*s. A key ingredient in de Finetti's representation theorem and the graph in Figure 3.5 is that of *conditional independence*. In the next section, we provide a brief overview of conditional independence; the following sections resume the discussion of the implications of exchangeability and conditional independence in modeling.

3.4.2 Conditional Independence

To develop the notions of conditional independence, let us begin with the construction of the joint distribution of two variables. In general, the joint distribution may be constructed as the product of the marginal probability distribution of one variable and the conditional probability distribution of the other, given the first. For two variables x_1 and x_2,

$$p(x_1, x_2) = p(x_1)p(x_2 \mid x_1). \tag{3.5}$$

If the variables are independent, the joint distribution simplifies to the product of the marginal distributions

$$p(x_1, x_2) = p(x_1)p(x_2). \tag{3.6}$$

The lone difference between (3.5) and (3.6) is that in the latter, the distribution for x_2 is not formulated as conditional on x_1. In other words, if the variables are independent then the conditional distribution for x_2 given x_1 is equal to the marginal distribution for x_2 (without regard or reference to x_1)

$$p(x_2 \mid x_1) = p(x_2). \tag{3.7}$$

This reveals the essence of what it means for the two variables to be independent or unrelated—the distribution of x_2 does not change depending on the value of x_1. From the epistemic probability perspective, independence implies that learning the value of x_1 does not change our beliefs about (i.e., the distribution for) x_2.

It is very common for variables to be dependent (related) in assessment applications. For example, consider variables derived from scoring student responses to tasks in the same domain. Performance on one task is likely related to performance on the other; students who tend to perform well on one task also tend to perform well on the other, while students who struggle with one task also tend to struggle with the other. In this case, the variables resulting from scoring the student responses will be dependent. In terms

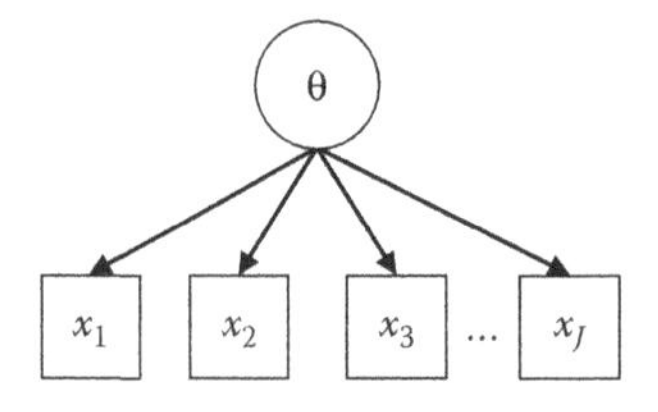

FIGURE 3.5
Graphical representation of the right-hand side of (3.3) or (3.4), illustrating conditional independence of the *x*s given θ, in line with exchangeability.

of "probability as an expression of beliefs," learning how a student performs on one task informs what we think about how they will perform on the other task. As a consequence of their dependence, (3.6) and (3.7) do not hold.

However, such variables may be *conditionally independent* given some other variable, θ, in which case

$$p(x_1, x_2 \mid \theta) = p(x_1 \mid \theta) p(x_2 \mid \theta). \tag{3.8}$$

Conceptually, (3.8) expresses that although x_1 and x_2 may be dependent (related), once θ is known they are rendered independent. More generally, a set of J variables $x_1, \ldots, x_J$ are conditionally independent given θ if

$$p(x_1, \ldots, x_J \mid \theta) = \prod_{j=1}^{J} p(x_j \mid \theta). \tag{3.9}$$

3.4.3 Exchangeability, Independence, and Random Sampling

Exchangeability is sometimes presented as the "Bayesian i.i.d." This is perhaps useful as a first step for an audience steeped in frequentist traditions and i.i.d. probability models, but from our perspective it misses the mark on several counts. If variables are i.i.d., they are exchangeable; however, the converse is not necessarily true as exchangeability is a weaker, more general condition. As that suggests, frequentist statistical modeling and analysis is grounded in assumptions of exchangeability as well (O'Neill, 2009). Moreover, the phrase potentially gives the impression that i.i.d. aligns only with frequentist approaches to inference, whereas the preceding developments have shown, and as we will see time and again, a Bayesian approach makes great use out of i.i.d. structures.

From a frequentist perspective—and hence, how it is commonly taught—i.i.d. is framed as arising from assumptions of a population and random sampling. From our perspective, random sampling is something in the world, specifically a process or mechanism that gives rise to data, and is particularly powerful in that it warrants independence models given appropriate conditioning variables. As we describe in Section 3.4.6, exchangeability, however, is an epistemic statement, reflecting an analyst's state of knowledge and purpose at a given point in time. Random sampling can certainly serve to justify exchangeability, and even as Bayesians we are quite happy to capitalize on such a circumstance, as it provides a strong justification for our useful but strong presumption of exchangeability and a corresponding conditional independence structure in some model we wish to build and reason through.

3.4.4 Conditional Exchangeability

It is often the case that exchangeability cannot be assumed for all units in the analysis. A paradigmatic example comes from grouping. Such situations may arise in nonexperimental settings, for example, when we study students that are organized into distinct classrooms. They may also arise in experimental settings, for example when we study subjects that have been randomly assigned to treatment or control conditions. In this case, the subjects may be exchangeable to us before the experiment but once the experiment is complete they are no longer exchangeable. Here, we may be reluctant to assert that we have the same beliefs for all units; indeed, we suspect that the grouping (treatment vs. control, teacher A vs. teacher B) makes a difference, even without articulating what we think such differences may be. Instead, we may be willing to assert that we have the same beliefs for all units in the treatment group, and the same beliefs for all units in the control group. That is, we may assert *partial* or *conditional exchangeability*, where exchangeability holds conditional on group membership. This motivates a multiple-group model, where all the units of the treatment group are specified as following a probability model and all the units of the control group are specified following a different probability model. More generally, we may have any number of covariates or explanatory variables that are deemed relevant. Once all such variables are included in the model, (conditional) exchangeability can be asserted.

3.4.5 Structuring Prior Distributions via Exchangeability

Noting that the exchangeability concept can be applied to parameters as well as observable variables, Lindley (1971; see also Lindley & Smith, 1972) developed a powerful approach to specifying prior distributions. Consider an assessment situation, the situation with n examinees, each of whom has a parameter representing proficiency on a set of tasks, which we collect in $\boldsymbol{\theta} = (\theta_1, \ldots, \theta_n)$. The prior distribution is now an n-variate distribution $p(\boldsymbol{\theta}) = p(\theta_1, \ldots, \theta_n)$. Treating the examinees as exchangeable, this joint distribution may be factored into the product of conditionally independent and identically distributed univariate distributions, formulated as conditional on parameters, say, θ_P,

$$p(\boldsymbol{\theta} \mid \theta_P) = p(\theta_1, \ldots, \theta_n \mid \theta_P) = \prod_{i=1}^{n} p(\theta_i \mid \theta_P). \tag{3.10}$$

The situation here is one in which the parameters of the original model, $\boldsymbol{\theta}$, have a structure in terms of other, newly introduced parameters θ_P, termed *hyperparameters** (Lindley & Smith, 1972). For example, a specification that the examinee variables are normally distributed may be expressed as $p(\theta_i \mid \theta_P) = N(\theta_i \mid \mu_\theta, \sigma_\theta^2)$, where $\theta_P = (\mu_\theta, \sigma_\theta^2)$ are the hyperparameters. If values of θ_P are known or chosen to be of a fixed value, the prior specification is complete. If they are unknown, they of course require a prior distribution, which may depend on other parameters. This sequence leads to a hierarchical specification, with parameters depending on hyperparameters, which may in turn depend on other, higher-order hyperparameters.

The implications of conditional exchangeability discussed in Section 3.4.4 hold analogously here. If the parameters are not deemed exchangeable, it is still possible that they be conditionally exchangeable. Returning to the example, if the examinees come from different groups, conditional exchangeability implies that we may construct the prior distribution as

$$p(\boldsymbol{\theta} \mid \theta_P) = \prod_{g=1}^{G} \prod_{i=1}^{n_g} p_g(\theta_{ig} \mid \theta_{Pg}), \tag{3.11}$$

where n_g is the number of examinees in group g, $g = 1, \ldots, G$, θ_{ig} refers to the parameter for examinee i in group g, and the subscripting of p and θ_P by g indicates that each group may have a different functional form and/or parameters governing the distribution. For example, we may have $p(\theta_{ig} \mid \theta_{Pg}) = N(\theta_{ig} \mid \mu_{\theta g}, \sigma_{\theta g}^2)$, where $\theta_{Pg} = (\mu_{\theta g}, \sigma_{\theta g}^2)$ are the hyperparameters that govern the prior distribution for examinees in group g.

* The use of the subscript 'P' in 'θ_P' is adopted to signal that these are the parameters that govern the *prior* distribution for $\boldsymbol{\theta}$. Similarly, the subscript indicates that, in a directed acyclic graph representation, the elements of θ_P are the *parents* of $\boldsymbol{\theta}$.

The upshot in this approach to building prior distributions is that exchangeability supports the specification of a common prior for each element conditional on some other, possibly unknown, parameters. This has become the standard approach to specifying prior distributions in highly parameterized psychometric models, and we will make considerable use of it throughout. Further, this connects with a particular conception of Bayesian modeling as "building out," described in Section 3.6. We will see in Chapter 15 that the purpose and context of inference can be an important factor in how, and indeed whether, we include these kinds of structures into a model. In educational testing, even with the same test and the same examinees, it can be ethical to do so in one situation and unethical in another; even legal in one situation and illegal in another!

3.4.6 Exchangeability, de Finetti's Theorem, Conditional Independence, and Model Construction

In probability-based reasoning, we construct a model by setting up a joint probability distribution for all the entities. It is here that exchangeability, de Finetti's representation theorem, and conditional independence may be brought to bear. Returning to the right-hand side of (3.4), we recognize a conditional independence relationship among the *x*s in the term $\prod_{j=1}^{J} p(x_j \mid \theta)$. De Finetti's theorem shows that the distribution of an exchangeable sequence of variables may be expressed as the expectation of a mixture of conditionally independent variables, where $p(\theta)$ denotes the mixing distribution.

In stressing the importance conditional independence plays in reasoning and statistical modeling, Pearl (1988, p. 44) argued that employing conditional independence structures has computational and psychological advantages:

> The conditional independence associated with [such variables] makes it a convenient anchoring point from which reasoning "by assumptions" can proceed effectively, because it decomposes the reasoning task into a set of independent subtasks. It permits us to use local chunks of information... and fit them together to form a global inference... in stages, using simple, local, vector operations. It is this role that prompts us to posit that conditional independence is not a grace of nature for which we must wait passively, but rather a psychological necessity which we satisfy actively by organizing our knowledge in a specific way. An important tool in such organization is the identification of intermediate variables that induce conditional independence among observables; if such variables are not in our vocabulary, we create them. In medical diagnosis, for instance, when some symptoms directly influence one another, the medical profession invents a name for that interaction (e.g., "syndrome," "complication," "pathological state") and treats it as a new auxiliary variable that induces conditional independence; dependency between any two interacting systems is fully attributed to the dependencies of each on the auxiliary variable.

De Finetti's exchangeability theorem shows that, by expressing our substantive beliefs within the generally applicable structure afforded by exchangeability, we can greatly simplify the cognitive task of expressing our beliefs, working with smaller, much more manageable parts. As Pearl's quote stresses, this strategy is so powerful that if variables that induce conditional independence cannot be found, we introduce them. In employing de Finetti's theorem and conditional independence of the *x*s given θ, we are therefore not asserting that θ necessarily exists *out there in the world*. Rather, θ exists in our model, as a tool for organizing our thinking and facilitating model-based reasoning (and when data are in hand, examining whether the structure we proposed needs to be revised).

The simplifications afforded by exchangeability and conditional independence assumptions have so much appeal that we seek them out. If exchangeability cannot be assumed, we pursue conditional exchangeability. Building probability models can then be thought of as the quest for exchangeability. We start with all the units and proceed to ask, what are the salient differences that we suspect may have bearing on the problem at hand? Do we think it matters which group you are in? If so, then the units are not exchangeable, at least until we condition on group. Does it additionally matter when the data were collected? If so, then the units are not exchangeable, at least until we condition on when the data were collected. Does it matter whether you are left handed or right handed? What the weather was like? What your favorite color is? Eventually, we reach the point where the answers to these questions become no, and we deem variance with respect to the remaining variables (e.g., the weather, your favorite color) as irrelevant. We arrive at exchangeability when we can declare that, for us, *further differences make no difference*. We may admit variability at this level, but no longer try to model it by entities within our modeling structure. Engaging in this quest allows us to declare what variables *are* important, so we can include them in the model. Achieving conditional exchangeability is important from this perspective not because it has been attained as a goal, but rather an indication that we have obtained the goal of identifying themes in the story our model represents that are necessary for the story to not be misleading in some way that matters.

Conditional exchangeability amounts to asserting that, for us, the relevant variables are included and that variation with respect to other variables can safely be ignored. The qualifier in the preceding sentence—"for us"—requires some unpacking. Our view is that exchangeability is an epistemic claim in that it concerns our thinking about the situation at hand, in service of setting up a necessarily imperfect model to serve our purposes (Lindley & Novick, 1981; Mislevy, 1994). Our thinking is necessarily localized in a variety of ways, chiefly in that it occurs for a particular person at a particular time. Asserting exchangeability is not akin to claiming that we have reached the final word on all relevant aspects of the real-world situation of interest. It is entirely coherent to assert exchangeability and simultaneously acknowledge that there may be—or perhaps even that there indeed are—further differences that are relevant for the problem at hand. Consider a scene that plays out countless times in our education systems. Students walk into a classroom and take a test. Once completed, the tests will be sent to the psychometricians for analysis. There may be any number of features that make some students more similar to other students and less similar to other students in ways that are surely relevant for their performance on the test, and the test scores that result (e.g., who their teachers were, study habits, home life experiences, motivation, amount of sleep the night before). Yet if we, the psychometricians, are not able to distinguish among these features, the scores are exchangeable to us. Technically, the joint distribution of test scores is invariant to a permutation in the elements; more conceptually, we do not think anything differently about one examinee's score as any other examinee's score.

Thus exchangeability is an expression of beliefs on the part of an analyst, situated in terms of who is conducting the analysis and what they know at the time. What is exchangeable for one person may not be exchangeable for another. The archetypical critique that an analysis is suspect because it has ignored a particular covariate may be seen as a statement that a variable important to condition on when asserting exchangeability has been ignored. Likewise, what we view to be exchangeable at an earlier time may not be deemed exchangeable at a later time, in light of what has been learned in the interim.

It is perfectly acceptable for different reasoning agents, be they different people or the same person at different points in time, to posit different exchangeability claims, which yield contradictory conditional independence relationships. Of course, it is impossible for any two, let alone all, such contradictory relationships to both (all) be true descriptions of the real-world mechanisms that give rise to the variables in question. This does not undercut their use because exchangeability, the resulting conditional independence relationships, and the parameters introduced to achieve those relationships, are expressions about *our thinking*, not about the *world*. In terms of Figure 1.4, exchangeability resides in the model-space, rather than the real-world situation.

The focus on the epistemic nature of exchangeability, and model construction more generally, is not to downplay the importance of the fidelity of our specifications to real-world mechanisms. The better able we are to model salient features and patterns of the real world, the better our resulting inferences are likely to be and the more useful our model will be. A central motivation for engaging in model-data fit analyses and model criticism, discussed in Chapter 10, is the desire to critique the plausibility of our specifications, so as to gain trust in the resulting inferences when they are reasonable, and improve our model when they are not.

3.4.7 Summary of Exchangeability: From Narrative to Structured Distributions to Parametric Forms

We can think about exchangeability in terms of different layers, representing increasing amounts of statistical specification. In the first layer, exchangeability is the term we use to convey that we think the same thing about multiple entities. Conditional exchangeability means that we think the same thing about entities but only after we have taken other things into account. In assessment, these entities are usually examinees, tasks, or examinee performances on tasks. For example, asserting that examinees are exchangeable means that we have the same beliefs about each examinee. Asserting that examinees are exchangeable conditional on what school they are in means that we have the same beliefs about each examinee at the same school. At this level, exchangeability is not really about probability distributions or statistical models, it is part of our narrative or story of the situation. We may even acknowledge that there may be differences or variation beyond those that we are explicitly considering, but that we not distinguishing them further in our story and eventually our model.

De Finetti's representation theorem helps move the concept of exchangeability from a narrative level to one of statistical modeling. It has a decidedly statistical form as it concerns the structure of joint distributions, and facilitates a shift from a hard-to-think-about joint distribution to easier-to-think-about univariate distributions.* Conceptually, it formalizes the intuition that if we think the same thing about multiple entities (i.e., exchangeability in the narrative sense), then we can treat them the same in our probability model. The theorem is very powerful in that it says we use the same thinking as i.i.d. for situations that we do not actually believe are truly i.i.d., but have no information otherwise. And we can express this with the marginalized conditional independence expression on the right-hand side of (3.4). This is tremendously freeing in how we are warranted to think about and tackle situations, applying the machinery of probability originally developed to deal with the more limited case of actual i.i.d.

* Really, they are easier-to-think-about lower dimensional distributions. We have considered the case of univariate distributions, but in principle they could multivariate, say, if each $p(x_j \mid \theta)$ was a joint distribution of a set of (fewer than J) entities.

Throughout, we will make repeated use of exchangeability in formulating our models. We will say that we are asserting (assuming, positing) exchangeability, and then we will write out the marginalized conditional independence form with particular distributional specifications. When we say this, it is really shorthand for a longer, two-part statement:

- *Part 1*: We are using the marginalized conditional independence structure, as sanctioned by the representation theorem (i.e., through exchangeability we move from the left-hand side to the right-hand side of Equation 3.4).
- *Part 2*: Further, we are populating that form with a particular distributional form for the conditionally independent distributional specification (i.e., a particular form for $p(x_j \mid \theta)$ in Equation 3.4) and a particular distributional form for the entities that are conditioned on (i.e., a particular form for $p(\theta)$ in Equation 3.4).

The exchangeability theorem is powerful in that it provides a *structure* for the distribution in terms of the marginalized conditional independence expression (part 1). In and of itself, the exchangeability representation theorem is just a piece of mathematical machinery. In terms of Figure 1.4, it resides in the model-space, and it is unassailable. Importantly, we cannot be wrong in adopting the marginalized conditional independence form, if we truly have no distinguishing information or have chosen not to use it in the model we are positing.

But the exchangeability theorem is vacuous because though this expression holds, no particular form is specified. Things get interesting when we come to actually make the specifications for the model, for the forms of the distributions (part 2). These vary considerably, and modelers face the challenge of positing models that reflect what they know substantively about the problem at issue. A goal of the book is to show how many psychometric models are variations on the theme, with different particular forms or extensions, motivated by what we know about people, psychology, and assessment.

When we posit particular functional forms for the distributions, however, we are taking an epistemic stance of using a more constrained probability model. Exchangeability qua exchangeability still holds as a necessarily-true feature of our reasoning. The same can be said about the conditional independence relationships, even when populated with particular parametric forms for the distributions. In terms of Figure 1.4, the exchangeability theorem, the marginalized conditional independence structure, and even the particular parametric forms reside in the model-space, and there they are unassailable as structures that represent our belief at a given point in time. Populating the marginalized conditional independence structure with particular parametric forms accomplishes two things. First, it provides *content* to the *structure* justified by the exchangeability theorem. The theorem is true, but it does not get us far enough. Particular parametric forms are needed for us to get work done, in terms of model-based implications that guide inferences and yield substantive conclusions. Second, it gets us to a place where we can critique the model, in terms of its match to the real-world situation. Included among the model-based implications generated by the use of particular distributional forms are implications for the data. In terms of Figure 1.4, when we add particular distributional specifications we can project from the model-space back down to the real-world situation in the lower level of the figure. And in this way the specifications can be wrong in that they yield or permit joint probabilities that are demonstrably at odds with the data. Techniques for investigating the extent and ways in which the implications of a model are at odds with the data are the subject of Chapter 10.

3.5 Why Bayes?

Readers steeped in frequentist traditions and new to Bayesian approaches to modeling and inference may feel that the use of probabilities in reference to parameters and associated probability statements about parameters is appealing, but perhaps foreign. And they may ask, how is it that we can treat parameters as random and use probabilistic expressions for them, both in terms of prior distributions needed to specify the model and the resulting posterior distribution? Sections 3.2 and 3.4.5 offered some strategies for *how* to specify prior distributions, and Bayes' theorem instructs us in *how* to arrive at posterior distributions. The balance of this section attempts to address *why* such practices are warranted and advantageous.

There has been considerable debate over the legitimacy or appropriateness of Bayesian statistical inference. Importantly, there is no controversy over the mechanics of Bayes' theorem, as it follows in a straightforward manner from the law of total probability. Rather, the controversy surrounds whether Bayesian methods, and in particular their use of distributions for parameters, should be applied to treat certain problems of inference. We say *certain* because despite the dominance of frequentist approaches to statistical inference over the last 100 years, some features and applications of Bayesian inference have gained widespread acceptance, even in the frequentist community. Examples include the specifications that may be viewed as prior distributions, and the use of Bayes' theorem to yield posterior probabilities of group membership in latent class and mixture models. Provocatively, a number of developments in frequentist modeling strategies can be seen as inherently Bayesian.

We recognize that condensing the debate concerning Bayesian inference to this point about the use of distributions for parameters is a gross oversimplification. Nevertheless, we suspect that readers new to Bayesian principles may be most apprehensive about this aspect. What grounds are there for treating parameters as random, which is the distinguishing feature of Bayesian approach? We endeavor to address that in the coming subsections.

It would be disingenuous to suggest that there is a single, monolithic philosophy or perspective that has attracted a consensus among those who engage in Bayesian modeling. To the contrary, we suspect that in the years since Good (1971) identified 46,656 possible varieties of Bayesians, the palette of Bayesian flavors has expanded, not shrunk. Debates carry on over fundamental issues such as the meaning of probability and probabilistic statements, the propriety of treating parameters as random, the meaning and role of prior and posterior distributions, and the nature of statistical modeling, particularly as it pertains to ontological commitments. These issues are not without consequence; they can have considerable implications for the practice of (Bayesian) statistical modeling and inference, including the specification of models, when certain procedures are warranted or even permitted, and the relevance of various activities that cut across Bayesian and non-Bayesian statistical modeling, such as model use, model-data fit, model criticism, model comparison, and model selection. Bayesian analyses have been motivated and similarly criticized from a number of perspectives on these issues. We do not aim to survey this debate or address all of its aspects. In this and the following subsections, we summarize the key elements as they relate to our position and the developments in the book.

3.5.1 Epistemic Probability Overlaid with Parameters

Once we commit to the use of statistical models with parameters, then the use of probability distributions for those parameters is consistent with the epistemic perspective on probability. In fact, distributional specifications for parameters are natural from the epistemic probability perspective. Prior distributions express our uncertainty about the parameters before the data are incorporated; posterior distributions express uncertainty about the parameters after the data are incorporated.

3.5.2 Prior Probability Judgments Are Always Part of Our Models

The epistemic probability perspective illuminates another argument, specifically how modeling parameters distributionally when specifying models may be viewed as a generalization of conventional approaches to modeling parameters in frequentist approaches (Box, 1980; Levy & Choi, 2013). In frequentist traditions, parameter specification typically amounts to indicating whether a parameter is or is not included in the model. If it is included, the parameter will be estimated using information in the data; in ML, via the likelihood. Excluding a parameter may often be seen as specifying it to be equal to a constant (usually 0). Each of these situations may be specified in a Bayesian analysis that models parameters distributionally. As the first column of Figure 3.1 illustrates, using a uniform prior over the support of the distribution* yields a posterior that is driven only by the likelihood, and is akin to the frequentist approach where the parameter is included in the model. A situation where the parameter is constrained to a constant (e.g., 0) may be modeled in a Bayesian approach by specifying a prior for this parameter with all of its mass at that value.

The conventional approaches thereby represent two ends of a distributional spectrum. At one end, the use of a prior distribution with all of its mass at one point can be viewed as corresponding to the conventional approach in which the parameter is fixed to that value. Viewing the prior distribution as an expression of uncertain beliefs, this suggests that fixing a parameter by concentrating all the mass of the prior at that value represents a particularly strong belief, namely one of complete certainty about the parameter. At the other end of the spectrum, the use of uniform prior distribution reflects a very weak assumption, one of maximal uncertainty, about the parameter.

Importantly, a Bayesian approach that specifies parameters distributionally affords the possibility of specification along the spectrum in between these endpoints. By choosing a prior distribution of a particular form (shape, central tendency, variability, etc.), we can encode assumptions of varying strengths or beliefs of various levels of certainty. This is particularly natural if a prior can be constructed where one of its parameters captures this strength. For example, the use of a normal prior distribution allows for the specification of uncertainty via the prior variance. A prior variance of 0 indicates the parameter is constrained to be equal to the value of the prior mean (e.g., 0); increasing the prior variance reflects an increase in uncertainty. Recently, such flexibility has been gainfully employed in fitting psychometric models reflecting differential amounts of uncertainty, including those that pose considerable challenges for frequentist approaches, as we discuss in the context of factor analysis in Chapter 9. The flexibility afforded by a distributional approach to specifying model parameters along a continuum of possibilities has been exploited to

* A uniform prior is an actual probability distribution in case of finite distributions, and in the absence of other considerations is the greatest uncertainty (maximal entropy) one can express. For variables that can take infinitely many values, a uniform distribution might have an infinite integral and thus not be an actual distribution. A uniform prior is proper on [0,1], but one on $(-\infty,\infty)$ is not. Sometimes the Bayesian machinery works anyway, producing a true posterior. A more strictly correct Bayesian approach is to use a proper but extremely diffuse prior, such as $N(0,10^5)$ for a real-valued parameter, or, with prior distributions that afford them, specifying parameter values that maximize entropy (Jaynes, 1988). Gelman et al. (2013) discussed approaches to specifying noninformative priors.

great advantage in other modeling contexts. For example, this same machinery allows for partial pooling among groups in multilevel modeling, which lies between the extremes of complete pooling of all the groups or no pooling among groups (Gelman & Hill, 2007; Novick et al., 1972).

This modeling flexibility is an important advantage, but is actually somewhat tangential to the current point, which is that prior probability judgments and assertions are always present, even when the analysis does not use Bayesian methods for inference. The inclusion of certain parameters and the exclusion of others (i.e., fixing them to be 0) in a frequentist analysis amounts to two particular probabilistic beliefs—complete uncertainty about the former and complete certainty about the latter. That is, *any and every* structure of the model amounts to an expression of prior beliefs, in ways that can be viewed as distributions. This holds for the other so-called model assumptions or model specifications, typically thought to be distinct from prior probabilistic beliefs. This can readily be seen by embedding a model into a larger one. For example, modeling a variable as normally distributed (e.g., errors in typical regression or factor analysis models) may be viewed as modeling the variable as member of a larger class of distributions with all the prior probability concentrated in such a way that yields the normal distribution (e.g., as a t distribution where the prior probability for the degrees of freedom has all its mass at a value of infinity). Generally, all of our model assumptions—effects/parameters that are included or excluded, functional forms of relationships, distributional specifications, and so on—are the result of prior probability judgments. Box tidily summarized this view (1980, p. 384), remarking that

> In the past, the need for probabilities expressing prior belief has often been thought of, not as a necessity for all scientific inference, but rather as a feature peculiar to Bayesian inference. This seems to come from the curious idea that an outright assumption does not count as a prior belief.... [I]t is impossible logically to distinguish between model assumptions and the prior distribution of the parameters. The model *is* the prior in the wide sense that it is a probability statement of all the assumptions currently to be tentatively entertained *a priori*.

Our view is that the distinction between "model specifications/assumptions," which are central features in conventional modeling, and "prior distributions," which are excluded from such traditions, is more terminological than anything else. The distinction may be useful for structuring and communicating modeling activities with analysts steeped in different traditions, but too strict adherence to this distinction may obscure that what is important is the model as a whole, and the boundaries between parts of the model may be false. Both "model specifications/assumptions" and "prior distributions" involve subjective decisions (Winkler, 1972); analysts who wish to employ the former but prohibit the latter on the grounds that the latter are less legitimate have their philosophical work cut out for them. We prefer to think about model specifications as prior specifications. They reflect our beliefs about the salient aspects of the real-world situation at hand before we have observed data, and are subject to possible revision as we learn from data and update our beliefs. Procedures for learning about the weaknesses (and strengths) in our model and the theories they represent are discussed in Chapter 10.

Historically, disagreement regarding the propriety of incorporating prior beliefs was *the* main source of contention between Bayesian and frequentist inference (Weber, 1973). This is less of an issue currently, partly because of the theoretical considerations such as those just discussed, and partly because of the recognition that the critique that a Bayesian approach with priors is subjective—used as a pejorative term in the critique—is undermined by the recognition that frequentist approaches hold no stronger claim to objectivity (Berger & Berry, 1988; Lindley & Philips, 1976).* It is also less of an issue because the use of very mild priors can be applied when we do not have information that would be brought in properly even in the eyes of frequentists, and because the real issue is seen now as not use-prior-information-or-not, but as whether to move into the framework where everything is in the same probability space, and we can make probability statements about any variables, with any conditioning or marginalizing we might want to do. In complex models this is really the only practical way to carry out inference in problems people need to tackle substantively (Clark, 2005; Stone, Keller, Kratzke, & Strumpfer, 2014).

That all modeled assumptions may be seen as prior probability judgments amounts to a connection between Bayesian approaches and the whole enterprise of statistical modeling. But as described next, there are still deeper connections between the formation of statistical models from first principles and a Bayesian approach that specifies parameters distributionally.

3.5.3 Exchangeability, de Finetti's Theorem, and Bayesian Modeling

The epistemic perspective on probabilistic expressions and modeling aligns well with the implications of de Finetti's representation theorem. Note that θ does not appear on the left-hand side of (3.4). In a sense, the parameter θ and the specification of the variables being conditionally independent given this parameter *emerge* from the assumption of exchangeability.

But that is not all that emerges; so too does $p(\theta)$. Thus, de Finetti's theorem may be seen in a Bayesian light where (a) $\prod_{j=1}^{J} p(x_j \mid \theta)$ is the conditional probability of the observed variables given a parameter, with the conditional independence relationship supporting the factoring into the product, and (b) $p(\theta)$ is the prior distribution for the parameter. Some scholars see the Bayesian approach as following from de Finetti's theorem (Jackman, 2009; Lindley & Phillips, 1976; Lunn et al., 2013). As Lindley and Phillips (1976, p. 115) gently put it

> The point is that exchangeability... produces $p(\theta)$ and demonstrates the soundness of the subsequent Bayesian manipulations... Thus, from a single assumption of exchangeability the Bayesian argument follows. This is one of the most beautiful and important results in modern statistics. Beautiful, because it is so general and yet so simple. Important, because exchangeable sequences arise so often in practice. If there are, and we are sure there will be, readers who find $p(\theta)$ distasteful, remember it is only as distasteful as exchangeability; and is that unreasonable?

At a minimum, de Finetti's theorem shows that, from an assumption of exchangeability, we can arrive at representation that is entirely consistent with the Bayesian approach to modeling in which the parameter θ is treated as random and modeled via a distribution. To be clear, we are *not* asserting that a Bayesian approach is the only justifiable approach. Indeed, we subscribe to the statistical ecumenism of Box (1983) that welcomes all methods. We are asserting that the use of prior distributions for parameters is consistent with an assumption of exchangeability.

* It may be further argued that Bayesian approaches offer something more than frequentist approaches on this issue, namely transparency of what is subjective (Berger & Berry, 1988).

We are also not asserting that the preceding developments and de Finetti's theorem implies that a parameter θ exists in any ontological sense; from exchangeability and de Finetti's theorem, we do not conclude that there is some previously unknown parameter, lurking out there in the world just waiting to be discovered. Viewing probability expressions as reflections of beliefs and uncertainty sheds a particular light on the meaning of de Finetti's theorem. The left-hand side is the joint distribution of the variables, which amounts to an expression of our beliefs about the full collection of variables. What de Finetti's theorem shows is that if we view the variables as exchangeable, we can parameterize this joint distribution—indeed, our beliefs—by specifying the variables as conditionally independent given a parameter, and a distribution for that parameter.

To summarize, probability-model-based reasoning begins with setting up the joint distribution (Gelman et al., 2013), which represents the analyst's beliefs about the salient aspects of the problem. If we can assert exchangeability, or once we condition on enough variables to assert exchangeability, de Finetti's theorem permits us to accomplish this daunting task by invoking conditional independence relationships—manufacturing entities to accomplish this if needed—which permits us to break the problem down into smaller, more manageable components. That is, we can specify a possibly high-dimensional joint distribution by specifying more manageable conditional distributions given parameters, and a distribution for the parameters.

Prior probability distributions for parameters are then just some of the building blocks of our models used to represent real-world situations, on par with other features we broadly refer to as *specifications* or *assumptions* common in Bayesian and frequentist modeling, including those regarding the distribution of the data or likelihood (e.g., linearity, normality), characteristics or relationships among persons (e.g., independence of persons or clustering of persons into groups), characteristics of the parameters (e.g., discrete or continuous latent variables, the number of such variables, variances being greater than 0), and among parameters (e.g., certain parameters are equal). What is important in the Bayesian perspective is modeling what features of situations we think are salient and how we think they might be related, using epistemic tools such as conditional exchangeability. In this light, the use of prior distributions is not a big deal. It is the admission price for being able to put all the variables in the framework of probability-based reasoning—not only from parameters to variables, like frequency-based statistics can do, but also from variables to parameters, or some variables to other variables, or current values of variables to future values, and so on—all in terms of sound and natural expression of beliefs of variables within the model framework through probability distributions.

We specify priors based on a mixture of considerations of substantive beliefs, computational demands, and ease of interpretability and communication. Importantly, the same can be said of specifying distributions of the data that induce the likelihood function. Those new to Bayesian methods are often quick to question the specification of prior distributions, or view the use of a particular prior distribution with a skeptical eye. We support this skepticism; like any other feature of a model, it and its influence should be questioned and justifications for it articulated. Analysts should not specify prior distributions without having reasons for such specifications.

In addition, sensitivity analyses in which solutions from models using different priors can be compared to reveal the robustness of the inferences to the priors or the unanticipated effects of the priors. Figure 3.1 illustrates how this could be done for the Beta-binomial model, where the analyst may see how the use of different prior distributions affects the substantive conclusions. The importance of the prior in influencing the posterior depends also on the information in the data as the posterior represents a balancing of the contributions of the data, in terms of the likelihood, and the prior. As the relative contribution of the data increases, the posterior typically becomes less dependent on the prior and more closely resembles the likelihood in shape, as illustrated in Figure 3.3.

The role of the prior distribution is often the focus of criticism from those skeptical of Bayesian inference, but the aforementioned issues of justification and examination apply to other features of the model, including the specification of the likelihood. We disagree with the perspective that views a prior as somehow on a different ontological or epistemological plane from the likelihood and other model assumptions. Viewed from the broader perspective of model-based reasoning, both the prior and the likelihood have the same status. They are convenient fictions—false but hopefully useful accounts we deploy to make the model reflect what is believed about the real-world situation—that reside in the model-space as part of the larger model used to enact reasoning to arrive at inferences. This equivalence is illuminated once we recognize that elements of a likelihood may be conceived of as elements of the prior, more broadly construed.

3.5.4 Reasoning through the Machinery of the Model

In probability models, reasoning through the machinery of the model amounts to deploying the calculus of probabilities. The model in (3.4) and Figure 3.5 is set up such that given the value of θ, we have (conditional) distributions for the observables. This naturally supports probabilistic *deductive* reasoning from the former to the latter. Of course, in practice we need to reason *inductively*, to reverse the direction of the flow of the model. Bayes' theorem enacts exactly this reversal, obtaining $p(\theta \mid \mathbf{x})$ by synthesizing $p(\mathbf{x} \mid \theta)$ with the prior distribution $p(\theta)$, enabling probability-based reasoning about the unknown θ.

3.5.5 Managing and Propagating Uncertainty

Conventional approaches to psychometric modeling often proceed in stages in such a way that the uncertainty at one stage is ignored at later stages. For example, modeling and inference in adaptive testing commonly proceeds by first obtaining estimates of parameters of item response theory models, which are then treated as known for estimating students' proficiencies (e.g., Wainer et al., 2000). As another example, many conventional approaches to model-data fit in psychometric modeling rely on point estimates of parameters obtained from samples. A Bayesian approach allows us to incorporate and propagate uncertainty throughout all aspects of modeling, including parameter estimation, model-data fit, and the management of missing data. For simple problems with large sample sizes, the failure to incorporate and propagate uncertainty may be relatively harmless; in complex problems, the failure to fully account for the uncertainty may prove consequential. We elaborate on and illustrate these points as they play out in psychometrics in several contexts throughout the second part of this book.

3.5.6 Incorporating Substantive Knowledge

Adopting a Bayesian approach can make it easier to incorporate substantive knowledge about the problem into the analysis and specify more flexible models, including those that pose considerable challenges to frequentist approaches to estimation (see Levy, 2009, Levy, Mislevy, & Behrens, 2011, and Levy & Choi, 2013 for discussions and references of examples in psychometrics; see also Clark, 2005). One way to incorporate substantive knowledge is via prior distributions, which can rein in estimates that based on the likelihood alone may be extreme. This is particularly salient in situations with small samples and/or sparse data, in which case sampling variability is high. Bayesian approaches, even with fairly diffuse prior distributions, have been shown to perform as well or better than frequentist methods with small samples (e.g., Ansari & Jedidi, 2000; Chung, 2003; Chung, Lanza, & Loken, 2008; Depaoli, 2013; Finch & French, 2012; Hox, van de Schoot, & Matthijsse, 2012; Kim, 2001; Lee & Song, 2004; Muthén & Asparouhov, 2012; Novick, Jackson, Thayer, & Cole, 1972; Sinharay, Dorans, Grant, & Blew, 2009). We discuss this issue in more detail in Chapter 11, where prior distributions can serve to regularize estimates or adjudicate between competing sets of estimates of parameters in item response theory models.

3.5.7 Accumulation of Evidence

Importantly, Bayesian methods reflect the accumulation of evidence (Jackman, 2009). The posterior distribution is a synthesis of the prior and the data. With little data, the solution is more heavily influenced by the prior and less so by the data. As more and more data arrive, they swamp the prior such that the solution becomes increasingly like what the data alone dictate (Section 3.1.2). As a result, analysts with different prior beliefs (and prior distributions) may have very different conclusions if there is no or little data. But as more and more data are increasingly incorporated their conclusions will converge to a common destination, that is, their posterior distributions will increasingly resemble each other as they increasingly resemble a spike at the point suggested by the data (Blackwell & Dubins, 1962).

A related point concerns the accumulation of evidence as data arrives. Let x_1 and x_2 be two datasets; we may think of their subscripts as indicating a temporal ordering, though this need not be case. Applying Bayes' theorem and assuming that x_1 and x_2 are conditionally independent given θ, we have

$$\begin{aligned} p(\theta | x_1, x_2) &\propto p(x_1, x_2 | \theta) p(\theta) \\ &= p(x_2 | \theta, x_1) p(x_1 | \theta) p(\theta) \\ &= p(x_2 | \theta) p(x_1 | \theta) p(\theta) \\ &\propto p(x_2 | \theta) p(\theta | x_1). \end{aligned} \tag{3.12}$$

The second line in (3.12) follows from a factorization of the conditional probability of the data. The third line follows from the assumption that x_1 and x_2 are conditionally independent given θ. The last line follows from recognizing that, from Bayes' theorem, $p(\theta | x_1) \propto p(x_1 | \theta)p(\theta)$. The right-hand side of the last line takes the form of the right-hand side of Bayes' theorem in (2.1), where $p(x_2|\theta)$ is the conditional probability of the (new) data, and $p(\theta | x_1)$ plays the role of the prior distribution for θ—that is, prior to having observed x_2. But $p(\theta | x_1)$ is just the posterior distribution for θ given x_1. Equation (3.12) reveals how a Bayesian approach naturally accommodates the arrival of data and updating beliefs about unknowns. We begin with a prior distribution for the unknowns, $p(\theta)$. Incorporating the first dataset, we have the posterior distribution $p(\theta | x_1)$, which in turn serves as the prior distribution when incorporating the second dataset, x_2. At any point, our "current" distribution is both a posterior distribution and prior distribution; it is posterior to the past data, and prior to future data. Today's posterior is just tomorrow's prior, and the updating is facilitated by Bayes' theorem. In psychometrics, this line of thinking supports adaptive testing discussed in Chapter 11, with θ modeled as constant over time, and models for student learning over time discussed in Chapter 14, with θ modeled as possibly changing over time.

3.5.8 Conceptual and Computational Simplifications

Finally, a Bayesian approach facilitates simplifications. In psychometrics, the entity we have so far denoted by θ has been variously referred to as a "parameter" and a "latent variable." In frequentist traditions, this difference has implications for what analyses and methods are permissible (Bartholomew, 1996; Maraun, 1996). In addition, it has been argued that θ should be conceived of as missing data (Bollen, 2002), which brings into the picture the possibilities of associated methods (Enders, 2010). Taken as a whole, we have a set of possible analyses and methods cobbled together from various corners of the frequentist world.

A fully Bayesian analysis offers considerable simplicity for how to proceed: if θ is unknown it gets a prior distribution, and once we condition on what we do know, we will have a posterior distribution. More broadly, though we may find terminological, epistemological, or ontological differences useful for conveying ideas or drawing distinctions among the roles that a latent variable, parameter, or missing data point may play, a fully Bayesian analysis offers considerable simplicity for how to proceed. They are all unknowns, in the sense that we have uncertainty about each of them. Accordingly, each such entity gets a prior distribution and once we condition on what we do know, we will have a posterior distribution for the ones that remain uncertain.

This conceptual simplicity translates to technical matters. Model fitting and estimating posterior distributions remains the same regardless of what we call θ, and in Markov chain Monte Carlo strategies, the process for obtaining a posterior distribution for unknown xs, often referred to as missing data, is the same as for unknown θs, regardless of what we call it (see Chapter 12). This unification and simplification represents in some ways a reduced burden for the student of quantitative methods—inference is just obtaining the conditional distribution for unknowns given knowns.

Moreover, it can aid in avoiding misconceptions. For example, categorical factor analysis and item response theory are two different paradigms for models with continuous latent variables and discrete observables. Traditions within these paradigms evolved over time in many ways independent from one another. The connections between and equivalence among models from these various traditions were formalized over a quarter century ago (Takane & de Leeuw, 1987). However, as noted by Wirth and Edwards (2007), the use of frequentist estimation routines associated with these modeling paradigms has yielded contradictory misconceptions about the weaknesses of the models. Adopting a Bayesian approach to modeling and estimation, the distinctions between the factor analytic and item response traditions disappear, and the conflicting misconceptions are recognized as illusory (Levy, 2009; Wirth & Edwards, 2007).

3.5.9 Pragmatics

One need not subscribe to the philosophy sketched here to employ Bayesian methods. One can adopt different orientations toward probability, parameters/latent variables, measurement, modeling, and inference, and still gainfully employ Bayesian methods. For those who adopt an alternative philosophical outlook, or do not believe philosophical positions have any bearing, there is another class of arguments that can be advanced in favor of adopting a Bayesian approach to inference. These are more practical in their focus, and advocate adopting a Bayesian approach to the extent that it is useful. These arguments advance the notion that adopting a Bayesian approach allows one to fit models and reach conclusions about questions more readily than can be done using frequentist approaches, and allows for the construction and use of complex models that are more aligned with features of the real world and the data (Andrews & Baguley, 2013; Clark, 2005). See Levy (2009); Levy et al. (2011); Levy and Choi (2013); and Rupp, Dey, and Zumbo (2004) for discussions and recent reviews of Bayesian approaches to psychometrics and related models that survey a number of applications that trade on the utility of Bayesian methods.

3.6 Conceptualizations of Bayesian Modeling

In the course of describing Bayesian modeling we have introduced several different conceptions on what it is we are doing when we engage in Bayesian modeling. These are briefly reviewed here for the purposes of calling them out as viable lenses through which we may proceed.

A first perspective views Bayesian inference as a belief updating process. We begin with what is initially believed about the parameters, expressed by the prior. We then update that in light of the information in the data, expressed in the likelihood, to yield the posterior. This conceptualization of Bayes as belief updating is strongly aligned with a subjective, epistemic view of probability. From this perspective, the data count as information to be folded into previously held beliefs, through a model that expresses our beliefs about how the data are related to the parameters. In this view, Bayes' theorem is a mechanism for updating distributions (beliefs) in light of observed data.

A second perspective gives more of a conceptual primacy to the data. Here, Bayesian inference is seen as taking what the data imply about the values of the parameters, which is contained in the likelihood, and tempering that in light of what was known, believed, or assumed about the parameters before observing the data, which is contained in the prior. This perspective views the prior as augmenting the information in data, contained in the likelihood. Readers steeped in frequentist traditions are likely to initially conceptualize Bayesian inference in this way. A similar perspective motivates constructing prior distributions that are mildly informative in that they contain some information to rein in or regularize results that would be obtained from a frequentist approach of a Bayesian approach with extremely diffuse priors (Gelman, Jakulin, Pittau, & Su, 2008). Examples of the use of priors in this capacity in psychometrics include Chung et al. (2008), Chung et al. (2013), Martin and McDonald (1975), Mislevy (1986), and Maris (1999).

Both of these conceptualizations trade on the distinction between the prior and the likelihood, which results from the factoring of the joint distribution. A third conceptualization does not rely on this distinction for conceptual grounding, and instead conceives of the joint distribution as the primary conceptual basis for analysis. This view is consistent with the recognition that it is not always easy, or necessary, to distinguish what is part of the likelihood and what is part of the prior (Box, 1980). Likewise, it aligns well with characterizations of Bayesian inference in which the first step is to set up the joint distribution of all entities and the second step is to condition on whatever is known or observed (i.e., the data) to yield the posterior distribution for all unknowns (Gelman et al., 2013). Although somewhat artificial, the distinction is useful in two ways. First, in practice, the specification of the joint distribution is typically accomplished by factoring into the likelihood and the prior components. Second, it is didactically useful for introducing Bayesian principles and procedures to those steeped in frequentist approaches, and accordingly we make use of it throughout the book.

A fourth conceptualization is one of model building or expansion, in which model construction occurs in stages. This begins not by assuming parameters, but instead by seeking to specify a distribution for the data, **x** (Bernardo & Smith, 2000). If we can assert exchangeability, we may structure the distribution based on parameters θ, as $p(\mathbf{x} \mid \theta)$. If θ is unknown, it requires a distribution. In simple cases, a distribution is specified as $p(\theta)$. This perspective does not seem to offer much in simple models, but is especially advantageous in specifying complex, multivariate models which can be framed as having multiple levels. That is, if $p(\theta)$ may be framed as conditional on parameters, say, θ_P, as $p(\theta \mid \theta_P)$ in (3.10). If θ_P are unknown, they too require a prior distribution $p(\theta_P)$, which itself may depend on parameters, possibly unknown, and so on. The distributions at the different levels may have their own covariates and exchangeability structures, such as student characteristics and school characteristics. This perspective of "building out" the model is strongly aligned with strategies in multilevel modeling (e.g., Gelman & Hill, 2007), and is often framed as a *hierarchical* specification of the model and prior distributions (Lindley & Smith, 1972). This perspective has come to dominate Bayesian modeling for wide variety of psychometric and complex statistical models (Jackman, 2009). At various points in our treatment of Bayesian psychometric modeling, we will adopt each of these perspectives on modeling to bring out certain inferential issues.

3.7 Summary and Bibliographic Note

This chapter has toured several key conceptual aspects of Bayesian approaches to psychometric modeling. Bayesian inference may be viewed as an instance of model-based reasoning, in which we construct a model to represent the salient aspects of a complicated real-world problem and reason through the machinery of the model. Exchangeability and conditional independence structures greatly facilitate the process of constructing the model, and Bayes' theorem is our workhorse for conducting inductive inference. In assessment, this will be in reasoning from observations of examinee behaviors to characterizations of them more broadly construed.

The distinctive feature of Bayesian approaches is the treatment of parameters, indeed, all entities, as random, with accompanying distributional specifications. Viewing probability as an expression of beliefs, the posterior distribution is an expression of beliefs arrived at by the synthesis of the prior beliefs represented in the prior distribution and the information in the data represented in the likelihood.

The perspective on Bayesian inference presented in Section 3.5 is one of many that have been advanced. See Barnett (1999), Berger (2006), Box (1983), Gelman (2011), Gelman and Shalizi (2013), Jaynes (2003), Senn (2011), and Williamson (2010) for recent expositions on varying perspectives on Bayesian inference, and discussions.

Our treatment has stressed exchangeability as a concept and its role in model construction. More formal textbook treatments of exchangeability can be found in Bernardo and Smith (2000) and Jackman (2009); Barnett (1999) provides a discussion of its historical origins and its role in inference.

Techniques and strategies for eliciting and codifying prior information and for eliciting probabilistic beliefs, including ways to calibrate the performance of individuals in their probability assessment, may be found in Almond (2010), Almond et al. (2015), Chow, O'Leary, and Mengersen (2009), Garthwaite, Kadane, and O'Hagan, 2005, Kadane and Wolfson (1998), Novick and Jackson (1974), O'Hagan (1998), O'Hagan et al. (2006), Press (1989), Savage (1971), and Winkler (1972). See De Leeuw and Klugkist (2012) for several alternatives for basing prior distributions on prior research in the context of regression modeling. We discuss strategies for specifying prior distributions based on substantive beliefs and previous research in psychometric models in the context of item response theory in Chapter 11.

4

Normal Distribution Models

This chapter provides a treatment of popular Bayesian approaches to working with normal distribution models. We do not attempt a comprehensive account, instead providing a more cursory treatment that has two aims. First, it is valuable to review a number of Bayesian modeling concepts in the context of familiar normal distributions. Second, normal distributions are widely used in statistical and psychometric modeling. As such, this chapter provides a foundation for more complex models; in particular, the development of regression, classical test theory, and factor analysis models will draw heavily from the material introduced here.

As a running example, suppose we have a test scored from 0 to 100, and we are interested in the distribution of test scores for examinees. We obtain scores from $n = 10$ examinees, and let those scores be $\mathbf{x} = (91, 85, 72, 87, 71, 77, 88, 94, 84, 92)$, where x_i is the score for examinee i. Assuming that the scores are independently and identically normally distributed, $x_i \sim N(\mu, \sigma^2)$, unbiased least-squares estimates of the mean and variance are 84.1 and 66.77; the MLEs for the mean and variance are 84.1 and 60.09. In what follows, we explore a series of Bayesian models for this setup. In Section 4.1, we model the unknown mean treating the variance as known. In Section 4.2, we model the unknown variance treating the mean as known. In Section 4.3, we consider the case where we model both the mean and variance as unknown. We conclude this chapter with a brief summary in Section 4.4.

4.1 Model with Unknown Mean and Known Variance

4.1.1 Model Setup

We begin by considering the situation where σ^2 is known and inference is focused on the unknown mean μ. Following Bayes' theorem,

$$p(\mu \mid \mathbf{x}, \sigma^2) \propto p(\mathbf{x} \mid \mu, \sigma^2) p(\mu). \tag{4.1}$$

The first term on the right-hand side of (4.1) is the conditional probability of the data. Treating the examinees' scores as exchangeable and normally distributed,

$$p(\mathbf{x} \mid \mu, \sigma^2) = \prod_{i=1}^{n} p(x_i \mid \mu, \sigma^2), \tag{4.2}$$

where

$$x_i \mid \mu, \sigma^2 \sim N(\mu, \sigma^2) \tag{4.3}$$

Note that (4.2) and (4.3) are akin to those in conventional frequentist approaches.

4.1.2 Normal Prior Distribution

In a Bayesian analysis, the unknown parameter is treated as a random variable and assigned a prior distribution, which is the second term on the right-hand side of (4.1). A popular choice for the prior distribution in this context is a normal distribution. That is,

$$\mu \sim N(\mu_\mu, \sigma_\mu^2), \tag{4.4}$$

where μ_μ and σ_μ^2 are hyperparameters. The subscript notation adopted here is used to reflect that these are features of the distribution of μ prior to observing **x**, in contrast to the posterior distribution.

4.1.3 Complete Model and Posterior Distribution

Figure 4.1 presents the directed acyclic graph (DAG) for the model. The portion of the DAG from the top to the node for μ in the middle represents the prior distribution and corresponds to (4.4). The portion of the DAG from the middle layer (i.e., μ and σ^2) down represents the conditional probability of the data and corresponds to (4.2) and (4.3). The plate over the index i indicates that there is a value of x_i for $i = 1,\ldots, n$. Importantly, the parents of x_i, namely μ and σ^2, fall outside the plate, indicating that they do not vary over i. Taken together, this is a graphical consequence of the exchangeability assumption that facilitates the factoring in (4.2) and the specification of each individual x_i on a common set of parent variables.

Putting the pieces together and substituting into (4.1), we have

$$\begin{aligned} p(\mu \mid \mathbf{x}, \sigma^2) &\propto p(\mathbf{x} \mid \mu, \sigma^2) p(\mu) \\ &= \prod_{i=1}^{n} p(x_i \mid \mu, \sigma^2) p(\mu), \end{aligned} \tag{4.5}$$

where

$$x_i \mid \mu, \sigma^2 \sim N(\mu, \sigma^2) \quad \text{for } i = 1,\ldots, n$$

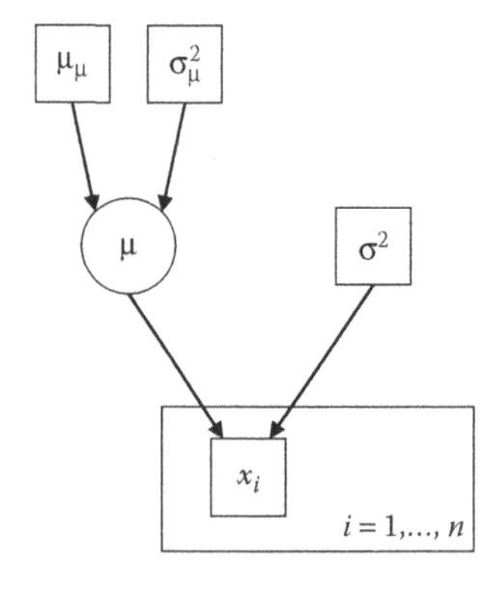

FIGURE 4.1
Directed acyclic graph for a normal distribution model with unknown mean μ and known variance σ^2, where μ_μ and σ_μ^2 are hyperparameters for the unknown mean μ.

and

$$\mu \sim N(\mu_\mu, \sigma_\mu^2).$$

The normal prior distribution is popular in part because it is a conjugate prior in this case. It can be shown that the posterior distribution is then itself normal (Lindley & Smith, 1972; see also Gelman et al., 2013; Gill, 2007; Jackman, 2009):

$$\mu \mid \mathbf{x}, \sigma^2 \sim N(\mu_{\mu|\mathbf{x}}, \sigma_{\mu|\mathbf{x}}^2), \quad (4.6)$$

where

$$\mu_{\mu|\mathbf{x}} = \frac{\left(\mu_\mu/\sigma_\mu^2\right)+\left(n\bar{x}/\sigma^2\right)}{\left(1/\sigma_\mu^2\right)+\left(n/\sigma^2\right)}, \quad (4.7)$$

$\bar{x} = n^{-1}\sum_i x_i$ is the mean of the observed data, and

$$\sigma_{\mu|\mathbf{x}}^2 = \frac{1}{\left(1/\sigma_\mu^2\right)+\left(n/\sigma^2\right)}. \quad (4.8)$$

The subscript notation adopted on the right-hand side of (4.6) and in (4.7) and (4.8) is used to reflect that these are features of the distribution of μ posterior to observing $\mathbf{x}$.* The posterior distribution also reveals that n and $\bar{x}$ are sufficient statistics for the analysis; they jointly capture all the relevant information in the data.

4.1.4 Precision Parameterization

A number of key features of the result here are more easily seen by a reparameterization of the normal distribution in terms of the *precision*, defined as the inverse of the variance. In this parameterization, we employ the precision in the distribution of the data

$$\tau = \sigma^{-2} \quad (4.9)$$

and the prior precision

$$\tau_\mu = \sigma_\mu^{-2}. \quad (4.10)$$

The posterior distribution is then written as

$$\begin{aligned} p(\mu \mid \mathbf{x}, \tau) &\propto p(\mathbf{x} \mid \mu, \tau)p(\mu) \\ &= \prod_{i=1}^{n} p(x_i \mid \mu, \tau)p(\mu), \end{aligned} \quad (4.11)$$

* As (4.7) and (4.8) reveal, these results are also conditional on the values for the hyperparameters and the variance of the data. In the current example, these are all known values. When we turn to models that treat the mean and variance as unknown, we will expand our notation for these expressions accordingly.

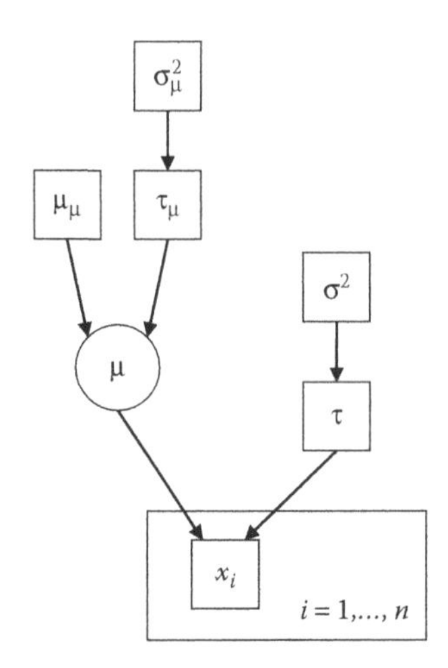

FIGURE 4.2
Directed acyclic graph for a normal distribution model with unknown mean μ in the precision parameterization, where τ is the known precision of the data, equal to the inverse of the variance σ^2, and μ_μ and τ_μ (the inverse of σ^2_μ) are hyperparameters for the unknown mean μ.

where

$$x_i \mid \mu, \tau \sim N(\mu, \tau) \text{ for } i = 1,\ldots,n$$

and

$$\mu \sim N(\mu_\mu, \tau_\mu).$$

Figure 4.2 contains the DAG for the model in the precision parameterization. It differs from the DAG in Figure 4.1 by modeling τ as a parent for x_i and τ_μ as a parent for μ. These precision terms are modeled as the children of the associated variance terms, reflecting the deterministic relations in (4.9) and (4.10).

Under this parameterization, the posterior distribution for μ is normal,

$$\mu \mid \mathbf{x}, \tau \sim N(\mu_{\mu|\mathbf{x}}, \tau_{\mu|\mathbf{x}}), \tag{4.12}$$

with posterior mean

$$\mu_{\mu|\mathbf{x}} = \frac{\tau_\mu \mu_\mu + n\tau\bar{x}}{\tau_\mu + n\tau} = \frac{\tau_\mu}{\tau_\mu + n\tau}\mu_\mu + \frac{n\tau}{\tau_\mu + n\tau}\bar{x} \tag{4.13}$$

and posterior precision

$$\tau_{\mu|\mathbf{x}} = \tau_\mu + n\tau. \tag{4.14}$$

The precision parameterization reveals several key features of the posterior distribtion. First, (4.14) indicates that the posterior precision ($\tau_{\mu|\mathbf{x}}$) is the sum of two components: the precision in the prior (τ_μ) and the precision in the data ($n\tau$). Conceptually, the variance of a distribution is a summary of our uncertainty—a distribution with a relatively large (small) variance indicates relatively high (low) uncertainty. In this light, the precision is a summary of our certainty—a distribution with a relatively large (small) precision indicates relatively high (low) certainty. In these terms, (4.14) states that our posterior certainty is the sum of certainty from two sources, namely the prior and the data.

The posterior mean ($\mu_{\mu|x}$) in (4.13) is a weighted average of the prior mean (μ_μ) and the mean of the data ($\bar{x}$). The weight for the prior mean ($\tau_\mu / [\tau_\mu + n\tau]$) is proportional to the prior's contribution to the total precision. Similarly, the weight for the mean of the data ($n\tau / [\tau_\mu + n\tau]$) is proportional to the data's contribution to the total precision. Viewing the posterior mean as a point summary of the posterior distribution, (4.13) also illustrates the general point that the posterior will be a synthesis of the information in the prior and the information in the data as expressed in the likelihood. In the current case, the relative contribution of the prior and the data in this synthesis is governed by the relative precision in each of these sources.

4.1.5 Example Analysis

To illustrate, we develop an analysis of the $n = 10$ examinee test scores listed at the outset of this chapter, which have a mean of $\bar{x} = 84.1$. The variance of the data is known, $\sigma^2 = 25$; equivalently the precision of the data is $\tau = .04$. We employ the following normal prior distribution for μ expressed in the variance parameterization

$$\mu \sim N(75, 50), \tag{4.15}$$

which implies the prior precision is $\tau_\mu = .02$. This prior, depicted in Figure 4.3, expresses the prior belief that μ is almost certainly larger than 50 and probably in the high 60s to low 80s. The likelihood is also depicted in Figure 4.3. It takes a maximum value at 84.1, the mean of the data. The posterior mean and precision are calculated through (4.13) and (4.14) as follows:

$$\mu_{\mu|x} = \frac{\tau_\mu \mu_\mu + n\tau\bar{X}}{\tau_\mu + n\tau} = \frac{(.02)(75) + (10)(.04)(84.1)}{.02 + (10)(.04)} \approx 83.67$$

and

$$\tau_{\mu|x} = \tau_\mu + n\tau = .02 + (10)(.04) = .42.$$

This last result implies that the posterior variance is $\sigma^2_{\mu|x} \approx 2.38$. The posterior distribution, expressed in the variance parameterization, is

$$\mu \mid \mathbf{x}, \sigma^2 \sim N(83.67, 2.38). \tag{4.16}$$

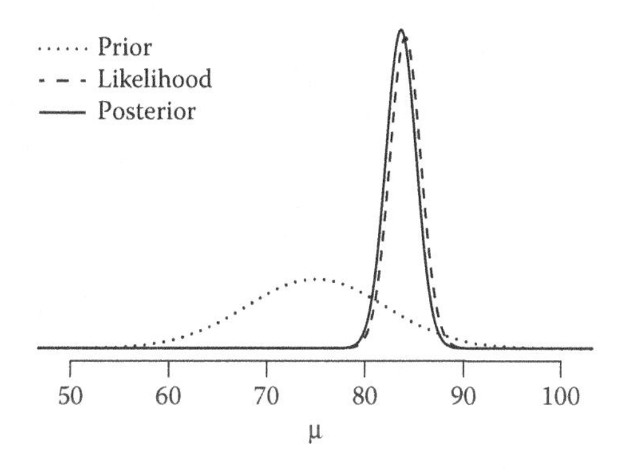

FIGURE 4.3
Prior distribution, likelihood, and posterior distribution for the example with an $N(75, 50)$ prior for the unknown mean μ, where $\sigma^2 = 25$, $\bar{x} = 84.1$, and $n = 10$.

The posterior is depicted in Figure 4.3. It may be summarized in terms of its central tendency, the posterior mean (and median and mode) is 83.67; and variability, the posterior standard deviation is 1.54. The 95% central and HPD interval is (80.64, 86.69).

4.1.6 Asymptotics and Connections to Frequentist Approaches

It is instructive to consider some properties of the posterior distribution in limiting cases as the information in the data increases or the information in the prior decreases. Consideration of (4.7) and (4.8) reveals that, holding the prior distribution constant, as $n \to \infty$,

$$\sigma^2_{\mu|x} \to \frac{\sigma^2}{n} \tag{4.17}$$

and

$$\mu_{\mu|x} \to \bar{x}. \tag{4.18}$$

Accordingly, as $n \to \infty$,

$$\mu \mid \mathbf{x}, \sigma^2 \to N\left(\bar{x}, \frac{\sigma^2}{n}\right). \tag{4.19}$$

Likewise, these limiting properties obtain as $\sigma^2_\mu \to \infty$, holding the features of the data constant.

These results illustrate a number of the principles discussed in Section 3.3. As the relative contribution of the data increases—either through the increased amount of data (n) or the decrease in the information in the prior (larger values of σ^2_μ)—the results become increasingly similar to what would be obtained from a frequentist analysis. Point summaries of the posterior (i.e., mean, median, and mode) get closer to $\bar{x}$, which is the MLE of μ. The posterior standard deviation of μ gets closer to the sampling variance of $\bar{x}$. And posterior credibility intervals resemble frequentist confidence intervals. Importantly, as discussed in Section 3.3, though the results may be numerically similar to their frequentist counterparts, their interpretations are different. Unlike a frequentist analysis, a Bayesian analysis yields probabilistic statements and reasoning about the parameters. The 95% central and HPD interval indicates that, according to the model, there is a .95 probability that μ is between 80.64 and 86.69.

4.2 Model with Known Mean and Unknown Variance

4.2.1 Model Setup

We turn to the situation where μ is known and inference is focused on the unknown variance σ^2. Following Bayes' theorem,

$$p(\sigma^2 \mid \mathbf{x}, \mu) \propto p(\mathbf{x} \mid \mu, \sigma^2) p(\sigma^2). \tag{4.20}$$

The first term on the right-hand side is the conditional probability of the data, which is the same as given in (4.2) and (4.3). The second term on the right-hand side is the prior distribution for σ^2.

4.2.2 Inverse-Gamma Prior Distribution

A popular choice for the prior distribution of the variance of a normal distribution is an inverse-gamma distribution. That is,

$$\sigma^2 \sim \text{Inv-Gamma}(\alpha, \beta), \tag{4.21}$$

where α and β are hyperparameters. Figure 4.4 depicts several inverse-gamma distributions, where it is seen that the support of the distribution is restricted to positive values and is positively skewed.

Inverse-gamma distributions are not as well known as normal distributions. Choosing the values of the hyperparameters is likely not natural to analysts not well steeped in them. One strategy for working with them involves a parameterization of the inverse gamma where $\alpha = \nu_0 / 2$ and $\beta = \nu_0\sigma_0^2 / 2$,

$$\sigma^2 \sim \text{Inv-Gamma}(\nu_0 / 2, \nu_0\sigma_0^2 / 2) \tag{4.22}$$

Under this parameterization, σ_0^2 has the interpretation akin to a best estimate for the variance and ν_0 has the interpretation akin to the degrees of freedom, or pseudo-sample size associated with that estimate. This could come from prior research, or from subject matter expert beliefs. For example, a subject matter expert may express their beliefs by saying "I think the variance is around 30. But I'm not very confident, it's only as if that came from observing 10 subjects." This could be modeled by setting $\sigma_0^2 = 30$ and $\nu_0 = 10$, yielding an Inv-Gamma(5,150) distribution. Then the inverse-gamma distribution may be plotted and inspected as to whether it represents prior beliefs as intended.

The inverse-gamma distribution has a positive skew shape like the more familiar χ^2 distribution. In fact, the χ^2 distribution is related to the inverse-gamma and the gamma distribution. Specifically, if $\sigma^2 \sim \text{Inv-Gamma}(\nu_0 / 2, \nu_0\sigma_0^2 / 2)$, then the quantity $\nu_0\sigma_0^2 / 2 \sim \chi_{\nu_0}^2$. Accordingly, some authors refer to this as a scaled inverse-χ^2 distribution (Gelman et al., 2013).

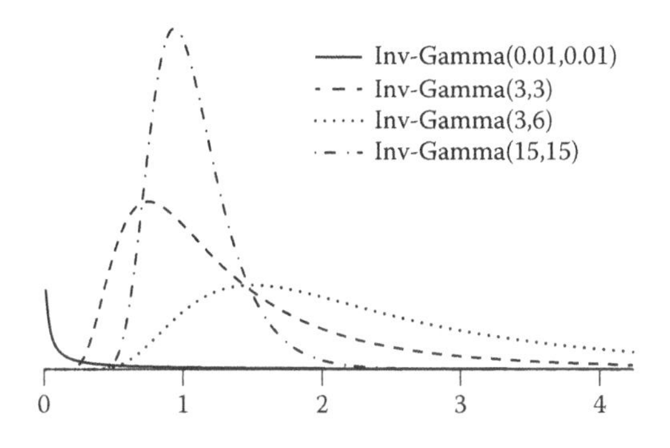

FIGURE 4.4
Inverse-gamma densities.

4.2.3 Complete Model and Posterior Distribution

Figure 4.5 presents the DAG for the model. The portion of the DAG from the top to the node for σ^2 in the middle represents the prior distribuion and corresponds to (4.22). The portion of the DAG from the middle layer down represents the conditional probability of the data and corresponds to (4.2) and (4.3).

Putting the pieces together and substituting into (4.20), the posterior distribution is

$$\begin{aligned} p(\sigma^2 \mid \mathbf{x}, \mu) &\propto p(\mathbf{x} \mid \mu, \sigma^2) p(\sigma^2) \\ &= \prod_{i=1}^{n} p(x_i \mid \mu, \sigma^2) p(\sigma^2), \end{aligned} \tag{4.23}$$

where

$$x_i \mid \mu, \sigma^2 \sim N(\mu, \sigma^2) \text{ for } i = 1, \ldots, n$$

and

$$\sigma^2 \sim \text{Inv-Gamma}(\nu_0 / 2, \nu_0 \sigma_0^2 / 2).$$

It can be shown that the posterior distribution is also an inverse-gamma distribution (Gill, 2007),

$$\sigma^2 \mid \mathbf{x}, \mu \sim \text{Inv-Gamma}\left(\frac{\nu_0 + n}{2}, \frac{\nu_0 \sigma_0^2 + SS(\mathbf{x} \mid \mu)}{2} \right), \tag{4.24}$$

where

$$SS(\mathbf{x} \mid \mu) = \sum_{i=1}^{n} (x_i - \mu)^2 \tag{4.25}$$

is the sum of squares from the data, with the conditioning notation indicating the sum of squares is taken about the known population mean μ. The posterior distribution also

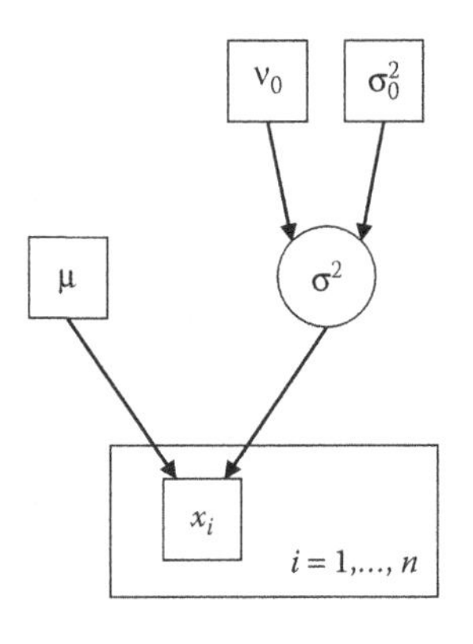

FIGURE 4.5
Directed acyclic graph for a normal distribution model with unknown variance σ^2 and known mean μ, where ν_0 and σ_0^2 are hyperparameters for the unknown variance σ^2.

reveals that n and $SS(\mathbf{x} \mid \mu)$ are sufficient statistics for the analysis; they jointly capture all the relevant information in the data.

4.2.4 Precision Parameterization and Gamma Prior Distribution

In the precision parameterization, the posterior distribution is

$$p(\tau \mid \mathbf{x}, \mu) \propto p(\mathbf{x} \mid \mu, \tau) p(\tau), \tag{4.26}$$

where $p(\tau)$ is the prior distribution for the precision. As discussed in Section 4.2.2, the inverse gamma is the conjugate prior for σ^2. Recognizing that τ is the inverse of σ^2, it is not surprising that the gamma distribution is the conjugate prior for τ. Thus, we specify the prior distribution for τ as

$$\tau \sim \text{Gamma}(\alpha, \beta), \tag{4.27}$$

where α and β are hyperparameters that govern the distribution. Figure 4.6 depicts several gamma distributions. Again, a convenient parameterization is one in which $\alpha = \nu_0 / 2$ and $\beta = \nu_0 \sigma_0^2 / 2$ with the interpretations of σ_0^2 and ν_0 as a prior best estimate for the variance and pseudo-sample size, respectively.

4.2.5 Complete Model and Posterior Distribution for the Precision Parameterization

Figure 4.7 contains the DAG for the model in the precision parameterization. It differs from the DAG in Figure 4.5 by modeling τ as the parent for x_i. The DAG also includes σ^2 as a child of τ, reflecting the deterministic relation between the two. σ^2 is included here to reflect that we might model things in terms of the precision, but for reporting we often prefer to employ the more familiar variance metric.

Putting the pieces together, the posterior distribution in the precision parameterization is

$$\begin{aligned} p(\tau \mid \mathbf{x}, \mu) &\propto p(\mathbf{x} \mid \mu, \tau) p(\tau) \\ &= \prod_{i=1}^{n} p(x_i \mid \mu, \tau) p(\tau), \end{aligned} \tag{4.28}$$

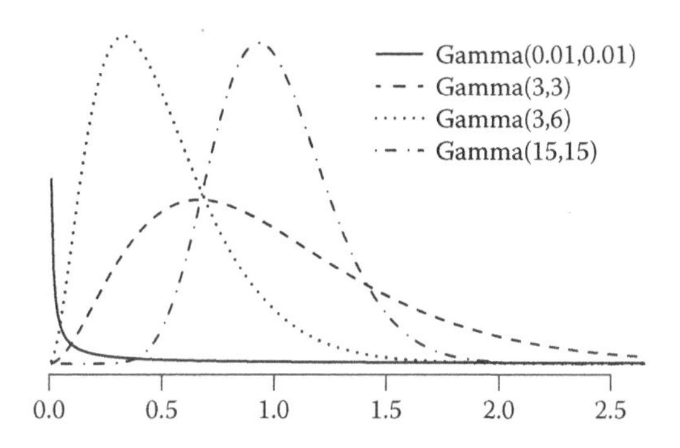

FIGURE 4.6
Gamma densities.

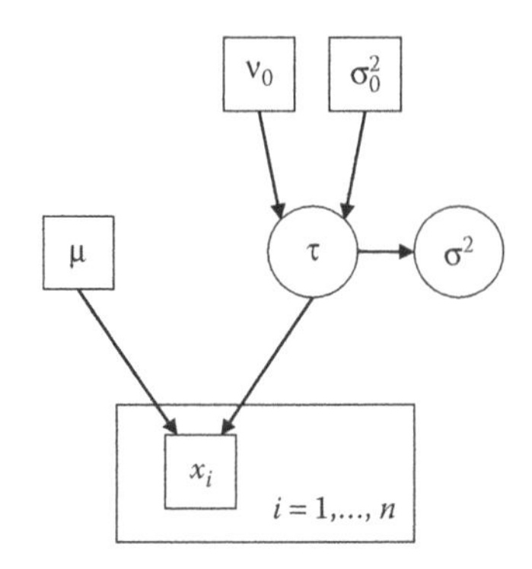

FIGURE 4.7
Directed acyclic graph for a normal distribution model in the precision parameterization, with unknown precision τ (inverse of variance, σ^2) and known mean μ, where ν_0 and σ_0^2 are hyperparameters for the unknown precision τ.

where

$$x_i \mid \mu, \tau \sim N(\mu, \tau) \text{ for } i = 1, \ldots, n$$

and

$$\tau \sim \text{Gamma}(\nu_0 / 2, \nu_0 \sigma_0^2 / 2).$$

The posterior distribution is then a gamma distribution

$$\tau \mid \mathbf{x}, \mu \sim \text{Gamma}\left(\frac{\nu_0 + n}{2}, \frac{\nu_0 \sigma_0^2 + SS(\mathbf{x} \mid \mu)}{2}\right). \tag{4.29}$$

4.2.6 Example Analysis

We return to the example with $n = 10$ scores, assuming a known mean $\mu = 80$. The sum of squares of the 10 scores around μ is 769. The prior distribution was developed based on modeling the belief that the variance is around 30, but expressing the amount of uncertainty as being akin to having observed that variance on the basis of 10 observations. Entering in values of $\sigma_0^2 = 30$ and $\nu_0 = 10$ into (4.22) yields the following prior distribution for σ^2,

$$\sigma^2 \sim \text{Inv-Gamma}(5, 150), \tag{4.30}$$

depicted in Figure 4.8.

The likelihood is also depicted in Figure 4.8. The posterior distribution, also depicted in Figure 4.8, is then

$$\begin{aligned} \sigma^2 \mid \mathbf{x}, \mu &\sim \text{Inv-Gamma}\left(\frac{\nu_0 + n}{2}, \frac{\nu_0 \sigma_0^2 + SS(\mathbf{x} \mid \mu)}{2}\right) \\ &= \text{Inv-Gamma}\left(\frac{10 + 10}{2}, \frac{(10)(30) + 769}{2}\right) \\ &= \text{Inv-Gamma}(10, 534.5). \end{aligned} \tag{4.31}$$

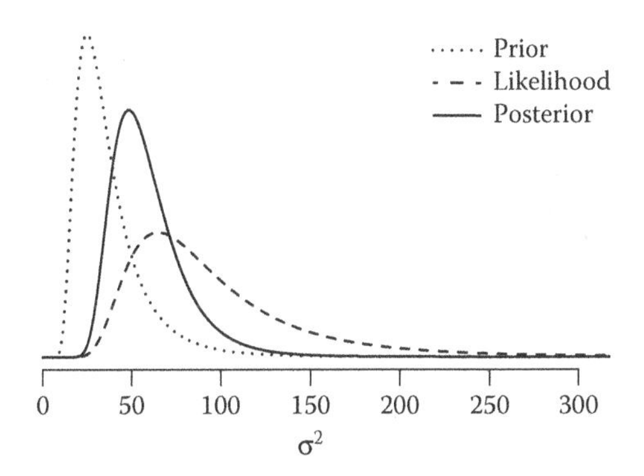

FIGURE 4.8
Prior distribution, likelihood, and posterior distribution for the example with an Inv-Gamma(5,150) prior for the unknown variance σ^2, where $\mu = 80$, $SS(\mathrm{x} \mid \mu) = 769$, and $n = 10$.

We may summarize the posterior numerically in terms of its central tendency, variability, and HPD interval: the posterior mean is 59.39 and the posterior mode is 48.59, the posterior standard deviation is 21.00, and the 95% HPD interval is (27.02,100.74).*

4.3 Model with Unknown Mean and Unknown Variance

4.3.1 Model Setup with the Conditionally Conjugate Prior Distribution

We now consider the situation where both μ and σ^2 are unknown. Following Bayes' theorem,

$$p(\mu,\sigma^2 \mid \mathbf{x}) \propto p(\mathbf{x} \mid \mu,\sigma^2)p(\mu,\sigma^2). \tag{4.32}$$

The first term on the right-hand side is the conditional probability of the data, which remains as given in (4.2) and (4.3).

The second term on the right-hand side is the joint prior distribution for μ and σ^2. To specify this distribution, we first assume independence between μ and σ^2, implying

$$p(\mu,\sigma^2) = p(\mu)p(\sigma^2). \tag{4.33}$$

This independence assumption is common in the specification of a number of models, particularly complex psychometric models. Of course, if there is substantive information to the contrary, it can be incorporated into the model by using a joint prior; we will see one way this can be done shortly. Importantly, using independent priors does not force the parameters to be independent in the posterior. Specifying the joint independence prior then comes to specifying the individual terms on the right-hand side of (4.33). Drawing from the previous developments, we employ a normal prior for μ and an inverse-gamma prior for σ^2 or, equivalently, a gamma prior for τ. The DAGs for the model in both the

* These values came from calculator for the inverse-gamma distribution. We will see shortly how to approximate them empirically. We do not need to do so in this problem, but it is a general approach we can use for more complicated posterior distributions.

variance and precision parameterizations are given in Figure 4.9. The posterior distribution is then

$$\begin{aligned} p(\mu,\sigma^2 \mid \mathbf{x}) &\propto p(\mathbf{x} \mid \mu,\sigma^2)p(\mu)p(\sigma^2) \\ &= \prod_{i=1}^{n} p(x_i \mid \mu,\sigma^2)p(\mu)p(\sigma^2), \end{aligned} \quad (4.34)$$

where the terms on the right-hand side of (4.34) are the conditional probability of the data and the prior distributions:

$$x_i \mid \mu,\sigma^2 \sim N(\mu,\sigma^2) \quad \text{for } i = 1,\ldots,n,$$

$$\mu \sim N(\mu_\mu,\sigma_\mu^2),$$

and

$$\sigma^2 \sim \text{Inv-Gamma}(\nu_0 / 2, \nu_0\sigma_0^2 / 2).$$

Though these prior distributions are conjugate priors in the cases where there is only one unknown, they do not constitute a conjugate joint prior in the current context and do not yield a closed-form joint posterior (Lunn et al., 2013). However, the posterior can be easily approximated empirically using simulation techniques illustrated in the next section and described more fully in Chapter 5. For the moment, it is sufficient to state that these procedures yield a series of values that, taken as a collection, constitute an empirical approximation to the posterior distribution. Following Jackman (2009), we refer to this situation as one where the priors are *conditionally* conjugate. They have also been termed generalized conjugate and semi-conjugate. All of these terms are aimed at reflecting that though this prior distribution does not yield a closed form for the posterior distribution, it does yield a manageable form, both conceptually and computationally using simulation strategies that involve conditioning.

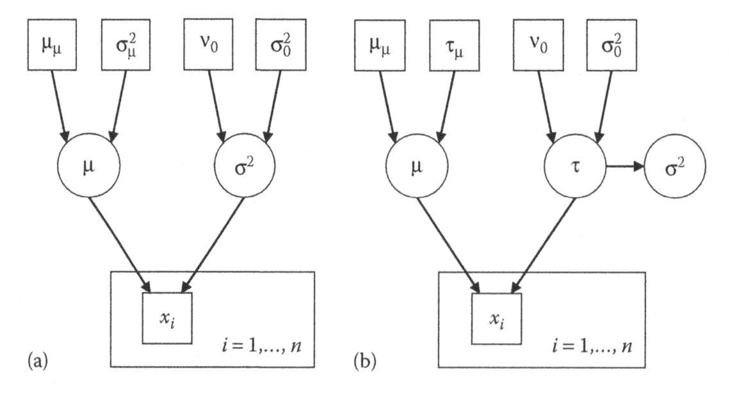

FIGURE 4.9
Directed acyclic graphs for a normal distribution model with unknown mean μ and unknown variance σ^2 or precision τ in the (a) variance parameterization and (b) precision parameterization.

4.3.2 Example Analysis

We return to the example with $n = 10$ scores and pursue inference for both μ and σ^2. Collecting all the pieces, the model is given in the variance parameterization by

$$p(\mu,\sigma^2 \mid \mathbf{x}) \propto p(\mathbf{x} \mid \mu,\sigma^2)p(\mu)p(\sigma^2)$$

$$= \prod_{i=1}^{n} p(x_i \mid \mu,\sigma^2)p(\mu)p(\sigma^2),$$

where

$$x_i \mid \mu,\sigma^2 \sim N(\mu,\sigma^2) \quad \text{for } i = 1,\ldots,10,$$

$$\mu \sim N(75, 50),$$

and

$$\sigma^2 \sim \text{Inv-Gamma}(5,150).$$

This model was analyzed in WinBUGS using the following code.

```
------------------------------------------------------------------------
########################################################################
# Model Syntax
########################################################################
model{

########################################################################
# Conditional distribution for the data
########################################################################

for(i in 1:n){
   x[i] ~ dnorm(mu, tau)
}

########################################################################
# Define the prior distributions for the unknown parameters
########################################################################

mu ~ dnorm(mu.mu, tau.mu)            # prior distribution for mu

mu.mu <- 75                          # mean of the prior for mu
sigma.squared.mu <- 50               # variance of the prior for mu
tau.mu <- 1/sigma.squared.mu         # precision of the prior for mu


tau ~ dgamma(alpha,beta)             # precision of the data
sigma.squared <- 1/tau               # variance of the data

nu.0 <- 10                           # hyperparameter for prior for tau
sigma.squared.0 <- 30                # hyperparameter for prior for tau
```

```
alpha <- nu.0/2                    # hyperparameter for prior for tau
beta <- nu.0*sigma.squared.0/2     # hyperparameter for prior for tau

}                                  # closes the model statement

########################################################################
Data statement
########################################################################
list(n=10, x=c(91, 85, 72, 87, 71, 77, 88, 94, 84, 92))
------------------------------------------------------------------------
```

Note that WinBUGS uses the precision parameterization for the normal distribution. In this example, the values for the observables and sample size are contained in the list statement for the data. The values for the hyperparameters are specified in the model portion of the code. As they are known values, they could have been specified as part of the data statement. We have specified them in the model portion to more clearly illustrate which hyperparameters go with which parameter, and how some of the hyperparameters are calculated as functions of others, to yield quantities used in WinBUGS's parameterization of the normal and (inverse-)gamma distributions. Figure 4.10 depicts the joint distribution for μ and σ^2 in the form of a scatterplot based on an analysis of the code above using 50,000 iterations.* As depicted there, the posterior correlation between the parameters is weak; in these 50,000 draws, the correlation was –.13. Figure 4.11 depicts empirical approximations to the marginal posterior distributions for μ and σ^2 produced by WinBUGS. Table 4.1 contains summary statistics for these marginal posterior distributions.

4.3.3 Alternative Prior Distributions

A variety of alternatives have been proposed for the joint prior distribution $p(\mu,\sigma^2)$. A conjugate prior is given by $p(\mu,\sigma^2) = p(\mu \mid \sigma^2)p(\sigma^2)$, where

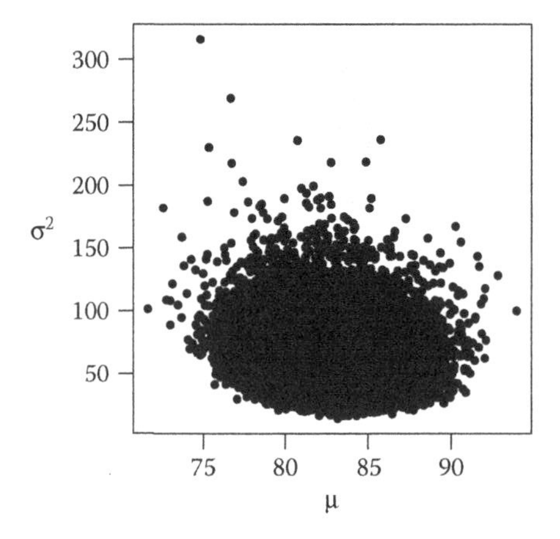

FIGURE 4.10
Scatterplot of the joint posterior distribution for the parameters of the normal distribution where both the mean μ and the variance σ^2 are unknown.

---

* We will have more to say about choices regarding the number of iterations to use in Chapter 5. Suffice it to say that 50,000 is plenty sufficient to characterize the posterior in the current case.

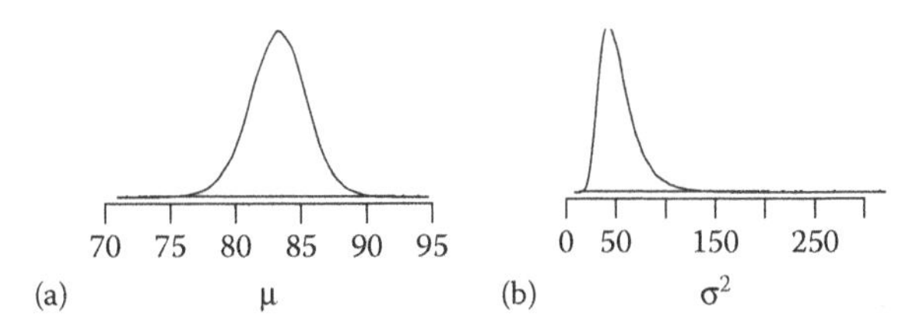

FIGURE 4.11
Marginal posterior densities for the parameters of the normal distribution: (a) the unknown mean μ and (b) the unknown variance σ^2.

TABLE 4.1
Summaries of the Marginal Posterior Distributions

Parameter	Mean	Median	Standard Deviation	95% Highest Posterior Density Interval
μ	83.23	83.27	2.20	(78.80, 87.48)
σ^2	53.19	49.32	19.45	(23.22, 91.34)

$$\sigma^2 \sim \text{Inv-Gamma}(\alpha, \beta) \tag{4.35}$$

and

$$\mu \mid \sigma^2 \sim N(\mu_\mu, \sigma^2 / \kappa), \tag{4.36}$$

with κ being a hyperparameter that aids in defining the prior variance for μ in terms of σ^2. κ may be thought of as a pseudo-sample size in that, as $\kappa \to 0$, the prior variance for μ gets larger, reflecting that the prior is expressing less certainty, as would be consistent with having fewer prior observations.

Though this approach affords conjugacy, it has seen limited application in psychometrics and latent variable models (see Lee, 2007, for examples) and we do not purse it further. The specification of independent priors for μ and σ^2 is attractive because (a) it is conceptually simpler to conceive of the mean and variance of the normal distribution as independent entities, and (b) the computation of the posterior using the independent, conditionally conjugate priors is fairly straightforward and usually quite fast, thereby minimizing the computational advantages of having full conjugacy.

Another choice for the prior distribution that is motivated by a desire to specify an increasingly diffuse prior distribution is the improper prior

$$p(\mu, \sigma^2) \propto \sigma^{-2}, \tag{4.37}$$

which is also uniform on $(\mu, \log \sigma)$. Provocatively, the posterior distribution in this case aligns closely with forms taken for sampling distributions in frequentist inference (Jackman, 2009). On reflection this is unsurprising and reinforces a key conceptual point: to the extent that there is less information in the prior—(4.37) is a limiting case of reducing the information in the prior—the posterior is increasingly similar to the likelihood.

4.4 Summary

This chapter has briefly described Bayesian approaches when the observable data is modeled via a normal distribution. Key ideas introduced include the precision τ as the inverse of σ^2, conjugate and conditionally conjugate prior distributions, and the posterior distribution for the mean as a precision-weighted synthesis of the prior mean and the mean of the data. Our treatment has been cursory, only going to a depth sufficient to support the development of models covered later. More comprehensive accounts can be found in any of a number of texts on Bayesian modeling, including those cited in the introduction to Section I. The specifications introduced here will re-emerge most prominently in our treatments of regression (Chapter 6), classical test theory (Chapter 8), and factor analysis (Chapter 9) models. The underlying principles will be used throughout the remaining chapters.

Exercises

4.1 An example for an inference about a mean (μ) when the variance is known was given in Section 4.1. A related analysis now conducting inference about the mean and the variance was given in Section 4.3. Compare the results for the two, in terms of the (marginal) posterior distribution for the mean. How do they differ? What explains why they differ in these ways? Under what circumstances would the results from such analyses yield increasingly similar results? Under what circumstances would the results from such analyses yield increasingly dissimilar results?

4.2 An example for an inference about a variance (σ^2) when the mean is known was given in Section 4.2. A related analysis now conducting inference about the mean and the variance was given in Section 4.3. Compare the results for the two, in terms of the (marginal) posterior distribution for the variance. How do they differ? What explains why they differ in these ways? Under what circumstances would the results from such analyses yield increasingly similar results? Under what circumstances would the results from such analyses yield increasingly dissimilar results?

4.3 In the WinBUGS code in Section 4.3.2, add a line defining the standard deviation `sigma` as the square root of the variance, and examine its posterior distribution. What is WinBUGS doing to produce this approximation of its posterior? Is the posterior mean of sigma equal to the square root of the posterior mean for sigma. squared? Why or why not?

4.4 (Advanced) Compare the posterior standard deviation of `sigma` with an approximation based on the posterior standard deviation of `sigma.squared`, transformed by the delta method. How close is the approximation, and why does it differ? Repeat the exercise with larger samples of data produced by creating multiple copies of the sample data, so the sample mean and dispersion remain constant as `n` increases. How large does `n` need to be for the asymptotic approximations in this exercise and the preceding one to be accurate?

5

Markov Chain Monte Carlo Estimation

In a Bayesian analysis, the posterior distribution is the solution obtained from fitting the model to the data. Analytical solutions are available for simple models with conjugate priors, such as the beta-binomial model, or a model for the mean of a normal distribution with a known variance. These situations arise when the likelihood is a member of the general exponential family of distributions (Bernardo & Smith, 2000), so that a special related distribution exists that can be used as a prior, and the posterior takes the same form with updated parameters.

Many complex statistical and psychometric models do not enjoy the benefit of having conjugate priors. When there is no closed form solution for the product of the prior and the likelihood, evaluating the possibly high-dimensional integrals in Bayes' theorem quickly become intractable, in practice preventing the analyst from obtaining the full posterior distribution. In some cases, Taylor series analytic approximations can be derived for moments (e.g., Tierney & Kadane, 1986), or posterior modes may be estimated via optimization techniques such as Newton–Raphson or the EM algorithm when derivatives are amenable (e.g., Mislevy, 1986). These resulting estimates and complementary estimates of the curvature of the posterior may be used to define an asymptotic approximation of the posterior. Such estimation procedures may be useful even in situations where the posterior is of a known form, say, in estimating the mode of the posterior when there is no closed form for it.

An attractive alternative when computing resources are available is to simulate values from distributions in such a way that the collection of such values forms an empirical approximation of the posterior distribution. Ideally these simulated values would be independent (or pseudo-independent). In some cases in psychometric modeling, certain choices for distributional forms (e.g., normality assumptions for the data and associated conjugate priors) yield posterior distributions of manageable form, in which case an empirical approximation may be obtained by simulating values using standard procedures (Lee, 2007).

However, in models that do not employ conjugate priors and/or have complicated features that preclude the specification of conjugate priors, the posterior distribution is often not of a form that facilitates independent sampling. An alternative is to simulate values that are dependent, which is sufficient if the values can be drawn throughout the posterior distribution and in the correct proportions (Gilks, Richardson, & Spiegelhalter, 1996a). A highly flexible framework for estimating distributions in this vein that supports Bayesian analyses in such complex models is provided by Markov chain Monte Carlo (MCMC) estimation (Brooks, 1998; Brooks, Gelman, Jones, & Meng, 2011; Gelfand & Smith, 1990; Gilks, Richardson, & Spiegelhalter, 1996b; Smith & Roberts, 1993; Spiegelhalter, Thomas, Best, & Lunn, 2007; Tierney, 1994). Although (possibly highly) dependent, if enough samples are simulated such that the values are obtained throughout the support of the distribution in the correct proportions, drawing such dependent samples provides an empirical approximation to the posterior distribution.

This chapter gives an overview of MCMC estimation. In Section 5.1, we give a broad description and comparison of MCMC to frequentist estimation. Next, we describe certain specific algorithms that have become the dominant approaches in Bayesian psychometric modeling with MCMC (Sections 5.2, 5.3, 5.5, and 5.6), and in doing so illuminate the conceptual alignment between MCMC and Bayesian statistical modeling (Section 5.4). We then briefly survey a number of matters pertaining to the practice of MCMC estimation (Section 5.7). We conclude in Section 5.8 with a summary and bibliographic note.

5.1 Overview of MCMC

MCMC routines are used to empirically approximate a distribution using a collection of values. The last two words, *Monte Carlo*, indicate that the process will involve simulating (sampling, generating, drawing) values from distributions. The word *chain* indicates that the drawn values are done so in ways that the values are linked, occurring in sequence. This stands in contrast to other sampling techniques in which the draws are independent (e.g., Rubin, 1988). To begin to formalize things, let $(\theta^{(1)},\ldots,\theta^{(T)})$ denote T values that constitute the chain, where the superscript denotes the iteration in the sequence. It is useful to think of this as an indicator of time. The dependence of the values in the chain follows the *Markov* property, which states that, given the value of the chain at any time t, the value of the chain at any *later* time is conditionally independent of the value at any *earlier* time. More formally,

$$p(\theta^{(t+a)} \mid \theta^{(t)}, \theta^{(t-b)}) = p(\theta^{(t+a)} \mid \theta^{(t)}), \quad a > 0, b > 0. \tag{5.1}$$

Given that certain general conditions hold (see, e.g., Jackman, 2009; Roberts, 1996; Tierney, 1994), a properly constructed chain is guaranteed to converge to a unique stationary distribution. Briefly, the transition kernel of the chain must be:

- Time-homogeneous: the transition probability from one state to another is constant;
- Irreducible: from any given point in the distribution, the chain can reach any other point with positive probability in some number of iterations; and
- Positive recurrent: in the long run, the chain will visit each state an infinite number of times.

Further, the chain should be aperiodic, which means it will not just oscillate between different states in a regular period. If these conditions hold, then the chain is said to be ergodic, and the chain will converge to its unique stationary distribution. This is accomplished by constructing chains that are reversible, discussed in more detail in Section 5.5.2.

To leverage Markov chains in Bayesian analysis, we set up the chain so that the desired posterior distribution is the chain's stationary distribution. If the chain is properly constructed, then $\theta^{(t)} \sim p(\theta \mid \mathbf{x})$ as $t \to \infty$. As such, MCMC estimation in a Bayesian analysis consists of drawing from a series of distributions that is, in the limit, equal to drawing from the posterior distribution. In practice, we seek to determine when the chain converges to its stationary distribution, which in our case is the posterior distribution. Let $\theta^{(C)}$ denote this point. The remaining draws $(\theta^{(C+1)},\ldots,\theta^{(T)})$ are considered draws from the posterior distribution $p(\theta \mid \mathbf{x})$.

The use of MCMC in Bayesian modeling aligns nicely with the Monte Carlo principle, which states that "anything we want to know about a random variable θ can be learned by sampling many times from $f(\theta)$, the density of θ" (Jackman, 2009, p. 133). The upshot of using MCMC in a Bayesian analysis is that the collection of values $(\theta^{(C+1)},\ldots,\theta^{(T)})$ from the chain constitute an empirical approximation to the posterior distribution $p(\theta \mid \mathbf{x})$. Features of the distribution of interest (e.g., the mean, standard deviation, intervals) are then empirically approximated by the corresponding features of the collection of draws, with increasing fidelity as more draws are taken.

Monte Carlo simulation offers an attractive approach when functions of interest do not have standard densities. Let $g(\theta)$ denote some function of the parameters θ. Even in situations where the posterior density $p(\theta \mid \mathbf{x})$ is well known, the posterior density $p(g(\theta) \mid \mathbf{x})$ might not be. Recall the situation in Section 4.2 where we sought inference for the variance of a normal distribution with known mean. In that case, $\theta = \sigma^2$ and (using the conjugate prior) the posterior distribution $p(\sigma^2 \mid \mathbf{x})$ was an inverse-gamma. However, suppose we were interested in the posterior distribution for the standard deviation defined as $\sigma = g(\sigma^2) = \sqrt{\sigma^2}$. It is important to note that in general the features of a distribution of a transformation of parameters *cannot* be obtained by applying the transformation to the features of the distribution of the parameters. For example, the mean of the posterior distribution of the standard deviation is *not* the square root of the mean of the posterior distribution for the variance; symbolically $E\left(\sqrt{\sigma^2} \mid \mathbf{x}\right) \neq \sqrt{E(\sigma^2) \mid \mathbf{x}}$. We can however sample values for σ^2 from $p(\sigma^2 \mid \mathbf{x})$ and then take the square root of each of those sampled values. This set of transformed values does indeed empirically approximate the posterior density $p\left(\sqrt{\sigma^2} \mid \mathbf{x}\right)$, which can be summarized by its mean or any other function as desired. More generally, we may empirically approximate the (posterior) density of a function $g(\theta)$ that transforms the parameters by sampling values from the (posterior) density of θ and then applying the transformation to those sampled values.

Before turning to descriptions of specific algorithms, we pause to clarify some points of departure between frequentist and Bayesian approaches as they play out in notions of estimation. The principal issue concerns what is being evaluated. In ML estimation, the function of interest is the likelihood. In a Bayesian analysis, what is of interest is the posterior distribution.

A second issue concerns the goals of estimation, and the methods of achieving those goals. In frequentist modeling that treats parameters as fixed, parameter estimation comes to arriving at a *point estimate* of the parameter. In ML, this amounts to finding the maximum of the likelihood function, often via iterative search methods (see, e.g., Süli & Mayers, 2003), akin to finding the highest point in a mountain range. In a fully Bayesian analysis, we do not seek a point, but rather a whole distribution. Continuing with the metaphor, Bayesian estimation works with a different mountain range than ML, and rather than seeking the highest peak, it seeks to map the entire terrain of peaks, valleys, and plateaus. The iterative simulation methods addressed in this chapter are flexible tools for accomplishing this mapping for a wide array of models.

5.2 Gibbs Sampling

5.2.1 Gibbs Sampling Algorithm

A popular algorithm for constructing the Markov chain is the Gibbs sampler, which relies on alternatively sampling components of θ conditional on the remaining components of θ and, in the Bayesian context, the data $\mathbf{x}$ as well (Gelfand & Smith, 1990; Geman & Geman, 1984; see also Brooks, 1998; Casella & George, 1992; Gilks et al., 1996b). Suppose θ has R components and let θ_r denote the rth component of θ. In the example below each component θ_r is a single parameter, but in general it may be a collection. Let θ_{-r} denote the remaining components in θ, that is, all components *except* θ_r. Then $p(\theta_r \mid \theta_{-r}, \mathbf{x})$ is the distribution of the rth component given the remaining components and the data, referred to as the *full conditional distribution* for θ_r. It can be shown that a joint distribution can be fully determined by the complete set of such full conditional distributions (Besag, 1974; Gelfand & Smith, 1990). That is, the posterior $p(\theta \mid \mathbf{x})$ is fully determined by $p(\theta_1 \mid \theta_{-1}, \mathbf{x}), \ldots, p(\theta_R \mid \theta_{-R}, \mathbf{x})$. A Gibbs sampler iteratively samples from these full conditional distributions. Letting $\theta_r^{(t)}$ denote the value of the model parameters in component r at iteration t, Gibbs sampling consists of proceeding with the following steps:

1. Assign initial values for all the components, yielding the collection $\theta_1^{(0)}, \ldots, \theta_R^{(0)}$ where the superscript of $t = 0$ conveys that these are initial values.
2. For $r = 1, \ldots, R$, draw values for component θ_r from its full conditional distribution given the observed data and the current values of all other components. In other words, for each component θ_r, we obtain the value of the chain at iteration $t + 1$ by drawing from its full conditional distribution $p(\theta_r \mid \theta_{-r}, \mathbf{x})$ using the current values for the remaining components θ_{-r}. One complete iteration is given by sequentially drawing values from

$$\theta_1^{(t+1)} \sim p(\theta_1 \mid \theta_2^{(t)}, \ldots, \theta_R^{(t)}, \mathbf{x})$$

$$\theta_2^{(t+1)} \sim p(\theta_2 \mid \theta_1^{(t+1)}, \theta_3^{(t)}, \ldots, \theta_R^{(t)}, \mathbf{x})$$

$$\vdots$$

$$\theta_r^{(t+1)} \sim p(\theta_r \mid \theta_1^{(t+1)}, \ldots, \theta_{r-1}^{(t+1)}, \theta_{r+1}^{(t)}, \ldots, \theta_R^{(t)}, \mathbf{x})$$

$$\vdots$$

$$\theta_R^{(t+1)} \sim p(\theta_R \mid \theta_1^{(t+1)}, \ldots, \theta_{R-1}^{(t+1)}, \mathbf{x}).$$

   Employing this process using the initial values ($t = 0$) yields the collection $\theta^{(1)} = (\theta_1^{(1)}, \ldots, \theta_R^{(1)})$ that constitutes the first draw for the components.
3. Increment t, by setting $t = t + 1$.
4. Repeat steps 2 and 3, for some large number T iterations.

5.2.2 Example: Inference for the Mean and Variance of a Normal Distribution

We return to the situation introduced in Chapter 4 for conducting a Bayesian analysis in the situation with data modeled as normally distributed with unknown mean and variance. For clarity of presentation, we collect and repeat the relevant expressions here. An exchangeability assumption regarding subjects supports the factorization of the conditional probability of the data as

$$p(\mathbf{x} \mid \mu, \sigma^2) = \prod_{i=1}^{n} p(x_i \mid \mu, \sigma^2), \tag{5.2}$$

where

$$x_i \mid \mu, \sigma^2 \sim N(\mu, \sigma^2). \tag{5.3}$$

We employ a conditionally conjugate prior, assuming independence of the parameters

$$p(\mu, \sigma^2) = p(\mu)p(\sigma^2), \tag{5.4}$$

where

$$\mu \sim N(\mu_\mu, \sigma_\mu^2) \tag{5.5}$$

and

$$\sigma^2 \sim \text{Inv-Gamma}(\nu_0 / 2, \nu_0\sigma_0^2 / 2). \tag{5.6}$$

The full conditional distributions are therefore $p(\mu \mid \sigma^2, \mathbf{x})$ and $p(\sigma^2 \mid \mu, \mathbf{x})$. It can be shown that the full conditionals are (Lunn et al., 2013)

$$\mu \mid \sigma^2, \mathbf{x} \sim N(\mu_{\mu\mid\sigma^2,\mathbf{x}}, \sigma^2_{\mu\mid\sigma^2,\mathbf{x}}), \tag{5.7}$$

where

$$\mu_{\mu\mid\sigma^2,\mathbf{x}} = \frac{(\mu_\mu/\sigma_\mu^2) + (n\bar{x}/\sigma^2)}{(1/\sigma_\mu^2) + (n/\sigma^2)}, \tag{5.8}$$

$$\sigma^2_{\mu\mid\sigma^2,\mathbf{x}} = \frac{1}{(1/\sigma_\mu^2) + (n/\sigma^2)}, \tag{5.9}$$

and

$$\sigma^2 \mid \mu, \mathbf{x} \sim \text{Inv-Gamma}\left(\frac{\nu_0 + n}{2}, \frac{\nu_0\sigma_0^2 + SS(\mathbf{x} \mid \mu)}{2}\right), \tag{5.10}$$

with

$$SS(\mathbf{x} \mid \mu) = \sum_{i=1}^{n} (x_i - \mu)^2 \tag{5.11}$$

being the sums of squares from the data taken about the population mean. These results for μ are the same as those obtained in the analysis where σ^2 was assumed known; compare (5.7)–(5.9) to (4.6)–(4.8). Similarly, the results for σ^2 are the same as those obtained in the analysis where μ was assumed known; compare (5.10) and (5.11) to (4.24) and (4.25). Working with the full conditionals puts us back in the contexts where, for each parameter, the other parameter is treated as known. Here lies the computational payoff of a conditionally conjugate prior specification. Although the joint posterior does not have a closed form, the full conditionals do. We are still capitalizing on the conjugacy, now just localized to each parameter.

This greatly eases the computational burden needed in the steps of Gibbs sampling. A Gibbs sampling algorithm for the current example is as follows:

1. Assign initial values for all the parameters: $\mu^{(0)}$ and $\sigma^{2^{(0)}}$, where the superscript indicates that $t = 0$.
2. For iteration $t + 1$, draw

$$\sigma^{2^{(t+1)}} \sim \text{Inv-Gamma}\left(\frac{\nu_0 + n}{2}, \frac{\nu_0\sigma_0^2 + SS^{(t)}(\mathbf{x} \mid \mu)}{2}\right)$$

where

$$SS^{(t)}(\mathbf{x} \mid \mu) = \sum_{i=1}^{n}(x_i - \mu^{(t)})^2$$

and

$$\mu^{(t+1)} \sim N\left(\frac{(\mu_\mu/\sigma_\mu^2) + (n\bar{x}/\sigma^{2^{(t+1)}})}{(1/\sigma_\mu^2) + (n/\sigma^{2^{(t+1)}})}, \frac{1}{(1/\sigma_\mu^2) + (n/\sigma^{2^{(t+1)}})}\right).$$

3. Increment t, by setting $t = t + 1$.
4. Repeat steps 2 and 3, for some large number T iterations.

We return to the example from Section 4.3 where $\mu_\mu = 75$, $\sigma_\mu^2 = 50$, $\sigma_0^2 = 30$, and $\nu_0 = 10$. Here, we briefly illustrate the computations for a few iterations. Let $\mu^{(0)} = 70$ and $\sigma^{2^{(0)}} = 10$ be the initial values for the parameters. These were arbitrarily selected; we will have more to say about strategies for selecting initial values in Section 5.7. These values are listed in the first row of Table 5.1 in the second and third columns, and represent the values for the iteration listed in the first column.

To conduct the first iteration, we work with the particular inverse-gamma full conditional distribution for σ^2 using the initial value for μ. For the first iteration, the first parameter of the inverse-gamma is

$$\frac{\nu_0 + n}{2} = \frac{10 + 10}{2} = 10.$$

In fact, this will be the first parameter for the inverse-gamma full conditional in every iteration, as its ingredients do not involve μ and therefore do not vary over iterations. However, the second parameter will vary with each iteration, as it depends on $SS(\mathbf{x} \mid \mu)$,

TABLE 5.1

Computations in the Gibbs Sampler for a Normal Distribution Model with Conditionally Conjugate Priors

	Parameters		Full Conditional for σ^2			Full Conditional for μ	
Iteration	μ	σ^2	$\frac{\nu_0+n}{2}$	$SS(\mathbf{x}\mid\mu)$	$\frac{\nu_0\sigma_0^2+SS(\mathbf{x}\mid\mu)}{2}$	$\frac{(\mu_\mu/\sigma_\mu^2)+(n\bar{x}/\sigma^2)}{(1/\sigma_\mu^2)+(n/\sigma^2)}$	$\frac{1}{(1/\sigma_\mu^2)+(n/\sigma^2)}$
0	70.00	10.00	–	–	–	–	–
1	81.19	110.89	10	2589.00	1444.50	82.45	9.08
2	82.78	31.70	10	685.76	492.88	83.56	2.98
3	82.38	43.39	10	618.42	459.21	83.37	3.99
4	81.82	51.72	10	630.32	465.16	83.25	4.69
5	83.16	76.20	10	652.69	476.34	82.90	6.61

which varies over iterations. We first compute the sums of squares of the data about the current value for the population mean, $\mu^{(0)}=70$:

$$SS^{(0)}(\mathbf{x}\mid\mu)=\sum_{i=1}^{n}(x_i-\mu^{(0)})^2=\sum_{i=1}^{n}(x_i-70)^2=2589.$$

Note that this is the value of SS that is used to define the full conditional for σ^2 in the *current* iteration; however, it is computed based on the value for μ from the *previous iteration*. Using this value for SS yields the following for the second parameter in the inverse gamma full conditional distribution:

$$\frac{\nu_0\sigma_0^2+SS^{(0)}(\mathbf{x}\mid\mu)}{2}=\frac{10\times 30+2589}{2}=1444.5.$$

We then take a draw from the full conditional: $\sigma^{2^{(1)}}\sim\text{Inv-Gamma}(10,1444.5)$. The value drawn from this distribution was 110.89, which is now the value for σ^2 for iteration 1. This value is reported in the second row of Table 5.1.

We then turn to the full conditional distribution for μ. The mean for this distribution is

$$\frac{(\mu_\mu/\sigma_\mu^2)+(n\bar{x}/\sigma^{2^{(1)}})}{(1/\sigma_\mu^2)+(n/\sigma^{2^{(1)}})}=\frac{(75/50)+(10\times 84.1/110.89)}{(1/50)+(10/110.89)}\approx 82.45$$

and the variance is

$$\frac{1}{(1/\sigma_\mu^2)+(n/\sigma^{2^{(1)}})}=\frac{1}{(1/50)+(10/110.89)}\approx 9.08.$$

Note the use of the just-drawn value of 110.89 for σ^2 in these computations. We then take a draw from the full conditional: $\mu^{(1)}\sim N(82.45,9.08)$. The drawn value was 81.19, which is now the value for μ for iteration 1, and is reported in the second row of Table 5.1. This completes one iteration of the Gibbs sampler.

To conduct the second iteration, we return to the full conditional distribution for σ^2. We compute the sums of squares of the data about the current value for the population mean

$$SS^{(1)}(\mathbf{x} \mid \mu) = \sum_{i=1}^{n}(x_i - \mu^{(1)})^2 = \sum_{i=1}^{n}(x_i - 81.19)^2 = 685.76$$

and the second parameter of the full conditional,

$$\frac{\nu_0\sigma_0^2 + SS^{(1)}(\mathbf{x} \mid \mu)}{2} = \frac{10 \times 30 + 685.76}{2} = 492.88.$$

We then take a draw from the full conditional: $\sigma^{2^{(2)}} \sim \text{Inv-Gamma}(10, 492.88)$. The value drawn from this distribution was 31.70, which is now the value for σ^2 for iteration 2.

Turning to μ, the mean for the full conditional distribution is

$$\frac{\left(\mu_\mu/\sigma_\mu^2\right) + \left(n\bar{x}/\sigma^{2^{(2)}}\right)}{\left(1/\sigma_\mu^2\right) + \left(n/\sigma^{2^{(2)}}\right)} = \frac{(75/50) + ((10 \times 84.1)/31.70)}{(1/50) + (10/31.70)} \approx 83.56$$

and the variance is

$$\frac{1}{\left(1/\sigma_\mu^2\right) + \left(n/\sigma^{2^{(2)}}\right)} = \frac{1}{(1/50) + (10/31.70)} \approx 2.98.$$

Accordingly, we then take a draw from the full conditional: $\mu^{(2)} \sim N(83.56, 2.98)$. The drawn value was 82.78, which is now the value for μ for iteration 2. These drawn values are listed in Table 5.1, as are values for three more iterations of the Gibbs sampler (see Exercise 5.6). Recalling that the goal is not to arrive at a point estimate, but rather a (posterior) distribution, five iterations are hardly sufficient to characterize the distribution. In applications, we would typically conduct many more iterations, and summarize them via densities and summary statistics. For this example, the results of running 50,000 iterations in WinBUGS were given in Figures 4.10 and 4.11 and Table 4.1.

5.2.3 Discussion

Many variations and extensions on the Gibbs sampling architecture are possible. One can work with subsets of the parameters, treating multiple parameters as a set and sampling from multivariate full conditionals (e.g., Patz & Junker, 1999b). One need not sample every parameter at each iteration, as long as each value will be visited infinitely many times in the long run. In cases where some parameters explore the parameter space more slowly than others, say due to poor mixing or high autocorrelation (see Section 5.7), it may be advantageous to devote more resources to sampling for those parameters, rather than others.

Note that the full conditional distribution for each parameter is formulated as conditional on all the remaining parameters. In complex models with many parameters, this potentially involves conditioning on a great many parameters. However, the full conditional distributions often simplify due to conditional independence specifications. It is here that the use of DAGs in conceptualizing the model can greatly aid in computation, as it can be shown that the full conditional distribution for any entity depends at most on its parents, its children, and the other parents of its children; all other parameters can be ignored (Lunn et al., 2009; Spiegelhalter & Lauritzen, 1990). In large problems built from simpler structures with conditional independence relationships, the simplification can be substantial.

Returning to the normal distribution example considered in Section 5.2.2, Figure 4.9a provides an expanded version of the DAG that includes all the entities in the model that are known or fixed, including the hyperparameters for the prior distributions. Working with the expanded DAG leads to an expanded notational scheme for the full conditionals, namely one that includes the hyperparameters of the prior distribution for the parameter in question. Reading off the DAG in Figure 4.9a, the full conditional for μ involves conditioning on its parents (μ_μ, σ_μ^2), its children ($\mathbf{x}$), and the other parents of its children (σ^2). The full conditional for σ^2 involves conditioning on its parents (ν_0, σ_0^2), its children ($\mathbf{x}$), and the other parents of its children (μ). The expanded DAG suggests writing the full conditionals with an expanded notation that includes the hyperparameters. This would involve rewriting (5.7)–(5.9) as

$$\mu \mid \sigma^2, \mu_\mu, \sigma_\mu^2, \mathbf{x} \sim N(\mu_{\mu|\sigma^2,\mu_\mu,\sigma_\mu^2,\mathbf{x}}, \sigma^2_{\mu|\sigma^2,\mu_\mu,\sigma_\mu^2,\mathbf{x}}),$$

where

$$\mu_{\mu|\sigma^2,\mu_\mu,\sigma_\mu^2,\mathbf{x}} = \frac{\left(\mu_\mu/\sigma_\mu^2\right) + \left(n\bar{x}/\sigma^2\right)}{\left(1/\sigma_\mu^2\right) + \left(n/\sigma^2\right)},$$

$$\sigma^2_{\mu|\sigma^2,\mu_\mu,\sigma_\mu^2,\mathbf{x}} = \frac{1}{\left(1/\sigma_\mu^2\right) + \left(n/\sigma^2\right)},$$

and rewriting (5.10) as

$$\sigma^2 \mid \mu, \nu_0, \sigma_0^2, \mathbf{x} \sim \text{Inv-Gamma}\left(\frac{\nu_0 + n}{2}, \frac{\nu_0\sigma_0^2 + SS(\mathbf{x} \mid \mu)}{2}\right).$$

This expanded notation aids in the derivation of full conditionals, as it highlights all the entities that are involved in computations. We employ this expanded notation in Appendix A where we develop the full conditional distributions for many of the models described in the rest of the book. For the rest of the chapters that comprise the main text, we use the briefer-if-not-fully-complete notation that ignores the hyperparameters on the right side of the conditioning bar.

If the full conditional distributions are of familiar form (as in the preceding example), sampling from them may proceed using Monte Carlo procedures. They can be particularly simple when we construct a larger model from constituent models and relationships that have conjugate or conditionally conjugate priors. However, in complex models, it might be the case that full conditional distributions are not of known form. In these cases, more complex sampling schemes are required. Sections 5.3, 5.5, and 5.6 describe such schemes.

5.3 Metropolis Sampling

In complex Bayesian models, the posterior distribution is often difficult to obtain through analytical methods, and not often easier to sample from directly. Metropolis sampling offers a powerful approach to handle such situations (Metropolis et al., 1953; see also

Brooks, 1998; Gilks et al., 1996b). The basic idea is that in lieu of drawing a value from the difficult-to-sample-from posterior distribution, we draw a value from a different distribution that we *can* easily sample from, and then decide to accept or reject that value as the next value in the chain. This decision rule is constructed in such a way that, after the chain reaches its stationary distribution, a subsequent draw has the same distribution as a draw from its posterior distribution. At this point, the frequency distribution of a large number of draws from the sequence converges to the posterior distribution. The procedure consists of conducting the following steps:

1. Initialize the parameters by assigning a value for $\theta^{(t)}$ for $t = 0$.
2. Draw a *candidate value* $\theta^* \sim q(\cdot \mid \theta^{(t)})$ from a *proposal* distribution q that we can easily sample from (e.g., a normal distribution).
3. Define the *acceptance probability*

$$\alpha(\theta^* \mid \theta^{(t)}) = \min\left[1, \frac{p(\theta^* \mid \mathbf{x})}{p(\theta^{(t)} \mid \mathbf{x})}\right],$$

   where $p(\theta^{(t)} \mid \mathbf{x})$ is the probability (or ordinate of the probability density) for the current value $\theta^{(t)}$ in the chain in the posterior distribution and $p(\theta^* \mid \mathbf{x})$ is the probability (or ordinate of the probability density) for the candidate value θ^* in the posterior distribution.
4. Set $\theta^{(t+1)} = \theta^*$ with probability $\alpha(\theta^* \mid \theta^{(t)})$. Set $\theta^{(t+1)} = \theta^{(t)}$ with probability $1 - \alpha(\theta^* \mid \theta^{(t)})$. Operationally this is typically accomplished by drawing a random variate $U \sim$ Uniform(0,1), and setting $\theta^{(t+1)} = \theta^*$ if $\alpha(\theta^* \mid \theta^{(t)}) > U$ and setting $\theta^{(t+1)} = \theta^{(t)}$ otherwise.
5. Increment t, by setting $t = t + 1$.
6. Repeat steps 2–5 for some large number T iterations.

The proposal distribution q in step 2 may be any symmetric distribution that is defined over the support of the stationary distribution, which in our case is the posterior distribution. Here, symmetry refers to a distribution being symmetric with respect to its arguments; that is, $q(\theta^* \mid \theta^{(t)}) = q(\theta^{(t)} \mid \theta^*)$. The most popular choice for q is a normal distribution centered at the current value of the chain. To simplify the presentation, first consider the situation where there is only one parameter so that q is a univariate normal distribution, and let $\sigma^2_{\theta^*}$ denote the variance of this distribution. Consideration of the normal probability density function

$$q(\theta^* \mid \theta^{(t)}) = N(\theta^* \mid \theta^{(t)}, \sigma^2_{\theta^*}) = \frac{1}{\sqrt{2\pi\sigma^2_{\theta^*}}} \exp\left[\frac{-1}{2\pi\sigma^2_{\theta^*}}(\theta^* - \theta^{(t)})^2\right] \tag{5.12}$$

reveals that interchanging the roles of θ^* and $\theta^{(t)}$ yields the same resulting value. That is, $N(\theta^* \mid \theta^{(t)}, \sigma^2_{\theta^*}) = N(\theta^{(t)} \mid \theta^*, \sigma^2_{\theta^*})$ and as such the normal distribution is symmetric with respect to its arguments.

Figure 5.1 illustrates the use of the situation for a model with a single parameter, where the posterior distribution takes on an irregular shape. Beginning with panel (a) the current value of the chain is $\theta^{(t)}$. Centered at that value is the normal proposal distribution q. Suppose that at this iteration, the value drawn from this proposal distribution is θ^*. The question then becomes whether to accept this as the next value for the chain, $\theta^{(t+1)}$.

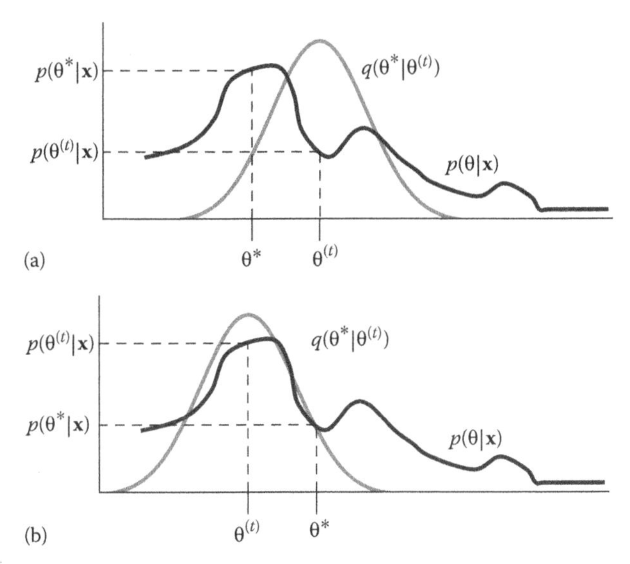

FIGURE 5.1
Illustration of two iterations of the Metropolis sampler: (a) a proposed value θ^* that will necessarily be accepted; once accepted it becomes the current value $\theta^{(t)}$ in (b), which shows a new proposed value that will possibly be accepted (with probability $[p(\theta^*|\mathbf{x})/p(\theta^{(t)}|\mathbf{x})]$). (Modified and used with permission from Educational Testing Service.)

The acceptance probability $\alpha(\theta^* \mid \theta^{(t)}) = \min\left[1, p(\theta^* \mid \mathbf{x}) / p(\theta^{(t)} \mid \mathbf{x})\right]$ in step 3 is so named because it is the probability that the candidate value θ^* is accepted as the value for the $(t + 1)$st iteration in the chain; accordingly, $1 - \alpha(\theta^* \mid \theta^{(t)})$ is the probability that the current value $\theta^{(t)}$ is retained as the value for the $(t + 1)$st iteration in the chain. This reveals a conceptual interpretation for the Metropolis sampler, which applies with slight modification to the Metropolis–Hastings sampler discussed in the Section 5.5. In Metropolis sampling, a symmetric proposal distribution q generates a candidate value θ^*. If θ^* has a higher probability (or ordinate of the probability density) than $\theta^{(t)}$ in the posterior, it is accepted as the next value in the chain. If θ^* has a lower probability (or ordinate of the probability density) than $\theta^{(t)}$ in the posterior, it is accepted as the next value in the chain with a probability determined by its relative probability to that of $\theta^{(t)}$. In the case of Figure 5.1 (panel a), the value of θ^* in the posterior density is greater than that associated with $\theta^{(t)}$. As a result, θ^* will be accepted as the next value (with probability 1).

The next iteration proceeds by centering the normal proposal distribution at this value. Figure 5.1 (panel b) depicts this situation, where the just-accepted value is now $\theta^{(t)}$ and the proposal distribution is centered at this point. Suppose that at this iteration, the value drawn from this proposal distribution is θ^* depicted in panel (b). The value of θ^* in the posterior density is less than that associated with $\theta^{(t)}$. As a result, θ^* will be accepted as the next value with probability $[p(\theta^* \mid \mathbf{x})/p(\theta^{(t)} \mid \mathbf{x})]$.

More casually, the acceptance probability in step 3 means that if θ^* is more likely in the posterior than the current value $\theta^{(t)}$, the chain will move to θ^* (Figure 5.1, panel a). On the other hand, if θ^* is less likely in the posterior than the current value $\theta^{(t)}$ (Figure 5.1, panel b), the chain will only move to θ^* sometimes, other times it will stay at the current value $\theta^{(t)}$. How often the chain moves to θ^* in this context is governed by how likely θ^* is in the posterior relative to $\theta^{(t)}$. Putting it together, the Metropolis sampler will always move to a candidate value if it is more likely than the current value, but only sometimes move if a candidate value is less likely than the current value. And it will do so in a way that the series of values in the chain produced from this process occur in a relative frequency dictated by the posterior distribution.

Remarkably, Metropolis sampling works to approximate a distribution for practically any proposal distribution as long as the mild regularity requirements are met. Some proposal distributions lead to more efficient estimation than others, though. Efficiency is governed by a number of factors including the dimension of the problem and conditional independence and conjugacy relationships, and proposal distributions that lead to a 30%–40% acceptance rate tend to be the most efficient (Gelman et al., 2013). As mentioned above, a common proposal distribution is a normal distribution centered at the previous value in the chain. Seeing that the rate of acceptance is too low suggests the proposal distribution is too wide, and the variance should be reduced. If the acceptance rate is too high, the variance should be increased.

5.4 How MCMC Facilitates Bayesian Modeling

As discussed previously, the denominator in Bayes' theorem for complex models is often analytically and computationally intractable. In these cases, the posterior distribution is usually only known up until a constant of proportionality (see Equation 2.1). As discussed in Section 2.2, this is sufficient to capture all the substantive aspects of the model, namely the synthesis of prior beliefs and the information in the data. The denominator serves only as a computational device to yield a proper distribution.

Inspection of the Metropolis sampler reveals that the posterior distribution appears in both the numerator and denominator of the acceptance probability and therefore only needs to be known up to a constant of proportionality. To see this, note that

$$\begin{aligned}
\alpha(\theta^* \mid \theta^{(t)}) &= \min\left[1, \frac{p(\theta^* \mid \mathbf{x})}{p(\theta^{(t)} \mid \mathbf{x})}\right] \\
&= \min\left[1, \frac{\left(p(\mathbf{x} \mid \theta^*)p(\theta^*) \Big/ \int_\theta p(\mathbf{x} \mid \theta)p(\theta)d\theta \right)}{\left(p(\mathbf{x} \mid \theta^{(t)})p(\theta^{(t)}) \Big/ \int_\theta p(\mathbf{x} \mid \theta)p(\theta)d\theta \right)}\right] \\
&= \min\left[1, \frac{p(\mathbf{x} \mid \theta^*)p(\theta^*)}{p(\mathbf{x} \mid \theta^{(t)})p(\theta^{(t)})}\right].
\end{aligned} \tag{5.13}$$

The second line results from applications of Bayes' theorem. The simplification from the second to third line reveals that the denominator in Bayes' theorem does not factor into the calculations. All that is required is the numerator: the prior and the likelihood.

As such MCMC alleviates the need to perform the integration over the parameter space in the denominator of Bayes' theorem to obtain the posterior distribution. This is the key feature of MCMC estimation that permits the estimation of complex Bayesian models. Prior to the advent of MCMC, applications of Bayesian modeling were limited because of the difficulty in analytically evaluating or empirically approximating the marginal probability of the data in the denominator of Bayes' theorem.* Practically speaking, MCMC is a general and now widely available way to avoid high-dimensional, prohibitively complex integration by repeatedly sampling from a distribution that is, in the limit, equal to the posterior distribution. The resulting drawn values thereby constitute an empirical approximation to the posterior.

5.5 Metropolis–Hastings Sampling

5.5.1 Metropolis–Hastings as a Generalization of Metropolis

Metropolis–Hastings sampling generalizes Metropolis sampling by allowing the proposal distribution to be asymmetric with respect to its arguments (Hastings, 1970; see also Brooks, 1998; Chib & Greenberg, 1995; Gilks et al., 1996b). The proposal distribution q may be any distribution that is defined over the support of the stationary distribution, which in our case is the posterior distribution. As such, the Metropolis–Hastings algorithm is an extremely flexible approach to estimating posterior distributions.

In Metropolis–Hastings, the acceptance probability outlined in Section 5.3 is replaced by $\alpha(\theta^* \mid \theta^{(t)}) = \min\left[1, p(\theta^* \mid \mathbf{x})q(\theta^{(t)} \mid \theta^*)/p(\theta^{(t)} \mid \mathbf{x})q(\theta^* \mid \theta^{(t)})\right]$, where, in addition to the terms previously defined, $q(\theta^{(t)} \mid \theta^*)$ is the probability (or ordinate of the probability density) for the current value $\theta^{(t)}$ in the chain in the proposal distribution, and $q(\theta^* \mid \theta^{(t)})$ is the probability (or ordinate of the probability density) for the candidate value θ^* in the proposal distribution.

The acceptance probability involves evaluating the posterior distribution and the proposal distribution q at both the current and candidate values. The Metropolis–Hastings algorithm generalizes the acceptance probability from the Metropolis algorithm to say that, if the proposal distribution by which the candidate values is generated is not symmetric with respect to its arguments, the relative probability of the current and candidate values in the proposal distribution must also be taken into account. Note that if the proposal distribution is indeed symmetric, the terms involving q in the numerator and denominator of $\alpha(\theta^* \mid \theta^{(t)})$ in the Metropolis–Hastings sampler cancel, yielding the Metropolis sampler in which $\alpha(\theta^* \mid \theta^{(t)}) = \min\left[1, p(\theta^* \mid \mathbf{x})/p(\theta^{(t)} \mid \mathbf{x})\right]$.

5.5.2 Explaining the Acceptance Probability

To facilitate an explanation of the acceptance probability, we consider discrete space models, though the logic generalizes. On the surface, it may not be obvious why the acceptance probability $\alpha(\theta^* \mid \theta^{(t)})$ accomplishes the goals of yielding values in concert with the desired distribution. To develop it conceptually, let us recognize that to iterate the chain we need a transition kernel to move from any point to any other point. In other words, when the chain is at a particular point, this transition kernel defines the probability of moving to another particular point, as well as its complement, the probability of not moving to that particular point. Letting $p(\theta^* \mid \theta^{(t)})$ denote the probability of moving from the current point $\theta^{(t)}$ to another point θ^*, the reversibility condition may be written as

* Conjugate priors are convenient, but they are limited to relatively simple models that can be expressed in the general exponential family. Graphical methods, analytic approximations, and numerical methods to find the maxima of posteriors often require specialized solutions (Bernardo & Smith, 2000), although in given problems, they can be an efficient choice for practical work.

$$p(\theta^{(t)} \mid \mathbf{x})p(\theta^* \mid \theta^{(t)}) = p(\theta^* \mid \mathbf{x})p(\theta^{(t)} \mid \theta^*). \tag{5.14}$$

The left-hand side of (5.14) is the probability of being at $\theta^{(t)}$ in the posterior distribution and then moving to θ^*. The right-hand side is the probability of being at θ^* in the posterior distribution and then moving to $\theta^{(t)}$. If the equality in (5.14) holds, it is said that the reversibility condition holds. This is sometimes referred as the equation being time reversible (Brooks, 1998), or that it exhibits detailed balance (Gilks et al., 1996a). Conceptually, Equation (5.14) states that the joint distribution of sampling the two points θ^* and $\theta^{(t)}$ is the same regardless of which comes first. One way to think of the reversibility condition is that it implies that if we ran the chain backwards, we would obtain the values we did with the same probability. If this condition is satisfied, the transition kernel will yield a sample from the distribution of interest, namely the posterior distribution $p(\theta \mid \mathbf{x})$.

The goal then becomes defining the transition kernel $p(\theta^* \mid \theta^{(t)})$ that makes (5.14) true. We begin by choosing a proposal distribution q. If it is the case that

$$p(\theta^{(t)} \mid \mathbf{x})q(\theta^* \mid \theta^{(t)}) = p(\theta^* \mid \mathbf{x})q(\theta^{(t)} \mid \theta^*), \tag{5.15}$$

then defining $p(\theta^* \mid \theta^{(t)}) = q(\theta^* \mid \theta^{(t)})$ makes (5.14) true. Generally, this will not be case; in general our proposal distribution q will be such that (5.15) does not hold. This means that the unconditional probability of being at a particular point (e.g., $\theta^{(t)}$, without loss of generality) and moving to the other point (θ^*) is greater than the unconditional probability of being at θ^* and moving to $\theta^{(t)}$. Relatively speaking (in terms of achieving the equality in Equation 5.14), we move from $\theta^{(t)}$ to θ^* too often and from θ^* to $\theta^{(t)}$ too rarely. To combat this, we devise a mechanism such that we reduce the probability of moving from $\theta^{(t)}$ to θ^*. More specifically, we only *accept* the move from $\theta^{(t)}$ to θ^* with a certain probability and remain at $\theta^{(t)}$ with the complement of that probability. We denote this acceptance probability as $\alpha(\theta^* \mid \theta^{(t)})$.

The probability of moving from $\theta^{(t)}$ to θ^* is then

$$p(\theta^* \mid \theta^{(t)}) = q(\theta^* \mid \theta^{(t)})\alpha(\theta^* \mid \theta^{(t)}). \tag{5.16}$$

This states that the probability of moving from $\theta^{(t)}$ to θ^* is defined as the probability of selecting θ^* from a proposal distribution multiplied by the probability of accepting that value of θ^*. Likewise, the probability of moving from θ^* to $\theta^{(t)}$ is defined as

$$p(\theta^{(t)} \mid \theta^*) = q(\theta^{(t)} \mid \theta^*)\alpha(\theta^{(t)} \mid \theta^*). \tag{5.17}$$

Recall that, relatively speaking, we move from θ^* to $\theta^{(t)}$ too rarely. To make sure that we never miss a chance to move from θ^* to $\theta^{(t)}$, set

$$\alpha(\theta^{(t)} \mid \theta^*) \equiv 1. \tag{5.18}$$

Substituting (5.16) and (5.17) into (5.14),

$$p(\theta^{(t)} \mid \mathbf{x})q(\theta^* \mid \theta^{(t)})\alpha(\theta^* \mid \theta^{(t)}) = p(\theta^* \mid \mathbf{x})q(\theta^{(t)} \mid \theta^*)\alpha(\theta^{(t)} \mid \theta^*). \tag{5.19}$$

Substituting (5.18) into (5.19) yields

$$p(\theta^{(t)} \mid \mathbf{x})q(\theta^* \mid \theta^{(t)})\alpha(\theta^* \mid \theta^{(t)}) = p(\theta^* \mid \mathbf{x})q(\theta^{(t)} \mid \theta^*). \tag{5.20}$$

A little algebra yields

$$\alpha(\theta^* \mid \theta^{(t)}) = \frac{p(\theta^* \mid \mathbf{x}) q(\theta^{(t)} \mid \theta^*)}{p(\theta^{(t)} \mid \mathbf{x}) q(\theta^* \mid \theta^{(t)})}, \tag{5.21}$$

which defines the probability for accepting a move from $\theta^{(t)}$ to θ^*. If the value of $\alpha(\theta^* \mid \theta^{(t)})$ exceeds one, the move from $\theta^{(t)}$ to θ^* is made. If the value of $\alpha(\theta^* \mid \theta^{(t)})$ is less than one, the move is made with that probability and not made with probability $1 - \alpha(\theta^* \mid \theta^{(t)})$.

5.6 Single-Component-Metropolis or Metropolis-within-Gibbs Sampling

It is easily seen that the Metropolis sampler is a special case of the Metropolis–Hastings sampler where $q(\theta^* \mid \theta^{(t)}) = q(\theta^{(t)} \mid \theta^*)$. It is somewhat less obvious that the Gibbs sampler may be viewed as a special case of the Metropolis sampler, namely where the proposal distribution for each parameter is the full conditional distribution, which implies that the acceptance probability $\alpha(\theta^* \mid \theta^{(t)}) = 1$ (Brooks, 1998). In this sense, Gibbs sampling is as an instance of Metropolis(–Hastings) sampling in which the candidate value is always accepted.

Recognizing the relation between Gibbs- and Metropolis-sampling facilitates ways to combine them (Brooks, 1998; Tierney, 1994). Conceptually, the Gibbs sampler is attractive because it decomposes the joint posterior into more manageable components. However, it is limited in that it requires the full conditionals to be of known form to facilitate sampling. The Metropolis (and Metropolis–Hastings) sampler is attractive because it can sample from distributions even when they are not of known form. However, it is sometimes limited in multivariate distributions in that it may be difficult to specify a proposal distribution that yields a candidate that is reasonably likely in the posterior with respect to all the components.

The *single-component-Metropolis* sampler, also termed the *Metropolis-within-Gibbs* sampler, combines the component decomposition approach of Gibbs sampling with the flexibility of Metropolis sampling (or, if asymmetric proposal distributions are used, with Metropolis–Hastings sampling). Specifically, the full conditionals are constructed as in Gibbs sampling. When they are of familiar form, they can be sampled from directly. When the full conditional for a parameter is not of known form, a Metropolis (or Metropolis–Hastings) step may be taken where a candidate value is drawn from a proposal distribution q and accepted with probability α as the next value in the chain for that parameter. This approach capitalizes on the advantages of each: we employ the one-component-at-a-time advantage of Gibbs, but take advantage of the flexibility of Metropolis(–Hastings) when the full conditionals are not of known form.

5.7 Practical Issues in MCMC

There are a number of important issues in the practice of MCMC. Popular software packages for Bayesian statistical and psychometric modeling hide or address these via defaults to varying degrees. Nevertheless, several of them are important for a few reasons. First, certain choices must be made by the analyst in many software programs. Second, they are useful for understanding some of the issues that arise in the use of such software. In the authors' experience, the fidelity and quality with which software programs implement Bayesian ideas and MCMC estimation vary considerably. Analysts are better off if they are prepared to spot results that should signal caution is needed before interpreting the output from software.

5.7.1 Implementing a Sampler

The first step in conducting MCMC is determining and implementing a sampling scheme. As examples, in Gibbs sampling this involves determining all the full conditional distributions and effective ways of sampling from them, and in Metropolis sampling with a normal proposal distribution centered at the current point this involves determining the variance of the proposal distribution for optimal acceptance rates (Gelman et al., 2013). These determinations and choices may be made and programmed by an analyst, but programs such as WinBUGS automate these procedures to the extent that usually the user just needs to specify the joint distribution of all the entities. From there, the program determines the structure of the full conditional distributions and an effective, if not necessarily optimal, computational scheme for a very wide range of models.

The following subsections address issues that are important to the analyst making these determinations and choices, or those using existing MCMC software such as WinBUGS. We focus our attention on analysts who use a program like WinBUGS rather than those who program their own samplers. Accordingly, we address these issues to the extent that most users of Bayesian psychometric models may need to consider them. Those wishing to program their own MCMC algorithms must consider these as well as several other aspects.

5.7.2 Assessing Convergence

The chain of draws from an MCMC process is guaranteed to converge to the target (posterior) distribution (see, e.g., chapter 4 of Jackman, 2009; Roberts, 1996; Tierney, 1994 for technical details of conditions that ensure convergence). However, there is no guarantee *when* that will occur. Assessing whether the chain has converged is therefore an important part of MCMC estimation.

To start, it is important to remember that, whereas in a frequentist paradigm we seek *convergence to a point*, in the fully Bayesian paradigm we seek *convergence to a distribution*. That is, we seek to a collection of points that occur with a specific relative frequency, namely that which is given by the posterior distribution. Returning to the mountain metaphor, whereas convergence in a frequentist approach means that we have reached the highest peak, convergence in the fully Bayesian approach means we are mapping the elevation of the correct terrain.

To illustrate the general idea of convergence, Figure 5.2 contains a trace plot of the draws from three chains for a parameter.* The trace plot is essentially a time-series plot of draws. In this case, there are three chains, run from disparate starting points. We see three chains start apart, and then come together at about iteration 600 and then continue to explore

* The parameter in the illustration is one from an item response theory model (Chapter 11). Here, we focus on the patterns of the behavior of the chains generally.

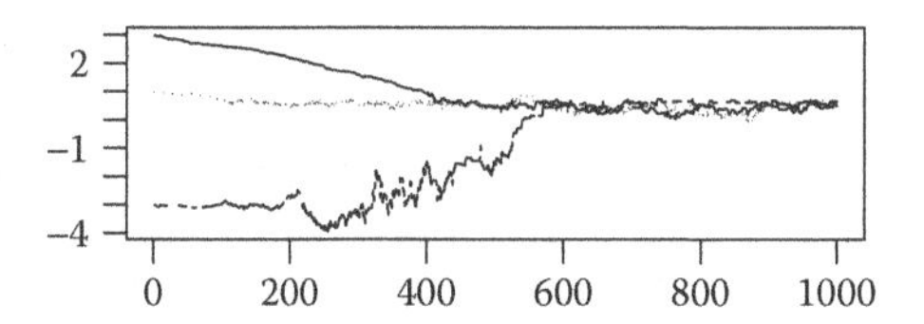

FIGURE 5.2
Trace plot from three chains illustrating monitoring convergence.

the same space. Because all chains eventually must converge to the same distribution, the coming together is taken as evidence, but not proof, of that convergence.

A number of approaches to diagnosing convergence have been proposed, and many can be viewed as comparing summaries of a subset of the draws to another. Examples include viewing trace plots as a graphical check of when the draws stabilize and related ANOVA-like analyses of multiple chains run in parallel examining the ratio of total-variability (i.e., from all chains) to within-chain variability (Brooks & Gelman, 1998; Gelman & Rubin, 1992). Following Gelman et al. (2013), these analyses may be conducted as follows. Run some number C chains from different starting points assumed to represent disparate locations in the posterior. For any parameter, the pooled within-chain variance obtained based on running T iterations in each of C chains is given by

$$W = \frac{1}{C(T-1)} \sum_{c=1}^{C} \sum_{t=1}^{T} (\theta_{(c)}^{(t)} - \bar{\theta}_{(c)})^2, \tag{5.22}$$

where $\theta_{(c)}^{(t)}$ is the value from iteration t in chain c and $\bar{\theta}_{(c)}$ is the mean of such values in chain c. The between chain variance is given by

$$B = \frac{T}{C-1} \sum_{c=1}^{C} (\bar{\theta}_{(c)} - \bar{\theta})^2. \tag{5.23}$$

The potential scale reduction factor is

$$\hat{R} = \sqrt{\frac{(T-1/T)W + (1/T)B}{W}}, \tag{5.24}$$

which decreases to 1 as $T \to \infty$. The interpretation is that if $\hat{R}$ is close to 1 (frequently operationalized as being < 1.1), this supports the inference that the sets of iterations from each chain are close to the target distribution. See Brooks and Gelman (1998) for a multivariate extension, and versions based on percentiles rather than variances. Other approaches suitable for evaluating chains individually include tests invoking normal statistical theory for whether the mean of the draws from a later part of the chain differs significantly from that from an earlier part of the chain (Geweke, 1992). The reader is referred to Cowles and Carlin (1996), Gill (2007), and Jackman (2009) for thorough treatments of MCMC convergence assessment.

Draws obtained before the chain converges are discarded as *burn-in* iterations. Upon convergence, subsequent iterations reflect draws from the stationary distribution and are retained and used to empirically approximate the posterior distribution. In practice using WinBUGS, the user typically must specify a number of iterations and then inspect whether there is evidence of convergence. If there is sufficient evidence of convergence, then the draws preceding the point of convergence are discarded as burn-in. If there is not sufficient evidence of convergence, more iterations can be conducted, and then the resulting draws can be inspected for evidence of convergence. In this way, the process of evaluating convergence and conducting iterations may itself be iterative, cycling between running more iterations and inspecting the resulting draws. Once there is sufficient evidence of convergence, the draws preceding the point of convergence are discarded as burn-in and the draws from subsequent iterations may be used to approximate the posterior.

5.7.3 Serial Dependence

The Markov property states that, given the current value of the chain, all future draws are conditionally independent from all past draws. However, successive draws are serially dependent, which induces a marginal dependence across the draws even with a lag greater than 1. The dependence between draws can be captured by the autocorrelation at various lags where values close to zero indicate approximate independence between draws. To illustrate, Figure 5.3 contains a plot of the autocorrelation by different lags for a chain. At lag 0, the autocorrelation is 1 (i.e., everything is perfectly correlated with itself). As the lag increases, the autocorrelation decreases. The presence of nonzero autocorrelation indicates that the values from the chain should not be considered independent draws from the posterior distribution.

One approach to managing high autocorrelations involves reparameterizing the model in some way (e.g., centering predictors in regression; see Dunson et al., 2005, for an application in the context of latent variable models).

The serial dependence in a set of draws may be mitigated by thinning the chain, in which only draws from a certain lag are retained. Figure 5.4 depicts the autocorrelation plot after thinning the chain by 5 (i.e., saving iterations 1, 6, 11, 16, etc.), 10 (i.e., saving iterations 1, 11, 21, 31, etc.), 30 (i.e., saving iterations 1, 31, 61, 91, etc.), and 40 (i.e., saving iterations 1, 41, 81, 121, etc.). As illustrated in Figure 5.4, increasing the thinning reduces the autocorrelation considerably. If approximately independent draws are desired, then thinning by 30 or certainly 40 gets us fairly close. As should be obvious, this is true *for the iterations that are retained*. Although it might be natural to prefer to retain a set of approximately independent draws, thinning does not provide a better portrait of the posterior. Thinning may be desirable if a fixed number of draws are to be retained, say for reasons of storage limitations. In that case it may be preferable to have the saved draws approximately independent. However, the discarded draws (assuming convergence has been reached) are legitimately considered draws from the posterior; thinning therefore represents a loss of information.

An alternative approach retains all the drawn values but takes their serial dependence into account as may be warranted for a particular analysis. For example, if the T draws are independent, the familiar equation for the standard error of the mean captures the

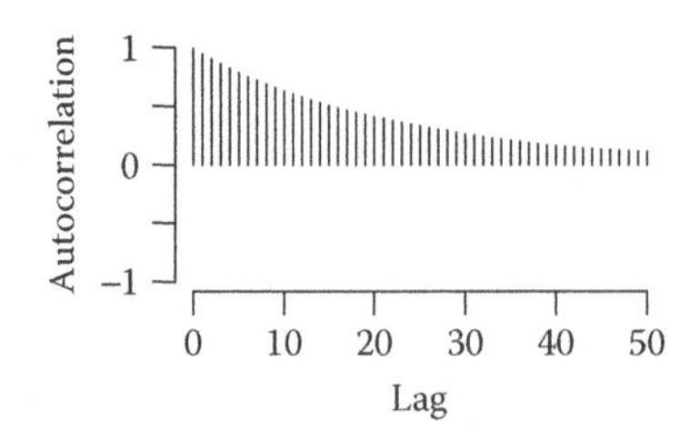

FIGURE 5.3
Autocorrelation plot.

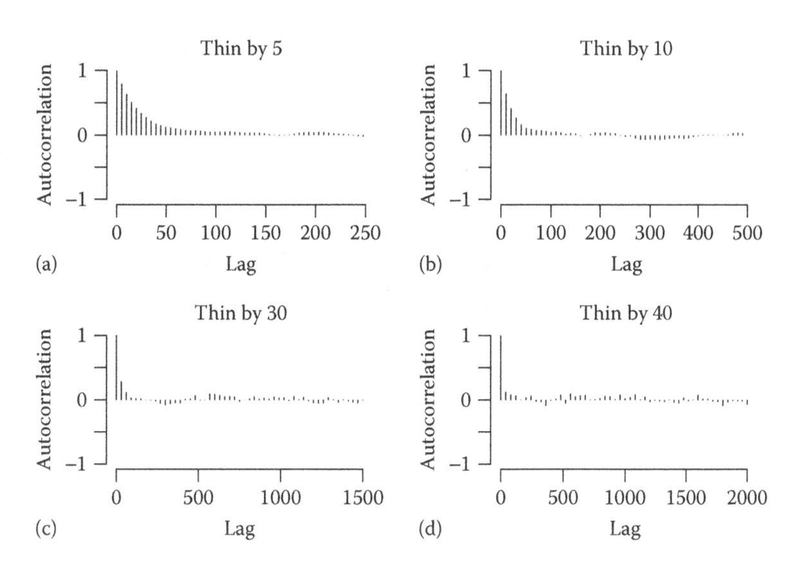

FIGURE 5.4
Autocorrelation plots when thinning by (a) 5, (b) 10, (c) 30, and (d) 40.

accuracy of the estimate of the posterior mean as a function of the number of draws and posterior standard deviation,

$$\sigma_{\bar{\theta}} = \frac{\sigma_\theta}{\sqrt{T}}, \tag{5.25}$$

where σ_θ is the posterior standard deviation for θ. But if there is autocorrelation, we have less information from a given number of draws. Borrowing from time series analysis, we can approximate the impact of autocorrelation with a corrected-for-autocorrelation standard error of the mean. If ρ is the autocorrelation in a first-order autocorrelation model, then the corrected standard error of the mean is

$$\sigma_{\bar{\theta}} = \frac{\sigma_\theta}{\sqrt{T}}\sqrt{\frac{1+\rho}{1-\rho}}. \tag{5.26}$$

The degree of autocorrelation can be very different for different variables in the same model, as it depends on the amount of information that data convey about a variable and its joint relationship with other variables.

One additional point about serial dependence is worth noting. Although draws may be dependent *within* a chain, they are independent *between* chains. Running multiple chains and then pooling the resulting iterations therefore also helps mitigate the effects of serial dependence.

5.7.4 Mixing

Informally, mixing refers to how well the chain "moves" throughout the support of the distribution. That is, we do not want the chain to get "stuck" in some region of the distribution, or ignore a certain area. Figure 5.2 illustrates how things look with relatively poor mixing of each of the chains for the first few hundred iterations. Each one appears "stuck" at a certain value. This could happen in the Metropolis(–Hastings) sampler if the candidate values are only rarely accepted. In that example, each chain mixes much better after about 500 iterations. This happens much faster in some problems and much slower in others.

A few comments on mixing are worthwhile. First, although the consideration of multiple chains (i.e., Figure 5.2) can be useful, mixing refers to how well an individual chain moves throughout the support of the distribution. Good mixing does not mean that the chain has converged, and poor mixing does not mean that the chain has not converged. But better mixing usually leads to faster convergence. Mixing is also distinct from serial dependence, but better mixing usually goes along with lower autocorrelations and cross-correlations between multiple parameters. Slow mixing is therefore often present when there is a high dependence among parameters, such as when high levels of multicollinearity are present in regression models. Reparameterizing the model (e.g., centering predictors in regression) can improve mixing. Alternatively, one can employ MCMC with block samplers in which highly correlated parameters are sampled jointly.

For a given number of draws, better mixing usually gives more information about the posterior, as suggested by the autocorrelation correction for the accuracy of the posterior mean. With worse mixing, more iterations are usually needed to achieve convergence and to achieve a desired level of precision for approximating the posterior or summaries of it.

5.7.5 Summarizing the Results from MCMC

Inferences are based on the draws obtained after the chain has converged. This is determined using the checks mentioned in the previous section. The draws prior to this point are referred to as burn-in iterations and are discarded. The draws obtained after convergence form an empirical approximation to the posterior, and inferences are made using familiar methods for summarizing distributions. Histograms or smoothed density plots for each parameter graphically convey the marginal posterior distributions. Figure 4.11 contained such plots for the example analysis of the unknown mean and variance of a normal distribution. Similarly, scatterplots, contour plots, and correlations between draws for parameters provide evidence about their dependence. Figure 4.10 contained the scatterplot from the normal distribution example; the correlation between the draws there was –.13. Numerical summaries of the posterior distribution are estimated via summaries of the series of draws; posterior means, standard deviations, percentiles, and credibility intervals are estimated by those quantities computed on the draws from the chain. Commonly, point estimates of posterior central tendency and variability (e.g., posterior means and standard deviations) and credibility intervals (e.g., a 95% central or HPD interval) are reported in documentation, for example as in Table 4.1.

5.8 Summary and Bibliographic Note

This chapter provided an overview of MCMC, focusing on the conceptual alignment of MCMC with the estimation of posterior distributions in Bayesian analyses, and on practical aspects of MCMC.

MCMC estimation is different from more traditional estimation procedures in terms of objectives and procedures. It is not *calculating* a solution, as can be done, say, in a closed form least-squares or ML solutions to regression. It is not *optimizing* any equation/condition, as is typically done, say, in iterative ML or least-squares analyses of complex models. Rather, it is *approximating* the posterior solution in a Monte Carlo framework by generating samples from the posterior.

However, we caution against simply associating iterative search approaches for optimization with frequentist estimation on one hand, and associating iterative simulation approaches for representing distributions with Bayesian estimation. The same optimization routines prevalent in ML estimation may be used to obtain estimates of the mode and curvature of the posterior, in support of an approximation of the full posterior (e.g., Mislevy, 1986). And the simulation methods described here in the context of estimating posterior distributions may be used to approximate, explore, and maximize likelihood functions (e.g., Cai, 2010; Fox, 2003; Geyer & Thompson, 1992; Lee & Zhu, 2002). This may be approximated even in the current context of Bayesian modeling by using completely diffuse priors, in which case the posterior is effectively a normalized likelihood. Here, the simulation techniques provide a survey of the normalized likelihood, and the highest peak in this mountain range is theoretically equivalent to the point estimate of traditional ML estimation methods. Similarly, as sample size increases the likelihood comes to dominate the posterior so that, asymptotically, posterior modes are equivalent to ML estimates.* The increasing similarity between the likelihood and the posterior distribution as sample size increases and/or the prior becomes more diffuse provides a connection between frequentist and Bayesian estimation. The methods reviewed in this chapter may therefore provide an attractive option for exploring, maximizing, or finding multimodality in likelihoods.

We briefly described Gibbs and Metropolis(–Hastings) samplers. These may be seen as ends of a spectrum of ease. When full conditionals are of a known form, Gibbs sampling is often an efficient and attractive option. When full conditionals are not of known form, Metropolis(–Hastings) is a very flexible approach, but comes at the price of additional computational demands. A number of algorithms reside "in between" these two in terms of complexity and efficiency, and may be gainfully employed in Bayesian analyses (see, e.g., Gelman et al., 2013; Jackman, 2009, for textbook treatments of a number of algorithms). Although there are differences between the versions of BUGS (Lunn et al., 2013), we may loosely describe its algorithm selection process as follows. If the full conditional is a known form, it will sample from it. If not, it will proceed to attempt to match features of the full conditional to a sampling scheme, and if no other sampling scheme is viable it will use a Metropolis sampler with a normal proposal distribution centered at the current value of the chain.

More comprehensive treatments of the theory and practice of MCMC estimation may be found in dedicated texts (Brooks et al., 2011; Gilks et al., 1996b) as well as texts on Bayesian modeling (Gelman et al., 2013; Jackman, 2009; Lynch, 2007). In addition, the reader is referred to Cowles and Carlin (1996), Gill (2007), and Jackman (2009) for thorough treatments of MCMC convergence assessment. Sinharay (2004) reported on the performance of a number of approaches to convergence assessment in the context of psychometric models.

It is difficult to overstate the importance of MCMC in the growth of Bayesian modeling generally, and in psychometrics. Gelfand and Smith's (1990) paper marked a turning point in the acceptance and application of MCMC in the broader statistics community, opening up possibilities for Bayesian modeling that were previously unobtainable due

* The relationship is not as straightforward in more complex models such as hierarchical models.

to computational intractability (Robert & Casella, 2011; McGrayne, 2011). We can identify a similar pivot point for MCMC and Bayesian modeling in psychometrics about a decade later with publications by Arminger and Muthén (1998) and Scheines, Hoijtink, and Boomsma (1999) in factor analysis and structural equation modeling, Hoijtink (1998) in latent class analysis, and particularly those of Patz and Junker (1999a, 1999b) in item response theory. Although seminal work by Albert (1992, Albert & Chib, 1993) had shown how normal-ogive item response theory parameters could be estimated via Gibbs sampling via data-augmentation strategies (Tanner & Wong, 1987), the full power of MCMC was not fully appreciated in the psychometric community until Patz and Junker (1999a, 1999b) described a Metropolis–Hastings-within-Gibbs sampling approach for the most common logistic IRT models without requiring conjugate relationships. The emergence of MCMC has precipitated the explosion in applications of Bayesian psychometrics (Levy, 2009), mirroring the effects of its recognition in other complex statistical modeling scenarios (Brooks et al., 2011; Gilks et al., 1996b). In the following chapters, we will employ MCMC estimation to empirically approximate posterior distributions, which is particularly powerful in light of psychometric models for which analytic solutions are intractable.

Exercises

5.1 Reconsider the beta-binomial model example introduced in Section 2.3. WinBUGS code for the model is given in Section 2.6. Conduct an analysis of the beta-binomial model in WinBUGS, monitoring θ.

a. List at least three possible dispersed initial values for θ that may be used gainfully in running multiple chains. How did you come up with them?

b. Using the initial values identified in (a), run three chains and monitor convergence. When do the chains appear to have converged, and what evidence do you have for that assessment?

c. Run 10,000 iterations for each chain past the point determined in (b). Obtain the density and summary statistics for θ. How does this representation of the posterior compare to the analytical solution depicted in Figure 2.4 and summarized in Section 2.6?

5.2 Conduct an analysis of the beta-Bernoulli version of the model in WinBUGS, monitoring θ.

a. Would the same initial values as used in Exercise 5.1 be sensible here? If not, why not and what values could be used instead?

b. Using the initial values identified in (a), run three chains and monitor convergence. When do the chains appear to have converged, and what evidence do you have for that assessment? How does this compare to what happened in the beta-binomial model in Exercise 5.1?

c. Run 10,000 iterations for each chain past the point determined in (b). Obtain the density and summary statistics for θ. How does this representation of the posterior compare to that from the beta-binomial model in Exercise 5.1?

5.3 Reconsider the example from Section 4.1 for inference about the mean of a normal distribution (μ) with the variance known.

a. Develop WinBUGS code for the model. (Hint: Consider modifying the code given in Section 4.3.2.)
b. List at least three possible dispersed initial values for μ that may be used gainfully in running multiple chains. How did you come up with them?
c. Using the initial values identified in (b), run three chains and monitor convergence. When do the chains appear to have converged, and what evidence do you have for that assessment?
d. Run 20,000 iterations for each chain past the point determined in (c). Obtain the density and summary statistics for μ. How does this representation of the posterior compare to the analytical solution depicted and summarized in Section 4.1?

5.4 Reconsider the example from Section 4.2 for inference about the variance of a normal distribution (σ^2) with the mean known.

a. Develop WinBUGS code for the model. (Hint: Consider modifying the code given in Section 4.3.2.)
b. List at least three possible dispersed initial values for σ^2 that may be used gainfully in running multiple chains. How did you come up with them?
c. Using the initial values identified in (b), run three chains and monitor convergence. When do the chains appear to have converged, and what evidence do you have for that assessment?
d. Run 20,000 iterations for each chain past the point determined in (c). Obtain the density and summary statistics for σ^2. How does this representation of the posterior compare to the analytical solution depicted and summarized in Section 4.2?

5.5 Reconsider the example from Section 4.3 and Section 5.2.2 for inference about the mean and variance of a normal distribution. Now suppose we are interested in reporting results for the variability in terms of the standard deviation.

a. Develop WinBUGS code for the model. (Hint: Consider modifying the code given in Section 4.3.2.)
b. Would the initial values for μ from Exercise 5.3 and the initial values for σ^2 from Exercise 5.4 be appropriate here? If not, why not and what values could be used instead?
c. Are additional initial values for the standard deviation needed? If so, what are some possible useful values? If not, why not?
d. Using the initial values identified in (b) and possibly (c), run three chains monitoring the mean, variance, and standard deviation. When do the chains appear to have converged, and what evidence do you have for that assessment?
e. Run 20,000 iterations for each chain past the point determined in (d). Obtain the marginal densities and summary statistics for all the parameters.
f. Consider an estimate of the standard deviation based on taking the square root of the posterior mean of the variance. How does this result compare to the marginal posterior distribution of the standard deviation from (e)?

5.6 Reconsider the example Section 5.2.2 for inference about the mean and variance of a normal distribution. The computations involved in the first two iterations of a Gibbs sampler were given in Section 5.2.2. The results for three additional iterations are given in Table 5.1. Show the computations for each of these additional iterations.

6

Regression

This chapter provides a treatment of popular Bayesian approaches to working with regression models. It builds on the normal distribution models discussed in Chapter 4. The current chapter once again offers a brief treatment; rather than seeking a comprehensive account, we aim to offer a depth sufficient to support the development of the psychometric models that are the focus of this book. In particular, this chapter focuses on the widely used normal linear model for a continuous outcome variable as it provides a foundation for the development of more complex classical test theory and factor analysis models.

In Section 6.1, we cover some background material, including a review of how regression models are conventionally perceived, as well as a high-level, abstract statement of the posterior distribution. In the following two sections, we fill in the details of the ingredients in the posterior, including the conditional probability of the data (Section 6.2) and the prior distribution (Section 6.3). We tie these developments together into a statement of the complete model and the posterior distribution in Section 6.4, and describe fitting the model using MCMC in Section 6.5. An example is analyzed in Section 6.6. In Section 6.7, we conclude the chapter with a brief summary and pointers to additional readings.

6.1 Background and Notation

6.1.1 Conventional Presentation

Regression models relate an outcome variable to a set of predictor variables. Let $\mathbf{y} = (y_1, \ldots, y_n)$ denote the collection of outcome variables from the n subjects. Let $\mathbf{x}_i = (x_{i1}, \ldots, x_{iJ})$ be the collection of J predictor variables for subject i. Further, let $\mathbf{x}$ be the full collection of J predictors from all n subjects.

We specify a linear model for the conditional expectation of the outcome variable given the predictors, a set of coefficients $\boldsymbol{\beta} = (\beta_1, \ldots, \beta_J)$ for those predictors, and an intercept term β_0. For an outcome variable at the level of an individual,

$$E(y_i \mid \beta_0, \boldsymbol{\beta}, \mathbf{x}) = \beta_0 + \beta_1 x_{i1} + \cdots + \beta_J x_{iJ}. \tag{6.1}$$

Letting

$$\varepsilon_i = y_i - E(y_i \mid \beta_0, \boldsymbol{\beta}, \mathbf{x}) \tag{6.2}$$

denote the error for subject i, then following assumptions of normality and homogeneity of variance for the εs, the regression model is given by

$$y_i = \beta_0 + \beta_1 x_{i1} + \cdots + \beta_J x_{iJ} + \varepsilon_i, \tag{6.3}$$

where

$$\varepsilon_i \sim N(0, \sigma_\varepsilon^2) \tag{6.4}$$

and σ_ε^2 is the error variance. The normal distribution of error terms assumed here is typical, although alternatives such as thicker-tailed distributions and variances that depend on the *x*s can be useful.

We suspect that readers are familiar with this manner of expressing a regression model. That is, if asked to write down the standard regression model, most readers would write down the contents of (6.3), or some notational variant of it, possibly including the distributional statement in (6.4). One theme of our approach to statistical modeling is that it is advantageous to shift from this sort of strategy of thinking *equationally* to one of thinking *probabilistically*. Combining the deterministic expression in (6.3) with the probabilistic expression in (6.4), the standard linear regression model is expressed distributionally as

$$y_i \mid \beta_0, \boldsymbol{\beta}, \mathbf{x}_i \sim N(\beta_0 + \beta_1 x_{i1} + \cdots + \beta_J x_{iJ}, \sigma_\varepsilon^2). \tag{6.5}$$

Rather than being an estimate or an answer to a particular question, the distributional expression in (6.5) tells us what we would believe about the value of y_i if we knew $\mathbf{x}_i$ and the model parameters—that is, it is a conditional expression of belief, which can be useful for a variety of inferential questions in its own right, and it can be a building block for managing evidence, belief, and uncertainty in a more complicated situation.

6.1.2 From a Full Model of All Observables to a Conditional Model for the Outcomes

As (6.5) makes explicit, the regression parameters β_0, $\boldsymbol{\beta}$, and σ_ε^2 describe the *conditional* distribution of the outcome given the predictors. The data in our regression situation include the predictors **x** as well as the outcomes **y**. A fully Bayesian analysis that views the data as observed values of random variables includes a distributional specification for **x** as well as **y**. If the values for **x** are fixed, such as when they are chosen in experimental studies, then it can be viewed as if their probability $p(\mathbf{x})$ is known (Gelman et al., 2013) or alternatively as though they are not drawn from a density at all (Jackman, 2009). More generally, letting $\boldsymbol{\Omega}$ denote the parameters that govern the distribution for **x**, $p(\mathbf{x} \mid \boldsymbol{\Omega})$, the posterior distribution for the full model is then

$$p(\beta_0, \boldsymbol{\beta}, \sigma_\varepsilon^2, \boldsymbol{\Omega} \mid \mathbf{y}, \mathbf{x}) \propto p(\mathbf{y}, \mathbf{x} \mid \beta_0, \boldsymbol{\beta}, \sigma_\varepsilon^2, \boldsymbol{\Omega}) p(\beta_0, \boldsymbol{\beta}, \sigma_\varepsilon^2, \boldsymbol{\Omega}). \tag{6.6}$$

The first term on the right-hand side is the conditional probability of all of the data. The second term is the prior distribution for all the parameters. Assuming prior independence of $(\beta_0, \boldsymbol{\beta}, \sigma_\varepsilon^2)$ and $\boldsymbol{\Omega}$ allows for the factorization $p(\beta_0, \boldsymbol{\beta}, \sigma_\varepsilon^2, \boldsymbol{\Omega}) = p(\beta_0, \boldsymbol{\beta}, \sigma_\varepsilon^2) p(\boldsymbol{\Omega})$. It can be shown (e.g., Jackman, 2009) that the posterior distribution can be then factored as

$$p(\beta_0, \boldsymbol{\beta}, \sigma_\varepsilon^2, \boldsymbol{\Omega} \mid \mathbf{y}, \mathbf{x}) = p(\beta_0, \boldsymbol{\beta}, \sigma_\varepsilon^2 \mid \mathbf{y}, \mathbf{x}) p(\boldsymbol{\Omega} \mid \mathbf{x}). \tag{6.7}$$

This implies that we can analyze the first term on the right-hand side—the elements of the standard regression model—by itself with no loss of information. As a consequence, the distinction between **x** being fixed or stochastic is irrelevant in the Bayesian analysis of the model (Jackman, 2009). Either way, the $p(\mathbf{x})$ and $\boldsymbol{\Omega}$ terms drop out of the model and subsequent analysis. This highlights the utility of thinking distributionally and expressing the model in (conditional) probabilistic terms in (6.5); namely, for inference about the standard regression model parameters β_0, β, and σ_ε^2, once we have conditioned on $\mathbf{x}$, there is no additional information gained from knowing Ω.

6.1.3 Toward the Posterior Distribution

The implication of these developments is that we can condition on $\mathbf{x}$ throughout our analysis of the regression model that specifies the conditional distribution of $\mathbf{y}$. The posterior distribution is therefore

$$p(\beta_0,\boldsymbol{\beta},\sigma_\varepsilon^2 \mid \mathbf{y},\mathbf{x}) = \frac{p(\mathbf{y} \mid \beta_0,\boldsymbol{\beta},\sigma_\varepsilon^2,\mathbf{x})p(\beta_0,\boldsymbol{\beta},\sigma_\varepsilon^2)}{p(\mathbf{y} \mid \mathbf{x})} \propto p(\mathbf{y} \mid \beta_0,\boldsymbol{\beta},\sigma_\varepsilon^2,\mathbf{x})p(\beta_0,\boldsymbol{\beta},\sigma_\varepsilon^2). \quad (6.8)$$

6.2 Conditional Probability of the Data

The first term on the right-hand side of (6.8) is the conditional probability of the outcomes $\mathbf{y}$ given the predictors $\mathbf{x}$ and the parameters $(\beta_0,\boldsymbol{\beta},\sigma_\varepsilon^2)$. Assuming exchangeability among subjects amounts to assuming that the outcome for any subject, y_i, is conditionally independent of the outcome from any other subject given $\mathbf{x}_i$ and $(\beta_0,\boldsymbol{\beta},\sigma_\varepsilon^2)$. This allows for $p(\mathbf{y} \mid \beta_0,\boldsymbol{\beta},\sigma_\varepsilon^2,\mathbf{x})$ to be factored as

$$p(\mathbf{y} \mid \beta_0,\boldsymbol{\beta},\sigma_\varepsilon^2,\mathbf{x}) = \prod_{i=1}^{n} p(y_i \mid \beta_0,\boldsymbol{\beta},\sigma_\varepsilon^2,\mathbf{x}_i). \quad (6.9)$$

To complete the specification, $p(y_i \mid \beta_0,\boldsymbol{\beta},\sigma_\varepsilon^2,\mathbf{x}_i)$ is the normal distribution given in (6.5).

Figure 6.1 contains a fragment of the DAG corresponding to the conditional probability of the data. The plate over i containing nodes x_{ij} and y_i indicates a repeating of the xs and y over subjects. The plate over j containing the x_{ij} and β_j nodes indicates a repeating over the various predictors and their associated coefficients. The DAG fragment in Figure 6.1 depicts the conditional probability of the data, and as such is a graphical representation of the regression model as conceived of in frequentist modeling traditions. To complete the Bayesian specification, we turn to the description and specification of the prior distribution for the unknown parameters.

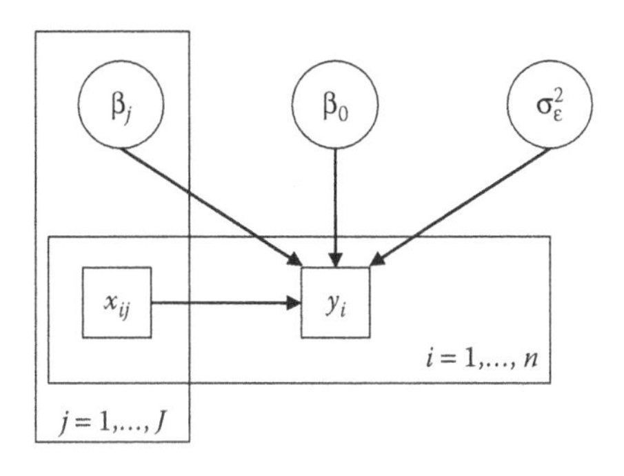

FIGURE 6.1
DAG fragment for the conditional probability of the outcomes (y_i) for the regression model that specifies predictors (x_{ij}), slopes (β_j), an intercept (β_0), and the error variance (σ_ε^2).

6.3 Conditionally Conjugate Prior

The second term on the right-hand side of (6.8) is the prior distribution for $(\beta_0, \boldsymbol{\beta}, \sigma_\varepsilon^2)$. Recognizing that the outcome has a conditional normal distribution, we employ the strategies and distributional forms introduced in Chapter 4 to yield a conditionally conjugate prior specification. The conditionally conjugate prior is adopted in part for its mathematical convenience, as it yields full conditionals of known form (see Appendix A). In addition, it simplifies things conceptually in that we can specify the prior one parameter at a time and it involves distributional forms that are familiar, with parameters that are reasonably easy to interpret when expressing prior beliefs. For alternative prior specifications in regression, see Gelman et al. (2013), Gill (2007), Jackman (2009), and Levy and Crawford (2009).

To begin, an assumption of a priori independence supports the factorization of the prior distribution $p(\beta_0, \boldsymbol{\beta}, \sigma_\varepsilon^2) = p(\beta_0)p(\boldsymbol{\beta})p(\sigma_\varepsilon^2)$. This does not restrict the parameters to be independent in the posterior. If the data indicate dependences, the posterior will reflect that. See Appendix A for an alternative approach that does not assume a priori independence between β_0 and $\boldsymbol{\beta}$. See Jackman (2001) for an approach that does not assume a priori independence among β_0, $\boldsymbol{\beta}$, and σ_ε^2.

We employ a normal prior distribution for the intercept β_0,

$$\beta_0 \sim N(\mu_{\beta_0}, \sigma_{\beta_0}^2), \tag{6.10}$$

where μ_{β_0} and $\sigma_{\beta_0}^2$ are hyperparameters. For the regression coefficients $\boldsymbol{\beta}$, an assumption of exchangeability motivates the factorization of their prior distribution,

$$p(\boldsymbol{\beta}) = \prod_{j=1}^{J} p(\beta_j). \tag{6.11}$$

Of course, if we had beliefs about the values of the β_js through our knowledge of the substance of an application—even for example just knowing the scales that the *x*s are measured on—we could forego the presumption of exchangeability for some or all of the β_js, or have subsets of exchangeable groups of them, as appropriate. It is quite common to treat the coefficients as exchangeable, or include enough conditioning variables to treat them as conditionally exchangeable. For ease of exposition, we continue with model assuming exchangeability. Each regression slope β_j, $j = 1, \ldots, J$, is specified as having a (common) normal distribution,

$$\beta_j \sim N(\mu_\beta, \sigma_\beta^2), \tag{6.12}$$

where μ_β and σ_β^2 are hyperparameters.

As we did in Chapter 4, we employ an inverse gamma for the prior distribution for the variance, specifically the error variance σ_ε^2,

$$\sigma_\varepsilon^2 \sim \text{Inv-Gamma}(\nu_0 / 2, \nu_0\sigma_{\varepsilon 0}^2/2), \tag{6.13}$$

where $\sigma_{\varepsilon 0}^2$ and ν_0 are hyperparameters that may be interpreted as an estimate for the error variance and degrees of freedom or pseudo-sample size associated with that estimate, respectively.

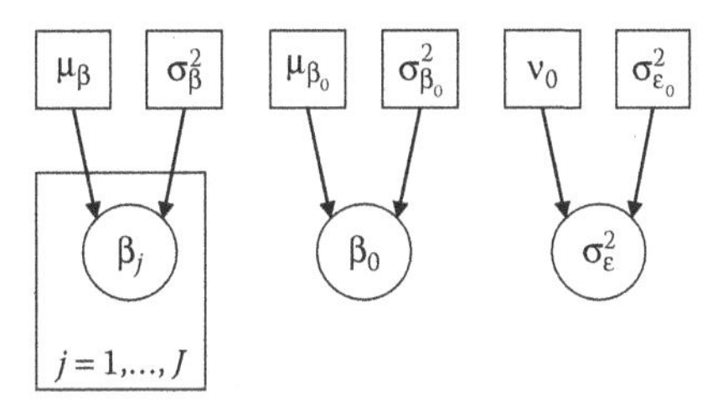

FIGURE 6.2
DAG fragment for the prior distribution for the regression model, with hyperparameters for the normal prior distributions for slopes (β_j), normal prior distribution for the intercept (β_0), and an inverse-gamma prior for the error variance (σ^2_ε).

Figure 6.2 contains a fragment of the DAG corresponding to the prior distribution for the parameters. The specification of independent priors for the parameters β_0, $\boldsymbol{\beta}$, and σ^2_ε is reflected in the graph by the absence of directed edges connecting the nodes for the parameters. Further, the plate over j containing the β_j nodes indicates replication over the coefficients for the J predictors. Note that the hyperparameters μ_β and σ^2_β lie *outside* the plate, indicating that the β_js have a common prior distribution, as is motivated by treating the β_js as exchangeable. As noted above, if we had beliefs about the values of the β_js through our knowledge of the substance of an application, we could forego the presumption of exchangeability for some or all of the β_js.

6.4 Complete Model and Posterior Distribution

Assembling the DAGs in Figures 6.1 and 6.2, the DAG for the complete model is given in Figure 6.3, depicting the dependence structure for each entity in the model. The corresponding distributional expressions are given in (6.5) and (6.10)–(6.13). We collect them here and state the posterior distribution:

$$
\begin{aligned}
p(\beta_0, \boldsymbol{\beta}, \sigma^2_\varepsilon \mid \mathbf{y}, \mathbf{x}) &\propto p(\mathbf{y} \mid \beta_0, \boldsymbol{\beta}, \sigma^2_\varepsilon, \mathbf{x}) p(\beta_0, \boldsymbol{\beta}, \sigma^2_\varepsilon) \\
&= \prod_{i=1}^{n} p(y_i \mid \beta_0, \boldsymbol{\beta}, \sigma^2_\varepsilon, \mathbf{x}_i)\, p(\beta_0) \prod_{j=1}^{J} p(\beta_j)\, p(\sigma^2_\varepsilon),
\end{aligned}
\tag{6.14}
$$

where

$$
\begin{aligned}
y_i \mid \beta_0, \boldsymbol{\beta}, \sigma^2_\varepsilon, \mathbf{x}_i &\sim N(\beta_0 + \beta_1 x_{i1} + \cdots + \beta_J x_{iJ}, \sigma^2_\varepsilon) \quad i = 1,\ldots,n, \\
\beta_0 &\sim N(\mu_{\beta_0}, \sigma^2_{\beta_0}), \\
\beta_j &\sim N(\mu_\beta, \sigma^2_\beta) \quad j = 1,\ldots, J,
\end{aligned}
$$

and

$$
\sigma^2_\varepsilon \sim \text{Inv-Gamma}(\nu_0 / 2, \nu_0 \sigma^2_{\varepsilon 0}/2).
$$

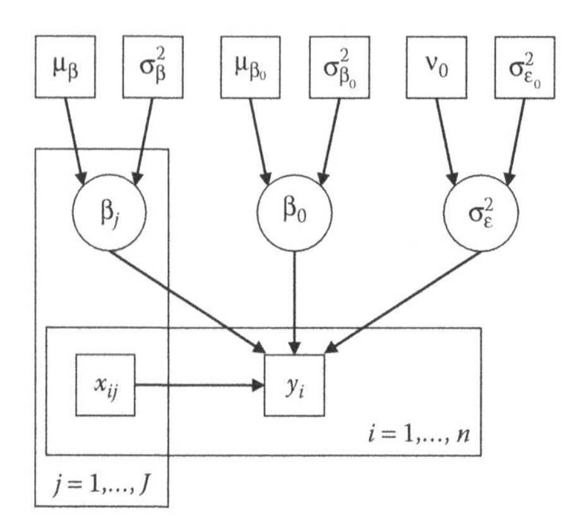

FIGURE 6.3
DAG for the regression model for outcomes (y_i) regressed on predictors (x_{ij}), with normal prior distributions for the slopes (β_j) and intercept (β_0), and an inverse-gamma prior for the error variance (σ^2_ε).

With the conditionally conjugate prior, the model is conceptually simple in that prior distributions are specified for each parameter individually. Furthermore, these specifications yield reasonably simple steps in MCMC techniques used to empirically approximate the posterior, to which we now turn.

6.5 MCMC Estimation

A Gibbs sampler is constructed by specifying initial values for all of the parameters $\beta_0^{(0)}, \beta_1^{(0)}, \ldots, \beta_J^{(0)}, \sigma_\varepsilon^{2(0)}$ where the parenthetical superscript refers to iteration t (here, $t = 0$) and sampling from the full conditional distributions. For iteration $t + 1$, we take a draw for each parameter from its full conditional distribution, which have the following structural forms:

$$\begin{aligned}
\beta_0^{(t+1)} &\sim p(\beta_0 \mid \beta_1^{(t)}, \ldots, \beta_J^{(t)}, \sigma_\varepsilon^{2(t)}, \mathbf{y}, \mathbf{x}), \\
\beta_1^{(t+1)} &\sim p(\beta_1 \mid \beta_0^{(t+1)}, \beta_2^{(t)}, \ldots, \beta_J^{(t)}, \sigma_\varepsilon^{2(t)}, \mathbf{y}, \mathbf{x}), \\
\beta_2^{(t+1)} &\sim p(\beta_2 \mid \beta_0^{(t+1)}, \beta_1^{(t+1)}, \beta_3^{(t)}, \ldots, \beta_J^{(t)}, \sigma_\varepsilon^{2(t)}, \mathbf{y}, \mathbf{x}), \\
&\vdots \\
\beta_J^{(t+1)} &\sim p(\beta_J \mid \beta_0^{(t+1)}, \beta_1^{(t+1)}, \ldots, \beta_{J-1}^{(t+1)}, \sigma_\varepsilon^{2(t)}, \mathbf{y}, \mathbf{x}),
\end{aligned} \tag{6.15}$$

and

$$\sigma_\varepsilon^{2(t+1)} \sim p(\sigma_\varepsilon^2 \mid \beta_0^{(t+1)}, \beta_1^{(t+1)}, \ldots, \beta_J^{(t+1)}, \mathbf{y}, \mathbf{x}).$$

The choice of the conditionally conjugate prior distribution for the parameters renders these full conditionals to be of known form (see Appendix A for derivations). We express the full conditional distributions in the following equations. On the left-hand side, the parameter in question is written as conditional on all the other relevant parameters and data.* On the right-hand side of each of the following equations, we give the parametric form for the full conditional distribution. In several places, we denote the arguments of the distribution (e.g., mean and variance for a normal distribution) with subscripts denoting that it refers to the full conditional distribution; the subscripts are then just the conditioning notation of the left-hand side.

To present all of them compactly, let $\mathbf{x}_A$ be the $(n \times [J+1])$ *augmented* predictor matrix obtained by combining an $(n \times 1)$ column vector of 1s to the predictor matrix $\mathbf{x}$, where $\mathbf{x}$ has been arranged such that $\mathbf{x}_i$ is in the ith row of $\mathbf{x}$. Let β_A be the $([J+1] \times 1)$ *augmented* vector of coefficients obtained by combining the intercept β_0 with the J coefficients in β. Let $\beta_{A(-j)}$ denote the $(J \times 1)$ vector obtained by omitting β_j from β_A. Similarly, let $\mathbf{x}_{A(-j)}$ denote the $(n \times J)$ matrix obtained by omitting the column of values from $\mathbf{x}_A$. When $j > 0$, $\mathbf{x}_j$ refers to column vector with values for the jth predictor; when $j = 0$, $\mathbf{x}_j$ refers to the column vector of 1s.

The full conditional distribution for β_j, $j = 0,\ldots, J$, is then

$$\beta_j \mid \beta_{A(-j)}, \sigma_\varepsilon^2, \mathbf{y}, \mathbf{x} \sim N(\mu_{\beta_j|\beta_{A(-j)},\sigma_\varepsilon^2,\mathbf{y},\mathbf{x}}, \sigma^2_{\beta_j|\beta_{A(-j)},\sigma_\varepsilon^2,\mathbf{y},\mathbf{x}}), \tag{6.16}$$

where

$$\mu_{\beta_j|\beta_{A(-j)},\sigma_\varepsilon^2,\mathbf{y},\mathbf{x}} = \left(\frac{1}{\sigma^2_{\beta_j}} + \frac{1}{\sigma^2_\varepsilon}\mathbf{x}_j'\mathbf{x}_j\right)^{-1}\left(\frac{1}{\sigma^2_{\beta_j}}\mu_{\beta_j} + \frac{1}{\sigma^2_\varepsilon}\mathbf{x}_j'(\mathbf{y} - \mathbf{x}_{A(-j)}\beta_{A(-j)})\right) \tag{6.17}$$

and

$$\sigma^2_{\beta_j|\beta_{A(-j)},\sigma_\varepsilon^2,\mathbf{y},\mathbf{x}} = \left(\frac{1}{\sigma^2_{\beta_j}} + \frac{1}{\sigma^2_\varepsilon}\mathbf{x}_j'\mathbf{x}_j\right)^{-1}. \tag{6.18}$$

When $j = 0$, μ_{β_j} and $\sigma^2_{\beta_j}$ in (6.17) and (6.18) refer to the prior mean and variance of the intercept, which we had denoted as μ_{β_0} and $\sigma^2_{\beta_0}$ in (6.10). When $j > 1$, μ_{β_j} and $\sigma^2_{\beta_j}$ in (6.17) and (6.18) refer to the prior mean and variance of the coefficient for the jth predictor, which under exchangeability was denoted as μ_β and σ^2_β in (6.12).

The full conditional distribution for σ^2_ε is

$$\sigma^2_\varepsilon \mid \beta_A, \mathbf{y}, \mathbf{x} \sim \text{Inv-Gamma}\left(\frac{\nu_0 + n}{2}, \frac{\nu_0\sigma^2_{\varepsilon 0} + SS(\mathbf{E})}{2}\right), \tag{6.19}$$

where

$$SS(\mathbf{E}) = (\mathbf{y} - \mathbf{x}_A\beta_A)'(\mathbf{y} - \mathbf{x}_A\beta_A). \tag{6.20}$$

A Gibbs sampler is constructed by iteratively drawing from the full conditional distributions defined by (6.16) and (6.19) using the just-drawn values for the conditioned parameters.

* We suppress the role of specified hyperparameters in this notation; see Appendix A for a presentation that formally includes the hyperparameters.

A summary of the Gibbs sampler, for one generic iteration $(t + 1)$ that occurs in a sequence of many such iterations, is as follows:

6.5.1 Gibbs Sampler Routine for Regression

1. *Sample the augmented regression coefficients.* For each coefficient β_j for $j = 0, \ldots, J$, let $\boldsymbol{\beta}_{A(-j)}^{(\text{current})}$ denote the current values of the $\boldsymbol{\beta}_A$ omitting β_j. $\boldsymbol{\beta}_{A(-j)}^{(\text{current})}$ will contain just-drawn values for coefficients that have already been sampled in the current iteration $(t + 1)$, and the values from the previous iteration for coefficients that have not yet been sampled in the current iteration. Let us sample the coefficients in the order given by their index j, $j = 0, \ldots, J$. In this case, $\boldsymbol{\beta}_{A(-j)}^{(\text{current})} = (\boldsymbol{\beta}_{j'<j}^{(t+1)}, \boldsymbol{\beta}_{j'>j}^{(t)})$, where $\boldsymbol{\beta}_{j'<j}$ is the collection of parameters with an index less than j and $\boldsymbol{\beta}_{j'>j}$ is the collection of parameters with an index greater than j. That is, for $\beta_0 : \boldsymbol{\beta}_{A(-0)}^{(\text{current})} = (\beta_1^{(t)}, \ldots, \beta_J^{(t)})$; for $\beta_1 : \boldsymbol{\beta}_{A(-1)}^{(\text{current})} = (\beta_0^{(t+1)}, \beta_2^{(t)}, \ldots, \beta_J^{(t)})$; and so on up until $\beta_J : \boldsymbol{\beta}_{A(-J)}^{(\text{current})} = (\beta_0^{(t+1)}, \ldots, \beta_{J-1}^{(t+t)})$. For each coefficient β_j for $j = 0, \ldots, J$, compute

$$\mu_{\beta_j|\boldsymbol{\beta}_{A(-j)}^{(\text{current})}, \sigma_\varepsilon^{2(t)}, \mathbf{y}, \mathbf{x}}^{(t+1)} = \left(\frac{1}{\sigma_{\beta_j}^2} + \frac{1}{\sigma_\varepsilon^{2(t)}} \mathbf{x}_j' \mathbf{x}_j \right)^{-1} \left(\frac{1}{\sigma_{\beta_j}^2} \mu_{\beta_j} + \frac{1}{\sigma_\varepsilon^{2(t)}} \mathbf{x}_j' (\mathbf{y} - \mathbf{x}_{A(-j)} \boldsymbol{\beta}_{A(-j)}^{(\text{current})}) \right) \tag{6.21}$$

and

$$\sigma_{\beta_j|\boldsymbol{\beta}_{A(-j)}^{(\text{current})}, \sigma_\varepsilon^{2(t)}, \mathbf{y}, \mathbf{x}}^{2(t+1)} = \left(\frac{1}{\sigma_{\beta_j}^2} + \frac{1}{\sigma_\varepsilon^{2(t)}} \mathbf{x}_j' \mathbf{x}_j \right)^{-1}, \tag{6.22}$$

and sample a value from (6.16) using these values for $\mu_{\beta_j|\boldsymbol{\beta}_{A(-j)}, \sigma_\varepsilon^2, \mathbf{y}, \mathbf{x}}$ and $\sigma^2_{\beta_j|\boldsymbol{\beta}_{A(-j)}, \sigma_\varepsilon^2, \mathbf{y}, \mathbf{x}}$:

$$\beta_j^{(t+1)} \mid \boldsymbol{\beta}_{A(-j)}^{(\text{current})}, \sigma_\varepsilon^{2(t)}, \mathbf{y}, \mathbf{x} \sim N\left(\mu_{\beta_j|\boldsymbol{\beta}_{A(-j)}^{(\text{current})}, \sigma_\varepsilon^{2(t)}, \mathbf{y}, \mathbf{x}}^{(t+1)}, \sigma_{\beta_j|\boldsymbol{\beta}_{A(-j)}^{(\text{current})}, \sigma_\varepsilon^{2(t)}, \mathbf{y}, \mathbf{x}}^{2(t+1)} \right). \tag{6.23}$$

2. *Sample the error variance.* Compute the sums of squares using the just-drawn values for the augmented vector of regression coefficients,

$$SS(\mathbf{E}^{(t+1)}) = (\mathbf{y} - \mathbf{x}_A \boldsymbol{\beta}_A^{(t+1)})'(\mathbf{y} - \mathbf{x}_A \boldsymbol{\beta}_A^{(t+1)}), \tag{6.24}$$

and sample from (6.19) using this value for $SS(\mathbf{E})$,

$$\sigma_\varepsilon^{2(t+1)} \mid \boldsymbol{\beta}_A^{(t+1)}, \mathbf{y}, \mathbf{x} \sim \text{Inv-Gamma}\left(\frac{\nu_0 + n}{2}, \frac{\nu_0 \sigma_{\varepsilon 0}^2 + SS(\mathbf{E}^{(t+1)})}{2} \right). \tag{6.25}$$

6.6 Example: Regressing Test Scores on Previous Test Scores

6.6.1 Model Specification

To illustrate, we describe analyses based on an example originally reported by Levy and Crawford (2009). The current analyses are conducted to facilitate the exposition of key ideas and differ slightly from that original source. The context is a prototypical one for education research, and involves the use of regression to model the relationships among scores among three end-of-chapter tests in a course. Specifically, scores from the tests in Chapters 1 and 2 are used to predict scores from that in Chapter 3. Test scores were formed by summing the number of correctly answered items. Table 6.1 contains summary statistics obtained from 50 students.

The posterior distribution is

$$p(\beta_0,\boldsymbol{\beta},\sigma_\varepsilon^2 \mid \mathbf{y},\mathbf{x}) \propto p(\mathbf{y} \mid \beta_0,\boldsymbol{\beta},\sigma_\varepsilon^2,\mathbf{x})p(\beta_0,\boldsymbol{\beta},\sigma_\varepsilon^2)$$

$$= \prod_{i=1}^{n} p(y_i \mid \beta_0,\boldsymbol{\beta},\sigma_\varepsilon^2,\mathbf{x}_i)p(\beta_0)\prod_{j=1}^{J} p(\beta_j)p(\sigma_\varepsilon^2),$$

where

$$y_i \mid \beta_0,\beta_1,\beta_2,\mathbf{x}_i,\sigma_\varepsilon^2 \sim N(\beta_0+\beta_1 x_{i1}+\beta_2 x_{i2},\sigma_\varepsilon^2) \quad i=1,\ldots,50,$$

$$\beta_0 \sim N(0,1{,}000),$$

$$\beta_j \sim N(0,1{,}000) \quad j=1,2,$$

and

$$\sigma_\varepsilon^2 \sim \text{Inv-Gamma}(1,1).$$

In addition, we obtain the posterior distribution for the proportion of variance in y explained by the model (R^2). There are a number of potential ways to formulate versions of R^2 for Bayesian models. Following Gelman and Hill (2006), we work with a version that computes a residual for each person:

$$\varepsilon_i = y_i - (\beta_0 + \beta_1 x_{i1} + \beta_2 x_{i2}). \tag{6.26}$$

The proportion of variance explained is then constructed as

$$R^2 = 1 - \frac{\text{var}(\varepsilon)}{\text{var}(y)}, \tag{6.27}$$

where $\text{var}(y) = \frac{1}{n-1}\sum_{i=1}^{n}(y-\bar{y})^2$ is the usual finite-sample variance formula, and analogously $\text{var}(\varepsilon) = \frac{1}{n-1}\sum_{i=1}^{n}(\varepsilon_i-\bar{\varepsilon})^2$. The first of these is straightforward because all ys are known and we are simply calculating a measure of their dispersion. The latter, var(ε), has

TABLE 6.1

Summary Statistics for the Three End-of-Chapter Tests for n = 50 Subjects

	Chapter 1	Chapter 2	Chapter 3
Number of items	16	18	15
Mean	14.10	14.34	12.22
Standard deviation	2.02	3.29	2.96
	Chapter 1	**Chapter 2**	
Chapter 2	.58		
Chapter 3	.69	.68	

Note: The bottom half of the table gives the correlations between the test scores.

a straightforward interpretation from a Bayesian perspective but would seem an odd mix of "parameter" and "estimate" from a frequentist perspective. Once the full joint distribution of all the variables in the regression model has been constructed and the ys and xs have been observed, we have posterior distributions conditional on them for the βs and, as functions of variables in the model, for the εs, var(ε), and R^2 as well. If the βs were known with certainty, R^2 would characterize the uncertainty remaining about the ys for the realized data when we condition on the xs. The posterior distribution for R^2 additionally takes the uncertainty about the βs into account for this descriptor of the predictive value of the model for the data in hand. This contrasts with the frequentist calculation of a single estimate of R^2 using point estimates of the model parameters, and interest in its distribution in repeated samples of y.

6.6.2 Gibbs Sampling

To illustrate the process of Gibbs sampling, we briefly present the computations for several iterations of the Gibbs sampler defined in Section 6.5. We begin by specifying initial values $\beta_0^{(0)} = 3$, $\beta_1^{(0)} = 1$, $\beta_2^{(0)} = 0.5$, $\sigma_\varepsilon^{2(0)} = 5$. These are given in the first row in each section of Table 6.2. To enact the Gibbs sampler, we cycle through taking draws from the full conditional distributions using the current values for the other parameters:

1. *Sample the augmented regression coefficients.*
   a. In general, when sampling the $(t+1)$st value for β_0, $\boldsymbol{\beta}_{A(-j)}^{(\text{current})} = \boldsymbol{\beta}_{A(-0)}^{(\text{current})} = (\beta_1^{(t)}, \beta_2^{(t)})'$. When sampling a value for the first iteration $\beta_0^{(1)}$, $\boldsymbol{\beta}_{A(-j)}^{(\text{current})} = \boldsymbol{\beta}_{A(-0)}^{(\text{current})} = (\beta_1^{(0)}, \beta_2^{(0)})'$. We compute

$$\mu^{(1)}_{\beta_0|\boldsymbol{\beta}_{A(-0)}^{(\text{current})}, \sigma_\varepsilon^{2(0)}, \mathbf{y}, \mathbf{x}} = \left(\frac{1}{\sigma_{\beta_0}^2} + \frac{1}{\sigma_\varepsilon^{2(0)}}\mathbf{x}_0'\mathbf{x}_0\right)^{-1}\left(\frac{1}{\sigma_{\beta_0}^2}\mu_{\beta_0} + \frac{1}{\sigma_\varepsilon^{2(0)}}\mathbf{x}_0'(\mathbf{y} - \mathbf{x}_{A(-0)}\boldsymbol{\beta}_{A(-0)}^{(\text{current})})\right)$$

$$= \left(\frac{1}{\sigma_{\beta_0}^2} + \frac{1}{\sigma_\varepsilon^{2(0)}}\mathbf{x}_0'\mathbf{x}_0\right)^{-1}\left(\frac{1}{\sigma_{\beta_0}^2}\mu_{\beta_0} + \frac{1}{\sigma_\varepsilon^{2(0)}}\mathbf{x}_0'(\mathbf{y} - \mathbf{x}_{A(-0)}(\beta_1^{(0)}, \beta_2^{(0)})')\right)$$

$$= \left(\frac{1}{1{,}000} + \frac{1}{5}\mathbf{1}'\mathbf{1}\right)^{-1}\left(\frac{1}{1{,}000}0 + \frac{1}{5}\mathbf{1}'(\mathbf{y} - \mathbf{x}_{A(-0)}(1, 0.5)')\right) \approx -9.05$$

   and

$$\sigma^{2(1)}_{\beta_0|\boldsymbol{\beta}_{A(-0)}^{(\text{current})}, \sigma_\varepsilon^{2(0)}, \mathbf{y}, \mathbf{x}} = \left(\frac{1}{\sigma_{\beta_0}^2} + \frac{1}{\sigma_\varepsilon^{2(0)}}\mathbf{x}_0'\mathbf{x}_0\right)^{-1} = \left(\frac{1}{1{,}000} + \frac{1}{5}\mathbf{1}'\mathbf{1}\right)^{-1} \approx 0.10.$$

   We draw a value from the full conditional for β_0 from its full conditional using these values for the arguments $\beta_0^{(1)} | \beta_1^{(0)}, \beta_2^{(0)}, \sigma_\varepsilon^{2(0)}, \mathbf{y}, \mathbf{x} \sim N(-9.05, 0.10)$. The drawn value was −9.17, as reported in the second row of the appropriate section in Table 6.2.

   b. In general, when sampling the $(t+1)$st value for β_1, $\boldsymbol{\beta}_{A(-j)}^{(\text{current})} = \boldsymbol{\beta}_{A(-1)}^{(\text{current})} = (\beta_0^{(t+1)}, \beta_2^{(t)})'$. Note that we are using the value drawn in step (a) for β_0. For sampling $\beta_1^{(1)}$, $\boldsymbol{\beta}_{A(-j)}^{(\text{current})} = \boldsymbol{\beta}_{A(-1)}^{(\text{current})} = (\beta_0^{(1)}, \beta_2^{(0)})'$. We compute

TABLE 6.2

Computations in the Gibbs Sampler for the Example Regressing Chapter 3 Test Scores on Chapter 1 Test Scores and Chapter 2 Test Scores

Iteration		Ingredients of the Full Conditional		
	β_0	$\dfrac{\left(\dfrac{\mu_{\beta_0}}{\sigma^2_{\beta_0}} + \dfrac{\mathbf{1}'(\mathbf{y} - \mathbf{x}_{A(-0)}\boldsymbol{\beta}_{A(-0)})}{\sigma^2_\varepsilon}\right)}{\left(\dfrac{1}{\sigma^2_{\beta_0}} + \dfrac{1}{\sigma^2_\varepsilon}\mathbf{1}'\mathbf{1}\right)}$		$\dfrac{1}{\left(\dfrac{1}{\sigma^2_{\beta_0}} + \dfrac{1}{\sigma^2_\varepsilon}\mathbf{1}'\mathbf{1}\right)}$
0	3.00	–		–
1	−9.17	−9.05		0.10
2	−9.13	−8.90		0.07
3	−8.89	−8.69		0.09
	β_1	$\dfrac{\left(\dfrac{\mu_{\beta_1}}{\sigma^2_{\beta_1}} + \dfrac{\mathbf{x}_1'(\mathbf{y} - \mathbf{x}_{A(-1)}\boldsymbol{\beta}_{A(-1)})}{\sigma^2_\varepsilon}\right)}{\left(\dfrac{1}{\sigma^2_{\beta_1}} + \dfrac{1}{\sigma^2_\varepsilon}\mathbf{x}_1'\mathbf{x}_1\right)}$		$\dfrac{1}{\left(\dfrac{1}{\sigma^2_{\beta_1}} + \dfrac{1}{\sigma^2_\varepsilon}\mathbf{x}_1'\mathbf{x}_1\right)}$
0	1.00	–		–
1	0.98	1.00		0.0005
2	0.99	0.98		0.0004
3	0.97	1.00		0.0004
	β_2	$\dfrac{\left(\dfrac{\mu_{\beta_2}}{\sigma^2_{\beta_2}} + \dfrac{\mathbf{x}_2'(\mathbf{y} - \mathbf{x}_{A(-2)}\boldsymbol{\beta}_{A(-2)})}{\sigma^2_\varepsilon}\right)}{\left(\dfrac{1}{\sigma^2_{\beta_2}} + \dfrac{1}{\sigma^2_\varepsilon}\mathbf{x}_2'\mathbf{x}_2\right)}$		$\dfrac{1}{\left(\dfrac{1}{\sigma^2_{\beta_2}} + \dfrac{1}{\sigma^2_\varepsilon}\mathbf{x}_2'\mathbf{x}_2\right)}$
0	0.50	–		–
1	0.51	0.52		0.0005
2	0.48	0.50		0.0003
3	0.49	0.51		0.0004
	σ^2_ε	$\dfrac{\nu_0 + n}{2}$	$SS(E)$	$\dfrac{\nu_0\sigma^2_{\varepsilon 0} + SS(E)}{2}$
0	5.00	–	–	–
1	3.55	26.00	220.26	111.13
2	4.33	26.00	221.38	111.69
3	3.92	26.00	219.61	110.81

$$\mu^{(1)}_{\beta_1|\boldsymbol{\beta}^{(current)}_{A(-1)},\sigma^{2(0)}_\varepsilon,\mathbf{y},\mathbf{x}} = \left(\frac{1}{\sigma^2_{\beta_1}} + \frac{1}{\sigma^{2(0)}_\varepsilon}\mathbf{x}_1'\mathbf{x}_1\right)^{-1}\left(\frac{1}{\sigma^2_{\beta_1}}\mu_{\beta_1} + \frac{1}{\sigma^{2(0)}_\varepsilon}\mathbf{x}_1'(\mathbf{y} - \mathbf{x}_{A(-1)}\boldsymbol{\beta}^{(current)}_{A(-1)})\right)$$

$$= \left(\frac{1}{\sigma^2_{\beta_1}} + \frac{1}{\sigma^{2(0)}_\varepsilon}\mathbf{x}_1'\mathbf{x}_1\right)^{-1}\left(\frac{1}{\sigma^2_{\beta_1}}\mu_{\beta_1} + \frac{1}{\sigma^{2(0)}_\varepsilon}\mathbf{x}_1'(\mathbf{y} - \mathbf{x}_{A(-1)}(\beta^{(1)}_0, \beta^{(0)}_2)')\right)$$

$$= \left(\frac{1}{1{,}000} + \frac{1}{5}\mathbf{x}_1'\mathbf{x}_1\right)^{-1}\left(\frac{1}{1{,}000}0 + \frac{1}{5}\mathbf{x}_1'(\mathbf{y} - \mathbf{x}_{A(-1)}(-9.17, 0.5)')\right) \approx 1.00$$

and

$$\sigma^{2(1)}_{\beta_1|\beta^{(\text{current})}_{A(-1)},\sigma^{2(0)}_{\varepsilon},\mathbf{y},\mathbf{x}} = \left(\frac{1}{\sigma^2_{\beta_1}} + \frac{1}{\sigma^{2(0)}_{\varepsilon}}\mathbf{x}_1'\mathbf{x}_1\right)^{-1} = \left(\frac{1}{1,000} + \frac{1}{5}\mathbf{x}_1'\mathbf{x}_1\right)^{-1} \approx 0.0005.$$

We draw a value from the full conditional for β_1 from its full conditional using these values for the arguments $\beta_1^{(1)} \mid \beta_0^{(1)}, \beta_2^{(0)}, \sigma_\varepsilon^{2(0)}, \mathbf{y}, \mathbf{x} \sim N(1.00, 0.0005)$. The drawn value was .98, as reported in Table 6.2.

c. In general, when sampling the $(t+1)$st value for β_2, $\beta^{(\text{current})}_{A(-j)} = \beta^{(\text{current})}_{A(-2)} = (\beta_0^{(t+1)}, \beta_1^{(t+1)})'$. Note that we are using the value drawn in steps (a) and (b) for β_0 and β_1, respectively. For sampling the first value, $\beta^{(\text{current})}_{A(-j)} = \beta^{(\text{current})}_{A(-2)} = (\beta_0^{(1)}, \beta_1^{(1)})'$. We compute

$$\begin{aligned}\mu^{(1)}_{\beta_2|\beta^{(\text{current})}_{A(-1)},\sigma^{2(0)}_{\varepsilon},\mathbf{y},\mathbf{x}} &= \left(\frac{1}{\sigma^2_{\beta_2}} + \frac{1}{\sigma^{2(0)}_{\varepsilon}}\mathbf{x}_2'\mathbf{x}_2\right)^{-1}\left(\frac{1}{\sigma^2_{\beta_2}}\mu_{\beta_2} + \frac{1}{\sigma^{2(0)}_{\varepsilon}}\mathbf{x}_2'(\mathbf{y} - \mathbf{x}_{A(-2)}\beta^{(\text{current})}_{A(-2)})\right)\\ &= \left(\frac{1}{\sigma^2_{\beta_2}} + \frac{1}{\sigma^{2(0)}_{\varepsilon}}\mathbf{x}_2'\mathbf{x}_2\right)^{-1}\left(\frac{1}{\sigma^2_{\beta_2}}\mu_{\beta_2} + \frac{1}{\sigma^{2(0)}_{\varepsilon}}\mathbf{x}_2'(\mathbf{y} - \mathbf{x}_{A(-2)}(\beta_0^{(1)}, \beta_1^{(1)})')\right)\\ &= \left(\frac{1}{1,000} + \frac{1}{5}\mathbf{x}_2'\mathbf{x}_2\right)^{-1}\left(\frac{1}{1,000}0 + \frac{1}{5}\mathbf{x}_2'(\mathbf{y} - \mathbf{x}_{A(-2)}(-9.17, 0.98)')\right) \approx 0.52\end{aligned}$$

and

$$\sigma^{2(1)}_{\beta_2|\beta^{(\text{current})}_{A(-2)},\sigma^{2(0)}_{\varepsilon},\mathbf{y},\mathbf{x}} = \left(\frac{1}{\sigma^2_{\beta_2}} + \frac{1}{\sigma^{2(0)}_{\varepsilon}}\mathbf{x}_2'\mathbf{x}_2\right)^{-1} = \left(\frac{1}{1,000} + \frac{1}{5}\mathbf{x}_2'\mathbf{x}_2\right)^{-1} \approx 0.0005.$$

We draw a value from the full conditional for β_2 from its full conditional using these values for the arguments $\beta_2^{(1)} \mid \beta_0^{(1)}, \beta_1^{(1)}, \sigma_\varepsilon^{2(0)}, \mathbf{y}, \mathbf{x} \sim N(0.52, 0.0005)$. The drawn value was .51, as reported in Table 6.2.

2. *Sample the error variance.* Turning to σ^2_ε, we compute the first parameter for the full conditional as $\frac{\nu_0 + n}{2} = \frac{2+50}{2} = 26$, which will not vary over iterations. To compute the second parameter, we compute the sums of squares using the values for the augmented vector of regression coefficients drawn in step 1:

$$\begin{aligned}SS(\mathbf{E}^{(1)}) &= (\mathbf{y} - \mathbf{x}_A\beta_A^{(1)})'(\mathbf{y} - \mathbf{x}_A\beta_A^{(1)})\\ &= (\mathbf{y} - \mathbf{x}_A(\beta_0^1, \beta_1^1, \beta_2^1))'(\mathbf{y} - \mathbf{x}_A(\beta_0^1, \beta_1^1, \beta_2^1))\\ &= (\mathbf{y} - \mathbf{x}_A(-9.17, 0.98, 0.51))'(\mathbf{y} - \mathbf{x}_A(-9.17, 0.98, 0.51)) \approx 220.26,\end{aligned}$$

and then compute the second parameter

$$\frac{\nu_0\sigma^2_{\varepsilon 0} + SS(\mathbf{E}^{(1)})}{2} = \frac{(2)(1) + 220.26}{2} = 111.13.$$

Accordingly, we draw a value from the full conditional $\sigma_\varepsilon^{2(1)} \mid \beta_0^{(1)}, \beta_1^{(1)}, \beta_2^{(1)}, \mathbf{y}, \mathbf{x} \sim$ Inv-Gamma(26, 111.13). As reported in Table 6.2, the drawn value was 3.55.

Table 6.2 lists the results from the first three iterations of a Gibbs sampler (see Exercise 6.1).

6.6.3 MCMC Using WinBUGS

The posterior distribution was empirically approximated using MCMC by fitting the model in WinBUGS, which employs the precision parameterization for the normal distribution, using the following code.

```
-----------------------------------------------------------------------
########################################################################
# Model Syntax
########################################################################
model{

########################################################################
# Prior distributions
########################################################################
beta.0 ~ dnorm(0, .001)     # prior for the intercept
beta.1 ~ dnorm(0, .001)     # prior for coefficient 1
beta.2 ~ dnorm(0, .001)     # prior for coefficient 2
tau.e ~ dgamma(1, 1)        # prior for the error precision
sigma.e <- 1/sqrt(tau.e)    # standard deviation of the errors


########################################################################
# Conditional distribution of the data
# Via a regression model
########################################################################
for(i in 1:n){
  y.prime[i] <- beta.0 + beta.1*x1[i] + beta.2*x2[i]
  y[i] ~ dnorm(y.prime[i], tau.e)
}


########################################################################
# Calculate R-squared
########################################################################
  for(i in 1:n){
  error[i] <- y[i] - y.prime[i]
}

var.error <- sd(error[])*sd(error[])
var.y <- sd(y[])*sd(y[])

R.squared <- 1 - (var.error/var.y)
} # closes the model statement
-----------------------------------------------------------------------
```

Three chains were run from dispersed start values. Convergence diagnostics including the history plots and the potential scale reduction factor (Section 5.7.2) suggested convergence within just a few iterations. The first 1,000 iterations were discarded as burn-in and then 10,000 iterations from each chain were obtained and saved, yielding 30,000 iterations that served to empirically approximate the posterior distribution.

Figure 6.4 contains density plots representing the marginal posterior distributions for the intercept (β_0), slope for the Chapter 1 test score (β_1), slope for the Chapter 2 test score (β_2),

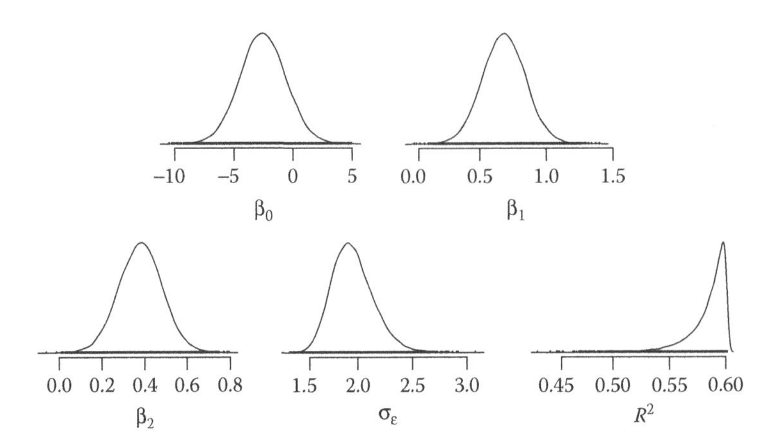

FIGURE 6.4
Marginal posterior densities for the parameters and R^2 from the example regressing tests scores on previous test scores.

standard deviation of the errors $\sigma_\varepsilon = \sqrt{\sigma_\varepsilon^2}$, and R^2. The marginal distributions are unimodal and fairly symmetric, with the distribution for σ_ε having a slight positive skew and that for R^2 having a negative skew. Accordingly, we report the posterior median as well as the posterior mean in Table 6.3, along with posterior standard deviations and 95% HPD intervals as summaries of the marginal posterior distributions for these parameters, as well as R^2.

Table 6.3 also contains the results from a frequentist (ML) solution to the model. Note the similarity in the values between the Bayesian and frequentist analysis. This illustrates the general point that, with diffuse priors, the posterior distribution strongly resembles the likelihood function and the results from the two approaches will be similar. We stress that these results are *numerically similar*, but *conceptually different*. The Bayesian analysis yields direct summary and probabilistic statements about the parameters themselves, which represent our uncertain beliefs about their values. For β_1, a posterior mean of 0.66 and posterior standard deviation of 0.17 are descriptions of the distribution for the parameter, not a parameter estimator as in frequentist analyses. Similarly, the credibility interval is

TABLE 6.3

Summary of the Results from Bayesian and Frequentist Analyses of the Example Regressing the Chapter 3 Test Scores on the Chapter 1 Test Scores and the Chapter 2 Test Scores

	Bayesian Analysis				Frequentist Analysis		
	Posterior Mean	Posterior Median	Posterior SD[a]	95% HPD[b] Int.	Est.	SE[c]	95% Conf. Int.
β_0	−2.53	−2.54	1.94	(−6.43, 1.15)	−2.54	1.93	(−6.41, 1.34)
β_1	0.66	0.66	0.17	(0.34, 0.99)	0.66	0.17	(0.33, 0.99)
β_2	0.38	0.38	0.10	(0.17, 0.57)	0.38	0.10	(0.18, 0.59)
σ_ε	1.91	1.90	0.20	(1.54, 2.31)	1.95	0.28	(1.60, 2.37)
R^2	0.58	0.59	0.02	(0.55, 0.60)	0.60		

Source: The frequentist results are from Levy, R., & Crawford, A. V. (2009). Incorporating substantive knowledge into regression via a Bayesian approach to modeling. *Multiple Linear Regression Viewpoints, 35*, 4–9. Reproduced here with permission of the publisher.

[a] SD = Standard Deviation.

[b] HPD = Highest Posterior Density.

[c] SE = Standard Error.

interpreted as a probabilistic statement about the parameter; there is a .95 probability that β_1 is between 0.34 and 0.99, according to this model. As noted above, in contrast to frequentist approaches, a Bayesian approach yields a distribution for R^2, which captures our uncertainty regarding the predictive power of the model.

6.7 Summary and Bibliographic Note

This chapter has briefly described Bayesian approaches to linear regression models rooted in normality assumptions. Our treatment has been cursory, aiming for an account sufficient to support the development of models covered later. Classic accounts can be found in Box and Tiao (1973), Zellner (1971), Lindley and Smith (1972), Smith (1973), and Goldstein (1976). More recent accounts far more comprehensive than that given here can be found in any of a number of texts on Bayesian modeling, including those cited in the introduction to Part I.

Regression models play a key role in classical test theory and factor analysis models (Chapters 8 and 9). A number of features of the model developed here associated with linear modeling and normal distributions will re-appear in those contexts. These include the use of conditionally conjugate normal prior distributions for coefficients and intercepts, and inverse-gamma prior distributions for error variances. Furthermore, several of the more abstract features of the model described here will appear throughout essentially all of the psychometric models. These include key assumptions of exchangeability and conditional independence of outcome variables conditional on predictors.

Jackson, Novick, and Thayer (1971) and Novick et al. (1972) discussed an extension to the regression models presented here that will be important for our developments later. Their work considered the case of fitting regression models simultaneously in multiple groups, and showed that a hierarchical approach to constructing the prior distribution for the regression parameters based on assumptions of exchangeability improved estimation, particularly in small sample sizes, by facilitating the borrowing of information across groups. This can be cast as an instance of partial pooling as in multilevel modeling (Gelman & Hill, 2006). We take up such hierarchical constructions in more detail in the second part of the book.

Exercises

6.1 Reconsider the multiple regression model example introduced in Section 6.6. The computations involved in the first iteration of a Gibbs sampler were given in Section 6.6.2. The results for two additional iterations are given in Table 6.2. Show the computations for each of these additional iterations.

6.2 Based on the results reported in Table 6.3, write a brief summary of the results of fitting the model using a Bayesian approach, including interpretations for each parameter.

6.3 The posterior implies that β_0 is most likely negative. Calculate the posterior probability of this, $p(\beta_0 < 0 \mid \mathbf{y}, \mathbf{x})$, by running a modified model in WinBUGS. This can be accomplished by adding the following lines to the WinBUGS code in Section 6.6.3.

```
--------------------------------------------------------------------
is.beta.0.greater.than.0 <- max(beta.0, 0)
probability.beta.0.less.than.0 <- equals(is.beta.0.greater.than.0, 0)
--------------------------------------------------------------------
```

The posterior probability that $p(\beta_0 < 0 \mid \mathbf{y}, \mathbf{x})$ is then given by the posterior mean of `probability.beta.0.less.than.0`.

6.4 Given the interpretation of a regression intercept in Exercise 6.2, is it sensible to think that β_0 could be negative? If not, why does this occur? What could be done to prevent this from happening, that is, how could the model be modified to build in the restriction that β_0 should not be negative?

6.5 Fit a modified model based on Exercise 6.4 in WinBUGS, monitoring convergence, and summarize the results. How do they compare to those for the original model reported in Table 6.3?

Section II

Psychometrics

Having reviewed foundational elements of Bayesian statistical modeling, we now turn our attention to psychometrics. Chapter 7 gives an overview of a Bayesian perspective on psychometric modeling, presenting core elements in a fairly abstract manner. This is then instantiated and extended in the context of several psychometric modeling paradigms: classical test theory in Chapter 8, factor analysis in Chapter 9, item response theory in Chapter 11, latent class analysis in Chapter 13, and Bayesian networks in Chapter 14. In each of these chapters, we present a conventional treatment of the modeling family in question. Such treatments are necessarily incomplete accounts of the conventional approaches, but serve two purposes: (1) for readers familiar with the model, they are a refresher; for readers new to the model, they are an overview sufficient to support our second purpose, namely that (2) they may be leveraged as launching points for Bayesian conceptualizations and developments. The treatments in these chapters operate within the traditional boundaries of these psychometric modeling paradigms as they have historically developed. Although these boundaries are useful as a didactic organizational tool, they are somewhat artificial. The possibilities for building models that cross these boundaries, and the advantages of adopting a Bayesian approach when doing so, are discussed in Chapter 15.

Chapters 10 and 12 cover model evaluation and missing data modeling. These are of course important topics in the broader statistical literature. Our treatments of them are situated in the contexts of the psychometric models that have been introduced at each point, but they apply generally. In Chapter 10, for example, the factor analysis models of Chapter 9 provide a first deeply substantive context to illustrate model evaluation techniques as they apply to particularly psychometric concerns. Similarly, in each of the chapters devoted to a modeling family, we discuss applications or ideas in the context of that family. Examples include small-variance priors for added flexibility in the context of factor analysis, adaptive testing in the context of item response theory, and models for learning over time in the context of Bayesian networks. Although our presentations occur in the context of a particular model, they are more general and apply with little difference to many of the psychometric models discussed in the rest of the book. These applications were selected in part because of their strong alignment with Bayesian thinking. Other applications and areas of psychometric modeling that benefit from a Bayesian approach are mentioned in Chapter 15.

One purpose of covering multiple psychometric modeling paradigms is to highlight similarities due to a shared structure. Our goal is to provide a coherent Bayesian account that is widely applicable and transcends the usual boundaries between these modeling families, as a counterpart to texts that achieve these ends from a predominantly frequentist perspective (e.g., Bartholomew et al., 2011; Skrondal & Rabe-Hesketh, 2004). This aids both in expanding the possibilities for conceiving of existing models and in service of the construction of new models. This emphasis on breadth comes at a price of depth. Each chapter is intended to provide an introductory account of Bayesian approaches to the models and their applications, and concludes with suggested additional readings.

7

Canonical Bayesian Psychometric Modeling

We now pivot from foundational issues that comprised the first half of the book to focus on psychometrics. We begin by introducing what we refer to as the canonical psychometric model, an abstraction of the popular psychometric models that are the foci of the second half of the book. We develop Bayesian approaches to common psychometric activities in the context of this model. We consider these approaches to be canonical as well, and they will shape much of the rest of this book. After introducing them, we briefly discuss the extent to which these approaches are prevalent in operational assessment. We conclude with a summary discussion that looks ahead to the rest of the book. The treatment is abstract throughout the chapter, and may be seen as the platform for instances and extensions of these core ideas that make up most of the balance of the book. To facilitate the developments in the chapter, we first introduce a distinction among different kinds of DAGs.

7.1 Three Kinds of DAGs

In our development of Bayesian approaches to psychometrics, we will find it useful to distinguish among three kinds of DAGs that differ in their level of granularity. Regression models introduced in the last chapter are a convenient means of illuminating the distinctions. Figure 7.1 contains three DAGs for a regression model with one predictor. The first (panel a) is the simplest, and expresses that there is a variable y that depends on another variable x. Here, only entities associated with persons (examinees) are included in the DAG. Accordingly, subscripting by i and an associated plate is implicit and need not be shown. This *just-persons* DAG is useful for thinking about what is going on vis-à-vis a single person. It also connects strongly with path diagrammatic representations from path analysis and structural equation modeling traditions, which we treat in more depth in Chapter 9.

In the just-persons DAG, the parameters β_1, β_0, and σ^2_ε are implicit, in a sense lurking behind the directed edge from x to y. Panel (b) contains an *all-unknowns* DAG that may be seen as bringing out these regression parameters, making their role in governing the distribution of y explicit. The subscripting by i and associated plate over persons is now needed to communicate that certain entities are hypothesized to vary for different persons. This second DAG connects strongly with fully Bayesian modeling in that it contains all the entities that are treated as random in our fully Bayesian model, and expresses how their joint distribution is structured in terms of parent–child relationships.

The third DAG (panel c) augments the second by including the hyperparameters that are used in specifying the prior distributions for the unknowns. This DAG is the most transparent in that it includes the greatest amount of information involved in model specification. As noted in Chapter 2, this representation is useful for triangulating among graphical representations of models, mathematical expressions of models, and software code or commands.

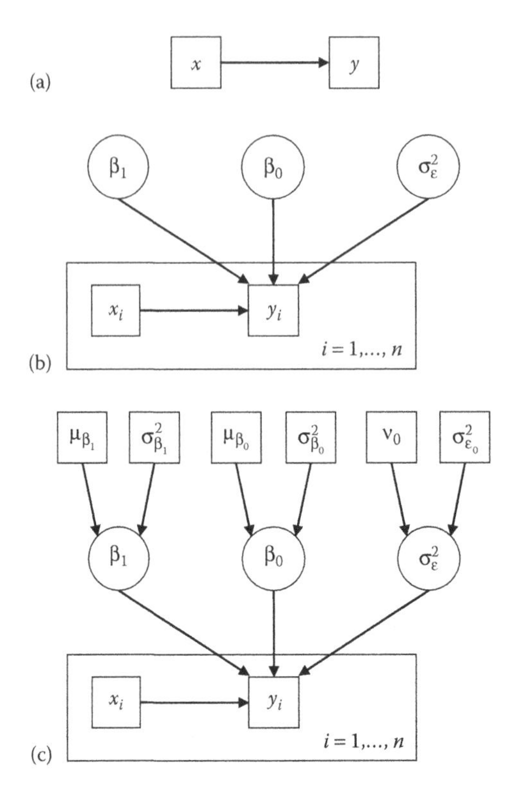

FIGURE 7.1
Three kinds of directed acyclic graphs for a regression model: (a) just-persons, (b) all-unknowns, and (c) with hyperparameters.

7.2 Canonical Psychometric Model

The psychometric or measurement model is the junction between the observable variables and the latent variables of inferential interest. A large number of psychometric models share a common strategy for the structure of this junction, varying principally in terms of the assumed continuous or discrete natures of the variables (see Table 1.3). We refer to this as the *canonical psychometric model*, and depict it in the just-persons DAG in Figure 7.2, where θ denotes a latent student model variable and $x_1, \ldots, x_J$ are J observables.*

We can recognize that, remarkably, this structure is the same as that in Figure 3.5, which expresses de Finetti's representation theorem in DAG form. Accordingly, the model for the

* In the psychometric model, "observables" x are the connection between the latent-variable space and persons' real-world actions. Recall that the nature of observable variables and how they come to take their values differ substantially across types of measuring instruments (e.g., multiple-choice tests, opinion-survey responses, evaluations of dance performances). The observables in applications of classical test theory are often test scores, while the observables in item response theory could be item scores on the same test. There can be several stages in evaluating complex responses, as when essays are scored by computers and distinct processes of lexical, syntactic, and semantic analyses are carried out and combined to approximate human raters. Although these considerations are integral to the *meaning* of the θs in any application of a latent variable model and essential to the design and interpretation of any assessment, they are not focus of the present volume. The reader is referred to Mislevy et al. (2003) for further discussion of these issues.

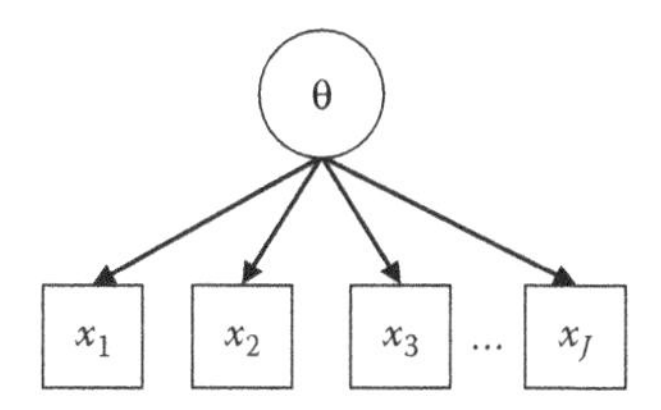

FIGURE 7.2
The just-persons directed acyclic graph for the canonical psychometric model.

observables may be seen as arising from an assumption that they are exchangeable and may be modeled as conditionally independent given the latent student model variable

$$p(\mathbf{x}) = p(x_1, \ldots, x_J) = \int_{\theta} \prod_{j=1}^{J} p(x_j \mid \theta) p(\theta) d\theta. \tag{7.1}$$

This conditional independence assumption is also referred to as *local independence* in the psychometric literature.* Many psychometric models may be seen as variations on this theme. Accordingly, time and again we will see our models take the following generic form implied by the representation theorem; letting $\mathbf{x}_i = (x_{i1}, \ldots, x_{iJ})$ be the full collection of observables for examinee i, the model for examinee i is

$$p(\mathbf{x}_i) = p(x_{i1}, \ldots, x_{iJ}) = \int_{\theta_i} \prod_{j=1}^{J} p(x_{ij} \mid \theta_i) p(\theta_i) d\theta_i. \tag{7.2}$$

Recall that the representation theorem says that a marginalized conditional independence structure exists to express beliefs about an exchangeable set of variables, but does not specify the forms of either the conditional distribution or the mixing distribution. In practical work, these must come from the analyst's beliefs about the substance of the problem. Based on the form of the data and theory about the nature of proficiency, the various psychometric models differ in terms of the chosen forms of the distributions and variables (e.g., continuous or discrete xs, continuous or discrete θs, the latter of which implies replacing the integral in (7.2) with a summation) and possibly additional parameters needed to warrant exchangeability and the conditional independence structures.

The importance of structuring psychometric models via conditional independence relationships has long been recognized for its computational benefits. We briefly rehash some of the discussion from Section 3.5 to additionally stress its psychological role in psychometric models. As Pearl (1988) noted, conditional independence relationships are so powerful for reasoning that we not only seek out variables that induce them, but we also invent these variables if they do not exist. This in line with an epistemic lens on probability models, in which the formal entities are abstractions defined in a model space to facilitate the use of the machinery of the model, namely probability calculus. De Finetti's representation theorem does not imply that θ necessarily exists in an ontological sense, out there in the world to be discovered. Rather, it is introduced as a tool in the model space to organize our thinking about the world.

The implication is that using a psychometric model does not require assuming that θ exists out there in the world. It is constructed or defined in the model space to be that

* In certain Bayesian and graphical modeling traditions the term *local independence* is used in a different sense than that used here (Spiegelhalter & Lauritzen, 1990).

which renders the observables conditionally independent, or actually, as close to conditionally independent as it needs to be, given additional features of the model. It is a piece of machinery we build to help us reason about complicated relations and patterns among what we can observe in the world. The role of θ as a modeling device subject to the beliefs and purposes of the analyst may be best seen in the common psychometric question regarding the number of latent variables used to characterize examinees. To fully render the observables as conditionally independent, we might need as many latent variables as there are observables. Unfortunately this does not offer much in the way of a simplification for model-based reasoning; we need a strategy for arriving at fewer latent variables in the model. In commenting on research aimed at determining the "true" number of latent variables, Reckase (2009, p. 181) articulated that the answer to the question about the number of latent variables (or dimensions in his terminology) to include comes not only from the world but also from the analyst's thinking and purposes at hand:

> The position taken here is that a true number of dimensions does not exist, but that a sufficient number of dimensions is needed to accurately represent the major relationships in the... data. If the fine details of the relationships in the data are of interest, more dimensions are needed to accurately show those details.

An analyst chooses the number of latent variables based on the desired purposes in concert with the dependencies in the data. Our goal in building a psychometric is not to be correct, but to be useful for the purposes at hand. We want the observables to be as close to conditionally independent as will balance the technical advantages of conditional independence, the practicality of implementation and understanding, and the modeling (or circumvention) of conditional dependence that would cause unacceptable errors in the inferences we make through the model.

In summary, when conducting model-based reasoning in educational assessment, we introduce latent variables as modeling devices. Latent variables lurk in the realm of the model space that we as analysts build, and are introduced to organize our thinking about examinee capabilities, proficiencies, and attributes. This reflects a *constructive* nature of modeling, in that models are based on certain beliefs and directed towards achieving certain purposes. The beliefs that are being modeled are better thought of as epistemological entities, rather than literal or ontological entities. The use of a conditional independence structure is not an ontological statement about the world, but rather an epistemological statement about our thinking about the world as it pertains to our beliefs and reasoning about the situation at hand. Though a latent variable is often interpreted as representing examinee capabilities, it arguably has more to do with what is going on in *our* heads than the examinees' heads. Our beliefs are informed by what we actually do know about the situation from theory and experience, but also tailored to the purposes we have in mind. This makes us more comfortable with using models of different grain sizes or different qualitative structures, when they highlight different aspects of situations that might be differentially relevant for different problems. To be sure, the fidelity of our thinking to the real-world situation is critical. We may of course decide that a simpler model is insufficient for desired inferences or find that it simply does not do sufficient justice to the data. Modeling involves balancing tradeoffs among fidelity to the real-world situation we aim to reason about, practicality of use and interpretability, computational ease, and consequences of inferential errors that may result from the model.

7.2.1 Running Example: Subtraction Proficiency

We have already seen the basis for a psychometric model in Chapter 1 when we sketched a situation where we had

- a dichotomous latent variable denoted θ corresponding to subtraction proficiency that can take on a value of Proficient or Not Proficient;
- a dichotomous observable variable x corresponding to the response of an examinee to an item that can take on a value of Correct or Incorrect;
- a probability structure modeling the dependence of the observable on the latent variable.*

Suppose we have multiple observable variables derived from responses to several subtraction items. Specifically, in this chapter we will employ a running example where there are two such subtraction items, yielding two observables corresponding to scored responses to those items. Treating the observables (items) as exchangeable, we may represent the joint distribution of the observables for any examinee via (7.2), which motivates the introduction of θ to render the observables conditionally independent. Its use in the model does not correspond to an ontological claim regarding any such variable. Rather, the specification of θ here is a choice made to frame and organize our thinking as tailored to our assessment purposes and desired inferences, informed by the fit of the proposed model to the data. Other choices could be made, reflecting interest in more and finer-grained distinctions regarding subtraction proficiency, a point we return to in an example in Chapter 14.

7.3 Bayesian Analysis

7.3.1 Bayesian Analysis for Examinees

Note that the edges in the DAG for the canonical psychometric model in Figure 7.2 are directed with a source at θ and destinations at the xs. This direction naturally facilitates the deductive reasoning, which flows from the latent variable (θ) thought of as the general characterization, to the observable variables (xs) that represent the particulars. This is expressed notationally in the conditioning of the latter on the former on the right-hand side of (7.1). The model expressed in the conditional distribution $p(x_j \mid \theta)$ explicitly articulates what we think or what we expect for an observable given the value of the latent variable.

In assessment we need to reason inductively—from the particular examinee behaviors we observe back to the general conception of their capabilities. To accomplish this we need to reverse the direction of the flow of the model. The deductive structures $p(x_j \mid \theta)$ express beliefs about behavior given proficiency, so we now need to reason "back through" this structure to arrive at beliefs about proficiency given behavior. Bayes' theorem engineers this reversal, yielding the posterior distribution that synthesizes the information in the data carried through $p(x_j \mid \theta)$ with the prior information expressed

* We had a similar example in the context of the medical diagnosis model in Chapter 2, with x being the result from a mammography screener and θ being cancer status.

in $p(\theta)$, enabling probability-based reasoning about the unknown θ. For examinee i, the posterior distribution for the latent variable given values for the J observables is

$$p(\theta_i \mid \mathbf{x}_i) = \frac{\prod_{j=1}^{J} p(x_{ij} \mid \theta_i) p(\theta_i)}{\int_{\theta_i} \prod_{j=1}^{J} p(x_{ij} \mid \theta_i) p(\theta_i) d\theta_i} \propto \prod_{j=1}^{J} p(x_{ij} \mid \theta_i) p(\theta_i). \tag{7.3}$$

where the integral for the normalizing constant in the denominator is replaced by a summation in the case of a discrete latent variable. The posterior summarizes our beliefs about the examinee after having observed her performance.

To complete the model specification in (7.3), we must specify $p(\theta_i)$. The form of the distribution is in part determined by the proposed nature of the latent variable. As we will see, a normal distribution is commonly specified when the latent variable is continuous, and Bernoulli or categorical distributions are commonly specified when the latent variable is discrete. For our current development, we let $\boldsymbol{\theta}_P$ generically denote the parameters of the prior distribution that is specified.* As examples, if θ_i is modeled as a normal random variable, $\boldsymbol{\theta}_P$ contains the mean and variance of the normal distribution; if θ_i is modeled as a Bernoulli random variable, $\boldsymbol{\theta}_P$ contains the probability that θ_i takes on one of its possible values.

To extend inference to the case of multiple examinees, an assertion of exchangeability implies the use of a common measurement model and prior distribution for all examinees. A hierarchical model results. Letting $\boldsymbol{\theta} = (\theta_i, \ldots, \theta_n)$ be the full collection of latent variables from n examinees, the joint posterior of all the examinees' latent variables is

$$p(\boldsymbol{\theta} \mid \mathbf{x}) \propto \prod_{i=1}^{n} \prod_{j=1}^{J} p(x_{ij} \mid \theta_i) p(\theta_i). \tag{7.4}$$

7.3.1.1 Running Example: Subtraction Proficiency

Returning to the subtraction proficiency example, θ_i is a Bernoulli random variable and suppose $p(\theta_i = \text{Proficient}) = .6$, indicating that our prior belief is that the examinee is most likely Proficient, but there is considerable uncertainty about that. Suppose we have two items, with conditional probabilities of response given in Table 7.1. It can be seen that

TABLE 7.1

Conditional Probabilities for the Responses to Two Items

	Response to Item 1		Response to Item 2	
Proficiency	**Correct**	**Incorrect**	**Correct**	**Incorrect**
Proficient	.70	.30	.65	.35
Not Proficient	.20	.80	.15	.85

* Recall that the use of the subscript "P" in "$\boldsymbol{\theta}_P$" is adopted to signal that these are the parameters that govern the *prior* distribution for θ. Similarly, the subscript indicates that, in a DAG representation, the elements of $\boldsymbol{\theta}_P$ are the *parents* of θ.

item 2 is slightly harder than item 1 by recognizing that, for each level of proficiency, the (conditional) probability of a correct response is lower for item 2 than item 1.

Now suppose we observe an examinee correctly answer the first item, and incorrectly answer the second item. The posterior probability for the examinee's latent variable taking on a value of c (in the current example: Proficient or Not Proficient) is given by

$$p(\theta_i = c \mid \mathbf{x}_i) = \frac{\prod_{j=1}^{J} p(x_{ij} \mid \theta_i = c) p(\theta_i = c)}{\sum_{g} \prod_{j=1}^{J} p(x_{ij} \mid \theta_i = g) p(\theta_i = g)}.$$

Working through the computations, the posterior probabilities are

$$p(\theta_i = \text{Proficient} \mid x_{i1} = \text{Correct}, x_{i2} = \text{Incorrect})$$

$$= \frac{p(x_{i1} = \text{Correct} \mid \theta_i = \text{Proficient})\ p(x_{i2} = \text{Incorrect} \mid \theta_i = \text{Proficient})\ p(\theta_i = \text{Proficient})}{\sum_{g=\text{Proficient,Not Proficient}} p(x_{i1} = \text{Correct} \mid \theta_i = g)\ p(x_{i2} = \text{Incorrect} \mid \theta_i = g)\ p(\theta_i = g)}$$

$$= \frac{(.70)(.35)(.60)}{(.70)(.35)(.60) + (.20)(.85)(.40)} = \frac{.147}{.147 + .068} \approx .68$$

and

$$p(\theta_i = \text{Not Proficient} \mid X_{i1} = \text{Correct}, X_{i2} = \text{Incorrect})$$

$$= \frac{p(x_{i1} = \text{Correct} \mid \theta_i = \text{Not Proficient})\, p(x_{i2} = \text{Incorrect} \mid \theta_i = \text{Not Proficient})\, p(\theta_i = \text{Not Proficient})}{\sum_{g=\text{Proficient,Not Proficient}} p(x_{i1} = \text{Correct} \mid \theta_i = g)\ p(x_{i2} = \text{Incorrect} \mid \theta_i = g)\ p(\theta_i = g)}$$

$$= \frac{(.20)(.85)(.40)}{(.70)(.35)(.60) + (.20)(.85)(.40)} = \frac{.068}{.147 + .068} \approx .32.$$

On the basis of the observed data, the posterior probability that the examinee is proficient is .68, which is slightly higher than the prior probability (.6). This indicates that the evidentiary bearing of the observed data serves to slightly increase our belief that the examinee is proficient (see Exercise 7.1). Note that the posterior probability that the examinee is not proficient is .32 = 1 – .68, which reflects the general relationship ensured by Bayes' theorem that our posterior belief is still expressed as a probability distribution; in this case, implying that $p(\theta_i = \text{Not Proficient} \mid X_{i1} = \text{Correct}, X_{i2} = \text{Incorrect}) = 1 - p(\theta_i = \text{Proficient} \mid x_{i1} = \text{Correct}, x_{i2} = \text{Incorrect})$.

7.3.2 Probability-Model-Based Reasoning in Assessment

To summarize, exchangeability, de Finetti's representation theorem, and the simplifications afforded by conditional independence support the construction of a psychometric model with a particular flow, namely from the latent variable to the observables. The latent variable need not exist in an ontological sense; it is an expression of our thinking, the result of articulating exchangeability assumptions, conditional independence relationships, and substantive beliefs. We then reverse the directional flow of the model, via Bayes' theorem, effecting probability-based reasoning from the observables to the latent variable.

We can now address, at an abstract level, the lingering question from Chapter 1 on how we can represent and construct an argument to bridge the divide between what is available, namely observations of an examinee's behaviors, and what is of inferential interest, namely the examinee's capabilities more broadly conceived. We engage in model-based reasoning, in particular using a probability model to represent the situation and probability calculus to reason within the model. This approach was advocated by Mislevy (1994), who summarized why such an enterprise is appealing (pp. 447–448)

> When it is possible to map the salient elements of an inferential problem into the probability framework, powerful tools become available to combine explicitly the evidence that various probans [facts offering proof] convey about probanda [facts to be proved], as to both weight and direction of probative force.... A properly-structured statistical model embodies the salient qualitative patterns in the application at hand, and spells out, within that framework, the relationship between conjectures and evidence. It overlays a substantive model for the situation with a model for our knowledge of the situation, so that we may characterize and communicate what we come to believe—as to both content and conviction—and why we believe it—as to our assumptions, our conjectures, our evidence, and the structure of our reasoning.

Operating in a probability model, Bayes' theorem is the key tool in probability calculus for reasoning from examinee performances, captured by observable variables, to their capabilities more broadly conceived, as captured by the latent variable.

7.3.3 Enter Measurement Model Parameters

The $p(x_{ij} \mid \theta_i)$ terms play a critical role in reasoning via (7.3) and (7.4). When we employ multiple observables (i.e., evaluated features of task performances) in assessment, we usually want to allow for them to vary in terms of their evidentiary relevance; that is, we wish to allow $p(x_{ij} \mid \theta_i)$ to vary over j. Many psychometric models commonly express the conditional distributions of the observables by making use of parameters (e.g., loadings in factor analysis, difficulty parameters in item response theory, slip and guessing parameters in diagnostic classification models; cf. nonparametric models, Ramsay, 1991; Sijtsma & Molenaar, 2002). We refer to these as *measurement model parameters*, as they govern the connection between the latent and observable variables within a posited functional form. We denote the measurement model parameters for observable j as ω_j. The DAG in Figure 7.3 expands on the earlier just-persons DAG and reflects that the conditional distribution for each observables is now expressed as $p(x_{ij} \mid \theta_i, \omega_j)$.

One way to proceed that does not involve much in the way of new activities is to specify the values for the measurement model parameters, say, on the basis of subject matter experts' beliefs about them. Such an approach may be sufficient in low-stakes assessment applications. However, this often will not suffice in higher stakes and research settings where we seek to learn about the measurement model parameters from data. To proceed in this setting, we treat the measurement model parameters as unknowns in our model. Letting $\omega = (\omega_1, \ldots, \omega_J)$ denote the full collection of measurement model parameters and assuming a priori independence between the latent variables and measurement model parameters, the posterior distribution is

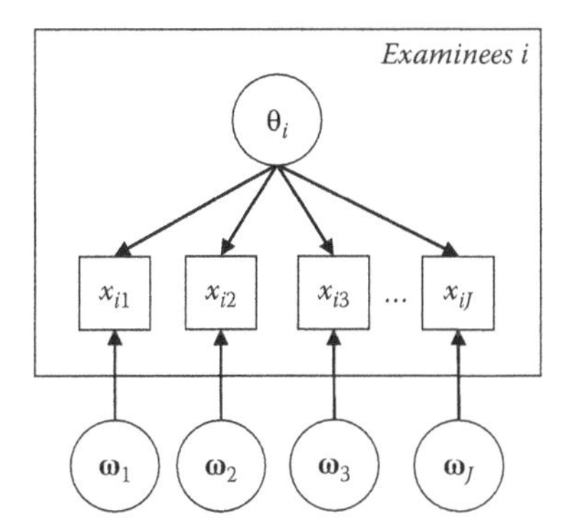

FIGURE 7.3
Directed acyclic graph for the canonical psychometric model including the measurement model parameters.

$$p(\theta, \omega \mid \mathbf{x}) \propto \prod_{i=1}^{n} \prod_{j=1}^{J} p(x_{ij} \mid \theta_i, \omega_j) p(\theta_i) p(\omega_j). \tag{7.5}$$

To specify $p(\omega_j)$, an assumption of exchangeability with regard to the tasks and the resulting observables supports the use of a common prior for the measurement model parameters for different observables. Let ω_P denote the parameters of the prior distribution that is specified for the ω_js. Again we will see that particular choices for the forms of the distribution will vary based on the nature of the psychometric model. Further, it is possible when substantive considerations merit to fall back to conditional exchangeability—for example, among the difficulty parameters of test items that share the same key features (Rijmen & de Boeck, 2002).

7.3.3.1 Running Example: Subtraction Proficiency

In the subtraction proficiency example, the observables had different conditional distributions, reflecting their difference in difficulty and relationship to the proficiency of inferential interest. This can be seen by defining the conditional distribution for each observable in terms of two measurement model parameters $\omega_j = (\pi_{\text{Proficient},j}, \pi_{\text{Not Proficient},j})$, where $\pi_{cj} = p(x_{ij} = \text{Correct} \mid \theta_i = c)$. In the previous analyses, these were treated as known with $\omega_1 = (\pi_{\text{Proficient},1}, \pi_{\text{Not Proficient},1}) = (.70, .20)$ and $\omega_2 = (\pi_{\text{Proficient},2}, \pi_{\text{Not Proficient},2}) = (.65, .15)$. When modeling these as unknown, they require prior distributions. As the elements of the ω_js are probabilities, a beta prior distribution is a natural choice (see Chapters 2 and 13).

7.3.4 Full Bayesian Model

Figure 7.4 contains a now-expanded DAG, using a plate over observables to more compactly represent the model. If θ_P and ω_P are unknown, they too require prior distributional specifications. Assuming prior independence, we denote the priors for θ_P and ω_P by $p(\theta_P)$ and $p(\omega_P)$, respectively. Accordingly, our posterior will now include θ_P and ω_P as well. Again, the particular forms used will depend on the nature of the psychometric model. The full Bayesian model is then

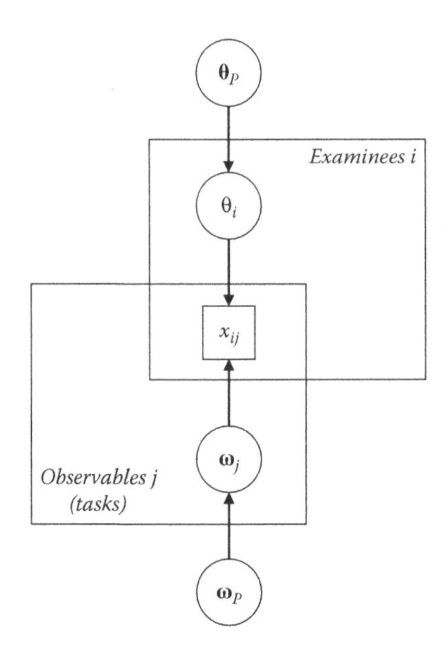

FIGURE 7.4
The all-unknowns directed acyclic graph for the canonical psychometric model including the hyperparameters.

$$p(\boldsymbol{\theta},\boldsymbol{\omega},\boldsymbol{\theta}_P,\boldsymbol{\omega}_P \mid \mathbf{x}) \propto \prod_{i=1}^{n}\prod_{j=1}^{J} p(x_{ij} \mid \boldsymbol{\theta}_i,\boldsymbol{\omega}_j)p(\boldsymbol{\theta}_i \mid \boldsymbol{\theta}_P)p(\boldsymbol{\omega}_j \mid \boldsymbol{\omega}_P)p(\boldsymbol{\theta}_P)p(\boldsymbol{\omega}_P). \tag{7.6}$$

7.4 Bayesian Methods and Conventional Psychometric Modeling

Although Bayesian approaches are not the dominant paradigm in psychometric modeling, they have become widely used in certain applications, including some where there is apparent unanimous consent as to their propriety and advantages. Further, the recognition of the power of MCMC for obtaining posterior distributions has led to an explosion in applications of Bayesian methods in psychometrics (Levy, 2009), a trend that is likely to continue as computing resources become more accessible. Over the next few chapters, we will discuss a number of both entrenched and more novel applications of Bayesian methods. In places where Bayesian approaches depart from conventional methods, we aim to highlight the relative strengths and weaknesses of adopting a Bayesian approach.

As a first example, the Bayesian methods described in the previous section are by no means ubiquitous in psychometrics. In operational assessment, a distinction is commonly drawn between two processes, often called scoring and calibration. *Scoring* refers to arriving at a representation for an examinee (in terms of $\boldsymbol{\theta}_i$) based on her performance on an assessment. In ECD terms, it is the result of working through the processes of: evidence identification, in which we extract key features of examinee work products and encode the results as observables (i.e., task scoring); and evidence accumulation, in which the information in the observables can be synthesized to yield updated values for the examinee's latent student model variables via the psychometric model (i.e., test scoring). Scoring is also commonly referred to as *estimating the latent variables* or *person parameters*, the latter being a term for the θs popular in item response theory traditions, or *factor scores*, a term popular in factor analysis traditions.

The second process, *calibration*, refers to arriving at a representation for the (other) parameters of the model: principally, the measurement model parameters (the ω_js), and possibly the hyperparameters (θ_P or ω_P) if they are unknown. In the Bayesian analysis described in Section 7.3.4, the representations come in the form of a posterior distribution. This could of course be used to construct other representations, such as using a measure of central tendency as a point representation. In frequentist analyses, the representations are point estimates, say, from maximum likelihood (ML) estimation. Calibration is also commonly referred to as *estimating the model parameters*, particularly in traditions that do not refer to θ as a person parameter. To avoid possible confusion over whether "estimating the model parameters" includes the θs, we avoid this terminology and refer to the distinct processes of scoring and calibration. Further, because the psychometric models we discuss start with observables that have already been identified, the focus here is on test scoring.

A few paradigmatic situations in operational assessment are as follows:

- *Scoring only*: If we have values for the measurement model parameters, say, from having set them in advance, or from some previous calibration, or from divine revelation, we can use them to conduct scoring for any examinee. The posterior analysis in the subtraction proficiency example in Section 7.3.1 is an instance of scoring.
- *Calibration only*: Conversely, if we have values for the examinees' latent variables, we can use them to conduct calibration for the tasks (observables), obtaining representations of the measurement model parameters and possibly hyperparameters. In the subtraction proficiency example, if we knew which examinees were proficient and which were not, we could estimate the conditional probabilities of correct response as the proportions of each group that correctly answer the items. A similar situation arises in operational assessment when some tasks have been previously calibrated, but others have not. Here, the previously calibrated tasks are used to conduct examinee scoring, and then the resulting values for the examinees' latent variables are used to conduct calibration for the new tasks.
- *Calibration and scoring*: If we do not have values for either the examinees' latent variables or the measurement model parameters, we need to conduct inference for both.

The first two scenarios may be viewed as special cases of the last scenario, which is typically what is meant when analysts talk about fitting psychometric models, and is the most complicated of the three from a statistical perspective.

Equation (7.6) provides an answer to this scenario from a Bayesian perspective, yielding a joint posterior distribution for all the unknowns. However, that is not how things are always handled in operational assessment. More commonly, analysts employ a two-stage strategy in which calibration is conducted first (e.g., from an initial sample of examinees) followed by scoring (e.g., for the examinees from that initial sample, or for a large number of subsequent examinees). We begin our development working from an ML framework, and mention Bayesian variations along the way.

In the first stage, the joint distribution of the observable and latent variables is constructed, conditional on the (fixed in the frequentist framework) measurement model parameters and hyperparameters that govern the distribution of the latent variables:

$$\begin{aligned} p(\mathbf{x},\boldsymbol{\theta} \mid \boldsymbol{\omega},\boldsymbol{\theta}_P) &= p(\mathbf{x} \mid \boldsymbol{\theta},\boldsymbol{\omega})p(\boldsymbol{\theta} \mid \boldsymbol{\theta}_P) \\ &= \prod_{i=1}^{n} p(\mathbf{x}_i \mid \theta_i,\boldsymbol{\omega})p(\theta_i \mid \boldsymbol{\theta}_P) \\ &= \prod_{i=1}^{n}\prod_{j=1}^{J} p(x_{ij} \mid \theta_i,\boldsymbol{\omega}_j)p(\theta_i \mid \boldsymbol{\theta}_P). \end{aligned} \tag{7.7}$$

To conduct calibration, the marginal distribution of the observables is obtained by integrating the latent variables out

$$\begin{aligned} p(\mathbf{x} \mid \boldsymbol{\omega},\boldsymbol{\theta}_P) &= \int_{\boldsymbol{\theta}} p(\mathbf{x},\boldsymbol{\theta} \mid \boldsymbol{\omega},\boldsymbol{\theta}_P)d\boldsymbol{\theta} \\ &= \int_{\boldsymbol{\theta}} p(\mathbf{x} \mid \boldsymbol{\theta},\boldsymbol{\omega})p(\boldsymbol{\theta} \mid \boldsymbol{\theta}_P)d\boldsymbol{\theta} \\ &= \prod_{i=1}^{n}\int_{\theta_i} p(\mathbf{x}_i \mid \theta_i,\boldsymbol{\omega})p(\theta_i \mid \boldsymbol{\theta}_P)d\theta_i \\ &= \prod_{i=1}^{n}\int_{\theta_i}\prod_{j=1}^{J} p(x_{ij} \mid \theta_i,\boldsymbol{\omega}_j)p(\theta_i \mid \boldsymbol{\theta}_P)d\theta_i. \end{aligned} \tag{7.8}$$

Viewed as a function of the model parameters, (7.8) is a marginal likelihood function, which can be maximized to yield estimates for $\boldsymbol{\omega}$ and $\boldsymbol{\theta}_P$. These estimates are referred to as ML estimates in factor analysis and latent class analysis, and marginal maximum likelihood (MML) estimates in item response theory (Bock & Aitkin, 1981).*

One Bayesian variation on this theme includes the use of prior distributions for some or all of the measurement model parameters; maximizing the resulting function yields Bayes modal estimates (Mislevy, 1986). Another Bayesian variation conducts the fully Bayesian analysis in (7.6), obtains the marginal posterior for each element of $\boldsymbol{\omega}$ and $\boldsymbol{\theta}_P$, and then uses a measure of central tendency (e.g., posterior mean) as a point summary or estimate. The key point for our purposes is that this first stage yields a set of point estimates for $\boldsymbol{\omega}$ and $\boldsymbol{\theta}_P$, be they ML estimates or some summary of the posterior distribution. Let $(\tilde{\boldsymbol{\omega}},\tilde{\boldsymbol{\theta}}_P)$ denote these values from the first stage.

The second stage proceeds by conducting scoring using $(\tilde{\boldsymbol{\omega}},\tilde{\boldsymbol{\theta}}_P)$ as the values for $\boldsymbol{\omega}$ and $\boldsymbol{\theta}_P$. An MLE for θ for each examinee may be obtained by defining a likelihood function for θ by viewing

* The terminological differences between psychometric modeling traditions are no doubt due to a several reasons (e.g., historical development, the stronger emphasis on estimating values for examinee latent variables in many applications of item response theory, where they are sometimes referred to as person parameters), and we leave such pursuits to professional etymologists. At present we merely wish to highlight that that despite the differences in terminology, there are core structures common to many paradigms, and suggest that the different terminology may be related to the more pervasive use of Bayesian approaches in some traditions as opposed to others.

$$p(\mathbf{x}_i \mid \theta_i) = \prod_{j=1}^{J} p(x_{ij} \mid \theta_i, \omega = \tilde{\omega}) \tag{7.9}$$

as a function of θ_i. A Bayesian perspective involves obtaining a posterior distribution for each examinee as an instance of (7.3),

$$p(\theta_i \mid \mathbf{x}_i) \propto \prod_{j=1}^{J} p(x_{ij} \mid \theta_i, \omega = \tilde{\omega}) p(\theta_i \mid \theta_P = \tilde{\theta}_P), \tag{7.10}$$

or taking a point summary of it such as a posterior mean, sometimes referred to as the expected a posteriori (EAP) estimate, or the mode, sometimes referred to as the maximum a posteriori (MAP) estimate. The Bayesian analysis in (7.10) taking point estimates from a prior stage as known are often called empirical Bayes estimates. We prefer the term *partially Bayesian* because (a) it better calls out that there is a distinction from fully Bayesian analyses where all unknowns are modeled stochastically, (b) the term *empirical Bayes* suggests that other (fully) Bayesian analyses are not empirical, and (c) the resulting estimates may not be Bayes estimates (Deely & Lindley, 1981).

An advantage of this two-stage approach is that it splits the inferences of unknown examinee latent variables from that of the remaining parameters. This has considerable practical advantages. Once a set of tasks (observables) are calibrated using a dataset, the estimates can be used not only to score the examinees in that dataset, but future examinees as well. In addition, it is generally easier to work with point estimates of measurement model parameters than distributions for them when we come to do all the other things that go in psychometrics, such as selecting tasks, creating comparable test forms, and equating multiple assessments.

However, this approach is potentially limited in that the uncertainty associated with the point estimates from calibration is ignored in scoring. A bit more formally, (7.10) amounts to specifying the posterior distribution for an examinee's latent variable using point values for the other parameters: $p(\theta_i \mid \mathbf{x}_i) = p(\theta_i \mid \mathbf{x}_i, \omega = \tilde{\omega}, \theta_P = \tilde{\theta}_P)$. In contrast, if we have the full posterior distribution for all unknowns, the marginal posterior distribution for an examinee's latent variable is $p(\theta_i \mid \mathbf{x}_i) = \int\int p(\theta_i \mid \mathbf{x}_i, \omega, \theta_P) p(\omega, \theta_P \mid \mathbf{x}) d\theta_P d\omega$, which will be more diffuse to the extent that there is posterior variability in ω and θ_P. By using the full posterior distribution for ω and θ_P, our uncertainty in those parameters is acknowledged and induces some additional uncertainty about the examinees' latent variables. This is not so if we use $(\tilde{\omega}, \tilde{\theta}_P)$ as the values for ω and θ_P. We would then be treating these parameters if they were known with certainty, on a par with us coming to know them through divine revelation, as opposed to based on fallible information in the forms of a finite dataset, subject matter expertise, or some combination thereof.

In many cases, the practical demands of an operational assessment dictate that this limitation is acceptable. Resources are limited, and decisions must be made to accomplish multiple, possibly misaligned or conflicting purposes. Maintaining and carrying through the uncertainty from one stage of an analysis to another may be a casualty of pragmatics. The consequences of doing so are likely minimal to the extent that the measurement model parameter estimates have relatively less uncertainty associated with them (which is subsequently ignored). The consequences are likely more severe the greater amount of uncertainty is being ignored. The fully Bayesian framework articulated here allows us to investigate just what the implications are when taking the simpler approach. In Section 13.4, we will examine a simple scenario in the context of latent class models where the differences are small, and the consequence are not dire. In other scenarios, the differences and consequences may be striking. Tsutakawa and Soltys (1988) and Tsutakawa and Johnson (1990) investigated paradigmatic applications of item response theory models for scoring, comparing the use of an approximation to the full posterior for measurement model parameters to the use of point estimates from ML and partially Bayes procedures. They found that the procedures that treated point estimates of measurement model parameters as known had substantially smaller standard errors and correspondingly narrower interval estimates.

In test construction, ignoring estimation error in measurement model parameters can lead to capitalizations on chance that causes items with higher magnitudes of errors to be selected in constructing fixed and adaptive tests (Hambleton & Jones, 1994; Hambleton, Jones, & Rogers, 1993; Patton, Cheng, Yuan, & Diao, 2013; van der Linden & Glas, 2000). This can lead in turn to overestimates of information and precision of estimation during scoring. Johnson and Jenkins (2005) investigated a complicated scenario in the context of the National Assessment of Educational Progress, comparing a fully Bayesian specification to typical operational procedures in which the measurement model parameters are estimated at one stage, and are then treated as known with certainty in later stages that estimate parameters of subpopulations of examinees (Mazzeo, Lazer, & Zieky, 2006; von Davier, Sinharay, Oranje, & Beaton, 2007). Using simulated and real data, they concluded that the procedures that treated point estimates from an earlier stage as known systematically underestimated posterior uncertainty for parameters estimated at a later stage. In addition, the fully Bayesian analysis provided more stable estimates of the sampling variability as compared to the typical approach.

7.5 Summary and Looking Ahead

In summary, we construct a probability model that encodes our beliefs about the situation directed toward our purposes and desired inferences. In assessment this typically involves, at a minimum, beliefs about the examinee's capability, their behavior, and the relationship between the two. In line with the implications of de Finetti's representation theorem and the power afforded by conditional independence relationships, modern psychometric models typically specify variables representing behaviors on tasks as stochastically dependent on variables representing proficiencies. Once the model is constructed, we employ the calculus of probabilities to "reason through" the model, arriving at model-based implications for features of inferential interest. The directionality of the dependence of the model quite naturally supports deductive reasoning, from the general (i.e., examinee proficiency) to the particulars (i.e., examinee behaviors on tasks). Of course, in most assessment situations we seek to reverse this directional flow, to reason inductively from the particulars (examinee behaviors on tasks) to the general (examinee proficiency). Operating in a probability model constructed this way, Bayes' theorem is the workhorse for accomplishing the reversal. This process amounts to a model-based synthesis of the implications of the data. It is particularly powerful in situations where multiple chunks of data are available, have complex relationships with the inferential target, and are possibly in conflict with one another.

If measurement model parameters are unknown, our probability model expands to accommodate them, specifying prior distributions for them, again often capitalizing on exchangeability assumptions. They too become subject to posterior inference, now jointly with examinees' latent variables in the full Bayesian model. Conventional practices in various psychometric modeling paradigms vary in the extent to which Bayesian approaches are employed. Commonly, they depart from the fully Bayesian perspective by estimating parameters in stages, often treating point estimates from earlier stages as known in later stages.

The model in (7.6) and Figure 7.4 serves as a cornerstone of Bayesian psychometric modeling. It is well targeted to what we may call the standard assessment paradigm (Mislevy, Behrens, DiCerbo & Levy, 2012), which is driven in part by twentieth century conceptions of psychology (Mislevy, 2006; Rupp & Mislevy, 2007) and technological capabilities (Cohen & Wollack, 2006; DiCerbo & Behrens, 2012; Dubois, 1970). Here, data from each examinee are fairly sparse, typically in the form of a single attempt on tasks that were chosen beforehand. Feedback does not occur during the assessment, so learning during the assessment is assumed to be negligible, and performance can then be interpreted as information regarding an examinee's static level of proficiency conceived in trait or behaviorist psychology.

Much of the material in the following chapters describes how this model and extensions of it play out in a variety of psychometric traditions. Along the way, we will also mention the practices that have emerged as the conventional approaches in these psychometric modeling paradigms. Provocatively, though many of these paradigms share the structure illustrated in Figure 7.4, we will see that the standard conventional practices in each tradition vary considerably in the degree to which they involve Bayesian approaches when conducting scoring and calibration.

Not all psychometric models can be framed as an instance of this model, but many can, and it is a useful starting point for building more complicated models that depart from the standard assessment paradigm, particularly in ways that harness shifts in psychology (Mislevy, 2006, 2008; Rupp & Mislevy, 2007) and technological capabilities to collect and process data that vary in type, quality, and quantity (Behrens & DiCerbo, 2013, 2014; Cohen & Wollack, 2006; Drasgow, Luecht, & Bennett, 2006). In the coming chapters, we discuss several such examples including: seeking inferences about multiple aspects of proficiency; acknowledging and modeling the possibility of learning during the assessment (which may be its very intent); leveraging collateral information about examinees and tasks and formulating exchangeability conditional on relevant covariates; and adding layers to the hierarchical specification by specifying θ_P or ω_P as dependent on other unknown parameters, thereby borrowing strength from the collective of examinees or tasks for estimating the parameters for individual members.

In our tour, we will use notation specific to each modeling paradigm. Table 7.2 lists some key elements of the notation for the various modeling paradigms, including elements not present in the canonical psychometric model discussed here. What is listed in the table is not broad enough to cover all of the variations and extensions of the basic model we will cover. It is intended to serve as a reference and illuminate how the basic model developed in this chapter will reappear across the modeling paradigms, list some of the directions for

TABLE 7.2

Notation for the Psychometric Models in This Part of Book

	Psychometric Modeling Paradigm				
Parameter	**General (Chap. 7)**	**Classical Test Theory (Chap. 8)**	**Factor Analysis (Chap. 9)**	**Item Response Theory (Chap. 11)**	**Latent Class Analysis, Bayesian Networks, Diag. Classification (Chap. 13–14)**
Examinee	θ	T	ξ	θ	θ
Hyperparameters for examinees (first level)	θ_P	μ_T, σ_T^2	κ, ϕ	$\mu_\theta, \sigma_\theta^2$	γ
Hyperparameters for examinees (second level)			$\mu_\kappa, \sigma_\kappa^2$	$\mu_{\mu_\theta}, \sigma_{\mu_\theta}^2$	α_γ
			d, Φ_0	d, Σ_{θ_0}	
Measurement model	ω	σ_E^2	τ	d (or b)	π
			λ	a	
			ψ	c	
Hyperparameters for observables (first level)	ω_P	$\nu_{\sigma_E^2}, \sigma_{\sigma_E^2}^2$	μ_τ, σ_τ^2	μ_d, σ_d^2	α_π
			$\mu_\lambda, \sigma_\lambda^2$	μ_a, σ_a^2	
			ν_ψ, ψ_0	α_c, β_c	
Hyperparameters for observables (second level)				$\mu_\omega, \Sigma_\omega$	

extensions to come, and aid readers familiar with some but not all of these traditions and notations in seeing the connections between them.

Exercises

7.1 Reconsider the subtraction proficiency example, where the prior probability that a student is proficient is $p(\theta_i = \text{Proficient}) = .6$ and the conditional probabilities of correct responses to the two items are given in Table 7.1. In Section 7.3.1, we obtained the posterior distribution for proficiency for a student who correctly answered the first question and incorrectly answered the second question:

$$p(\theta_i = \text{Proficient} \mid x_{i1} = \text{Correct}, x_{i2} = \text{Incorrect}) \approx .68$$

$$p(\theta_i = \text{Not Proficient} \mid x_{i1} = \text{Correct}, x_{i2} = \text{Incorrect}) \approx .32.$$

a. Explain how this posterior distribution can be interpreted from the Bayes-as-using-data-to-update-prior perspective.

b. Explain how this posterior can be interpreted from the Bayes-as-prior-reining-in-data perspective.
c. Obtain the posterior distributions for the remaining possible response patterns (both responses correct, only the second correct, and both incorrect).

7.2 Reconsider the subtraction proficiency example, where the prior probability that a student is proficient is $p(\theta_i = \text{Proficient}) = .6$ and the conditional probabilities of correct responses to the items on a two-item test are given in Table 7.1.

a. Suppose we observe a student respond correctly to the first item, and at this point that is all we have observed. What is the posterior distribution for her proficiency?
b. We now plan to administer the second item, and will update our distribution for her proficiency based on her response. What will our prior distribution be for this analysis?
c. Suppose we observe that she responds incorrectly to this second item. What is the posterior distribution for her proficiency?
d. Suppose we have another student and decide to administer the second item first this time. Suppose we observe he incorrectly answers this item and at this point that is all we have observed. What is the posterior distribution for his proficiency?
e. We now plan to administer the first item, and will update our distribution for his proficiency based on his response. What will our prior distribution be for this analysis?
f. Suppose we observe that he responds correctly to this item. What is the posterior distribution for his proficiency?
g. Compare the results from (c), (f), and the computations shown in Section 7.3.1. What does this indicate about Bayesian inference?
h. In general terms, how would we proceed if we thought that θ might change from the first response to the second? What inferential errors might result if θ did indeed change but we used a standard model assuming it was constant during the time of observation?

7.3 Reconsider the subtraction proficiency example, where the prior probability that a student is proficient is $p(\theta_i = \text{Proficient}) = .6$

a. Suppose we have an item with the following conditional distribution of response:

	Response to Item	
Proficiency	**Correct**	**Incorrect**
Proficient	.90	.10
Not Proficient	.20	.80

and we observe a student responds correctly to the item. What is the posterior for his proficiency?

b. Suppose we have an item with the following conditional distribution of response:

	Response to Item	
Proficiency	**Correct**	**Incorrect**
Proficient	.90	.10
Not Proficient	.90	.10

What is the posterior distribution for the examinee's proficiency given a correct response? Given an incorrect response? (Hint: you can answer these without doing any calculations.)

c. Compare the results from (a), (b), and Exercise 7.2(a). Why do the posteriors differ? What does that indicate about the discrimination of the items?

7.4 In Section 7.4, we discussed how it is common to first conduct calibration and then, using the results of calibration, scoring for both the examinees employed in the calibration analysis and other examinees. Explain how the use of the results from a calibration for scoring other examinees (i.e., not those used in the calibration) may be seen as relying on notions of exchangeability.

8

Classical Test Theory

We begin our tour of psychometric modeling paradigms with classical test theory (CTT). Textbook treatments and overviews of CTT from a conventional perspective can be found in Crocker and Algina (1986), Lewis (2007), Lord and Novick (1968), and McDonald (1999). The CTT model has been conceived of in several distinct ways, differing in terms of whether an author develops the model from single or multiple observables, and from one of several different distributional notions of error (cf. Bollen, 1989; Haertel, 2006; Lewis, 2007; Lord & Novick, 1968; McDonald, 1999). What follows is a cursory treatment of the model focusing on consensus notions. Readers are referred to the above cited works for considerably more details on CTT, and for constructing assessments that lead to the "scores" that our look at Bayesian modeling for CTT takes as a starting point. We begin our development of CTT in Section 8.1 by considering a single observable (test) assuming that the measurement model parameters and the hyperparameters are known. In Section 8.2, we generalize to the case where there are multiple observables (tests), still assuming that the measurement model parameters and hyperparameters are known. In Section 8.3, we treat the case where the measurement model parameters and hyperparameters are unknown. Section 8.4 concludes this chapter with a summary and bibliographic note.

8.1 CTT with Known Measurement Model Parameters and Hyperparameters, Single Observable (Test or Measure)

We begin with the situation in which there is a single test or measure, the resulting score on which will serve as the observable. We describe the model as it is conventionally conceived, and then turn to a Bayesian treatment, which is developed in stages. We then extend the model to the case of multiple observables (tests or measures) in Section 8.2. Throughout, our focus will be on inference about examinees, or scoring, in the parlance of Chapter 7.

8.1.1 Conventional Model Specification

Consider first the situation where examinees from a population are administered a single measure or test. For each examinee i, x_i denotes the resulting observable score on the test. In most applications the test is made up by several items and the test score is the sum of the scored responses to the individual items. In this chapter, the observable x_i may be taken as a test score formed in this way. Much the same can be said for the factor analysis models of Chapters 9 and 10. We return to situations where the observables are item responses in Chapter 11.

The CTT model specifies x_i as an additive combination of two components that may vary over examinees,

$$x_i = T_i + E_i, \tag{8.1}$$

where T_i is a true score (for examinee i) with mean μ_T and variance σ_T^2 and E_i is an error (for examinee i) with mean 0 and variance σ_E^2. Errors are uncorrelated with true scores in the population,

$$\rho_{TE} = 0, \tag{8.2}$$

and similarly in all subpopulations.

The CTT model implies that the mean of the observed scores μ_x is equal to the mean of the true scores,

$$\mu_x = \mu_T. \tag{8.3}$$

The observed score variance σ_x^2 is given by

$$\begin{aligned} \sigma_x^2 &= \sigma_T^2 + \sigma_E^2 + 2\sigma_{TE} \\ &= \sigma_T^2 + \sigma_E^2, \end{aligned} \tag{8.4}$$

where $\sigma_{TE} = \rho_{TE}\sigma_T\sigma_E = 0$ is the covariance between true scores and errors.

On this basis, the classical definition of reliability of the test is given as the proportion of observed score variance that is accounted for by the true score variance,

$$\rho = \frac{\sigma_T^2}{\sigma_x^2} = \frac{\sigma_T^2}{\sigma_T^2 + \sigma_E^2}, \tag{8.5}$$

which is also the squared correlation between observable and true scores, ρ_{xT}^2.

True scores for examinees may be estimated via Kelley's formula, which dates to at least Kelley (1923, p. 214). In the current notation,

$$\begin{aligned} T_i' &= \rho x_i + (1-\rho)\mu_x \\ &= \mu_x + \rho(x_i - \mu_x), \end{aligned} \tag{8.6}$$

where μ_x is the mean of the observed scores and T_i' is the estimated true score for examinee i. The use of the prime notation in T_i' to stand for an estimate is derived from noting that the formula in (8.6) may be framed as the regression of true scores on error scores (e.g., Crocker & Algina, 1986; Haertel, 2006; Lewis, 2007; Lord & Novick, 1968; the same idea appears in hierarchical modeling as well, as in Raudenbush, 1988). The second formulation on the right-hand side in (8.6) yields an interpretation of Kelley's formula as one in which we get an estimate for an examinee's true score by starting with the mean, and then moving away from the mean in the direction of their observed score. We do not move all the way—rather, the amount we move is in proportion to the reliability.

Importantly, the last interpretation reveals that Kelley's formula contradicts the intuitive notion that we should use the examinee's observed score as the estimate of the true score (i.e., $T_i' = x_i$). The reasoning for the latter approach is that x_i is obviously related to the true score, only differing by a random error component and that, after all, x_i is all we have for the examinee. This line of reasoning is quite appealing on the surface, so it is instructive to consider how Kelley's formula departs from this, and what it reveals about assessment. Note that Kelley's formula will indeed yield the observed score as the estimate of the true score when $\rho = 1$. At the other end of the spectrum, when $\rho = 0$, Kelley's formula will yield μ_x as the estimate of the true score.* In this case, Kelley's formula instructs us to ignore the actual observed score for the examinee. The reason is that the variance of x is all due to error variance. The test is essentially worthless as an inferential tool to differentiate examinees—we would do just as well if did not administer the test at all. Kelley's formula therefore instructs us to ignore the examinee's observed score for this purpose.†

Thankfully, in practice it is highly unlikely that $\rho = 0$.‡ If ρ falls between the extremes of 0 and 1, the estimated true score will fall between x_i and μ_x. To understand what this captures, we can hardly do better than Kelley himself, who remarked (1947, p. 409)

> This is an interesting equation in that it expresses the estimate of true ability as the weighted sum of two separate estimates,–one based upon the individual's observed score, X_1 [x_i in the current notation], and the other based upon the mean of the group to which he belongs, M_1 [μ_x in the current notation]. If the test is highly reliable, much weight is given to the test score and little to the group mean, and vice versa.

Finally, the variability associated with estimation of the true score is captured by the standard deviation of the errors of estimation of the true score,

$$\sigma_{T|x} = \sigma_T \sqrt{(1-\rho)}, \tag{8.7}$$

known as the standard error of the true score (Guilford, 1936) or the standard error of estimation (Lord & Novick, 1968). Note that this refers to the variability of the true scores given the observed scores, and differs from the standard error of measurement that refers to the variability of observed scores given true scores (Dudek, 1979).

8.1.2 Bayesian Modeling

In the present development, let us assume that the mean and variance of the distribution of the true scores, μ_T and σ_T^2, and the variance of the errors, σ_E^2, are known. In Section 8.3, we will relax this assumption of known parameters. For the moment, the only stochastic entities are the observed scores and the true scores of individual examinees, of which only the latter are unknown.

Following Bayes' theorem, the posterior distribution for the full collection of true scores, $\mathbf{T} = (T_1, \ldots, T_n)$, given the full collection of the observed scores, $\mathbf{x} = (x_1, \ldots, x_n)$, and the other, assumed known entities is

* In the rest of this section and Section 8.2, we assume that μ_x is known. If it is unknown, a conventional approach estimates μ_x as the mean of the test scores from a sample of examinees. We treat the situation of unknown μ_x in Section 8.3.

† In this case, we might have reservations about estimating true scores at all, because our best estimate for everyone is the population mean. If it comes to pass that $\rho = 0$, the next step is not really to estimate true scores to facilitate interpretations and decisions, but rather to revisit and revise the assessment process.

‡ But it is about as likely as $\rho = 1$. If ρ was indeed 1, that would indicate that there was no measurement error, obviating the need for much of psychometrics. In practice, many achievement tests have reliability estimates that exceed .8.

$$p(\mathbf{T} \mid \mathbf{x}, \mu_T, \sigma_T^2, \sigma_E^2) \propto p(\mathbf{x} \mid \mathbf{T}, \sigma_E^2) p(\mathbf{T} \mid \mu_T, \sigma_T^2). \tag{8.8}$$

The first term on the right-hand side of (8.8) is the conditional distribution of the data given the true scores and the error variance. To structure this distribution, we assume that, given the examinee's true score, the observed score from the examinee is conditionally independent of all other observed scores. This assumption allows for the factorization of the conditional distribution of the data as

$$p(\mathbf{x} \mid \mathbf{T}, \sigma_E^2) = \prod_{i=1}^{n} p(x_i \mid T_i, \sigma_E^2). \tag{8.9}$$

CTT, as formulated so far, defines the model in terms of the first two moments. In our development of a Bayesian analysis, we will specify the full distributions.* For each examinee, we now model the observed score as being normally distributed around the examinee's true score,

$$x_i \mid T_i, \sigma_E^2 \sim N(T_i, \sigma_E^2). \tag{8.10}$$

The second term on the right-hand side of (8.8) is the prior distribution for the true scores. An assumption of exchangeability regarding the examinees implies the use of a common prior distribution for each, supporting the factorization of the joint prior of all n examinees' true scores as

$$p(\mathbf{T} \mid \mu_T, \sigma_T^2) = \prod_{i=1}^{n} p(T_i \mid \mu_T, \sigma_T^2). \tag{8.11}$$

For each examinee, we now also model the true score as being normally distributed,

$$T_i \mid \mu_T, \sigma_T^2 \sim N(\mu_T, \sigma_T^2). \tag{8.12}$$

Note that these normal distribution specifications for the observables in (8.10) and for the true scores in (8.12) are common in conventional presentations of the model rooted in frequentist paradigms (e.g. Bollen, 1989; Crocker & Algina, 1986) as well as in Bayesian developments (Lindley, 1970; Novick, 1969; Novick, Jackson, & Thayer, 1971). Other assumptions can be made, such as assuming normal distributions for errors, but estimating the distribution of true scores (Mislevy, 1984).

A DAG representation of the model is given in Figure 8.1. Each examinee's observable is modeled with the associated true score and the error variance as parents, in accordance with (8.10). Similarly, each true score is modeled with the hyperparameters μ_T and σ_T^2 as parents, in accordance with (8.12).

Substituting into (8.8), the posterior distribution is

$$\begin{aligned} p(\mathbf{T} \mid \mathbf{x}, \mu_T, \sigma_T^2, \sigma_E^2) &\propto p(\mathbf{x} \mid \mathbf{T}, \sigma_E^2) p(\mathbf{T} \mid \mu_T, \sigma_T^2) \\ &= \prod_{i=1}^{n} p(x_i \mid T_i, \sigma_E^2) p(T_i \mid \mu_T, \sigma_T^2), \end{aligned} \tag{8.13}$$

* An alternative development might involve just the first- and second-order moments, capitalizing on Bayesian analyses of linear models without invoking a full distributional specification (Hartigan, 1969).

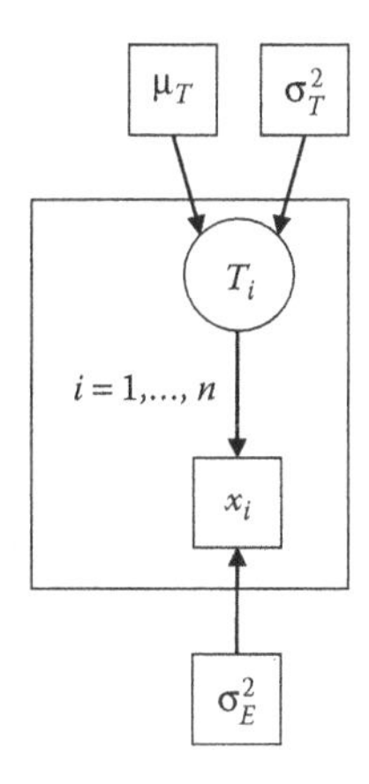

FIGURE 8.1
Directed acyclic graph for the classical test theory model for one observable variable, with known true score mean, true score variance, and error variance.

where

$$x_i \mid T_i, \sigma_E^2 \sim N(T_i, \sigma_E^2) \quad \text{for } i = 1,\ldots,n,$$

and

$$T_i \mid \mu_T, \sigma_T^2 \sim N(\mu_T, \sigma_T^2) \quad \text{for } i = 1,\ldots,n.$$

Note that the right-hand side of (8.13) has a form where all the elements are inside the product operator. This reveals that the conditional independence assumptions imply that the model for all the examinees can be viewed as a model for an individual examinee, which is then repeated over all examinees. The DAG illustrates this in that the only stochastic entities, namely the true scores and the observables, lie inside the plate for examinees.

This implies that the model is one where, for each examinee, the data x_i is normally distributed with unknown mean T_i and known variance σ_E^2. The unknown mean T_i is normally distributed with known mean μ_T and variance σ_T^2. This is exactly the situation discussed in Section 4.1. The expression in (8.13) is just an instantiation of (4.5), as can be seen by framing the notation of the former in terms of the latter:

Notation in (8.13)		Notation in (4.5)
T_i	$\leftrightarrow$	μ
σ_E^2	$\leftrightarrow$	σ^2
μ_T	$\leftrightarrow$	μ_μ
σ_T^2	$\leftrightarrow$	σ_μ^2

Thus, the posterior distribution for each T_i is an instantiation of (4.6) and is a normal distribution,

$$T_i \mid x_i, \mu_T, \sigma_T^2, \sigma_E^2 \sim N(T_i \mid \mu_{T_i|x_i}, \sigma_{T_i|x_i}^2), \tag{8.14}$$

where

$$\mu_{T_i|x_i} = \frac{\left(\mu_T/\sigma_T^2\right)+\left(x_i/\sigma_E^2\right)}{\left(1/\sigma_T^2\right)+1/\sigma_E^2} \tag{8.15}$$

and

$$\sigma_{T_i|x_i}^2 = \frac{1}{\left(1/\sigma_T^2\right)+\left(1/\sigma_E^2\right)}. \tag{8.16}$$

A little algebra reveals that the posterior mean can be expressed as

$$\mu_{T_i|x_i} = \frac{\sigma_T^2}{\sigma_T^2+\sigma_E^2} x_i + \frac{\sigma_E^2}{\sigma_T^2+\sigma_E^2} \mu_T. \tag{8.17}$$

The coefficient for x_i is recognized as ρ. A little more algebra reveals that the coefficient for μ_T is $1-\rho$. Thus, the posterior mean for an examinee's true score is

$$\begin{aligned} \mu_{T_i|x_i} &= \rho x_i + (1-\rho)\mu_T \\ &= \mu_T + \rho(x_i - \mu_T). \end{aligned} \tag{8.18}$$

Recalling that the means of the true and observed scores are equal (see Equation 8.3), we see from the right-hand side of (8.18) that, remarkably, the posterior mean is just an expression of Kelley's formula as given in (8.6). In the Bayesian analysis, the posterior mean is a precision-weighted combination of the mean of the data and the mean of the prior. For inferences about an examinee's true score in CTT, the precision of the data is captured by the reliability of the test, ρ, and the precision in the prior for the true score is captured by $1-\rho$. Echoing similar statements made in Chapter 4, this reveals that to the extent that the test is reliable (i.e., the variation in observed scores is due to variation in true scores), the observed test score should drive the solution and our beliefs about the examinee's true score. To the extent that the test is unreliable (i.e., the variation in observed scores is due to something *other* than variation in true scores), the influence of the data ought to be reduced. Thus, we may echo Kelley's sentiments quoted in Section 8.1.1 with some slight revisions in our phrasing: Kelley's formula, as it appears in (8.18), is indeed an interesting equation in that it expresses a point summary of our posterior beliefs about an examinee's true ability as the weighted sum of two separate sources of information: one being the examinees's observed score, x_i, and the other being what was believed about the true ability prior to observing any data, captured by μ_T. If the test is highly reliable, much weight is given to the observed score and little is given to our prior beliefs, and vice versa.

In Bayesian terms, Kelley's formula amounts to saying that the best estimator for an examinee's true score is the posterior mean. For several reasons, it is important to note that Kelley did not derive his formula in Bayesian terms, but rather as the regression of true score on observed score (Kelley, 1923, 1947). What's more, it can be shown that the posterior standard deviation,

$$\sigma_{T_i|x_i} = \sqrt{\sigma^2_{T_i|x_i}} = \sqrt{\frac{1}{\left(1/\sigma_T^2\right)+\left(1/\sigma_E^2\right)}}, \tag{8.19}$$

is equal to the standard error of the estimate of the true score in (8.7), which in conventional approaches is framed as the unexplained variation of the regression of true scores on observed scores.

The conventional framing of Kelley's formula as the regression of true score on observed scores, and the standard error of the estimate of true scores as the associated variation from that regression (Haertel, 2006; Lord & Novick, 1968), brings to light the alignment between the goals of assessment and the Bayesian approach to inference. To begin, we can recognize that the formulation of CTT in (8.1) has the appearance of regressing the observables x on the true scores T. The "flow" of the model here and in the associated DAG in Figure 8.1 is *from the true score to the observable*, supporting deductive reasoning. Kelley's formula reverses this flow, regressing the true scores T on the observables x in support of inductive reasoning. The model is constructed with x conditional on T, but once a value for x is observed, the goal is to *reason back from the observed score to the true score*. Kelley's formula facilitates this reasoning partially, by regressing T on x, giving the expectation $E(T \mid x)$. The standard error of the estimate of the true scores complements this partial reasoning, providing a first inkling of variation. A Bayesian perspective makes explicit this reversal of flow and reasoning back, recognizes Kelley's formula as the posterior mean and the standard error of the true scores as the posterior standard deviation, and goes beyond these to give the full posterior distribution in (8.14).

Confusion over the appropriate way to reason back from observed scores to true scores has been with us almost as long as CTT itself (Dudek, 1979) and does not appear to be disappearing, at least in some disciplines (McManus, 2012). Generally, a Bayesian approach to inference constructs a joint distribution for all entities and then conditions on what is known to yield a conditional distribution for all unknown entities. As such it offers a framework well suited to reasoning in either direction. If true scores are known, we have the conditional distribution for observables in (8.10). If observed scores are known, we have the conditional distribution for true scores in (8.14). No entity has any special status that prevents us from formulating conditional distributions for it. Our construction of the joint distribution emerges from structuring the observables as conditional on true scores. In the current situation with one observable, this choice does not seem of great importance. When we expand our scope to include multiple observables, we will model the multiple observables for each examinee as conditionally (locally) independent given their true score, in line with the implications of exchangeability and the utility in organizing our beliefs around conditional independence specifications. Bayes' theorem is then the machinery for taking this construction of the joint distribution and turning it into a conditional distribution for unknown true scores given observed scores.

8.1.3 Example

Suppose we have a test scored from 0 to 100.* We are interested in the posterior distributions of true scores for individuals, where it is known that the mean of the true scores is $\mu_T = 80$ the standard deviation of the true scores is $\sigma_T = 6$, and the standard deviation of the error scores is $\sigma_E = 4$. The reliability of the test is calculated via (8.5); rounding to two decimal places, $\rho \approx .69$.

8.1.3.1 Analytical Solution

For each examinee, the posterior distribution is normal with the mean given by (8.17),

$$\begin{aligned}\mu_{T_i|x_i} &= \frac{\sigma_T^2}{\sigma_T^2 + \sigma_E^2} x_i + \frac{\sigma_E^2}{\sigma_T^2 + \sigma_E^2} \mu_T \\ &= \frac{36}{36+16} x_i + \frac{16}{36+16} 80 \\ &\approx .69 x_i + 24.62.\end{aligned}$$

Alternatively, using the equivalent re-expression in (8.18) and substituting in for ρ,

$$\begin{aligned}\mu_{T_i|x_i} &= \mu_T + \frac{\sigma_T^2}{\sigma_T^2 + \sigma_E^2}(x_i - \mu_T) \\ &= 80 + \frac{36}{36+16}(x_i - 80) \\ &\approx 80 + .69(x_i - 80).\end{aligned}$$

The latter expression affords the interpretation that the posterior mean for an examinee's true score can be viewed as departing from the mean for all examinees (80) in the direction of the examinee's score (x_i), and in an amount proportional to the reliability (.69). Again, we note that each examinee's posterior mean is exactly what is obtained by the conventional application of Kelley's formula. The posterior variance is constant for all examinees,

$$\sigma_{T_i|x_i}^2 = \frac{1}{\left(1/\sigma_T^2\right) + \left(1/\sigma_E^2\right)} = \frac{1}{\left(1/36\right) + \left(1/16\right)} \approx 11.08.$$

To illustrate, we compute the posterior distribution for examinees, with scores 70, 80, and 96, rounding the values obtained to two decimal places. For $x_i = 70$, $p(T_i \mid x_i = 70) = N(73.08, 11.08)$; for $x_i = 80$, $p(T_i \mid x_i = 80) = N(80, 11.08)$; for $x_i = 96$, $p(T_i \mid x_i = 96) = N(91.08, 11.08)$. Figure 8.2 depicts these results.

For each examinee, the posterior is a synthesis of the information in the prior and the information in the data. In CTT, the information in the data is imperfect, as reflected

* As noted above, tests usually contain multiple items, and the test score is the aggregation over the items, say, by summing the scored responses for the individual items. Estimates of ρ are obtained as functions of variation among items. But for the purpose of laying out a Bayesian treatment of CTT, the starting point is a single observed (test) score for each examinee, however, obtained, such that the CTT structure applies.

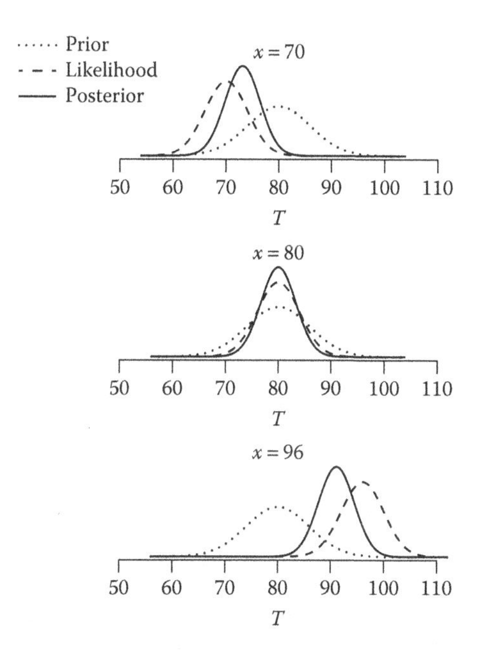

FIGURE 8.2
Plotting the components of a Bayesian analysis for selected examinees' true scores, with the prior distribution (dotted), likelihood (dashed), and posterior distribution (solid).

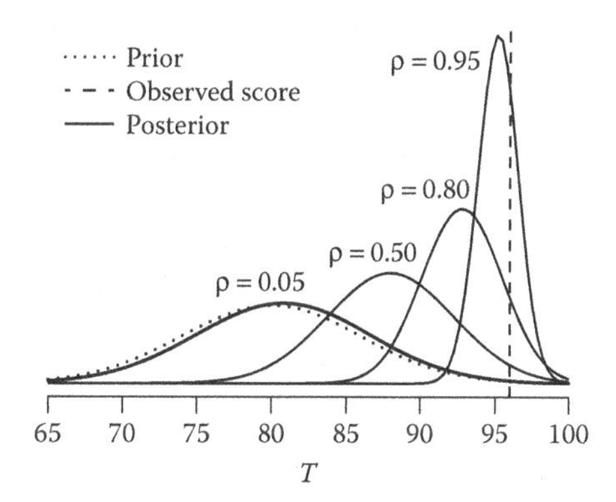

FIGURE 8.3
Posterior densities (solid lines) for the true score for an $N(80, 36)$ prior (dotted line) and an observed score of 96 (dashed line) as the reliability (ρ) changes.

by the reliability being less than 1. To further illustrate this point, consider the effect of different values for the reliability. Figure 8.3 depicts analyses where $\mu_T = 80$, $\sigma_T = 6$, and $x_i = 96$. The curves depict different posterior distributions corresponding to different values of the reliability of the test. When the reliability is near 0, the posterior distribution is very similar to the prior distribution. As the reliability increases, the posterior distribution "moves" closer to the observed test score of 96 and becomes more narrowly distributed.

8.1.3.2 MCMC Estimation via WinBUGS

The WinBUGS code for the model, using the precision parameterization for the normal distribution, and the data statement are given as follows.

```
-------------------------------------------------------------------------
#########################################################################
# Model Syntax
#########################################################################
model{

#########################################################################
# Classical Test Theory
# With Known
#    True Score Mean, True Score Variance
#    Error Variance
#########################################################################

#########################################################################
# Known Parameters
#########################################################################
mu.T <- 80                        # Mean of the true scores
sigma.squared.T <- 36             # Variance of the true scores
sigma.squared.E <- 16             # Variance of the errors

tau.T <- 1/sigma.squared.T        # Precision of the true scores
tau.E <- 1/sigma.squared.E        # Precision of the errors


#########################################################################
# Model for True Scores and Observables
#########################################################################

for (i in 1:n){
   T[i] ~ dnorm(mu.T, tau.T)      # Distribution of true scores
   x[i] ~ dnorm(T[i], tau.E)      # Distribution of observables
}

}                                 # closes the model statement


#########################################################################
# Data statement
#########################################################################
list(n=3, x=c(70, 80, 96))
-------------------------------------------------------------------------
```

The model was run with three examinees. Three chains were run from dispersed starting values. Convergence diagnostics, including the history plots and the potential scale reduction factor (Section 5.7.2), indicated convergence within a few iterations. This is unsurprising given that the posterior is of a known form (see Equation 8.14) that is recognized as such by WinBUGS. To be conservative, we discarded the first 100 iterations and 1,000 additional iterations were run, yielding 3,000 iterations for use in summarizing the posterior distribution.

Figure 8.4 contains plots of the analytical marginal posterior distributions for the true scores (solid line, akin to the solid lines in Figure 8.2) and empirical approximations from WinBUGS (dashed line) for the three examinees. Table 8.1 gives, for each examinee, the posterior mean and standard deviation from the analytical solution, and the estimated posterior mean and standard deviation from the empirical approximation from WinBUGS. The empirical approximations come quite close to the analytical solutions in terms of shape, location, and scale.

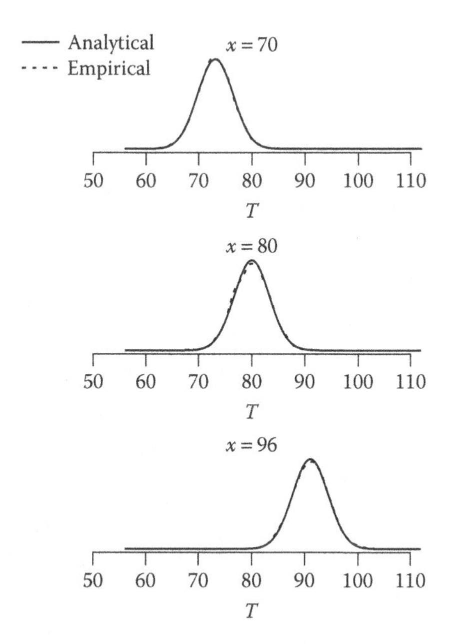

FIGURE 8.4
Analytical (solid) and empirical (dashed) marginal posterior densities for true scores for three examinees. The empirical density closely mirrors the analytical density, such that it is obscured by the latter when plotted.

TABLE 8.1
Analytical Solutions and Empirical Approximations to the Posterior Distributions for Three Examinees

	Analytical Solution		**Empirical Solution from MCMC**[a]	
x_i	**Posterior Mean**	**Posterior SD**[b]	**Posterior Mean**	**Posterior SD**[b]
70	73.08	3.33	73.07	3.28
80	80.00	3.33	79.96	3.34
96	91.08	3.33	91.13	3.36

[a] MCMC = Markov chain Monte Carlo.
[b] SD = Standard deviation.

8.2 CTT with Known Measurement Model Parameters and Hyperparameters, Multiple Observables (Tests or Measures)

We now extend the CTT model to the case of multiple observables (tests or measures) for examinees. Conceptually, we may think of administering the same exam to the examinees on multiple occasions absent the effects of learning or fatigue. Alternatively, we may think of administering parallel tests or forms, which are different tests (or test forms) where each examinee's true score on one test (form) is the same as their true score on another test (form), and the error variances for the observable test scores are the same for each test (form). As before, we consider the case where μ_T, σ_T^2, and σ_E^2 are known; we turn to the case where these are unknown in Section 8.3.

8.2.1 Conventional Model Specification

Let x_{ij} denote the observable variable corresponding to examinee i on observable j, $i=1,\ldots,n$ denoting the examinee, and $j = 1, \ldots, J$ denoting the observable (test, test form, measure). The CTT model specifies x_{ij} as an additive combination of the examinee's true score and an error score for the examinee,

$$x_{ij} = T_i + E_{ij}, \tag{8.20}$$

where E_{ij} is now the error for examinee i on observable j. The means of the errors for all observables are 0, and the variances of the errors for the observables are equal, denoted by σ_E^2. Letting E_j denote the error scores for observable j, the correlation between the true scores and the errors for observable j is

$$\rho_{TE_j} = 0 \tag{8.21}$$

for all j. The correlation between errors for observable j and observable j' is

$$\rho_{E_j E_{j'}} = 0 \tag{8.22}$$

for $j \neq j'$.

Letting μ_{xj} and $\sigma_{x_j}^2$ denote the mean and variance of the scores for observable j, the CTT model implies that, for $j = 1, \ldots, J$,

$$\mu_{x_j} = \mu_T \tag{8.23}$$

and

$$\sigma_{x_j}^2 = \sigma_T^2 + \sigma_E^2. \tag{8.24}$$

As μ_{xj} and $\sigma_{x_j}^2$ are the same for all j, we drop the second subscript and use μ_x and σ_x^2 to denote the mean and variance for any observable.

Under these specifications, the reliability of any observable is given by (8.5). The reliability of an equally weighted composite of the J observables, such as their sum or mean, is given by the Spearman–Brown prophecy formula,

$$\rho_c = \frac{J\rho}{(J-1)\rho+1} = \frac{J\sigma_T^2}{J\sigma_T^2+\sigma_E^2}, \tag{8.25}$$

where ρ_c denotes the reliability of the composite of the J observables.

8.2.2 Bayesian Modeling

8.2.2.1 Model Specification

The DAG for the model is given in Figure 8.5. The lone difference between this and the DAG in Figure 8.1 is the existence of multiple observables, represented here by the plate over j. Implicit in this plate structure is the conditional independence assumptions that allow for the factorization of the conditional probability of the data.

Following Bayes' theorem, the posterior distribution is

$$p(\mathbf{T} \mid \mathbf{x}, \mu_T, \sigma_T^2, \sigma_E^2) \propto p(\mathbf{x} \mid \mathbf{T}, \sigma_E^2) p(\mathbf{T} \mid \mu_T, \sigma_T^2), \tag{8.26}$$

where now $\mathbf{x}$ is the full collection of $(n \times J)$ observables. The second term on the right-hand side of (8.26) is the prior distribution for examinees' true scores. As before, following assumptions of exchangeability and normality, the prior distribution is given in (8.11) and (8.12).

The first term on the right-hand side of (8.26) is the conditional distribution of the data given the true scores and the error variance. As before, we assume conditional independence between the values of the observables from different examinees. This supports the factorization

$$p(\mathbf{x} \mid \mathbf{T}, \sigma_E^2) = \prod_{i=1}^{n} p(\mathbf{x}_i \mid T_i, \sigma_E^2), \tag{8.27}$$

where $\mathbf{x}_i = (x_{i1}, \ldots, x_{iJ})$ is the collection of J observable variables for examinee i. Assuming that, for each examinee, the observables are conditionally (locally) independent given their true score allows for the specification of a common distribution for each observable,

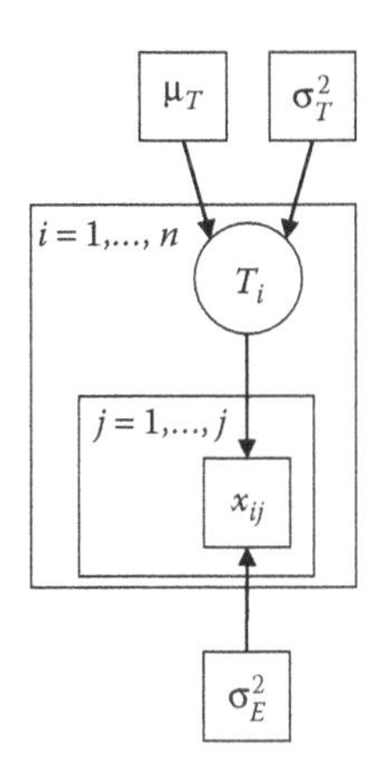

FIGURE 8.5
Directed acyclic graph for the classical test theory model for multiple observable variables, with known true score mean, true score variance, and error variance.

$$p(\mathbf{x}_i \mid T_i, \sigma_E^2) = \prod_{j=1}^{J} p(x_{ij} \mid T_i, \sigma_E^2). \tag{8.28}$$

Combining these two expressions, we have the factorization of the conditional distribution of the data as

$$p(\mathbf{x} \mid \mathbf{T}, \sigma_E^2) = \prod_{i=1}^{n} \prod_{j=1}^{J} p(x_{ij} \mid T_i, \sigma_E^2). \tag{8.29}$$

For each examinee, we model the observables as being normally distributed around the examinee's true score,

$$x_{ij} \mid T_i, \sigma_E^2 \sim N(T_i, \sigma_E^2). \tag{8.30}$$

Substituting into (8.26), the posterior distribution is then

$$\begin{aligned} p(\mathbf{T} \mid \mathbf{x}, \mu_T, \sigma_T^2, \sigma_E^2) &\propto p(\mathbf{x} \mid \mathbf{T}, \sigma_E^2) p(\mathbf{T} \mid \mu_T, \sigma_T^2) \\ &= \prod_{i=1}^{n} \prod_{j=1}^{J} p(x_i \mid T_i, \sigma_E^2) p(T_i \mid \mu_T, \sigma_T^2), \end{aligned} \tag{8.31}$$

where

$$x_{ij} \mid T_i, \sigma_E^2 \sim N(T_i, \sigma_E^2) \quad \text{for } i = 1, \ldots, n, j = 1, \ldots, J$$

and

$$T_i \mid \mu_T, \sigma_T^2 \sim N(\mu_T, \sigma_T^2) \quad \text{for } i = 1, \ldots, n.$$

8.2.2.2 Hierarchical Specification

At this point it is worth taking a step back to develop the model from the perspective of "building out" from the data. We begin with the $(n \times J)$ collection of observed values. For any examinee i, we have a collection of J values in $\mathbf{x}_i$. To specify the distribution for $\mathbf{x}_i$, we invoke an assumption of exchangeability. This allows for the specification of the distribution as a product of conditional distributions, given a possibly vector-valued parameter. This is expressed in (8.28), where the conditioning parameter consists of the known error variance and the unknown true score. The latter of these, being unknown, requires a distributional specification. An assumption of exchangeability with respect to the examinees implies that the same conditional distributions for observables can be used for all examinees, as expressed in (8.29). This process yields n true scores in need of distributional specification. Again, the exchangeability assumption with respect to examinees allows for the specification of a common conditional distribution, as in (8.11). The conditional distribution here is specified with respect to a parameter that consists of the known mean and variance of the true scores. This completes the specification of the model. To recap, we begin with the data and "build out" by invoking exchangeability assumptions to simplify the specification of the distribution of the data. But that introduces new unknown parameters (here, true scores) that stand in need of distributional specification. We accomplish this by again invoking exchangeability assumptions to simplify the specification of the distribution of the true scores.

8.2.2.3 Sufficient Statistics

The full model in (8.31) has the structure where, essentially, there is a model for an individual examinee that is repeated over examinees. For each examinee, the model is one where there are multiple normally distributed observables, $x_{i1},\ldots, x_{iJ}$, with known variance σ_E^2 and unknown mean T_i. This is therefore an instance of the model described in Section 4.1. As discussed there, the model may be formulated in terms of the sufficient statistics, which in this case is the sample mean, $\bar{x}_i = \sum_{j=1}^{J} x_{ij}/J$, and the number of observables, J.

We may therefore model the observables in terms of the observable sample means for examinees conditional on the true scores and error variance,

$$\bar{x}_i \mid T_i, \sigma_E^2 \sim N\left(T_i, \frac{\sigma_E^2}{J}\right). \tag{8.32}$$

The DAG for this version of the model is given in Figure 8.6.

8.2.2.4 Posterior Distribution

Collecting the preceding developments, the posterior distribution is

$$\begin{aligned} p(\mathbf{T} \mid \mathbf{x}, \mu_T, \sigma_T^2, \sigma_E^2) &\propto p(\mathbf{x} \mid \mathbf{T}, \sigma_E^2) p(\mathbf{T} \mid \mu_T, \sigma_T^2) \\ &= \prod_{i=1}^{n} p(\bar{x}_i \mid T_i, \sigma_E^2) p(T_i \mid \mu_T, \sigma_T^2), \end{aligned} \tag{8.33}$$

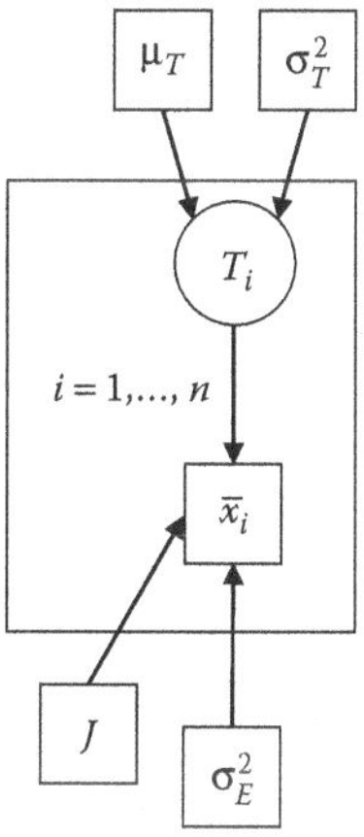

FIGURE 8.6
Directed acyclic graph for the classical test theory model for multiple observable variables formulated using the mean of the observables, with known true score mean, true score variance, and error variance.

where

$$\bar{x}_i \mid T_i, \sigma_E^2 \sim N\left(T_i, \frac{\sigma_E^2}{J}\right) \quad \text{for } i = 1, \ldots, n,$$

and

$$T_i \mid \mu_T, \sigma_T^2 \sim N(\mu_T, \sigma_T^2) \quad \text{for } i = 1, \ldots, n.$$

The posterior distribution for each T_i is again an instantiation of (4.6),

$$T_i \mid \mathbf{x}_i, \mu_T, \sigma_T^2, \sigma_E^2 \sim N(T_i \mid \mu_{T_i|\mathbf{x}_i}, \sigma^2_{T_i|\mathbf{x}_i}), \tag{8.34}$$

where

$$\mu_{T_i|\mathbf{x}_i} = \frac{\left(\mu_T/\sigma_T^2\right) + \left(J\bar{x}_i/\sigma_E^2\right)}{\left(1/\sigma_T^2\right) + \left(J/\sigma_E^2\right)} \tag{8.35}$$

and

$$\sigma^2_{T_i|\mathbf{x}_i} = \frac{1}{\left(1/\sigma_T^2\right) + \left(J/\sigma_E^2\right)}. \tag{8.36}$$

A little algebra reveals that the posterior mean can be expressed as

$$\mu_{T_i|\mathbf{x}_i} = \frac{J\sigma_T^2}{J\sigma_T^2 + \sigma_E^2}\bar{x}_i + \frac{\sigma_E^2}{J\sigma_T^2 + \sigma_E^2}\mu_T. \tag{8.37}$$

The coefficient for $\bar{x}_i$ is recognized as the reliability ρ_c. A little more algebra reveals that the coefficient for μ_T is $1 - \rho_c$. Thus, the posterior mean for an examinee's true score is

$$\begin{aligned} \mu_{T_i|\mathbf{x}_i} &= \rho_c \bar{x}_i + (1 - \rho_c)\mu_T \\ &= \mu_T + \rho_c(\bar{x}_i - \mu_T). \end{aligned} \tag{8.38}$$

The right-hand side of (8.38) is an instantiation of Kelley's formula expressed in terms of the sample mean (Lewis, 2007). In light of the connection between Kelley's formula and Bayesian inference, this result is unsurprising. To restate, the posterior mean is the weighted composite of the mean of the observables for the examinee and the mean of the prior, where the weights are the reliability of the observables and its complement, respectively. This concept is even more clearly seen in the precision parameterization of the normal distribution. Recalling that the precision is the inverse of the variance, the posterior distribution for an examinee's true score may be written as

$$T_i \mid \mathbf{x}_i, \mu_T, \tau_T, \tau_E \sim N(T_i \mid \mu_{T_i|\mathbf{x}_i}, \tau_{T_i|\mathbf{x}_i}), \tag{8.39}$$

where $\tau_T = 1/\sigma_T^2$ is the precision of the true scores, $\tau_E = 1/\sigma_E^2$ is the precision of the error scores,

$$\mu_{T_i|\mathbf{x}_i} = \frac{J\tau_E}{\tau_T + J\tau_E}\bar{x}_i + \frac{\tau_T}{\tau_T + J\tau_E}\mu_T, \tag{8.40}$$

and

$$\tau_{T_i|\mathbf{x}_i} = \tau_T + J\tau_E \tag{8.41}$$

is the posterior precision. Equation (8.41) reveals how the precision in the data and therefore the total precision in the posterior increases as the number of observables J increases. This reflects the principles of the Spearman–Brown prophecy formula in CTT and the notion of Bayesian inference as the mechanism for the accumulation of evidence and the representation of uncertainty as evidence accumulates. Equation (8.40) reveals that the posterior mean is a precision-weighted average of the mean of the observables and the prior mean. This lends the following interpretation to the reliability; namely, the proportion of the posterior precision that is due to the precision in the observables. The complement of the reliability is the proportion of the posterior precision due to the prior.

8.2.3 Example

We continue the example scenario, where we have a test scored from 0 to 100 and we are interested in true scores for students where it is known that the mean of the true scores is $\mu_T = 80$, the standard deviation of the true scores is $\sigma_T = 6$, and the standard deviation of the error scores for each observable is $\sigma_E = 4$. The reliability of each test is calculated via (8.5), $\rho \approx .69$. The reliability of an equally weighted composite is calculated via (8.25), $\rho_c \approx .92$. We consider a dataset of 10 examinees and 5 tests, given in Table 8.2.

8.2.3.1 Analytical Solution

For each examinee, the posterior distribution is normal with mean

$$\begin{aligned}\mu_{T_i|\mathbf{x}_i} &= \frac{J\sigma_T^2}{J\sigma_T^2 + \sigma_E^2}\bar{x}_i + \frac{\sigma_E^2}{J\sigma_T^2 + \sigma_E^2}\mu_T \\ &= \frac{(5)36}{(5)36 + 16}\bar{x}_i + \frac{16}{(5)36 + 16}80 \\ &\approx .92\bar{x}_i + 6.53.\end{aligned}$$

Again, we note that each examinee's posterior mean is exactly what is obtained by the conventional application of Kelley's formula. The posterior variance is constant for all examinees,

$$\sigma^2_{T_i|\mathbf{x}_i} = \frac{1}{\left(1/\sigma_T^2\right) + \left(J/\sigma_E^2\right)} = \frac{1}{(1/36) + (5/16)} \approx 2.94.$$

For example, the posterior distribution for the first examinee's true score is $p(T_1 \mid \mathbf{x}_1) = N(76.88, 2.94)$. Table 8.2 lists the posterior means and standard deviations for all examinees under the columns headed by Posterior (Analytical).

TABLE 8.2

Dataset with Observed Scores from 10 Examinees to 5 Tests, and the Results from a Bayesian Classical Test Theory Analysis

	Test					Mean	Posterior (Analytical)		Posterior (MCMC[a])	
Examinee	1	2	3	4	5	$\bar{x}$	Mean	SD[b]	Mean	SD[b]
1	80	77	80	73	73	76.6	76.88	1.71	76.87	1.73
2	83	79	78	78	77	79.0	79.08	1.71	79.12	1.70
3	85	77	88	81	80	82.2	82.02	1.71	81.91	1.72
4	76	76	76	78	67	74.6	75.04	1.71	75.01	1.72
5	70	69	73	71	77	72.0	72.65	1.71	72.65	1.74
6	87	89	92	91	87	89.2	88.45	1.71	88.47	1.70
7	76	75	79	80	75	77.0	77.24	1.71	77.26	1.67
8	86	75	80	80	82	80.6	80.55	1.71	80.54	1.74
9	84	79	79	77	82	80.2	80.18	1.71	80.22	1.67
10	96	85	91	87	90	89.8	89.00	1.71	88.99	1.71

[a] MCMC = Markov chain Monte Carlo.
[b] SD = Standard deviation.

8.2.3.2 MCMC Estimation via WinBUGS

The WinBUGS code for the model is given as follows.

```
-----------------------------------------------------------------------
#########################################################################
# Model Syntax
#########################################################################
model{

#########################################################################
# Classical Test Theory
# With Known
#    True Score Mean, True Score Variance
#    Error Variance
#########################################################################

#########################################################################
# Known Parameters
#########################################################################
mu.T <- 80                        # Mean of the true scores
sigma.squared.T <- 36             # Variance of the true scores
sigma.squared.E <- 16             # Variance of the errors

tau.T <- 1/sigma.squared.T        # Precision of the true scores
tau.E <- 1/sigma.squared.E        # Precision of the errors


#########################################################################
# Model for True Scores and Observables
#########################################################################
```

```
for (i in 1:n) {
  T[i] ~ dnorm(mu.T, tau.T)                    # Distribution of true scores
  for(j in 1:J){
    x[i,j] ~ dnorm(T[i], tau.E)                # Distribution of observables
  }
}
}                                               # closes the model statement


##########################################################################
# Data statement
##########################################################################
list(n=10, J=5, x=structure(.Data= c(
80, 77, 80, 73, 73,
83, 79, 78, 78, 77,
85, 77, 88, 81, 80,
76, 76, 76, 78, 67,
70, 69, 73, 71, 77,
87, 89, 92, 91, 87,
76, 75, 79, 80, 75,
86, 75, 80, 80, 82,
84, 79, 79, 77, 82,
96, 85, 91, 87, 90), .Dim=c(10, 5))
)
--------------------------------------------------------------------------
```

The model was run with the data contained in Table 8.2. Three chains were run from dispersed starting values. Convergence diagnostics including the history plots and the potential scale reduction factor (Section 5.7.2) indicated convergence within a few iterations. This is unsurprising given that the posterior is of a known form (see Equation 8.39) that is recognized as such by WinBUGS. To be conservative, 100 iterations were discarded as burn-in and 1,000 additional iterations were run, yielding 3,000 iterations for use in summarizing the posterior distribution. The marginal posterior distributions for the true scores for the examinees were all approximately normal. The final two columns in Table 8.2 report the estimated posterior mean and standard deviation for each examinee from the MCMC estimation. It is clear that these closely approximate the analytically derived values.

8.3 CTT with Unknown Measurement Model Parameters and Hyperparameters

8.3.1 Bayesian Model Specification and Posterior Distribution

In the preceding development of CTT, the mean and variance of the true scores were assumed known, as was the variance of the errors. In the parlance of Chapter 7, our focus was on scoring. In this section, we generalize to conditions where these quantities are not assumed known and are therefore incorporated into the model as random variables, moving to a situation where we are conducting calibration as well as scoring.

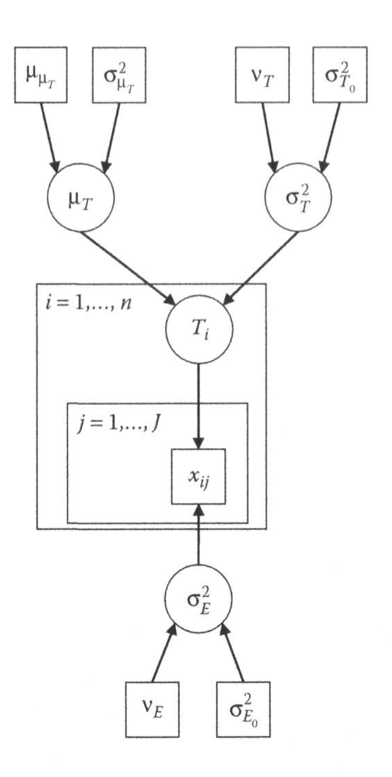

FIGURE 8.7
Directed acyclic graph for the classical test theory model for multiple observable variables, with unknown hyperparameters: true score mean, true score variance, and error variance.

The DAG for the model is given in Figure 8.7. In contrast to the DAGs for the previous models, the mean and variance of the true scores and the error variance are depicted as circles, reflecting that they are unknown. Accordingly, they are modeled with a higher structure needed to specify their distributions. As stochastic entities, they will need to be incorporated into the distributional specification.

Following Bayes' theorem, the posterior distribution for the full collection of unknowns given the full collection of the observed scores $\mathbf{x}$ is

$$p(\mathbf{T}, \mu_T, \sigma^2_T, \sigma^2_E \mid \mathbf{x}) \propto p(\mathbf{x} \mid \mathbf{T}, \sigma^2_E) p(\mathbf{T} \mid \mu_T, \sigma^2_T) p(\mu_T, \sigma^2_T, \sigma^2_E). \tag{8.42}$$

The first two terms on the right-hand side of (8.42) are (a) the conditional distribution of the data given the true scores and the error variance and (b) the prior distribution for true scores given the mean and variance of the true scores. These terms are unchanged from the situation in which μ_T, σ^2_T, and σ^2_E are assumed known and we adopt the same distributional specifications in (8.11), (8.12), (8.29), and (8.30).

The third term on the right-hand side of (8.42) is the prior distribution for μ_T, σ^2_T, and σ^2_E. As the structure of the DAG in Figure 8.7 conveys, we proceed by assuming independence in the prior distribution,

$$p(\mu_T, \sigma^2_T, \sigma^2_E) = p(\mu_T) p(\sigma^2_T) p(\sigma^2_E). \tag{8.43}$$

We complete the task by specifying conditionally conjugate priors:

$$\mu_T \sim N(\mu_{\mu T}, \sigma^2_{\mu T}), \tag{8.44}$$

$$\sigma^2_T \sim \text{Inv-Gamma}(\nu_T/2, \nu_T \sigma^2_{T0}/2), \tag{8.45}$$

and

$$\sigma^2_E \sim \text{Inv-Gamma}(\nu_E/2, \nu_E \sigma^2_{E0}/2), \tag{8.46}$$

where $\mu_{\mu T}$ and $\sigma^2_{\mu T}$ are hyperparameters that govern the distribution of μ_T, ν_T, and σ^2_{T0} are hyperparameters that govern the distribution of σ^2_T, and ν and $\sigma^2_{E0}$are hyperparameters that govern the distribution of σ^2_E.

Substituting into (8.42), the posterior distribution is then

$$p(\mathbf{T}, \mu_T, \sigma^2_T, \sigma^2_E \mid \mathbf{x}) \propto \prod_{i=1}^{n} \prod_{j=1}^{J} p(x_{ij} \mid T_i, \sigma^2_E) p(T_i \mid \mu_T, \sigma^2_T) p(\mu_T) p(\sigma^2_T) p(\sigma^2_E), \tag{8.47}$$

where

$$x_{ij} \mid T_i, \sigma^2_E \sim N(T_i, \sigma^2_E) \text{ for } i = 1, \ldots, n, j = 1, \ldots, J,$$

$$T_i \mid \mu_T, \sigma^2_T \sim N(\mu_T, \sigma^2_T) \text{ for } i = 1, \ldots, n,$$

$$\mu_T \sim N(\mu_{\mu T}, \sigma^2_{\mu T}),$$

$$\sigma^2_T \sim \text{Inv-Gamma}(\nu_T/2, \nu_T \sigma^2_{T0}/2),$$

and

$$\sigma^2_E \sim \text{Inv-Gamma}(\nu_E/2, \nu_E \sigma^2_{E0}/2).$$

8.3.2 MCMC Estimation

With the conditionally conjugate prior, the full conditional distributions are of known form, and we express them in the following equations (see Appendix A for derivations). On the left-hand side, the parameter in question is written as conditional on all the other relevant parameters and data.* On the right-hand side of each of the following equations, we give the parametric form for the full conditional distribution. In several places, we denote the arguments of the distribution (e.g., mean and variance for a normal distribution) with subscripts denoting that it refers to the full conditional distribution; the subscripts are then just the conditioning notation of the left-hand side.

Beginning with the examinees' latent variables, we present the full conditional for any examinee i. The same structure applies to all the examinees. The full conditional distribution for the latent variable for examinee i is

* We suppress the role of specified hyperparameters in this notation; see Appendix A for a presentation that formally includes the hyperparameters.

$$T_i \mid \mu_T, \sigma_T^2, \sigma_E^2, \mathbf{x}_i \sim N(\mu_{T_i \mid \mu_T, \sigma_T^2, \sigma_E^2, \mathbf{x}_i}, \sigma^2_{T_i \mid \mu_T, \sigma_T^2, \sigma_E^2, \mathbf{x}_i}), \tag{8.48}$$

where

$$\mu_{T_i \mid \mu_T, \sigma_T^2, \sigma_E^2, \mathbf{x}_i} = \frac{\left(\mu_T / \sigma_T^2\right) + \left(J\bar{x}_i / \sigma_E^2\right)}{\left(1/\sigma_T^2\right) + \left(J/\sigma_E^2\right)}$$

and

$$\sigma^2_{T_i \mid \mu_T, \sigma_T^2, \sigma_E^2, \mathbf{x}_i} = \frac{1}{\left(1/\sigma_T^2\right) + \left(J/\sigma_E^2\right)}.$$

The full conditional for the mean of the true scores is

$$\mu_T \mid \mathbf{T}, \sigma_T^2 \sim N(\mu_{\mu_T \mid \mathbf{T}, \sigma_T^2}, \sigma^2_{\mu_T \mid \mathbf{T}, \sigma_T^2}), \tag{8.49}$$

where

$$\mu_{\mu_T \mid \mathbf{T}, \sigma_T^2} = \frac{\left(\mu_{\mu_T} / \sigma^2_{\mu_T}\right) + \left(n\bar{T} / \sigma_T^2\right)}{\left(1/\sigma^2_{\mu_T}\right) + \left(n/\sigma_T^2\right)}$$

and

$$\sigma^2_{\mu_T \mid \mathbf{T}, \sigma_T^2} = \frac{1}{\left(1/\sigma^2_{\mu_T}\right) + \left(n/\sigma_T^2\right)}.$$

The full conditional for the variance of the true scores is

$$\sigma_T^2 \mid \mathbf{T}, \mu_T \sim \text{Inv-Gamma}\left(\frac{\nu_T + n}{2}, \frac{\nu_T \sigma^2_{T_0} + SS(\mathbf{T})}{2}\right), \tag{8.50}$$

where

$$SS(\mathbf{T}) = \sum_{i=1}^{n} (T_i - \mu_T)^2.$$

Finally, the full conditional for the error variance is

$$\sigma_E^2 \mid \mathbf{T}, \mathbf{x} \sim \text{Inv-Gamma}\left(\frac{\nu_E + nJ}{2}, \frac{\nu_E \sigma^2_{E_0} + SS(\mathbf{E})}{2}\right), \tag{8.51}$$

where

$$\text{SS}(\mathbf{E}) = \sum_{i=1}^{n}\sum_{j=1}^{J}(x_{ij} - T_i)^2.$$

A Gibbs sampler can be constructed by iteratively drawing from these full conditionals, using the just-drawn values for the conditioned parameters. We begin by setting $t = 0$ and specify initial values for the parameters: $T_1^{(0)},\ldots,T_n^{(0)}, \mu_T^{(0)}, \sigma_T^{2(0)}, \sigma_E^{2(0)}$, where the parenthetical superscript of 0 indicates that this is the initial value. We then proceed with a Gibbs sampler, repeating the following steps, generically written below for iteration $(t + 1)$:

8.3.2.1 Gibbs Sampler Routine for Classical Test Theory

1. *Sample the latent variables for examinees.* For each examinee $i = 1, \ldots, n$, sample a value for the true score from (8.48) using the current values for the other parameters,

$$T_i^{(t+1)} \mid \mu_T^{(t)}, \sigma_T^{2(t)}, \sigma_E^{2(t)}, \mathbf{x}_i \sim N(\mu_{T_i^{(t+1)}|\mu_T^{(t)},\sigma_T^{2(t)},\sigma_E^{2(t)},\mathbf{x}_i}, \sigma^2_{T_i^{(t+1)}|\mu_T^{(t)},\sigma_T^{2(t)},\sigma_E^{2(t)},\mathbf{x}_i}), \tag{8.52}$$

where

$$\mu_{T_i^{(t+1)}|\mu_T^{(t)},\sigma_T^{2(t)},\sigma_E^{2(t)},\mathbf{x}_i} = \frac{\left(\mu_T^{(t)}/\sigma_T^{2(t)}\right)+\left(J\bar{x}/\sigma_E^{2(t)}\right)}{\left(1/\sigma_T^{2(t)}\right)+\left(J/\sigma_E^{2(t)}\right)}$$

and

$$\sigma^2_{T_i^{(t+1)}|\mu_T^{(t)},\sigma_T^{2(t)},\sigma_E^{2(t)},\mathbf{x}_i} = \frac{1}{\left(1/\sigma_T^{2(t)}\right)+\left(J/\sigma_E^{2(t)}\right)}.$$

Note that this uses values of the other parameters from the previous iteration (t).

2. Sample the parameters for the latent variable distribution. We conduct this by sampling in a univariate fashion.
   a. Sample a value for the mean of the true scores from (8.49) using the current values for the remaining parameters,

$$\mu_T^{(t+1)} \mid \mathbf{T}^{(t+1)}, \sigma_T^{2(t)} \sim N(\mu_{\mu_T^{(t+1)}|\mathbf{T}^{(t+1)},\sigma_T^{2(t)}}, \sigma^2_{\mu_T^{(t+1)}|\mathbf{T}^{(t+1)},\sigma_T^{2(t)}}), \tag{8.53}$$

where

$$\mu_{\mu_T^{(t+1)}|\mathbf{T}^{(t+1)},\sigma_T^{2(t)}} = \frac{\left(\mu_{\mu T}/\sigma^2_{\mu T}\right)+\left(n\bar{T}^{(t+1)}/\sigma_T^{2(t)}\right)}{\left(1/\sigma^2_{\mu T}\right)+\left(n/\sigma_T^{2(t)}\right)}$$

and

$$\sigma^2_{\mu_T^{(t+1)}|\mathbf{T}^{(t+1)},\sigma_T^{2(t)}} = \frac{1}{\left(1/\sigma^2_{\mu_T}\right)+\left(n/\sigma_T^{2(t)}\right)}.$$

Note that this includes the just-sampled values for the true scores (i.e., from iteration $t + 1$) along with the values of the true score variance from the previous iteration (t).

b. Sample a value for the variance of the true scores from (8.50) using the current values for the remaining parameters,

$$\sigma_T^{2(t+1)} \mid \mathbf{T}^{(t+1)}, \mu_T^{(t+1)} \sim \text{Inv-Gamma}\left(\frac{\nu_T + n}{2}, \frac{\nu_T\sigma^2_{T_0} + SS(\mathbf{T}^{(t+1)})}{2}\right), \tag{8.54}$$

where

$$SS(\mathbf{T}^{(t+1)}) = \sum_{i=1}^{n}(T_i^{(t+1)} - \mu_T^{(t+1)})^2.$$

Note that this includes the just-sampled values for the true scores and the mean of the true scores (i.e., from iteration $t + 1$).

3. *Sample the measurement model parameters.* Sample a value for the error variance from (8.51) using the current values of the remaining parameters,

$$\sigma_E^{2(t+1)} \mid \mathbf{T}^{(t+1)}, \mathbf{x} \sim \text{Inv-Gamma}\left(\frac{\nu_E + nJ}{2}, \frac{\nu_E\sigma^2_{E_0} + SS(\mathbf{E}^{(t+1)})}{2}\right), \tag{8.55}$$

where

$$SS(\mathbf{E}^{(t+1)}) = \sum_{i=1}^{n}\sum_{j=1}^{J}(x_{ij} - T_i^{(t+1)})^2.$$

Note that this includes the just-sampled values for the true scores, the mean of the true scores, and the variance of the true scores (i.e., from iteration $t + 1$).

8.3.3 Example

8.3.3.1 Model Specification and Posterior Distribution

We continue with the example scenario and the data in Table 8.2 containing observed values from $n = 10$ examinees for $J = 5$ observables, now with μ_T, σ_T^2, and σ_E^2 treated as unknown. We construct a prior distribution for the mean of the true scores based on the prior beliefs that the mean is likely around 80, and almost certainly between 60 and 100. Accordingly, we specify a normal prior with mean $\mu_{\mu_T} = 80$ and variance $\sigma^2_{\mu_T} = 100$,

$$\mu_T \sim N(80,100).$$

We construct a prior distribution for the true score variance expressing the beliefs that the variance of the true scores is likely about 36 (i.e., the standard deviation is about 6), but we are not very confident, it is as if our beliefs were based on having observed two examinees. Accordingly, we specify $\sigma^2_{\sigma_T^2} = 36$ and $\nu_{\sigma_T^2} = 2$, yielding

$$\sigma_T^2 \sim \text{Inv-Gamma}(1,36).$$

We construct a prior distribution for the error variance expressing the beliefs that the variance of the errors is likely about 16 (i.e., the standard deviation is about 4), but we are not very confident, as if that were based on two examinees. Accordingly, we specify $\sigma^2_{\sigma_E^2} = 16$ and $\nu_{\sigma_E^2} = 2$, yielding

$$\sigma_E^2 \sim \text{Inv-Gamma}(1,16).$$

Writing out the posterior distribution for the model using these chosen values for the hyperparameters, we have

$$p(\mathbf{T},\mu_T,\sigma_T^2,\sigma_E^2 \mid \mathbf{x}) \propto \prod_{i=1}^{n}\prod_{j=1}^{J} p(x_{ij} \mid T_i,\sigma_E^2)p(T_i \mid \mu_T,\sigma_T^2)p(\mu_T)p(\sigma_T^2)p(\sigma_E^2),$$

where

$$x_{ij} \mid T_i,\sigma_E^2 \sim N(T_i,\sigma_E^2) \quad \text{for } i = 1,\ldots,10, j = 1,\ldots,5,$$

$$T_i \mid \mu_T,\sigma_T^2 \sim N(\mu_T,\sigma_T^2) \quad \text{for } i = 1,\ldots,10,$$

$$\mu_T \sim N(80,100),$$

$$\sigma_T^2 \sim \text{Inv-Gamma}(1,36),$$

and

$$\sigma_E^2 \sim \text{Inv-Gamma}(1,16).$$

8.3.3.2 Gibbs Sampling

The computations for several iterations of a Gibbs sampler defined in (8.52)–(8.55) are summarized in Table 8.3. We briefly step through the computations for the first iteration. We begin by specifying initial values $(T_1^{(0)},\ldots,T_{10}^{(0)}) = (80,\ldots,89)$, $\mu_T^{(0)} = 75$, $\sigma_T^{2(0)} = 50$, $\sigma_E^{2(0)} = 10$. To enact the Gibbs sampler, we cycle through (8.52)–(8.55):

TABLE 8.3

Computations in the Gibbs Sampler for the Classical Test Theory Example with Unknown True Scores (T), Mean of the True Scores (μ_T), Variance of the True Scores (σ_T^2), and Error Variance (σ_E^2)

	True Scores										Full Conditional for T_1	
Iteration	T_1	T_2	T_3	T_4	T_5	T_6	T_7	T_8	T_9	T_{10}	$\frac{(\mu_T/\sigma_T^2)+(J\bar{x}_1/\sigma_E^2)}{(1/\sigma_T^2)+(J/\sigma_E^2)}$	$\frac{1}{(1/\sigma_T^2)+(J/\sigma_E^2)}$
0	80.00	81.00	82.00	83.00	84.00	85.00	86.00	87.00	88.00	89.00	76.54	1.92
1	77.65	78.62	81.72	71.70	74.38	89.17	75.79	82.19	76.41	89.00	76.54	1.92
2	76.31	80.28	82.03	74.94	72.78	89.21	76.92	77.15	78.12	88.70	76.63	1.76
3	77.40	78.83	82.41	74.83	71.12	86.99	76.36	79.56	79.06	85.85	76.66	1.80
4	77.90	79.24	82.17	76.21	73.26	88.37	74.51	80.27	81.96	87.62	77.04	1.85
5	76.07	79.72	82.37	70.90	74.66	85.72	78.16	79.94	80.52	89.44	76.79	2.62

	Parameters			Full Conditional for μ_T		Full Conditional for σ_T^2			Full Conditional for σ_E^2		
Iteration	μ_T	σ_T^2	σ_E^2	$\frac{(\mu_{\mu T}/\sigma_{\mu T}^2)+(n\bar{T}/\sigma_T^2)}{(1/\sigma_{\mu T}^2)+(n/\sigma_T^2)}$	$\frac{1}{(1/\sigma_{\mu T}^2)+(n/\sigma_T^2)}$	$\frac{\nu_T+n}{2}$	$SS(T)$	$\frac{\nu_T\sigma_{T_0}^2+SS(T)}{2}$	$\frac{\nu_E+nJ}{2}$	$SS(E)$	$\frac{\nu_E\sigma_{E_0}^2+SS(E)}{2}$
0	75.00	50.00	10.00	84.29	4.76	6.00	985.00	528.50	26.00	2625.00	1328.50
1	76.99	26.11	9.45	79.68	4.76	6.00	381.46	226.73	26.00	641.87	336.94
2	77.23	19.96	9.88	79.66	2.54	6.00	334.38	203.19	26.00	568.27	300.13
3	80.07	14.49	10.57	79.26	1.96	6.00	218.09	145.05	26.00	592.83	312.42
4	79.84	45.44	13.91	80.15	1.43	6.00	232.51	152.25	26.00	572.36	302.18
5	77.38	31.60	14.35	79.76	4.35	6.00	313.59	192.80	26.00	647.43	339.72

1. *Sample the latent variables for examinees.* For the first examinee, we compute

$$\mu_{T_1^{(1)}|\mu_T^{(0)},\sigma_T^{2(0)},\sigma_E^{2(0)},\mathbf{x}_i} = \frac{\left(\mu_T^{(0)}/\sigma_T^{2(0)}\right)+\left(J\bar{x}_1/\sigma_E^{2(0)}\right)}{\left(1/\sigma_T^{2(0)}\right)+\left(J/\sigma_E^{2(0)}\right)} = \frac{(75/50)+((5)(76.6)/10)}{(1/50)+(5/10)} \approx 76.54$$

and

$$\sigma^2_{T_1^{(1)}|\mu_T^{(0)},\sigma_T^{2(0)},\sigma_E^{2(0)},\mathbf{x}_i} = \frac{1}{(1/50)+(5/10)} \approx 1.92,$$

and draw a value from the full conditional $T_1^{(1)} \mid \mu_T^{(0)},\sigma_T^{2(0)},\sigma_E^{2(0)},\mathbf{x}_1 \sim N(76.54,1.92)$. The drawn value was 77.65. Repeat this process for the remaining examinees $i = 2, \ldots, 10$.

2. *Sample the parameters for the latent variable distribution.*
   a. Turning to the full conditional for μ_T, we compute

$$\mu_{\mu_T^{(1)}|\mathbf{T}^{(1)},\sigma_T^{2(0)}} = \frac{\left(\mu_{\mu T}/\sigma^2_{\mu T}\right)+\left(n\bar{T}^{(1)}/\sigma_T^{2(0)}\right)}{\left(1/\sigma^2_{\mu T}\right)+\left(n/\sigma_T^{2(0)}\right)} = \frac{(80/100)+((10)79.67/50)}{(1/100)+(10/50)} \approx 79.68$$

   and

$$\sigma^2_{\mu_T^{(1)}|\mathbf{T}^{(1)},\sigma_T^{2(0)}} = \frac{1}{(1/100)+(10/50)} \approx 4.76,$$

   and draw a value from the full conditional $\mu_T^{(1)} \mid \mathbf{T}^{(1)},\sigma_T^{2(0)} \sim N(79.68,4.76)$. The drawn value was 76.99.

   b. Turning to σ_T^2, we compute the first parameter of the full conditional as $\nu_T + n/2 = 2 + 10/2 = 6$, which will not vary over iterations. To compute the second parameter for the first iteration, we compute

$$SS(\mathbf{T}^{(1)}) = \sum_{i=1}^{n}(T_i^{(1)} - \mu_T^{(1)})^2 = \sum_{i=1}^{n}(T_i^{(1)} - 76.99)^2 \approx 381.46$$

   and then

$$\frac{\nu_T\sigma^2_{T_0} + SS(\mathbf{T}^{(1)})}{2} = \frac{(2)(36)+381.46}{2} \approx 226.73.$$

   Accordingly, we draw a value from the full conditional $\sigma_T^{2(1)} \mid \mathbf{T}^{(1)},\mu_T^{(1)} \sim$ Inv-Gamma (6,226.73). The drawn value was 26.11.

3. *Sample the measurement model parameters.* Turning to σ_E^2 we compute the first parameter of the full conditional as $\nu_E + nJ/2 = 2 + (10)(5)/2 = 26$, which will not vary over iterations. To compute the second parameter for the first iteration, we compute

$$SS(\mathbf{E}^{(1)}) = \sum_{i=1}^{n}\sum_{j=1}^{J}(x_{ij} - T_i^{(1)})^2 \approx 641.87$$

and then

$$\frac{\nu_E \sigma_{E_0}^2 + SS(\mathbf{E}^{(1)})}{2} = \frac{(2)(16) + 641.87}{2} \approx 336.94.$$

Accordingly, we draw a value from the full conditional $\sigma_E^{2(1)} \mid \mathbf{T}^{(1)}, \mathbf{x} \sim$ Inv-Gamma (26,336.94). The drawn value was 9.45.

Table 8.3 lists the results from five iterations of the Gibbs sampler, including the relevant computations for the parameters and the true score for the first examinee.

8.3.3.3 WinBUGS

The WinBUGS code for the model, including the reliability of a single observable and the reliability of the composite of the observables, is given as follows. The data statement is the same as that given in the example in Section 8.2.3, and will not be repeated here.

```
-------------------------------------------------------------------------
#########################################################################
# Model Syntax
#########################################################################
#model{

#########################################################################
# Classical Test Theory
# With Unknown
#    True Score Mean, True Score Variance
#    Error Variance
#########################################################################

#########################################################################
# Prior Distributions for Parameters
#########################################################################
mu.T ~ dnorm(80,.01)              # Mean of the true scores,
                                  # in terms of its mean and precision

tau.T ~ dgamma(1, 36)             # Precision of the true scores
tau.E ~ dgamma(1, 16)             # Precision of the errors

sigma.squared.T <- 1/tau.T        # Variance of the true scores
sigma.squared.E <- 1/tau.E        # Variance of the errors

#########################################################################
# Model for True Scores and Observables
#########################################################################
```

```
for (i in 1:n) {
  T[i] ~ dnorm(mu.T, tau.T)          # Distribution of true scores
  for(j in 1:J){
    x[i,j] ~ dnorm(T[i], tau.E)      # Distribution of observables
  }
}

#########################################################################
# Reliability
#########################################################################
reliability <- sigma.squared.T/(sigma.squared.T+sigma.squared.E)
   reliability.of.composite <- J*reliability/((J-1)*reliability+1)

}                                    # closes the model statement
-------------------------------------------------------------------------
```

The model was run with the data contained in Table 8.2. Three chains were run from dispersed starting values. Convergence diagnostics including the history plots and the potential scale reduction factor (Section 5.7.2) suggested convergence within a few iterations. The quick convergence here is in part due to the use of the conditionally conjugate prior specification, which yields full conditional distributions that are of known form (Section 8.3.2) and recognized as such by WinBUGS. To be conservative, 100 iterations were discarded as burn-in and 10,000 additional iterations were run, yielding 30,000 iterations for use in summarizing the posterior distribution. The marginal posterior densities for the parameters are given in Figure 8.8. The reliability of each test (ρ) and the reliability of the composite (ρ_c) are also included. Table 8.4 gives the summary statistics describing these marginal posterior distributions. The posterior mean and median are given as measures of central tendency. These are quite similar for all parameters, despite the skewness in the posterior distributions for the variances and the reliability terms. The Bayesian approach results support probabilistic expressions of our uncertainty about the parameters. For example, the 95% HPD interval for ρ is (.53, .90), indicating that, based on our data and the model specifications, we are 95% sure that the reliability of each test is between .53 and .90.

8.4 Summary and Bibliographic Note

This chapter has detailed a Bayesian approach to analysis in CTT, the first of several psychometric modeling paradigms to be considered. We have seen the role of the exchangeability and conditional independence in constructing the models, a theme that will reappear throughout.

Early treatments of elements of Bayesian CTT were given by Novick (1969) and Lindley (1970), and crystallized in Novick et al.'s (1971) more comprehensive treatment focusing on the variance components, with model specification and estimation strategies reflecting the then-available foci and methods. Our development has taken a slightly different approach that directly exploits connections to the foundational material in Chapters 4 and 6 and serves as a foundation for the treatment of factor analysis models, which we turn to in the next chapter.

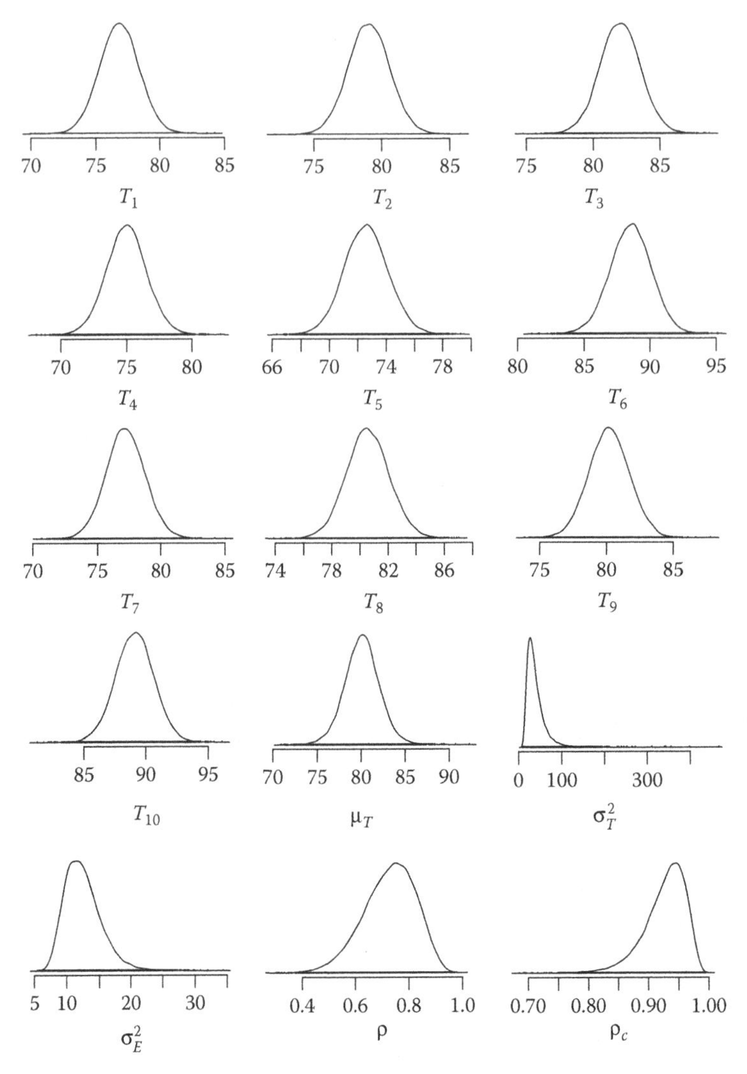

FIGURE 8.8
Marginal posterior densities for the classical test theory example: the 10 examinees' true scores ($T_1, \ldots, T_{10}$), the mean of the true scores (μ_T), true score variance (σ_T^2), error variance (σ_E^2), reliability of each test (ρ), and the reliability of the equally weighted composite (ρ_c).

Before turning to other models, let us review the developments in light of the assessment goal of making inferences about examinees. In CTT, the inferential target is the examinee's true score, so it is worthwhile to consider the various options in estimating an examinee's true score. An obvious candidate is the observed score for the examinee. In the case of one observable, this is x_i. In the case of multiple observables, this may be the mean for the observables for the examinee, $\bar{x}_i$. However, this fails to account for the uncertainty due to the (un)reliability of the tests. Kelley's formula for estimating true scores overcomes this, involving two parameters, the reliability (ρ) and the population mean of the observed scores (μ_x). (Note that if the unit of analysis is the mean over observables for an examinee, $\bar{x}_i$, the second parameter is the population mean of the $\bar{x}_i$, denoted as $\mu_{\bar{x}}$.

TABLE 8.4

Summaries of the Marginal Posterior Distributions for the Classical Test Theory Example

Parameter	Mean	Median	Standard Deviation	95% Highest Posterior Density Interval
T_1	76.85	76.85	1.53	(73.87, 79.83)
T_2	79.07	79.07	1.53	(76.00, 82.03)
T_3	82.04	82.05	1.52	(79.04, 85.03)
T_4	75.00	75.00	1.55	(71.97, 78.07)
T_5	72.6	72.59	1.55	(69.51, 75.60)
T_6	88.54	88.56	1.56	(85.50, 91.60)
T_7	77.23	77.22	1.54	(74.23, 80.26)
T_8	80.57	80.55	1.53	(77.48, 83.49)
T_9	80.2	80.19	1.52	(77.18, 83.13)
T_{10}	89.09	89.11	1.55	(86.02, 92.10)
μ_T	80.1	80.11	1.98	(76.23, 84.11)
σ_T^2	38.78	33.72	21.30	(10.83, 78.59)
σ_E^2	12.54	12.17	2.86	(7.59, 18.21)
ρ	0.73	0.74	0.10	(0.53, 0.90)
ρ_c	0.93	0.93	0.04	(0.86, 0.98)

It can easily be shown that $\mu_{\bar{x}} = \mu_x$.) In applications, these parameters are unknown. A frequentist approach employs point estimates of these parameters. A point estimate of reliability ρ may be obtained via any number of methods (Crocker & Algina, 1986; Haertel, 2006; Lewis, 2007). The population mean of the observed scores may be estimated via the sample mean of the observables. (Similarly, if $\bar{x}_i$ is the unit of analysis, $\mu_{\bar{x}}$ may be estimated as the mean of the $\bar{x}_i$). However, this ignores the uncertainty in the parameters. A fully Bayesian approach—one that models parameters as random, using distributions to represent uncertainty—incorporates the uncertainty about ρ and μ_x (alternatively, $\mu_{\bar{x}}$). (Note that in our Bayesian formulations this appears as μ_T; this is simply a change in notation, as $\mu_x = \mu_T = \mu_{\bar{x}}$.) This incorporation of uncertainty is accomplished by using the posterior distribution for the parameters, rather than point estimates, in the estimation of true scores. Of course, the uncertainty in the resulting estimation of true scores is also represented by the posterior distribution for a true score, rather than a point estimate. Finally, the same logic applies to the other parameters of interest, such as reliability. Here again, we have the posterior distribution as a synthesis of our prior beliefs and the information in the data. This affords the expression of beliefs and uncertainty about the reliability in probabilistic terms.

We close by noting that Kelley's formula is sometimes formulated using the mean of the observed scores for the group rather than a population parameter. Results of a Bayesian analysis of normal distributions using the observed mean in this way date to Lindley's discussion of Stein (1962). A more complete development in the context of evaluating true scores was given by Lindley and reported in Novick (1969). It is here that the connection between Kelley's formula and Bayes' theorem was recognized, culminating with Novick et al. (1971) explicitly stating the connection. This is certainly not the only time that that a methodological development that is well accepted as part of conventional statistical practice has subsequently been seen to be an instance of, or aligned with, Bayesian inference. Indeed, Bayesian inference has often been found to be sweeter by any other name (McGrayne, 2011).* Novick et al.'s (1971) development of the Bayesian analysis of true scores that reflects the emphasis of those times on maximally diffuse prior distributions, which may be obtained as a limiting case in the hierarchical formulation adopted here.

Exercises

8.1 Consider again the model in Section 8.3.1, depicted in the DAG in Figure 8.7. Based on the DAG, for each entity, list the entities that need to be conditioned on in the full conditional distribution.

8.2 Section 8.3.3 demonstrated the computations involved in the first iteration of a Gibbs sampler for the CTT model with unknown hyperparameters. The results for this and four additional iterations are given in Table 8.3. Show the computations for each of these additional iterations.

8.3 Consider again a CTT model with a test scored from 0 to 100 and $\sigma_E = 4$ is the standard deviation of the error scores. There are two groups of examinees: In Class A, $\mu_{T(A)} = 85$, in Class B, $\mu_{T(B)} = 75$, and in both groups the standard deviation of the true scores is $\sigma_T = 6$. Both within-group true-score distributions and the error distributions are normal. Generate 200 observed scores for both groups, in each case by first drawing a true score from the group distribution and then drawing an error term to add to it, so you know each simulee's θ and x. Round each observed score to the closest integer.

a. For each *observed* score between 70 and 90, calculate the mean of the true scores of simulees who obtained that x, within each group separately and combining across groups.
b. Calculate the posterior mean that corresponds to each observed score between 70 and 90, for each group separately and combining across groups. (Hint: Use Kelley's formula for the combined-group answers. Calculate the combined-group variance using the within-group variance and the squared difference between group means.)
c. Compare the results of (a) and (b).
d. What is the correct posterior mean for an individual with an observed score of 80? For an observed score of 70? (Hint: You may want to revisit this question after you have answered the next one.)

8.4 Some of the students in the classes from the previous problem are going on a field trip to the museum. There is one more place left, and the determination will

* For another instance of the issue considered here, see Efron and Morris (1977). For examples of other situations where advances in conventional approaches may be seen as Bayesian, see Good (1965); Goldstein (1976); Clogg, Rubin, Schenker, Schultz, and Weidman (1991); and Galindo-Garre, Vermunt, and Bergsma (2004).

be made on the basis of their proficiency. It comes down to deciding between Prayoon, from Class A, for whom $x = 79$, and Pat in Class B, for whom $x = 81$.

a. Who do you think should get to go on the trip?
b. If the students took a parallel form of the same test, meaning the true score variance, error variance, and individuals' respective true scores would be the same on two forms, who do you think would be likely to get a higher score?
c. Are your answers to (a) and (b) the same or different? Why?
d. What would your answers to (a), (b), and (c) be if we knew Prayoon's and Pat's scores, but not which classes they were in?
e. Suppose Pat is selected, and Prayoon sues the school district claiming that he is more deserving to go on the trip. What is your argument if you are Prayoon's lawyer? What is your argument if you are the school district's lawyer?

9

Confirmatory Factor Analysis

In factor analysis, continuous observable variables are modeled as dependent on continuous latent variables or factors, as the inventor of the technique, Charles Spearman (1904), called them. This chapter treats flavors of factor analysis that have come to be termed confirmatory factor analysis (CFA), in which the analyst specifies the number of latent variables (factors) and the pattern of dependence of observables on those latent variables. This stands in contrast to exploratory factor analysis (EFA; Gorsuch, 1983), in which the central goals are determining the number of latent variables and the pattern of dependence of observables on them. This distinction is not as sharp as some believe, as models typically viewed as belonging to one may be seen as belonging to the other, and because practices involved in CFA have exploratory elements and practices involved in EFA have confirmatory elements. Our view is that the terminological distinction reflects less of a difference in models and more of one in the strength of a priori beliefs and purposes at hand. The distinction is further blurred by a Bayesian approach to modeling, which can soften some of the restrictions involved in CFA models (Section 9.8).

Modern treatments typically couch CFA in a structural equation modeling (SEM) framework (e.g., Bollen, 1989, Brown & Moore, 2012; Kline, 2010; Mulaik, 2009), and we will do the same in this development. Following a conventional presentation of CFA in Section 9.1, we turn to describing a Bayesian approach (Section 9.2) and illustrative examples (Sections 9.3 and 9.4). In Section 9.5, we describe a Bayesian approach to CFA using summary statistics rather than individual data points. In the next several sections, we discuss additional practical and conceptual issues, including comparing DAGs to the path diagrams that are popular in conventional approaches to CFA and SEM (Section 9.6), model formulation via hierarchical structuring (Section 9.7), the flexibility of Bayesian approaches to modeling (Section 9.8), and Bayesian approaches to resolving the indeterminacies in psychometric models with continuous latent variables (Section 9.9). We conclude the chapter in Section 9.10 with a brief summary and bibliographic note.

9.1 Conventional Factor Analysis

9.1.1 Model Specification

Let $\mathbf{x}$ be an $(n \times J)$ matrix containing the potentially observable values from n examinees to J observable variables.* At the individual level, $\mathbf{x}_i = (x_{i1}, \ldots, x_{iJ})'$ is the $(J \times 1)$ vector of

* As was the case for CTT in Chapter 8, these observables may be the result of taking aggregates or other functions of other variables. And that is indeed the case for the example introduced in Section 9.3, where we have five observables, each of which is an average of the scored responses to items. For our purposes of laying out a Bayesian treatment of CFA, the starting point is a set of observed scores for each examinee, however obtained, such that the CFA structure applies.

observed values from examinee i (i.e., the contents of the ith row of $\mathbf{x}$, arranged as a column vector). These observables are related to a set of latent variables via a factor analytic measurement model,

$$\mathbf{x}_i = \boldsymbol{\tau} + \boldsymbol{\Lambda}\boldsymbol{\xi}_i + \boldsymbol{\delta}_i, \tag{9.1}$$

where

- $\boldsymbol{\xi}_i = (\xi_{i1},\ldots,\xi_{iM})'$ is the $(M \times 1)$ vector of M latent variables for examinee i; these may be collected in an $(n \times M)$ matrix $\boldsymbol{\Xi}$ (i.e., $\boldsymbol{\xi}_i$ has the contents of the ith row of $\boldsymbol{\Xi}$, arranged as a column vector), where $\boldsymbol{\kappa} = (\kappa_1,\ldots,\kappa_M)' = E(\boldsymbol{\xi}_i)$ is an $(M \times 1)$ mean vector and $\boldsymbol{\Phi} = \text{var}(\boldsymbol{\xi}_i)$ is an $(M \times M)$ covariance matrix of the latent variables;
- $\boldsymbol{\tau} = (\tau_1,\ldots,\tau_J)'$ is a $(J \times 1)$ vector of observable–variable intercepts;
- $\boldsymbol{\Lambda}$ is a $(J \times M)$ matrix of factor loadings, the jth row of which is denoted as $\boldsymbol{\lambda}_j = (\lambda_{j1},\ldots,\lambda_{jM})$, containing the loadings for observable j on the M latent variables;
- $\boldsymbol{\delta}_i$ is a $(J \times 1)$ vector of errors, where $E(\boldsymbol{\delta}_i) = \mathbf{0}$ and $\text{var}(\boldsymbol{\delta}_i) = \boldsymbol{\Psi}$ is a $(J \times J)$ covariance matrix.

It is also usually assumed that all the sources of covariation among the observables are expressed in the latent variables and therefore that $\boldsymbol{\Psi}$ is diagonal; this assumption may easily be relaxed in ways discussed later.* Here $\boldsymbol{\tau}$, $\boldsymbol{\Lambda}$, and $\boldsymbol{\Psi}$ are the measurement model parameters as they govern the dependence of the observables on the latent variables, which is made more explicit in Section 9.2.

The expression in (9.1) is a compact way to represent a system of equations. For expository reasons, we write them out more fully as

$$\begin{aligned}
x_{i1} &= \tau_1 + \boldsymbol{\xi}_i'\boldsymbol{\lambda}_1 + \delta_{i1} = \tau_1 + \lambda_{11}\xi_{i1} + \lambda_{12}\xi_{i2} + \cdots + \lambda_{1M}\xi_{iM} + \delta_{i1} \\
x_{i2} &= \tau_2 + \boldsymbol{\xi}_i'\boldsymbol{\lambda}_2 + \delta_{i2} = \tau_2 + \lambda_{21}\xi_{i1} + \lambda_{22}\xi_{i2} + \cdots + \lambda_{2M}\xi_{iM} + \delta_{i2} \\
&\vdots \\
x_{ij} &= \tau_j + \boldsymbol{\xi}_i'\boldsymbol{\lambda}_j + \delta_{ij} = \tau_j + \lambda_{j1}\xi_{i1} + \lambda_{j2}\xi_{i2} + \cdots + \lambda_{jM}\xi_{iM} + \delta_{ij} \\
&\vdots \\
x_{iJ} &= \tau_J + \boldsymbol{\xi}_i'\boldsymbol{\lambda}_J + \delta_{iJ} = \tau_J + \lambda_{J1}\xi_{i1} + \lambda_{J2}\xi_{i2} + \cdots + \lambda_{JM}\xi_{iM} + \delta_{iJ}.
\end{aligned} \tag{9.2}$$

EFA models commonly estimate a loading for each observable on each latent variable. CFA models typically specify the pattern of loadings, meaning that they express which observables load on which latent variables. To communicate this, path diagrams are commonly employed when specifying and communicating CFA models within the SEM framework (Ho et al., 2012). Figure 9.1 contains representations of one-factor and two-factor models for five observables in panels (a) and (b), respectively. The path diagrams correspond to

* The model may be viewed as simultaneously regressing the xs on the $\boldsymbol{\xi}$s, with $\boldsymbol{\tau}$, $\boldsymbol{\Lambda}$, and the diagonal elements of $\boldsymbol{\Psi}$ playing the roles of regression intercepts, coefficients, and error variances, respectively, only now the predictors in the regression model (the $\boldsymbol{\xi}$s) are unknown. See Exercise 9.1.

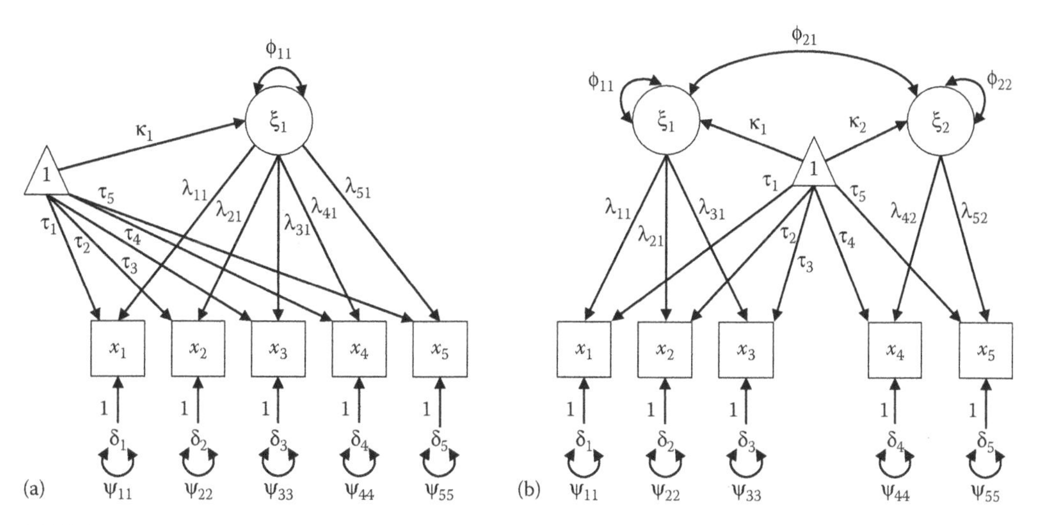

FIGURE 9.1
Path diagrams for confirmatory factor analysis models for five observables with (a) one latent variable (factor) and (b) two latent variables (factors).

systems of structural equations. Figure 9.1a corresponds to a one-factor model expressed in the following equations:

$$
\begin{aligned}
x_{i1} &= \tau_1 + \lambda_{11}\xi_{i1} + \delta_{i1} \\
x_{i2} &= \tau_2 + \lambda_{21}\xi_{i1} + \delta_{i2} \\
x_{i3} &= \tau_3 + \lambda_{31}\xi_{i1} + \delta_{i3} \\
x_{i4} &= \tau_4 + \lambda_{41}\xi_{i1} + \delta_{i4} \\
x_{i5} &= \tau_5 + \lambda_{51}\xi_{i1} + \delta_{i5}.
\end{aligned}
\tag{9.3}
$$

The path diagram depicted in Figure 9.1b corresponds to a two-factor model expressed in the following equations

$$
\begin{aligned}
x_{i1} &= \tau_1 + \lambda_{11}\xi_{i1} + \delta_{i1} \\
x_{i2} &= \tau_2 + \lambda_{21}\xi_{i1} + \delta_{i2} \\
x_{i3} &= \tau_3 + \lambda_{31}\xi_{i1} + \delta_{i3} \\
x_{i4} &= \tau_4 + \lambda_{42}\xi_{i2} + \delta_{i4} \\
x_{i5} &= \tau_5 + \lambda_{52}\xi_{i2} + \delta_{i5}.
\end{aligned}
\tag{9.4}
$$

For each model, the path diagram conveys the same information contained in the equations, and additionally communicates the mean, variance, and covariance terms for the latent variables (κ_1, κ_2, ϕ_{11}, ϕ_{22},ϕ_{21}) and variance terms for errors ($\psi_{11},\ldots,\psi_{55}$).*

* Although not typically done, the systems of expressions in (9.3) and (9.4) could be expanded to better include these elements formally. The path diagrams contain a "1" in a triangle that does not seem to appear in the equations. As discussed in Section 9.6, this is a modeling device to aid in connecting path diagrams to equations. For now, it suffices to say that this "1" is a constant, and may be seen in structural equations as the often-not-written value for which the parameter on the path emanating from it serves as a coefficient.

The observant reader may have noticed that the path diagrams in Figure 9.1 do not include subscripts for examinee variables: xs, ξs, and δs. This follows conventional use of path diagrams for CFA from within the SEM framework. Correspondingly, most presentations do not include such subscripts in the equations (e.g., Bollen, 1989). We have included them in (9.1)–(9.4) to make explicit that these entities are examinee variables and for continuity with our presentation throughout the book. It is worth more fully contrasting the structural equation and path diagrammatic representation of models with the probabilistic and DAG representation, a task we defer until Section 9.6.

9.1.2 Indeterminacies in Factor Analysis

Several indeterminacies exist in FA models, the implications of which are that transformations of the parameters result in the same conditional probability for the data. Conceptually, these indeterminacies may be seen as arising because the continuous latent variables do not possess a natural metric. As a result there are indeterminacies associated with location (origin), scale, and orientation, which ought to be resolved before interpreting the results. This is typically accomplished by fixing various parameters.

To resolve the indeterminacy in location, a common choice in CFA is to fix the means of the latent variables, usually to 0. Other strategies include fixing the intercepts or their sum, as may be preferred when the model is constructed such that the latent variable means are of inferential interest, as in growth modeling (Bollen & Curran, 2006). Similarly, the indeterminacy in scale may be resolved by fixing the variance of the latent variables, usually to 1. Another common choice to resolving the indeterminacy in scale is to fix a loading for each latent variable to a value, usually 1.

This latter approach also resolves an indeterminacy in the orientation of each of the latent variables that may otherwise manifest itself in a *reflection* of the loadings, as the model-implied features of the data are invariant to a transformation that multiplies each loading by –1. Fixing a loading to a particular value obviates this concern. Put another way, fixing λ_{11} to 1 resolves the scale of ξ_1 (by associating it with the scale of x_1); fixing λ_{11} to *positive* 1 resolves the orientation of ξ_1, orienting it such that larger values of ξ_1 are associated with larger values of x_1.

This orientation indeterminacy may be seen as akin to that of rotation; the latent variable may be rotated 180° or "flipped." For models with multiple latent variables, the rotational indeterminacy looms larger. This is a primary concern in EFA models in which there is a loading estimated for each observable on each latent variable, and it is typically resolved by defining a rotational criterion to be optimized (Gorsuch, 1983). It can be also be addressed by fixing loadings; this is commonly done in CFA by fixing certain loadings to be 0 (Bollen, 1989; for related discussions in item response theory, see Davey, Oshima, & Lee, 1996, and Jackman, 2001). For example, the model in (9.3) and Figure 9.1b does not include loadings for x_4 and x_5 on ξ_1 or loadings for x_1, x_2, and x_3 on ξ_2; these can be conceived of as being present in the model but fixed to 0, a view we discuss in greater detail in Section 9.8.

In the frequentist framework, these indeterminacies are commonly viewed as a matter of identification, because failing to adequately resolve them will result in the lack of a unique point estimate. We revisit these considerations from a Bayesian perspective in Section 9.9.

9.1.3 Model Fitting

The systems of equations and the path diagrams play a central role in conventional CFA modeling. Inference typically proceeds by working out the model-implied first- and second-order moments for the observables in terms of the model parameters through the algebra of expectations or via path-tracing rules for path diagrams (Mulaik, 2009; Wright, 1934), which in general yields

$$E(\mathbf{x}_i \mid \boldsymbol{\kappa}, \boldsymbol{\Phi}, \boldsymbol{\tau}, \boldsymbol{\Lambda}, \boldsymbol{\Psi}) = \boldsymbol{\tau} + \boldsymbol{\Lambda}\boldsymbol{\kappa} \tag{9.5}$$

and

$$\text{var}(\mathbf{x}_i \mid \boldsymbol{\kappa}, \boldsymbol{\Phi}, \boldsymbol{\tau}, \boldsymbol{\Lambda}, \boldsymbol{\Psi}) = \boldsymbol{\Lambda}\boldsymbol{\Phi}\boldsymbol{\Lambda}' + \boldsymbol{\Psi}. \tag{9.6}$$

The notation on the left-hand side of (9.5) and (9.6) stresses conditioning variables, which aids in contrasting these with representations that come later.

Frequentist approaches dominate conventional approaches to model fitting, which amounts to fitting these model-implied moments to data via least-squares or ML routines assuming normality of the observables to yield estimates for $\boldsymbol{\tau}$, $\boldsymbol{\Lambda}$, $\boldsymbol{\Psi}$, $\boldsymbol{\kappa}$, and $\boldsymbol{\Phi}$. In the parlance of Chapter 7, estimating these parameters amounts to model calibration. Following the introduction of some necessary terms in Section 9.2, we sketch an overview of ML estimation to draw out connections to Bayesian approaches (for treatments of these and other options in the frequentist framework, see Bollen, 1989, chapter 4; Mulaik, 2009, chapter 7; Yanai & Ichikawa, 2007; Lei & Wu, 2012). As we will see, the popular normal-theory ML routines can be grounded in distributional assumptions of normality for the latent variables and errors.

Interestingly, arriving at representations of the examinees' latent variables—referred to as scoring in Chapter 7—is less common in CFA as compared to other psychometric modeling traditions. This is no doubt in part due to the purposes of those applications of CFA where values for examinees' latent variables are not of inferential interest. We suspect that it is also in part due to the so-called problem of factor score indeterminacy (not to be confused with the indeterminacies discussed earlier) that has resulted in a number of proposed ways to represent examinees' latent variables (Bartholomew et al., 2011; Yanai & Ichikawa, 2007) and considerable debate (see, e.g., Maraun, 1996, and surrounding discussion articles).

Working from the principle that the latent variables are random variables, Bartholomew (1981) argued for the use of Bayesian inference for scoring in CFA, in which the posterior distribution $p(\boldsymbol{\xi}_i \mid \mathbf{x}_i)$ is obtained using point estimates for the model parameters from frequentist calibration, akin to (7.10). However, this approach is hardly one of consensus. Bartholomew (1981) also noted that what are typically called "regression" scores for latent variables correspond to the expectation of this posterior distribution (see also Aitkin & Aitkin, 2005; Bartholomew et al., 2011). The situation here is akin to what was observed for CTT, in which Kelley's formula for arriving at a point estimate of a true score coincided with the posterior mean from a Bayesian analysis.

9.2 Bayesian Factor Analysis

In a Bayesian analysis, fitting a model involves obtaining the posterior distribution for all unknowns, given the observed data. In CFA, the collection of unknowns are values of the latent variables ($\boldsymbol{\Xi}$), means and (co)variances for the latent variables ($\boldsymbol{\kappa}$ and $\boldsymbol{\Phi}$, respectively), and parameters for the measurement model relating the observables to the latent variables (τ,Λ, and Ψ). Once the data are observed, the posterior distribution following Bayes theorem is

$$p(\Xi,\kappa,\Phi,\tau,\Lambda,\Psi \mid \mathbf{x}) \propto p(\mathbf{x} \mid \Xi,\kappa,\Phi,\tau,\Lambda,\Psi)p(\Xi,\kappa,\Phi,\tau,\Lambda,\Psi). \quad (9.7)$$

The first term on the right-hand side of (9.7) is the conditional probability distribution of the observables. The second term is the prior distribution for the unknowns. Specifying the model involves specifying these terms. In the following sections, we will treat each in turn, and discuss some connections and departures from frequentist approaches to CFA.

9.2.1 Conditional Distribution of the Observables

The first term on the right-hand side of (9.7) is the probability distribution of the observables, expressed as conditional on all the parameters in the model. This may be simplified to express the conditional probability distribution of the observables given the latent variables (Ξ) and the measurement model parameters (τ, Λ, Ψ), as

$$p(\mathbf{x} \mid \Xi,\kappa,\Phi,\tau,\Lambda,\Psi) = p(\mathbf{x} \mid \Xi,\tau,\Lambda,\Psi). \quad (9.8)$$

This expression reflects that, given the values of the latent variables (Ξ) of examinees, the observables $\mathbf{x}$ are conditionally independent of the parameters that govern the distribution of the latent variables, namely the means (κ) and (co)variances (Φ). Put another way, once we know the values of the latent variables, knowing their means and (co)variances does not tell us anything new about the observables.

Looking at the right-hand side of (9.8), $\mathbf{x}$ contains the values for the observables for examinees, and Ξ contains the values for the latent variables for examinees. These vary over examinees. The remaining parameters are the measurement model parameters that govern the conditional distribution of the observables given the latent variables: the intercepts τ, loadings Λ, and error (co)variances Ψ. That these parameters do not vary over examinees reflects an exchangeability assumption with respect to the measurement model: we think the measurement quality of the observables is the same for each examinee. When exchangeability of measurement is not assumed, different measurement model parameters are specified for different examinees. This usually occurs by way of specifying group-specific parameters when group membership is known or unknown (as in latent class or finite mixture models; Pastor & Gagné, 2013), with applications in invariance or differential functioning analyses (Millsap, 2011; Verhagen & Fox, 2013).

Assuming exchangeability with respect to the measurement model, the joint conditional distribution for the observables may be factored into a product over examinees,

$$p(\mathbf{x} \mid \Xi,\tau,\Lambda,\Psi) = \prod_{i=1}^{n} p(\mathbf{x}_i \mid \xi_i,\tau,\Lambda,\Psi). \quad (9.9)$$

The assumption that Ψ is diagonal embodies the local independence assumption and reveals why local independence is best thought of with respect to the model as a whole, not just the latent variables (Levy & Svetina, 2011). To see why, consider what happens if we do not condition on the errors. If there are no error covariances (i.e., Ψ is diagonal), local independence will hold. In the more general case where error covariances are present, failing to condition on them will not render the observables independent; a nonzero error covariance indicates that the observables are dependent above and beyond that which can be accounted for by the latent variables (and other measurement model parameters).

As a result of the local independence assumption, the conditional probability of the observables for any examinee can be further factored into the product over observables,

$$p(\mathbf{x}_i \mid \xi_i, \tau, \Lambda, \Psi) = \prod_{j=1}^{J} p(x_{ij} \mid \xi_i, \tau_j, \lambda_j, \psi_{jj}). \tag{9.10}$$

The form of the distribution is dictated by the factor analytic model and the distributional assumption of the errors. The system of equations in (9.1) and the assumption that the errors have expectation of 0 for all variables and covariance matrix Ψ imply that the conditional expectation and variance of the observables for an examinee given the latent variables and the measurement model parameters are

$$E(\mathbf{x}_i \mid \xi_i, \tau, \Lambda, \Psi) = \tau + \Lambda \xi_i \tag{9.11}$$

and

$$\text{var}(\mathbf{x}_i \mid \xi_i, \tau, \Lambda, \Psi) = \Psi. \tag{9.12}$$

Following the assumption of normality of errors, the conditional distribution of the observables for each examinee is normal. Working at the level of the vector of observables for any examinee (i.e., the right-hand side of 9.9)

$$\mathbf{x}_i \mid \xi_i, \tau, \Lambda, \Psi \sim N(\tau + \Lambda \xi_i, \Psi). \tag{9.13}$$

Drilling down further to the level of each observable for any examinee (i.e., the right-hand side of 9.10)

$$x_{ij} \mid \xi_i, \tau_j, \lambda_j, \psi_{jj} \sim N(\tau_j + \xi_i \lambda_j', \psi_{jj}). \tag{9.14}$$

Collecting the preceding developments in (9.8)–(9.10), (9.13) and (9.14), we are now ready to state the conditional distribution of all the observables, expressed by drilling down to the level of each observable for each examinee.* The conditional distribution is

$$p(\mathbf{x} \mid \Xi, \kappa, \Phi, \tau, \Lambda, \Psi) = p(\mathbf{x} \mid \Xi, \tau, \Lambda, \Psi) = \prod_{i=1}^{n} \prod_{j=1}^{J} p(x_{ij} \mid \xi_i, \tau_j, \lambda_j, \psi_{jj}) \tag{9.15}$$

where

$$x_{ij} \mid \xi_i, \tau_j, \lambda_j, \psi_{jj} \sim N(\tau_j + \xi_i \lambda_j', \psi_{jj}).$$

9.2.1.1 Connecting Distribution-Based and Equation-Based Expressions

We pause in our development of the model to emphasize that though these last two expressions differ on the surface from the usual presentation of CFA as a system of equations, they express the same model. Working at the level of the vector of observables for

* As an exercise, we encourage the reader to create a DAG corresponding to the specifications described in this section.

any examinee, coupling the deterministic expression in (9.1) with the assumption that $\delta_i \sim N(\mathbf{0}, \mathbf{\Psi})$ implies the probabilistic expression in (9.13). Working at the level of any individual observable for any examinee, coupling the deterministic expression for observable j in (9.2) with the assumption that $\delta_{ij} \sim N(0, \psi_{jj})$ implies the probabilistic expression in (9.14).

Note that we have been working with a model for the conditional distribution of the data, structuring the data as dependent on model parameters. There is nothing particularly unique to Bayesian modeling here. The conditional distribution for data given the parameters plays a role in Bayesian inference of course, but it also plays a role in frequentist inference, namely by defining the likelihood function.

The conventional approach to expressing statistical and psychometric models is equation oriented. For CFA, the standard approach is to use *structural equation* modeling, as in (9.1)–(9.3); path diagrams such as those in Figure 9.1 closely correspond to those structural equations. This engenders us to *think equationally*. We value this perspective, and it serves psychometrics quite well in many respects. However, we find that it is often preferable to express models using distributional forms, as in (9.13) or (9.14). The equations of course play a role in specifying the distribution, as they yield the parameters of the resulting distribution in (9.11) and (9.12). Formulating the model in this latter way is an important step towards our aim of *thinking probabilistically*. Formulating and explicitly expressing the model in terms of probability distributions more naturally coheres with probability-based reasoning to conduct inference, and probabilistic expressions of beliefs. In addition, it aids with making conditioning explicit in notation, which allows us to more easily express local and other conditional independence relationships. This approach is well aligned with hierarchical strategies to building models, which we flesh out more fully in the current context in Section 9.7. The distributional form is also better connected to the substantive stories about the data we come to see. We also suspect that this approach may better encourage the specification of models that depart from or generalize those covered here, such as the specification of thicker-tailed distributions in support of modeling rather than discarding outlying observations (Lee & Xia, 2008; Zhang, Lai, Lu, & Tong, 2013).

9.2.2 Prior Distribution

The second term on the right-hand side of the posterior distribution in (9.7) is the joint prior distribution for all the unknowns. We factor this joint prior as

$$\begin{aligned} p(\Xi, \mathbf{\kappa}, \mathbf{\Phi}, \mathbf{\tau}, \mathbf{\Lambda}, \mathbf{\Psi}) &= p(\Xi \mid \mathbf{\kappa}, \mathbf{\Phi}) p(\mathbf{\kappa}, \mathbf{\Phi}, \mathbf{\tau}, \mathbf{\Lambda}, \mathbf{\Psi}) \\ &= p(\Xi \mid \mathbf{\kappa}, \mathbf{\Phi}) p(\mathbf{\kappa}, \mathbf{\Phi}) p(\mathbf{\tau}, \mathbf{\Lambda}, \mathbf{\Psi}). \end{aligned} \tag{9.16}$$

The first line reflects that the prior distribution for the examinee latent variables Ξ is governed by the parameters $\mathbf{\kappa}$ and $\mathbf{\Phi}$, and is independent of the measurement model parameters ($\mathbf{\tau}$, $\mathbf{\Lambda}$, and $\mathbf{\Psi}$). The second line reflects that these parameters that govern the distribution of the latent variables ($\mathbf{\kappa}$ and $\mathbf{\Phi}$) are independent of the measurement model parameters ($\mathbf{\tau}$, $\mathbf{\Lambda}$, and $\mathbf{\Psi}$). We now step through the further specifications of the three terms on the right-hand side of the second line.

9.2.2.1 Prior Distribution for Latent Variables

Beginning with the examinee latent variables (i.e., the first term on the right-hand side of 9.16), assuming exchangeability with regard to the examinees allows for the joint prior to be factored into the product of a common prior distribution

$$p(\Xi \mid \kappa, \Phi) = \prod_{i=1}^{n} p(\xi_i \mid \kappa, \Phi). \tag{9.17}$$

Assuming normality for the latent variables, the prior for each examinee is then

$$\xi_i \mid \kappa, \Phi \sim N(\kappa, \Phi). \tag{9.18}$$

9.2.2.2 Prior Distribution for Parameters That Govern the Distribution of the Latent Variables

Turning to the parameters that characterize the distribution of the latent variables in the second term of the right-hand side of the last line of (9.16), we pursue a conditionally conjugate prior specification for (κ, Φ), which are the mean vector and covariance matrix of a multivariate normal. As was the case with the parameters of the univariate normal distribution (Chapter 4), the conditionally conjugate prior in the multivariate normal case of interest here involves specifying the mean vector and covariance matrix to be independent in the prior,

$$p(\kappa, \Phi) = p(\kappa)p(\Phi). \tag{9.19}$$

Specifying the prior for (κ, Φ) now comes to specifying a prior for κ and a prior for Φ.

An assumption of exchangeability regarding the latent variables allows for the specification of a common prior for all elements of κ,

$$p(\kappa) = \prod_{m=1}^{M} p(\kappa_m). \tag{9.20}$$

As the κ_m are means of normal distributions, a common choice for the prior is a normal distribution

$$\kappa_m \sim N(\mu_\kappa, \sigma_\kappa^2), \tag{9.21}$$

where μ_κ and σ_κ^2 are parameters that govern these prior distributions, specified by the analyst.

For Φ, we employ an inverse-Wishart distribution, denoted by Inv-Wishart( ),

$$\Phi \sim \text{Inv-Wishart}(\Phi_0, d), \tag{9.22}$$

where Φ_0 is a matrix of values reflecting prior expectation for Φ and $d \geq M$ is a specified weight, with smaller values for d yielding a more diffuse prior distribution. The inverse-Wishart is a multivariate generalization of the inverse-gamma. If we were to take a univariate approach to modeling the latent variables, such as in a model with one latent variable, we might employ an inverse-gamma prior distribution for the univariate latent variable.

9.2.2.3 Prior Distribution for Measurement Model Parameters

Turning to the measurement model parameters in the third term on the right-hand side of the second line of (9.16), we present procedures associated with the common practice of assuming exchangeability and conditionally conjugate prior specifications. An exchangeability assumption with respect to the observables allows for the joint prior distribution for the observables' measurement model parameters to be factored into the product of a common prior distribution,

$$p(\boldsymbol{\tau}, \boldsymbol{\Lambda}, \boldsymbol{\Psi}) = \prod_{j=1}^{J} p(\tau_j, \boldsymbol{\lambda}_j, \psi_{jj}). \tag{9.23}$$

To specify $p(\tau_j, \boldsymbol{\lambda}_j, \psi_{jj})$, we employ conditionally conjugate priors. In this case, all the elements, including the possibly multiple loadings for each observable, are modeled as independent in the prior,

$$p(\tau_j, \boldsymbol{\lambda}_j, \psi_{jj}) = p(\tau_j)\prod_{m=1}^{M} p(\lambda_{jm})p(\psi_{jj}). \tag{9.24}$$

The intercepts are assigned a normal prior distribution,

$$\tau_j \sim N(\mu_\tau, \sigma_\tau^2), \tag{9.25}$$

as are the loadings along each latent variable,

$$\lambda_{jm} \sim N(\mu_\lambda, \sigma_\lambda^2), \tag{9.26}$$

and error variances are assigned an inverse-gamma distribution,

$$\psi_{jj} \sim \text{Inv-Gamma}(\nu_\psi/2, \nu_\psi\psi_0/2). \tag{9.27}$$

where μ_τ, σ_τ^2, μ_λ, σ_λ^2, ν_ψ, and ψ_0 are hyperparameters that govern these prior distributions, typically specified by the analyst.

9.2.2.4 The Use of a Prior Distribution in ML Estimation

We pause our development of a Bayesian treatment to highlight an often-overlooked but important shared feature common to Bayesian and ML approaches to CFA, namely that the latter can also be justified by the specification of a (normal) prior distribution for latent variables. For the readers steeped in frequentist approaches to CFA, the notion of a prior distribution for the latent variables may seem strange, as conventional presentations of ML estimation of CFA do not typically refer to a distribution of latent variables (e.g., Bollen, 1989; Lei et al., 2012; Yanai & Ichikawa, 2007), and when a distribution is mentioned, it is not typically referred to as a prior distribution (Skrondal & Rabe-Hesketh, 2004; Wall, 2009).

To see that this distributional specification is present in ML estimation, it is worth pursuing another difference between Bayesian and frequentist estimation, namely how the dependence of observables on the parameters is typically expressed. The formulation in the previous section expressed the observables as conditional on the latent variables as in

(9.11) through (9.14). Typical frequentist presentations do not include the latent variables in expressing the distribution of the data, instead presenting them as in (9.5) and (9.6). The differences in formulations can be reconciled by recognizing what is being conditioned on, and what each approach considers to be parameters.

We begin by constructing the joint distribution of the observable and latent variables:

$$\begin{aligned} p(\mathbf{x},\Xi \mid \kappa,\Phi,\tau,\Lambda,\Psi) &= p(\mathbf{x} \mid \Xi,\tau,\Lambda,\Psi)p(\Xi \mid \kappa,\Phi) \\ &= \prod_{i=1}^{n} p(\mathbf{x}_i \mid \xi_i,\tau,\Lambda,\Psi)p(\xi_i \mid \kappa,\Phi). \end{aligned} \tag{9.28}$$

The first term on the right-hand side of (9.28) is the conditional distribution of the observables given the latent variables (and measurement model parameters), defined in (9.15). The second term on the right-hand side of (9.28) is the distribution of the latent variables for examinees, defined in (9.17) and (9.18). The marginal distribution of the observables, without reference to the latent variables, is obtained by integrating (9.28) over the distribution of the latent variables as an instance of (7.8),

$$\begin{aligned} p(\mathbf{x} \mid \kappa,\Phi,\tau,\Lambda,\Psi) &= \int_{\Xi} p(\mathbf{x},\Xi \mid \kappa,\Phi,\tau,\Lambda,\Psi)d\Xi \\ &= \int_{\Xi} p(\mathbf{x} \mid \Xi,\tau,\Lambda,\Psi)p(\Xi \mid \kappa,\Phi)d\Xi \\ &= \prod_{i=1}^{n} \int_{\xi_i} p(\mathbf{x}_i \mid \xi_i,\tau,\Lambda,\Psi)p(\xi_i \mid \kappa,\Phi)d\xi_i. \end{aligned} \tag{9.29}$$

Given normality assumptions about the conditional distribution of the data (the first term in the second and third lines of (9.29) and of the prior distribution for the latent variables (the second term on these lines), the marginal distribution that results from integrating ξ_i out is also normal:

$$\mathbf{x}_i \mid \kappa,\Phi,\tau,\Lambda,\Psi \sim N(\tau+\Lambda\kappa, \Lambda\Phi\Lambda'+\Psi). \tag{9.30}$$

The resulting joint conditional distribution of all the observables on the left-hand side of (9.29) is the collection of i.i.d. normal $\mathbf{x}_i$ with mean vector given by $(\tau+\Lambda\kappa)$ and covariance matrix given by $(\Lambda\Phi\Lambda'+\Psi)$. Thus, we see from (9.30) that the conventional frequentist description of CFA in (9.5) and (9.6) may be derived from a probabilistic approach that integrates the latent variables out. The result expresses the dependence of the observables on the parameters that characterize the distribution of the latent variables, κ and Φ, as well as the measurement model parameters, τ, Λ, and Ψ. Once values for the observables are known, the conditional distribution in (9.30) induces a likelihood function for these parameters. It is this likelihood that yields the (normal-theory) ML fit function that is optimized in ML estimation of CFA models (e.g., Bollen, 1989).

Of course, that a multivariate normal specification for a prior for the latent variables yields a convenient marginal distribution might not have quite the same sway in a fully Bayesian analysis as it might in a frequentist analysis. However, the use of a normal distribution does aid the computational tractability of a fully Bayesian analysis (Section 9.2.4), it is a distributional form that many analysts are familiar with, and it has a long track record of successful use in the analysis of data in psychometric contexts generally and specifically for continuous latent variables. What's more, like any other feature, we may critically evaluate it using follow-up analyses such as those described in Chapter 10.

9.2.3 Posterior Distribution and Graphical Model

Collecting the preceding developments, we are now ready to state the posterior distribution for all the unknowns:

$$\begin{aligned} p(\Xi,\kappa,\Phi,\tau,\Lambda,\Psi \mid \mathbf{x}) &\propto p(\mathbf{x} \mid \Xi,\kappa,\Phi,\tau,\Lambda,\Psi)p(\Xi,\kappa,\Phi,\tau,\Lambda,\Psi) \\ &= p(\mathbf{x} \mid \Xi,\tau,\Lambda,\Psi)p(\Xi \mid \kappa,\Phi)p(\kappa)p(\Phi)p(\tau)p(\Lambda)p(\Psi) \\ &= \prod_{i=1}^{n}\prod_{j=1}^{J}\prod_{m=1}^{M} p(x_{ij} \mid \xi_i,\tau_j,\lambda_j,\psi_{jj}) \\ &\quad \times p(\xi_i \mid \kappa,\Phi)p(\kappa_m)p(\Phi)p(\tau_j)p(\lambda_{jm})p(\psi_{jj}), \end{aligned} \tag{9.31}$$

where

$$\begin{aligned} x_{ij} \mid \xi_i,\tau_j,\lambda_j,\psi_{jj} &\sim N(\tau_j+\xi_i\lambda_j',\psi_{jj}) \text{ for } i=1,\ldots,n, j=1,\ldots,J, \\ \xi_i \mid \kappa,\Phi &\sim N(\kappa,\Phi) \text{ for } i=1,\ldots,n, \\ \kappa_m &\sim N(\mu_\kappa,\sigma_\kappa^2) \text{ for } m=1,\ldots,M, \\ \Phi &\sim \text{Inv-Wishart}(\Phi_0,d), \\ \tau_j &\sim N(\mu_\tau,\sigma_\tau^2) \text{ for } j=1,\ldots,J, \\ \lambda_{jm} &\sim N(\mu_\lambda,\sigma_\lambda^2) \text{ for } j=1,\ldots,J, m=1,\ldots,M, \end{aligned}$$

and

$$\psi_{jj} \sim \text{Inv-Gamma}(\nu_\psi/2,\nu_\psi\psi_0/2) \text{ for } j=1,\ldots,J.$$

Figure 9.2 contains the DAG for the model, expressing the joint distribution of all the entities. We emphasize the correspondence between the DAG and elements of (9.31). Taken as a whole, the model is fairly complex, relative to those covered previously in the book. However, a modular approach recognizes the use of familiar structures developed earlier: (1) normal-theory regression-like structures for the observables as dependent on the latent variables and measurement model parameters, (2) associated prior distributions for those regression-like structures, (3) a normal model for the latent variables, (4) a normal prior on the means, and (5) an inverse-Wishart as a generalization of the inverse-gamma for the covariance matrix.

9.2.4 MCMC Estimation

The use of conditionally conjugate prior distributions makes for full conditional distributions of known form. It is possible to construct the full conditionals for the intercepts and loadings individually. A more compact representation is afforded by framing the intercept and loadings for an observable as a single entity. Both approaches are developed in Appendix A. Here, we present the latter, where λ_{jA} denotes the $([M+1] \times 1)$ *augmented* vector of loadings obtained by combining the intercept τ_j with the M loadings in λ_j, and Ξ_A denotes the $(n \times [M+1])$ *augmented* matrix of latent variables obtained by combining

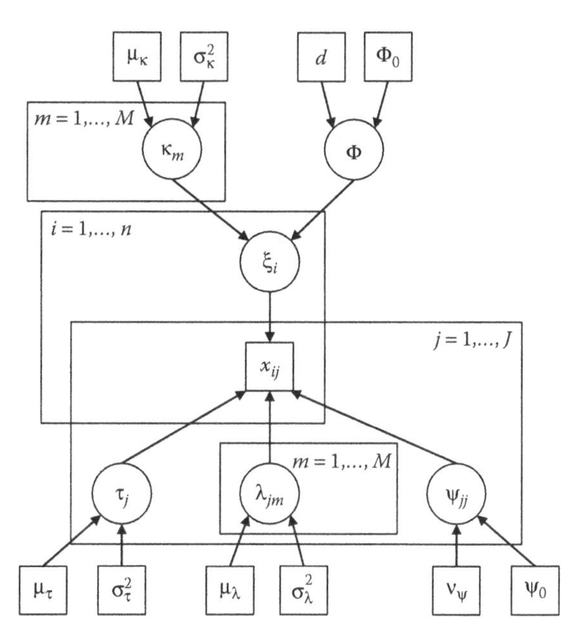

FIGURE 9.2
Directed acyclic graph for a confirmatory factor analysis model supporting multiple observables loading on multiple latent variables.

an ($n \times 1$) column vector of 1s to the matrix of latent variables Ξ. We employ a multivariate normal prior for the intercept and loadings for each observable j,

$$\lambda_{jA} \sim N(\mu_{\lambda_A}, \Sigma_{\lambda_A}). \tag{9.32}$$

where μ_{λ_A} is an ($[M + 1] \times 1$) prior mean vector and Σ_{λ_A} is an ($[M + 1] \times [M + 1]$) prior covariance matrix. Note that the univariate specification of the intercept and the loadings obtains as a special case when Σ_{λ_A} is diagonal. All other prior distributions remain the same.

We express each full conditional distribution in the following equations. On the left-hand side, the parameter in question is written as conditional on all the other relevant parameters and data.* On the right-hand side of each of the following equations, we give the parametric form for the full conditional distribution. In several places we denote the arguments of the distribution (e.g., mean vector and covariance matrix for a normal distribution) with subscripts denoting that it refers to the full conditional distribution; the subscripts are then just the conditioning notation of the left-hand side.

Beginning with the examinees' latent variables, we present the full conditional for any examinee i. The same structure applies to all the examinees. The full conditional distribution for the latent variables for examinee i is

$$\xi_i \mid \boldsymbol{\kappa}, \Phi, \Lambda, \tau, \Psi, \mathbf{x}_i \sim N(\mu_{\xi_i|\kappa,\Phi,\Lambda,\tau,\Psi,x_i}, \Sigma_{\xi_i|\kappa,\Phi,\Lambda,\tau,\Psi,x_i}), \tag{9.33}$$

* We suppress the role of specified hyperparameters in this notation; see Appendix A for a presentation that formally includes the hyperparameters.

where

$$\mu_{\xi_i|\kappa,\Phi,\Lambda,\tau,\Psi,\mathbf{x}_i} = (\Phi^{-1} + \Lambda'\Psi^{-1}\Lambda)^{-1}(\Phi^{-1}\kappa + \Lambda'\Psi^{-1}(\mathbf{x}_i - \tau)),$$

$$\Sigma_{\xi_i|\kappa,\Phi,\Lambda,\tau,\Psi,\mathbf{x}_i} = (\Phi^{-1} + \Lambda'\Psi^{-1}\Lambda)^{-1},$$

and $\mathbf{x}_i = (x_{i1}, \ldots, x_{iJ})$ is the $(J \times 1)$ vector of J observed values from examinee i. The subscript of "$\xi_i \mid \kappa,\Phi,\Lambda,\tau,\Psi,\mathbf{x}_i$" in (9.33) is rather unwieldy, but by writing it out fully explicitly indicate what the mean vector and covariance matrix in the full conditional for ξ_i do and do not depend on. They *do* depend on the structural parameters $\kappa,\Phi,\Lambda,\tau,$ and ψ. They *do* depend on examinee i's vector of observed scores, $\mathbf{x}_i$. Given these variables, however, they *do not* depend on the latent variables or observed scores of other examinees, ξ_{i^*} and $\mathbf{x}_{i^*}$ for $i^* \neq i$. The same scheme applies to interpreting the subscripts in the rest of the full conditional distributions that follow. The full conditional for the mean vector for the latent variables is

$$\kappa \mid \Xi,\Phi \sim N(\mu_{\kappa|\Xi,\Phi}, \Sigma_{\kappa|\Xi,\Phi}), \tag{9.34}$$

where

$$\mu_{\kappa|\Xi,\Phi} = (\Sigma_\kappa^{-1} + n\Phi^{-1})^{-1}(\Sigma_\kappa^{-1}\mu_\kappa + n\Phi^{-1}\bar{\xi}),$$

$$\Sigma_{\kappa|\Xi,\Phi} = (\Sigma_\kappa^{-1} + n\Phi^{-1})^{-1},$$

and $\bar{\xi}$ is the $(M \times 1)$ vector of means of the treated-as-known latent variables over the n examinees. The full conditional for the covariance matrix for the latent variables is

$$\Phi \mid \Xi,\kappa \sim \text{Inv-Wishart}(d\Phi_0 + n\mathbf{S}_\xi, d+n), \tag{9.35}$$

where

$$\mathbf{S}_\xi = \frac{1}{n}\sum_i (\xi_i - \kappa)(\xi_i - \kappa)'.$$

Turning to the measurement model parameters, we present the full conditionals for any observable j. The same structure applies to all the observables. The full conditional for the augmented loadings for observable j is

$$\lambda_{jA} \mid \Xi,\psi_{jj},\mathbf{x}_j \sim N(\mu_{\lambda_{jA}|\Xi,\psi_{jj},\mathbf{x}_j}, \Sigma_{\lambda_{jA}|\Xi,\psi_{jj},\mathbf{x}_j}), \tag{9.36}$$

where

$$\mu_{\lambda_{jA}|\Xi,\psi_{jj},\mathbf{x}_j} = \left(\Sigma_{\lambda_A}^{-1} + \frac{1}{\psi_{jj}}\Xi_A'\Xi_A\right)^{-1}\left(\Sigma_{\lambda_A}^{-1}\mu_{\lambda_A} + \frac{1}{\psi_{jj}}\Xi_A'\mathbf{x}_j\right),$$

$$\Sigma_{\lambda_{jA}|\Xi,\psi_{jj},\mathbf{x}_j} = \left(\Sigma_{\lambda_A}^{-1} + \frac{1}{\psi_{jj}}\Xi_A'\Xi_A\right)^{-1},$$

and $\mathbf{x}_j = (x_{1j}, \ldots, x_{nj})'$ is the $(n \times 1)$ vector of observed values for the n examinees for observable j.

The full conditional distribution for the error variance for observable j is

$$\psi_{jj} \mid \Xi,\lambda_{jA},\mathbf{x}_j \sim \text{Inv-Gamma}\left(\frac{\nu_\psi + n}{2}, \frac{\nu_\psi\psi_0 + SS(\mathbf{E}_j)}{2}\right), \tag{9.37}$$

where

$$SS(\mathbf{E}_j) = (\mathbf{x}_j - \Xi_A \boldsymbol{\lambda}_{jA})'(\mathbf{x}_j - \Xi_A \boldsymbol{\lambda}_{jA})$$

and $\mathbf{x}_j = (x_{1j}, \ldots, x_{nj})'$ is the $(n \times 1)$ vector of observed values for the n examinees for observable j. A Gibbs sampler is constructed by iteratively drawing from these full conditional distributions using the just-drawn values for the conditioned parameters, generically written below (for iteration $t + 1$).

9.2.4.1 Gibbs Sampler Routine for CFA

1. *Sample the latent variables for examinees.* For each examinee $i = 1, \ldots, n$, using the values of measurement model parameters and parameters that govern the latent distribution from the previous iteration (t), compute

$$\boldsymbol{\mu}_{\xi_i^{(t+1)}|\boldsymbol{\kappa}^{(t)},\boldsymbol{\Phi}^{(t)},\boldsymbol{\Lambda}^{(t)},\boldsymbol{\tau}^{(t)},\boldsymbol{\Psi}^{(t)},\mathbf{x}_i} = (\boldsymbol{\Phi}^{(t)^{-1}} + \boldsymbol{\Lambda}'^{(t)}\boldsymbol{\Psi}^{(t)^{-1}}\boldsymbol{\Lambda}^{(t)})^{-1}(\boldsymbol{\Phi}^{(t)^{-1}}\boldsymbol{\kappa}^{(t)} + \boldsymbol{\Lambda}'^{(t)}\boldsymbol{\Psi}^{(t)^{-1}}(\mathbf{x}_i - \boldsymbol{\tau}^{(t)}))$$

and

$$\boldsymbol{\Sigma}_{\xi_i^{(t+1)}|\boldsymbol{\kappa}^{(t)},\boldsymbol{\Phi}^{(t)},\boldsymbol{\Lambda}^{(t)},\boldsymbol{\tau}^{(t)},\boldsymbol{\Psi}^{(t)},\mathbf{x}_i} = (\boldsymbol{\Phi}^{(t)^{-1}} + \boldsymbol{\Lambda}'^{(t)}\boldsymbol{\Psi}^{(t)^{-1}}\boldsymbol{\Lambda}^{(t)})^{-1},$$

where $\mathbf{x}_i = (x_{i1}, \ldots, x_{iJ})'$ is the $(J \times 1)$ vector of observed values for examinee i (i.e., the ith row of $\mathbf{x}$). Sample a value for the latent variables from

$$\xi_i^{(t+1)} \mid \boldsymbol{\kappa}^{(t)}, \boldsymbol{\Phi}^{(t)}, \boldsymbol{\Lambda}^{(t)}, \boldsymbol{\tau}^{(t)}, \boldsymbol{\Psi}^{(t)}, \mathbf{x}_i \sim N(\boldsymbol{\mu}_{\xi_i^{(t+1)}|\boldsymbol{\kappa}^{(t)},\boldsymbol{\Phi}^{(t)},\boldsymbol{\Lambda}^{(t)},\boldsymbol{\tau}^{(t)},\boldsymbol{\Psi}^{(t)},\mathbf{x}_i}, \boldsymbol{\Sigma}_{\xi_i^{(t+1)}|\boldsymbol{\kappa}^{(t)},\boldsymbol{\Phi}^{(t)},\boldsymbol{\Lambda}^{(t)},\boldsymbol{\tau}^{(t)},\boldsymbol{\Psi}^{(t)},\mathbf{x}_i}).$$

2. *Sample the parameters for the latent variable distribution.*
   a. Using the just-sampled values for the latent variables (from iteration $t + 1$) and the latent variable covariance matrix from the previous iteration (t), compute

$$\boldsymbol{\mu}_{\boldsymbol{\kappa}^{(t+1)}|\Xi^{(t+1)},\boldsymbol{\Phi}^{(t)}} = (\boldsymbol{\Sigma}_{\boldsymbol{\kappa}}^{-1} + n\boldsymbol{\Phi}^{(t)^{-1}})^{-1}(\boldsymbol{\Sigma}_{\boldsymbol{\kappa}}^{-1}\boldsymbol{\mu}_{\boldsymbol{\kappa}} + n\boldsymbol{\Phi}^{(t)^{-1}}\bar{\xi}^{(t+1)})$$

   and

$$\boldsymbol{\Sigma}_{\boldsymbol{\kappa}^{(t+1)}|\Xi^{(t+1)},\boldsymbol{\Phi}^{(t)}} = (\boldsymbol{\Sigma}_{\boldsymbol{\kappa}}^{-1} + n\boldsymbol{\Phi}^{(t)^{-1}})^{-1},$$

   where $\bar{\xi}^{(t+1)}$ is the $(M \times 1)$ vector of means of the treated-as-known values of the latent variables $\Xi^{(t+1)}$ over the n examinees. Sample a value for the mean of the latent variables from

$$\boldsymbol{\kappa}^{(t+1)} \mid \Xi^{(t+1)}, \boldsymbol{\Phi}^{(t)} \sim N(\boldsymbol{\mu}_{\boldsymbol{\kappa}^{(t+1)}|\Xi^{(t+1)},\boldsymbol{\Phi}^{(t)}}, \boldsymbol{\Sigma}_{\boldsymbol{\kappa}^{(t+1)}|\Xi^{(t+1)},\boldsymbol{\Phi}^{(t)}}).$$

   b. Using the just-sampled values for the latent variables and the means of the latent variables (from iteration $t + 1$), compute

$$\mathbf{S}_{\xi}^{(t+1)} = 1/n\textstyle\sum_i(\xi_i^{(t+1)} - \boldsymbol{\kappa}^{(t+1)})(\xi_i^{(t+1)} - \boldsymbol{\kappa}^{(t+1)})'.$$

   Sample a value for the covariance matrix for the latent variables from

$$\boldsymbol{\Phi} \mid \Xi^{(t+1)}, \boldsymbol{\kappa}^{(t+1)} \sim \text{Inv-Wishart}(d\boldsymbol{\Phi}_0 + n\mathbf{S}_{\xi}^{(t+1)}, d + n).$$

3. *Sample the measurement model parameters.*
   a. For each observable, $j = 1,\ldots, J$, using the just-drawn values for the latent variables (from iteration $t + 1$) and the error variance from the previous iteration (t), compute

$$\boldsymbol{\mu}_{\lambda_{jA}^{(t+1)}|\Xi^{(t+1)},\psi_{jj}^{(t)},\mathbf{x}_j} = \left(\boldsymbol{\Sigma}_{\lambda A}^{-1} + \frac{1}{\psi_{jj}^{(t)}}\Xi_A'^{(t+1)}\Xi_A^{(t+1)}\right)^{-1}\left(\boldsymbol{\Sigma}_{\lambda A}^{-1}\boldsymbol{\mu}_{\lambda A} + \frac{1}{\psi_{jj}^{(t)}}\Xi_A'^{(t+1)}\mathbf{x}_j\right)$$

and

$$\boldsymbol{\Sigma}_{\lambda_{jA}^{(t+1)}|\Xi^{(t+1)},\psi_{jj}^{(t)},\mathbf{x}_j} = \left(\boldsymbol{\Sigma}_{\lambda A}^{-1} + \frac{1}{\psi_{jj}^{(t)}}\Xi_A'^{(t+1)}\Xi_A^{(t+1)}\right)^{-1},$$

where $\mathbf{x}_j = (x_{1j},\ldots, x_{nj})'$ is the $(n \times 1)$ vector of observed values for observable j (i.e., the jth column of $\mathbf{x}$). Sample a value for the augmented vector of loadings from

$$\boldsymbol{\lambda}_{jA}^{(t+1)} \mid \Xi^{(t+1)}, \psi_{jj}^{(t)}, \mathbf{x}_j \sim N(\boldsymbol{\mu}_{\lambda_{jA}^{(t+1)}|\Xi^{(t+1)},\psi_{jj}^{(t)},\mathbf{x}_j}, \boldsymbol{\Sigma}_{\lambda_{jA}^{(t+1)}|\Xi^{(t+1)},\psi_{jj}^{(t)},\mathbf{x}_j}).$$

   b. For each observable, $j = 1, \ldots, J$, using the just sampled values for the latent variables and the augmented vector of loadings (from iteration $t + 1$), compute

$$SS(\mathbf{E}_j^{(t+1)}) = (\mathbf{x}_j - \Xi_A^{(t+1)}\boldsymbol{\lambda}_{jA}^{(t+1)})'(\mathbf{x}_j - \Xi_A^{(t+1)}\boldsymbol{\lambda}_{jA}^{(t+1)})$$

where $\mathbf{x}_j = (x_{1j},\ldots, x_{nj})'$ is the $(n \times 1)$ vector of observed values for observable j (i.e., the jth row of $\mathbf{x}$). Sample a value for the error variance from

$$\psi_{jj}^{(t+1)} \mid \Xi^{(t+1)}, \boldsymbol{\lambda}_{jA}^{(t+1)}, \mathbf{x}_j \sim \text{Inv-Gamma}\left(\frac{\nu_\psi + n}{2}, \frac{\nu_\psi\psi_0 + SS(\mathbf{E}_j^{(t+1)})}{2}\right).$$

Note that the joint posterior distribution includes all unknowns, including the latent variables. As a result, a Gibbs sampler will yield draws for the latent variables from the posterior distribution. When working within a probability model, Bayes' theorem provides a mechanism for arriving at a representation of the examinees' latent variables, be it scoring using point estimates for other parameters (Bartholomew, 1981, 1996) or in a fully Bayesian analysis such as that described here (see also Aitkin & Aitkin, 2005). Within the probability model, the latent variables are unknowns, and Bayes' theorem provides the mechanism for yielding a posterior distribution for all the unknowns. Mechanically, this shows up in the Gibbs sampler taking draws for the latent variables for each examinee from the full conditional. For each examinee, the collection of the draws is an empirical approximation to the marginal distribution for their latent variables, which may be summarized in the usual ways (e.g., density plots, point and interval summaries, and standard deviations).

9.3 Example: Single Latent Variable (Factor) Model

We illustrate a Bayesian approach to CFA via an analysis on a subset of data previously reported in Levy (2011) using an example drawn from Mitchell (2009), who examined the factor structure of a web-based version of the Institutional Integration Scale (Pascarella &

Terenzini, 1980). The scale measures students' collegiate experiences with and perceptions of peers, faculty, intellectual growth, and academic goals. It contains 31 five-point Likert-type items with rating categories of strongly disagree, disagree, neutral, agree, and strongly agree, coded as integers from 1 to 5. The items are organized into five subscales: *Peer Interaction, Faculty Interaction, Academic and Intellectual Development, Faculty Concern,* and *Institutional Goal Commitment.* The current analysis is based on a subset of 500 examinees from the sample analyzed in Levy (2011). For each examinee, scores on the subscales are obtained by averaging the scores for the items in that subscale. Summary statistics for the subscales are given in Table 9.1.

In this example we pursue a model with a single latent variable (factor, i.e., $M = 1$) for the five observable subscale scores. In Section 9.4, we pursue a model with two latent variables (French & Oakes, 2004; Mitchell, 2009), which is but one of several possible alternatives.* The path diagram for the model is depicted in Figure 9.1a. For simplicity we can drop the subscripting associated with multiple latent variables, denoting the latent variable for an examinee by ξ_i, the mean of the latent variable by κ, and the variance of the latent variable by ϕ. In the current illustration, we resolve the indeterminacies by fixing the latent variable mean κ to be 0 and fixing the loading for the first observable (for *Peer Interaction*) λ_1 to be 1.

9.3.1 DAG Representation

The general DAG given in Figure 9.2 is fairly "open" in the sense that it does not contain any of the restrictions typically seen in applications, including those needed to resolve indeterminacies. For any parameter that is fixed to a particular value, the associated node would not be specified as a child of any other node, and in the conventions adopted here, it would be specified as a rectangle rather than a circle.

Figure 9.3a presents a DAG for the model for this example where the indeterminacies are resolved by specifying a value for the mean of the latent variable, κ, and the loading for the first observable, λ_1. As a result, these nodes are rectangles, rather than circles.

TABLE 9.1

Summary Statistics for the Institutional Integration Scale Subscales Data, Including Means, Standard Deviations, and the Correlations

Subscale	*PI*	*AD*	*IGC*	*FI*	*FC*
IGC	.41	.64			
FI	.56	.56	.31		
FC	.54	.63	.46	.67	
Mean	3.33	3.90	4.60	3.03	3.71
Standard Deviation	0.83	0.60	0.46	0.89	0.79

PI = Peer Interaction; AD = Academic and Intellectual Development; IGC = Institutional Goal Commitment; FI = Faculty Interaction; FC = Faculty Concern.

* One set of alternatives analyzes the individual items rather than the subscale aggregates, and we pursue an example analyzing the items of one subscale using the tools of item response theory in Section 11.4. This could be extended, for instance to simultaneously modeling all the items, using five latent variables—one for every subscale—using the tools of multidimensional item response theory discussed in Section 11.5.

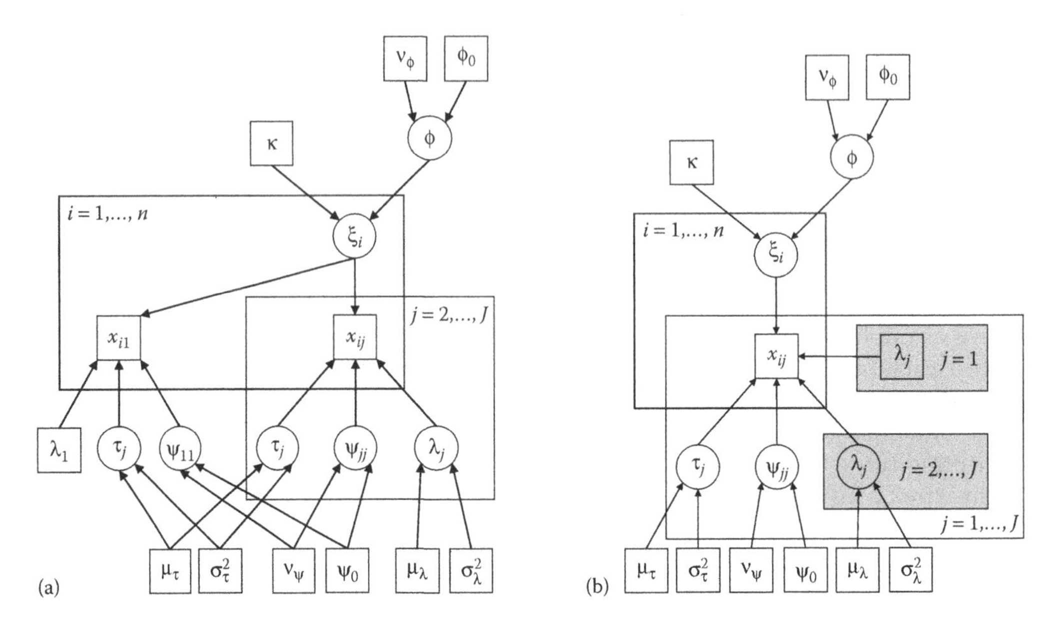

FIGURE 9.3
Two representations for the model with one latent variable with fixed latent variable mean and fixed loading. (a) Exact directed acyclic graph representation. (b) Inexact but less cluttered representation, where the shaded box indicates that there is a subset of the larger plate for j with a common structure.

λ_1 is now specified differently than the remaining loadings, and so resides outside the plate for observables, which is now amended to start at $j = 2$. What's more, the intercept (τ_1) and error variance (ψ_{11}) for the first observable must now lie outside the plate even though they are assigned the same prior distributions as the remaining intercepts and error variances. This is exact but somewhat unfortunate, as it produces some additional clutter at the bottom of the graph. On the positive side, it makes it more clear that there is something importantly different going on with the loading for the first observable, but not the intercept or error variance.

A visually simpler representation is given in Figure 9.3b. Here, there is a single plate for all observables $j = 1, \ldots, J$ indicating the basic structure. The shaded plates inside that plate also define values for the observables indexed by j. These indicate that the first loading is fixed to a specified value and the loadings for the remaining observables are unknown and follow a prior distribution governed by μ_λ and σ^2_λ. Though technically inexact and not a representation that directly corresponds to the probability distribution, this type of presentation is somewhat more elegant in communicating the big idea. Generally speaking, it combats the tendency for DAGs to get increasingly messy as additional restrictions are placed on certain parameters.

9.3.2 Model Specification and Posterior Distribution

For completeness, we write out the posterior distribution for the model with the chosen values for the fixed parameters and the hyperparameters:

$$p(\Xi,\phi,\tau,\Lambda,\Psi \mid \mathbf{x}) \propto \prod_{i=1}^{n}\prod_{j=1}^{J} p(x_{ij} \mid \xi_i,\tau_j,\lambda_j,\psi_{jj})p(\xi_i \mid \kappa,\phi)p(\phi)p(\tau_j)p(\psi_{jj})\prod_{j=2}^{J} p(\lambda_j),$$

where

$$
\begin{aligned}
x_{ij} \mid \xi_i, \tau_j, \lambda_j, \psi_{jj} &\sim N(\tau_j + \lambda_j \xi_i, \psi_{jj}) \text{ for } i = 1, \ldots, 500,\ j = 1, \ldots, 5, \\
\xi_i \mid \kappa, \phi &\sim N(\kappa, \phi) \text{ for } i = 1, \ldots, 500, \\
\kappa &= 0, \\
\phi &\sim \text{Inv-Gamma}(5, 10), \\
\tau_j &\sim N(3, 10) \text{ for } j = 1, \ldots, 5, \\
\lambda_1 &= 1, \\
\lambda_j &\sim N(1, 10) \text{ for } j = 2, \ldots, 5.
\end{aligned}
\tag{9.38}
$$

and

$$\psi_{jj} \sim \text{Inv-Gamma}(5, 10) \text{ for } j = 1, \ldots, 5.$$

The priors are reasonably diffuse, and we wish to draw attention to the choice of the priors for the loadings and the intercepts. An assumption of exchangeability reflects that we do not a priori believe anything different about the measurement quality of the observables. As a result, the loadings $\lambda_2, \ldots, \lambda_5$ are assigned a common prior. That this common prior is centered at 1 reflects the exchangeability assumption with respect to λ_1, which was fixed to 1. The use of a prior with a mean of 1 reflects that we do not believe anything different about the measurement quality of *Faculty Interaction*, *Academic and Intellectual Development*, *Faculty Concern*, and *Institutional Goal Commitment*, whose loadings are unknowns in the model, as we do about *Peer Interaction*, whose loading is fixed to 1 in the model. The prior mean of 3 for the (common) prior distribution for the intercepts is based on the knowledge of the response scale. The intercept is the expected value for the observable when the value of the latent variable is 0, which is also the mean of the latent variable, given by the choice of the fixed value for κ. Thus, for each observable, the intercept represents the expected response for an examinee at the mean of the latent variable. With items scored in integers from 1 to 5, it is sensible to choose a value in that range. We opt for a value in the middle, and use a prior variance that yields a diffuse density over 1–5.

9.3.3 WinBUGS

WinBUGS code for the model and list statements for three sets of initial values are given as follows.

```
---------------------------------------------------------------------------
###########################################################################
# Model Syntax
###########################################################################
model{

###########################################################################
# Specify the factor analysis measurement model for the observables
###########################################################################
for (i in 1:n){
  for(j in 1:J){
    mu[i,j] <- tau[j] + ksi[i]*lambda[j]
    x[i,j] ~ dnorm(mu[i,j], inv.psi[j])
  }
}
```

```
#########################################################################
# Specify the prior distribution for the latent variables
#########################################################################
for (i in 1:n){
  ksi[i] ~ dnorm(kappa, inv.phi)
}


#########################################################################
# Specify the prior distribution for the parameters that govern
# the latent variables
#########################################################################
kappa <- 0                          # Mean of factor
inv.phi ~ dgamma(5, 10)             # Precision of factor
phi <- 1/inv.phi                    # Variance of factor


#########################################################################
# Specify the prior distribution for the measurement model parameters
#########################################################################
for(j in 1:J){
  tau[j] ~ dnorm(3, .1)             # Intercepts for observables
  inv.psi[j] ~ dgamma(5, 10)        # Precisions for observables
  psi[j] <- 1/inv.psi[j]            # Variances for observables
}

lambda[1] <- 1.0                    # Loading fixed to 1.0
for (j in 2:J){
     lambda[j] ~ dnorm(1, .1)       # Prior for remaining loadings
}

}                                   # closes the model

#########################################################################
# Initial values for three different chains
#########################################################################
list(tau=c(.1, .1, .1, .1, .1), lambda=c(NA, 0, 0, 0, 0), inv.phi=1,
inv.psi=c(1, 1, 1, 1, 1))

list(tau=c(3, 3, 3, 3, 3), lambda=c(NA, 3, 3, 3, 3), inv.phi=2,
inv. psi=c(2, 2, 2, 2, 2))

list(tau=c(5, 5, 5, 5, 5), lambda=c(NA, 6, 6, 6, 6), inv.phi=.5,
inv. psi=c(.5, .5, .5, .5, .5))
-------------------------------------------------------------------------
```

A few comments about the code are warranted. First, note that the error variances are defined in the code as psi[j]. The use of a single index of j here contrasts with the double subscripting in "ψ_{jj}." This double subscripting reflects that the error variances are the diagonal of the $(J \times J)$ error covariance matrix $\mathbf{\Psi}$. A matrix representation is useful when error covariances on the off-diagonal are included, as discussed in Section 9.8. In the current setting where $\mathbf{\Psi}$ is diagonal, we can conceive of the diagonal being a vector of error variances. This simplifies the coding of the model and the inputting of initial values in WinBUGS. Second, we call out the use of "NA" for the first loading in the list statements for initial values. The first loading is fixed to 1 and therefore not a stochastic node. As a result, no initial value (not even 1) can be supplied to WinBUGS for this node. Finally, the three sets of initial values were specified to contain values that were anticipated to represent values for the parameters that would be fairly dispersed in the posterior distribution.*

9.3.4 Results

The model was fit in WinBUGS using three chains with these start values for the intercepts, loadings, error precisions, and the precision of the latent variable. The start values for the latent variables were generated by WinBUGS. Based on convergence diagnostics (Section 5.72), there was strong evidence of convergence within a few iterations. To be conservative, 500 iterations were discarded as burn-in, and 5000 subsequent iterations from each chain were used, totaling 15,000 draws used to empirically approximate the posterior distribution. The marginal densities were unimodal and fairly symmetric. Summary statistics are given in Table 9.2.

The loadings for *Faculty Interaction* and *Faculty Concern* are comparable to each other and the fixed loading for *Peer Interaction*, and the error variances for these variables are

TABLE 9.2

Summary of the Posterior Distribution for the Single Latent Variable Model, Including Summaries for Two Examinees' Latent Variables (ξ_1 and ξ_8)

Parameter	Mean	Standard Deviation	95% Highest Posterior Density Interval
λ_{PI}	1.00	–	–
λ_{AD}	0.73	0.04	(0.64, 0.82)
λ_{IGC}	0.42	0.04	(0.35, 0.49)
λ_{FI}	1.05	0.07	(0.92, 1.18)
λ_{FC}	0.98	0.06	(0.87, 1.10)
τ_{PI}	3.33	0.04	(3.25, 3.41)
τ_{AD}	3.90	0.03	(3.84, 3.95)
τ_{IGC}	4.60	0.02	(4.55, 4.64)
τ_{FI}	3.03	0.04	(2.95, 3.12)
τ_{FC}	3.71	0.04	(3.65, 3.79)
ψ_{PI}	0.37	0.03	(0.32, 0.43)
ψ_{AD}	0.18	0.01	(0.16, 0.21)
ψ_{IGC}	0.18	0.01	(0.16, 0.20)
ψ_{FI}	0.38	0.03	(0.32, 0.44)
ψ_{FC}	0.27	0.02	(0.22, 0.31)
ϕ	0.44	0.04	(0.35, 0.52)
ξ_1	−0.20	0.26	(−0.70, 0.30)
ξ_8	0.86	0.26	(0.35, 1.38)

PI = Peer Interaction; FI = Faculty Interaction; AD = Academic and Intellectual Development; FC = Faculty Concern; IGC = Institutional Goal Commitment.

* The data statement is not shown here on space considerations. Like all the other code used in the examples, it is available from the website for the book.

comparable, suggesting that the observables have comparable relations to the latent variable. The loading for *Academic and Intellectual Development* is a bit lower, as is its error variance. Most notably, the loading for *Institutional Goal Commitment* is quite a bit lower than the others, suggesting it has a weaker relationship to the latent variable. Table 9.2 includes results for two examinees, where it is easily seen that examinee 8 has a higher standing on the latent variable than examinee 1. These results do not speak to the viability of this model for this data, and we defer a treatment of techniques for evaluating models until Chapter 10, where this example forms the basis for discussing a variety of techniques for criticizing and comparing models.

9.4 Example: Multiple Latent Variable (Factor) Model

We reanalyze the data on the subscales of the Institutional Integration Scale, illustrating a model with two latent variables. More specifically, we consider an organization where the first three observables, *Peer Interaction*, *Academic and Intellectual Development*, and *Institutional Goal Commitment*, load on one latent variable interpreted as pertaining specifically to *Students*, and the other two observables, *Faculty Interaction* and *Faculty Concern*, load on a second latent variable interpreted as pertaining to *Faculty* (French & Oakes, 2004; Mitchell, 2009). The path diagram for the model was given in Figure 9.1(b). We resolve the indeterminacies by fixing the latent variable means, $\boldsymbol{\kappa} = \mathbf{0}$, and fixing the loading for the first observable (*Peer Interaction*) along the first latent variable, $\lambda_{11} = 1$, and the loading for the fourth latent variable (*Faculty Interaction*) along the second latent variable, $\lambda_{42} = 1$.

9.4.1 DAG Representation

When there are multiple latent variables, we often specify many fixed loadings (e.g., to 1 to resolve indeterminacies, to 0 to reflect the lack of loading of an observable on a latent variable), which can make for a cluttered DAG. We opt to present the model via the technically inexact but more elegant presentation given in Figure 9.4. Here, there is a single plate for all the observables, $j = 1, \ldots, J$, indicating the general structure as it pertains to most of the variables and parameters. Within this larger plate, we have introduced two smaller plates that indicate the structure for the loadings. One plate indicates that λ_{11} and λ_{42} are fixed and the other plate indicates that λ_{21}, λ_{31}, and λ_{52} are assigned a common prior distribution. There are a number of possible loadings that are not specified at all: namely λ_{12}, λ_{22}, λ_{32}, λ_{41}, and λ_{51}. These loadings are not part of the model. As discussed more in detail in Section 9.8, we could conceive of them as being fixed to 0, but there is little gained by doing so in the DAG.

9.4.2 Model Specification and Posterior Distribution

We write out the posterior distribution for the model using these chosen values for the fixed parameters and for the hyperparameters:

$$p(\Xi, \Phi, \tau, \Lambda, \Psi \mid \mathbf{x}) \propto \prod_{i=1}^{n} \prod_{j=1}^{J} \prod_{m=1}^{M} p(x_{ij} \mid \xi_i, \tau_j, \lambda_j, \psi_{jj}) p(\xi_i \mid \kappa, \Phi) p(\Phi)$$

$$\times p(\tau_j) p(\psi_{jj}) p(\lambda_{21}) p(\lambda_{31}) p(\lambda_{52}),$$

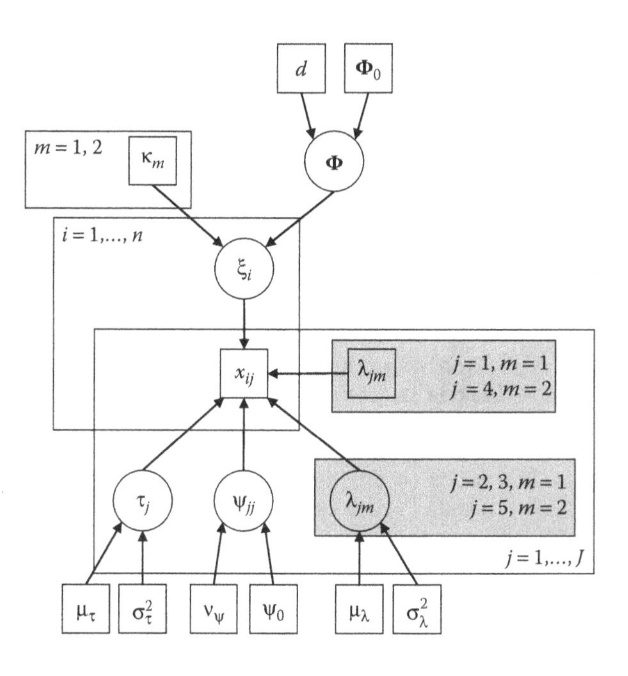

FIGURE 9.4
An inexact directed acyclic graph representation for the model with two latent variables with fixed latent variable means and fixed loadings.

where

$$x_{ij} \mid \xi_i, \tau_j, \lambda_j, \psi_{jj} \sim N(\tau_j + \lambda_{j1}\xi_{i1}, \psi_{jj}) \text{ for } i = 1, \ldots, 500, j = 1,2,3,$$

$$x_{ij} \mid \xi_i, \tau_j, \lambda_j, \psi_{jj} \sim N(\tau_j + \lambda_{j2}\xi_{i2}, \psi_{jj}) \text{ for } i = 1, \ldots, 500, j = 4, 5,$$

$$\xi_i \mid \boldsymbol{\kappa}, \boldsymbol{\Phi} \sim N(\boldsymbol{\kappa}, \boldsymbol{\Phi}) \text{ for } i = 1, \ldots, 500,$$

$$\boldsymbol{\kappa} = \mathbf{0},$$

$$\boldsymbol{\Phi} \sim \text{Inv-Wishart}\left(\begin{bmatrix} 1 & .3 \\ .3 & 1 \end{bmatrix}, 2\right),$$

$$\lambda_{11} = 1,$$

$$\lambda_{42} = 1,$$

$$\lambda_{j1} \sim N(1,10) \text{ for } j = 2, 3,$$

$$\lambda_{52} \sim N(1,10),$$

$$\tau_j \sim N(3,10) \text{ for } j = 1, \ldots, 5,$$

and

$$\psi_{jj} \sim \text{Inv-Gamma}(5,10) \text{ for } j = 1, \ldots, 5.$$

The priors are similar to those for the single-latent variable (factor) model. One key difference is that with multiple latent variables comes the specification of covariance matrix $\boldsymbol{\Phi}$. The adopted prior reflects beliefs that the latent variables are probably positively correlated, in the area of .3, but there is a lot of uncertainty.

9.4.3 WinBUGS

WinBUGS code for the model and list statements for three sets of initial values is given as follows.

```
-------------------------------------------------------------------------
#########################################################################
# Model Syntax
#########################################################################
model{

#########################################################################
# Specify the factor analysis measurement model for the observables
#########################################################################
for (i in 1:n){

   # expected value for each examinee for each observable
   mu[i,1] <- tau[1] + lambda[1,1]*ksi[i,1]
   mu[i,2] <- tau[2] + lambda[2,1]*ksi[i,1]
   mu[i,3] <- tau[3] + lambda[3,1]*ksi[i,1]
   mu[i,4] <- tau[4] + lambda[4,2]*ksi[i,2]
   mu[i,5] <- tau[5] + lambda[5,2]*ksi[i,2]

   for(j in 1:J){
     x[i,j] ~ dnorm(mu[i,j], inv.psi[j])

   }
}


#########################################################################
# Specify the prior distribution for the latent variables
#########################################################################
for (i in 1:n){
  ksi[i, 1:M] ~ dmnorm(kappa[], inv.phi[,])
}


#########################################################################
# Specify the prior distribution for the parameters that govern
# the latent variables
#########################################################################

# Means of latent variables
for(m in 1:M){
  kappa[m] <- 0
}

# precision matrix for the latent variables
inv.phi[1:M,1:M] ~ dwish(dxphi.0[ , ], d)

# the covariance matrix for the latent vars
phi[1:M,1:M] <- inverse(inv.phi[ , ])

phi.0[1,1] <- 1
phi.0[1,2] <- .3
phi.0[2,1] <- .3
phi.0[2,2] <- 1
d <- 2
```

```
for (m in 1:M){
  for (mm in 1:M){
    dxphi.0[m,mm] <- d*phi.0[m,mm]
  }
}


#########################################################################
# Specify the prior distribution for the measurement model parameters
#########################################################################
for(j in 1:J){
  tau[j] ~ dnorm(3, .1)          # Intercepts for observables
    inv.psi[j] ~ dgamma(5, 10)   # Precisions for observables
    psi[j] <- 1/inv.psi[j ]      # Variances for observables
}

lambda[1,1] <- 1.0               # Loading fixed to 1.0
lambda[4,2] <- 1.0               # Loading fixed to 1.0

for (j in 2:3){
  lambda[j,1] ~ dnorm(1, .1)     # Prior for the loadings
}
lambda[5,2] ~ dnorm(1, .1)       # Prior for the loadings

}                                # closes the model


#########################################################################
# Initial values for three different chains
#########################################################################
list(tau=c(.1, .1, .1, .1, .1), lambda= structure(.Data= c( NA, NA, 2,
NA, 2, NA, NA, NA, NA, 2), .Dim=c(5, 2)), inv.phi= structure(.Data=
c(1, 0, 0, 1), .Dim=c(2, 2)), inv.psi=c(1, 1, 1, 1, 1))

list(tau=c(3, 3, 3, 3, 3), lambda= structure(.Data= c( NA, NA, 5.00E-01,
NA, 5.00E-01, NA, NA, NA, NA, 5.00E-01), .Dim=c(5, 2)), inv.phi=
structure(.Data= c(1.33, -.667, -.667, 1.33), .Dim=c(2, 2)), inv.psi=c(2,
2, 2, 2, 2))

list(tau=c(5, 5, 5, 5, 5), lambda= structure(.Data= c( NA, NA, 1, NA, 1,
NA, NA, NA, NA, 1), .Dim=c(5, 2)), inv.phi= structure(.Data= c(1.96,
-1.37, -1.37, 1.96), .Dim=c(2, 2)), inv.psi=c(.5, .5, .5, .5, .5))
------------------------------------------------------------------------
```

Much of the code mimics code introduced earlier, including the specification of initial values for the chains that were anticipated to represent values for the parameters that would be fairly dispersed in the posterior distribution. We concentrate on what is new here, specifically the portion of the code that specifies the prior distribution for the parameters that govern the latent variables. The first section of that code specifies that the means of the latent variables are all 0 by means of a loop over the latent variables. The loop goes from 1 up to `M`, the number of latent variables. The value of `M` could be supplied in the code, but in this example it is part of our data statement, much like the other constants `n` and `J`. The rest of this section pertains to the covariance matrix of the latent variables.

The specification of the node `inv.phi` is the precision matrix, which is what WinBUGS requires for specifying the multivariate normal distribution, in this case for latent variables. The specification of the node `phi` uses the `inverse` function in WinBUGS to transform the precision matrix into the covariance matrix.

The prior for the precision matrix (`inv.phi`) is a Wishart distribution; the rest of the code in this section details the hyperparameters of that prior distribution. First, we define a matrix `inv.phi.0`, corresponding to $\boldsymbol{\Phi}_0$, by plugging in a value for each element in the matrix. Next we define `d`, corresponding to d. The next portion of code loops over the number of latent variables twice, which defines a particular row and column of the covariance matrix, and for each element multiplies the corresponding element of $\boldsymbol{\Phi}_0$ by d, and places it in a corresponding place in a node `dxphi.0`. The double looping effectively carries out the multiplication of the matrix $\boldsymbol{\Phi}_0$ by the scalar d. This process is conducted because the result is used in WinBUGS parameterization of the Wishart distribution (Lunn et al., 2013).

9.4.4 Results

The model was fit in WinBUGS using three chains with these start values for the intercepts, loadings, error precisions, and the precision of the latent variable. The start values for the latent variables were generated by WinBUGS. Based on convergence diagnostics (Section 5.72), there was strong evidence of fast convergence. To be conservative, 500 iterations were discarded as burn-in, and 5000 subsequent iterations from each chain were used, totaling 15,000 draws used to empirically approximate the posterior distribution. Density plots for the marginal distributions were unimodal and fairly symmetric. Summary statistics for the marginal posterior distributions are given in Table 9.3.

9.5 CFA Using Summary Level Statistics

An alternative approach to Bayesian CFA conceives of the data in terms of the first- and second-order moments as the data (Hayashi & Arav, 2006; Press & Shigemasu, 1997; Scheines, Hoijtink, & Boomsma, 1999). To accomplish this, we begin by conceiving of the model in terms of the conditional distribution of the data given $\boldsymbol{\kappa}, \boldsymbol{\Phi}, \boldsymbol{\tau}, \boldsymbol{\Lambda}$, and $\boldsymbol{\Psi}$, marginalized over the latent variables $\boldsymbol{\Xi}$. In this formulation, the distribution of the data is multivariate normal with mean vector $\boldsymbol{\tau} + \boldsymbol{\Lambda}\boldsymbol{\kappa}$ and covariance matrix $\boldsymbol{\Lambda}\boldsymbol{\Phi}\boldsymbol{\Lambda}' + \boldsymbol{\Psi}$ (see Equation 9.30). In this case, it can be shown that the distribution of the mean vector of the observables, $\bar{\mathbf{x}}$, and the covariance matrix of the observables, $\mathbf{S}$, have the following distribution:

$$p(\bar{\mathbf{x}}, \mathbf{S} \mid \boldsymbol{\kappa}, \boldsymbol{\Phi}, \boldsymbol{\tau}, \boldsymbol{\Lambda}, \boldsymbol{\Psi}) = p(\bar{\mathbf{x}} \mid \boldsymbol{\kappa}, \boldsymbol{\Phi}, \boldsymbol{\tau}, \boldsymbol{\Lambda}, \boldsymbol{\Psi}) p(\mathbf{S} \mid \boldsymbol{\Phi}, \boldsymbol{\Lambda}, \boldsymbol{\Psi}), \tag{9.39}$$

where

$$\bar{\mathbf{x}} \sim N\left(\boldsymbol{\tau} + \boldsymbol{\Lambda}\boldsymbol{\kappa}, \frac{1}{n}(\boldsymbol{\Lambda}\boldsymbol{\Phi}\boldsymbol{\Lambda}' + \boldsymbol{\Psi})\right)$$

and (9.40)

$$\mathbf{S} \sim \text{Wishart}(\boldsymbol{\Lambda}\boldsymbol{\Phi}\boldsymbol{\Lambda}' + \boldsymbol{\Psi}, n-1).$$

TABLE 9.3

Summary of the Posterior Distribution for the Two Latent Variable Models, Including Summaries for Two Examinees' Latent Variables (ξ_{11}, ξ_{12}, ξ_{81}, and ξ_{82})

Parameter	Mean	Standard Deviation	95% Highest Posterior Density Interval
λ_{PI}	1.00	–	–
λ_{AD}	0.77	0.05	(0.67, 0.88)
λ_{IGC}	0.46	0.04	(0.38, 0.54)
λ_{FI}	1.00	–	–
λ_{FC}	0.92	0.05	(0.81, 1.02)
τ_{PI}	3.33	0.04	(3.26, 3.41)
τ_{AD}	3.90	0.03	(3.84, 3.95)
τ_{IGC}	4.60	0.02	(4.55, 4.64)
τ_{FI}	3.04	0.04	(2.96, 3.12)
τ_{FC}	3.71	0.04	(3.64, 3.78)
ψ_{PI}	0.36	0.03	(0.31, 0.42)
ψ_{AD}	0.17	0.01	(0.14, 0.20)
ψ_{IGC}	0.17	0.01	(0.15, 0.20)
ψ_{FI}	0.34	0.03	(0.28, 0.40)
ψ_{FC}	0.24	0.02	(0.20, 0.28)
ϕ_{11}	0.39	0.04	(0.30, 0.47)
ϕ_{22}	0.49	0.05	(0.39, 0.59)
ϕ_{21}	0.38	0.04	(0.31, 0.45)
ξ_{11}	−0.15	0.27	(−0.70, 0.38)
ξ_{12}	−0.31	0.3	(−0.90, 0.27)
ξ_{81}	0.88	0.27	(0.36, 1.43)
ξ_{82}	0.84	0.30	(0.26, 1.45)

PI = Peer Interaction; FI = Faculty Interaction; AD = Academic and Intellectual Development; FC = Faculty Concern; IGC = Institutional Goal Commitment.

The posterior distribution is then

$$p(\boldsymbol{\kappa}, \boldsymbol{\Phi}, \boldsymbol{\tau}, \boldsymbol{\Lambda}, \boldsymbol{\Psi} \mid \bar{\mathbf{x}}, \mathbf{S}) \propto p(\bar{\mathbf{x}}, \mathbf{S} \mid \boldsymbol{\kappa}, \boldsymbol{\Phi}, \boldsymbol{\tau}, \boldsymbol{\Lambda}, \boldsymbol{\Psi}) p(\boldsymbol{\kappa}, \boldsymbol{\Phi}, \boldsymbol{\tau}, \boldsymbol{\Lambda}, \boldsymbol{\Psi}), \quad (9.41)$$

where the second term on the right-hand side is the prior distribution, which may be specified in the same manner as described in Section 9.2.2.

This approach using summary statistics has a number of advantages over the individual-level approach. It can be employed when summary statistics are available but the raw data are not (e.g., in a secondary analysis of published summary statistics; Hayashi & Arav, 2006). Note that in contrast to the individual-level approach, the specification here does not involve the examinees' values of the latent variables, as they are marginalized over in construction of the conditional distribution of the data. This can be advantageous in a simulation-based estimation environment, such as Markov chain Monte Carlo (MCMC), where the computational burden and time needed increase with each additional parameter to be estimated. Because the summary-level approach takes the mean vector and covariance matrix as the data, the number of parameters and the associated computational burden does not increase as sample size increases. In contrast, in the individual-level approach, each additional examinee brings with them another M latent variables. More concretely, in Gibbs sampling a full conditional distribution must be defined and sampled from for each examinee's latent variables at each iteration. Thus for the same model, the time needed to fit a model using the summary-level approach is likely to be less than in the individual-level approach, especially in large samples with many latent variables (Choi, Levy, & Hancock, 2006). The estimation-time advantage of the summary-level approach is likely to be diminished in the future as technological advances increase the computational power available to analysts. A related point is that the full conditional distributions for the examinee latent variables are normal distributions, and relatively easy to sample from. In the authors' experience, it is usually not prohibitively time-consuming to analyze individual-level models in WinBUGS. And in situations where it is prohibitively time-consuming, the analyst may turn to other software that is faster than WinBUGS for CFA and related models, such as M*plus* (Muthén & Muthén, 1998–2012) and Amos (Arbuckle, 2007).

However, the summary-level approach suffers relative to the individual-level approach in several respects. It does not yield a posterior distribution for the values of the examinees' latent variables. If inferences about the examinees' latent variables are desired, auxiliary analyses are required. Options here include taking draws from the full conditional distributions of the latent variables given the drawn values for the parameters (Aitkin & Aitkin, 2005), or partially Bayesian solutions using point estimates of parameters (Bartholomew et al., 2011) that could be applied here using a point summary of the parameters. Importantly, the typical summary-level approach relies on the assumption of multivariate normality of the observables, which was derived from an assumption of conditional normality of the observables given the latent variables. Hayashi and Yuan (2003) proposed a robust analysis to account for non-normality, though it is possible that greater flexibility is afforded by adopting an approach that specifies non-normal distributions for the individual-level data (e.g., Lee & Xia, 2008; Zhang et al., 2013).

9.6 Comparing DAGs and Path Diagrams

There are a number of similarities between the graphical modeling approach in the DAGs and the path diagrammatic approach from conventional SEM traditions. Importantly, however, there are also a number of differences. We briefly highlight key similarities and differences between these two approaches by way of comparing the structure of Figure 9.1b and Figure 9.2. Both approaches use nodes to represent observable and latent variables for examinees. The use of rectangles for observables and circles for latent variables is standard for path diagrams but not DAGs; we have adopted this convention for the DAGs in the book. For both approaches, this convention is useful, but imperfect in more complicated situations (e.g., modeling missing data, specifying fixed values for latent variables). Path diagrams additionally use triangles to represent constants. The "1" in Figure 9.1b is in a triangle to represent that this node does *not* vary over examinees. As discussed below, this is really a modeling trick that derives from path diagrams being representations of structural equations. Traditionally, DAGs do not model constants as nodes, as they are not needed for structuring dependence relationships in multivariate distributions. We have used expanded DAGs (see Sections 2.5 and 7.1) to aid in exposition, in which case constants are denoted by rectangles.

The need to indicate a constant via a different shape in path diagrams brings into relief a key distinction between path diagrams and DAGs. Path diagrams implicitly express that the nodes—the rectangles and circles—vary over examinees. As a result, the parameters—which do not vary over examinees—are *not* nodes in path diagrams. They are placed along the paths in the diagram. For directed paths (i.e., one-headed arrows), the parameter residing along the path serves as a coefficient for the variable at the source of the path for the structural equation for the variable at the destination of the path. When the source of the path is a constant (i.e., the "1" in the triangle) this just expresses that the parameter along the path is an intercept in the structural equation. For nondirected paths (i.e., two-headed arrows), the parameter residing along the path refers to the covariance between the variables connected by the path; when the connection is between a variable and itself, this is just the variance of the variable. Conventions for what are and are not represented as nodes align well with a frequentist perspective on inference: nodes correspond to variables associated with examinees that are assigned distributions; because parameters are treated as fixed entities in a frequentist framework, they are not depicted as nodes.

In contrast, DAGs make explicit the replication of variables over examinees, and other replications, via plate structures. In a fully probabilistic framework, unknown parameters are assigned distributions and appear as nodes in the graph. Path diagrams are well aligned with a frequentist perspective in which distributions are invoked only where a sampling concept is applied, namely for examinee variables. DAGs are well aligned with a Bayesian perspective in which all entities are treated as random variables, with distributions assigned to all. In this light, examinee variables do not have a distinct status; other entities appear as nodes in the graph and so the replication over examinees must be made explicit. This underscores a general point about a Bayesian approach; in contrast to the conventional frequentist approach, the latent variables have the same status as other parameters.

Importantly, path diagrams and DAGs differ in how they represent correlational structures. Path diagrams employ nondirectional (two-headed) arrows to indicate a correlation between variables, such as the correlation between the latent variables in Figure 9.1b. DAGs only employ directed (one-headed) arrows, and take a multivariate approach to specifying correlational structures. As the DAG in Figure 9.2 illustrates, we specify the vector of latent variables (for each examinee i) and express the correlation via the specification of a multivariate distribution with covariance structure $\boldsymbol{\Phi}$. Path diagrams also allow for cyclic dependences, where setting out on a path of directed arrows from a certain node, we can return to that node. The simplest example of this is when there two variables and between them are two directed arrows facing in opposite direction, indicating a feedback or reciprocal dependence (Kline, 2013).

We stress that though path diagrams are not as closely aligned with the probability-based modeling approach as DAGs, they pose certain advantages over DAGs. Path models are especially adept at communicating a number of key features: how many observable and latent variables are in the model; which observables load on which latent variables; and the means, variances, and correlations among the latent variables and possibly other exogenous entities (indicated by a bidirectional arrow). As such they are a powerful knowledge representation for analysts when formulating models and communicating the model to others. As path diagrams expressing the dependence and conditional independence relationships among the entities associated with examinee variables, they more easily support the derivation of model-implied mean and covariance structures via path tracing (Mulaik, 2009; Wright, 1934), which in turn supports other analyses such as determining how models are related (Levy & Hancock, 2007, 2011). For these reasons, path diagrams and DAGs may complement each other in how they express and facilitate reasoning using models, and both may be gainfully called into service simultaneously, as warranted by the needs of the analyst.

9.7 A Hierarchical Model Construction Perspective

9.7.1 Hierarchical Model Based on Exchangeability

In the preceding sections, we have developed Bayesian procedures to CFA by first presenting the CFA model and representations in terms that readers familiar with conventional approaches should recognize, and then shifting to a Bayesian perspective by overlaying prior distributions on unknown parameters. This has been the didactic approach taken throughout most of the book, as we suspect that most readers will find this approach to be the most accessible for learning Bayesian approaches to psychometrics. As noted in Section 3.6, the perspective that views Bayes as using priors to augment what is done in frequentist procedures is but one of many perspectives on Bayesian analyses, and it is worth viewing Bayesian procedures through other lenses to highlight key principles of Bayesian modeling and illustrate complementary interpretations.

Here, we develop the Bayesian CFA model from the perspective of a hierarchical model construction, in which we build a model through a layering of distributional specifications for unknowns introduced at lower levels (see Section 3.6). We begin with the joint distribution of the data, $\mathbf{x}$. This distribution is specified in (9.15). This structuring of the distribution of $\mathbf{x}$ introduces a number of parameters: $\boldsymbol{\tau}$, $\boldsymbol{\Lambda}$, $\boldsymbol{\Psi}$, and $\boldsymbol{\Xi}$. As these are unknown entities, they require prior distributions.

Starting with the first of these, we tackle the challenge of specifying a J-variate prior distribution for $\boldsymbol{\tau}$ by assuming exchangeability, which allows for the specification of a common univariate (normal) prior for all elements of $\boldsymbol{\tau}$, given in (9.25), with values of the hyperparameters μ_τ and σ^2_τ specified by the analyst. Similarly, an exchangeability assumption allows for the specification of a common univariate (normal) prior for the elements in $\boldsymbol{\Lambda}$ included in the model, given in (9.26), with hyperparameters μ_λ and σ^2_λ, the values of which are specified by the analyst. Likewise, an exchangeability assumption allows for the specification of a common prior for the diagonal elements of $\boldsymbol{\Psi}$, given in (9.27), with hyperparameters $\nu_{\psi0}$ and ψ_0, the values of which are specified by the analyst.

Turning to the latent variables, $\boldsymbol{\Xi}$, an assumption of exchangeability about the examinees supports the specification of a common (normal) prior for all examines in (9.18), with hyperparameters $\boldsymbol{\kappa}$ and $\boldsymbol{\Phi}$. As unknown entities, these also require prior distributions. An assumption of exchangeability with respect to the latent variables allows for a common univariate (normal) prior to be specified for the elements of $\boldsymbol{\kappa}$ in (9.21) with its own hyperparameters μ_κ and σ^2_κ, with values specified by the analyst. The prior distribution for $\boldsymbol{\Phi}$ in (9.22), with its own hyperparameters $\boldsymbol{\Phi}_0$ and d specified by the analyst, completes the model specification.

To recap, $\mathbf{x}$ is an $(n \times J)$ matrix of observables and we seek to specify a probabilistic model to facilitate desired inferences about the examinees and the tasks that yield the observables. Following de Finetti's representation theorem, we accomplish modeling this joint distribution by assuming exchangeability, introducing parameters that induce conditional independence, and specifying distributions for the newly introduced parameters. To specify the joint (prior) distribution for the newly introduced parameters, we follow the same steps. If this introduces new unknown parameters, we repeat the process until we reach the point where there are no more unknowns. Equation (9.9) expresses the first layer in the hierarchy, where we have structured the $(n \times J)$-variate distribution of observables in terms of the $(n \times M)$ latent variables Ξ, $(J \times M)$ loadings Λ, J intercepts τ, and J error variances in Ψ. As M is usually far less than J, this represents a considerable simplification.

We are now tasked with specifying a (prior) distribution for the newly introduced parameters, again by capitalizing on exchangeability, introducing parameters that induce conditional independence, and specifying distributions for them as needed. For the $(n \times M)$ matrix of unknowns in Ξ, (9.17) contains the next level of this hierarchy and introduces hyperparameters κ and Φ, for which (hyperprior) distributions are needed. These are specified in the next level of the hierarchy, in (9.20) and (9.22). To specify the former, again an assumption of exchangeability is invoked to allow for a common univariate prior for the elements of κ, in (9.21). In short, by invoking exchangeability assumptions and approaching the problem in a hierarchical manner, we have solved the problem of specifying an $(n \times M)$-variate prior distribution with the following (collecting Equations 9.18, 9.21, and 9.22):

$$\xi_i \mid \kappa, \Phi \sim N(\kappa, \Phi) \text{ for all } i,$$

$$\kappa_m \sim N(\mu_\kappa, \sigma_\kappa^2) \text{ for all } m, \tag{9.42}$$

and

$$\Phi \sim \text{Inv-Wishart}(\Phi_0, d).$$

We adopt the same strategies for the measurement model parameters. Assuming exchangeability allows for the specification of common priors for τ in (9.25), Λ in (9.26), and Ψ in (9.27). Note that we could specify a hierarchical structure on the measurement model parameters, that is, where the (hyper)parameters of their prior distributions are unknown and modeled via their own (hyper)priors. This specification is not common in CFA, but has been used more in the context of item response theory (Section 11.7.4).

This line of reasoning develops the model in a slightly different manner than CFA has typically been conceived. It is distinctly probabilistic in nature in that it (a) focuses on the specifications of (conditional) distributions and (b) for every parameter that is introduced, a (prior) distribution—itself possibly conditional on other parameters—is specified. Combining this approach with assumptions of exchangeability allows for the efficient construction not only of the distribution of the data (the first layer in the hierarchy) but of the prior distributions as well (the subsequent layers). The process concludes when there are no more unknown parameters in need of distributional specification.

9.7.2 Hierarchical Modeling under Conditional Exchangeability

Assumptions of exchangeability greatly facilitate the efficient construction of joint distributions, including prior distributions, via specifying a common distribution for its elements. This is not a requirement of Bayesian modeling. If assumptions of exchangeability are not warranted, the model can still be formulated. For example, if we cannot assume exchangeability between the measurement quality of two observables that load on a latent variable, we need not assign their loadings the same prior distribution. Rather, each loading can have a different prior.

In general, every parameter can have a unique prior. However, it is often the case that we need not resort to this extreme. Rather, if exchangeability cannot be assumed, we construct the model to allow for an assumption of conditional exchangeability (Jackman, 2009; Lindley & Novick, 1981), which means elements are exchangeable conditional on some other relevant entity. For example, in multiple-group modeling, examinees from *different* groups might not be assumed to be exchangeable, but examinees may be assumed to be exchangeable *within* groups, that is, *conditional* on the grouping variable. In that case, a conditionally exchangeable prior for Ξ is given by

$$\xi_{ig} \mid \kappa_g, \Phi_g \sim N(\kappa_g, \Phi_g),$$

$$\kappa_{mg} \sim N(\mu_{\kappa_g}, \sigma^2_{\kappa_g}),$$

and (9.43)

$$\Phi_g \sim \text{Inv-Wishart}(\Phi_{0_g}, d_g),$$

where the additional subscripting by g indicates group-specific parameters. Models that assume conditional exchangeability about examinee parameters are commonly used in multiple-group modeling and in examinations of differences in latent means (Thompson & Green, 2013).

In addition, we might assume that the measurement parameters are exchangeable within groups, but not across groups. In that case, a conditionally exchangeable model specification is given by

$$x_{ijg} \mid \xi_{ig}, \tau_{jg}, \boldsymbol{\lambda}_{jg}, \psi_{jjg} \sim N(\tau_{jg} + \xi_{ig}\boldsymbol{\lambda}'_{jg}, \psi_{jjg}),$$

$$\lambda_{jmg} \sim N(\mu_{\lambda_g}, \sigma^2_{\lambda_g}),$$

$$\tau_{jg} \sim N(\mu_{\tau_g}, \sigma^2_{\tau_g}), \quad (9.44)$$

and

$$\psi_{jjg} \sim \text{Inv-Gamma}(\nu_{\psi_g}/2, \nu_{\psi_g}\psi_{0_g}/2).$$

Models that assume the measurement model parameters are conditionally exchangeable are commonly used in measurement invariance and differential item functioning analyses where group membership is known (Millsap, 2011; Verhagen & Fox, 2013) or unknown as in latent class or mixture models (Pastor & Gagné, 2013).

Note that in (9.43) and (9.44) each group has distinct hyperparameters, reflecting different a priori beliefs about the group-specific parameters. This need not be the case. For example, we may specify the model as

$$\lambda_{jmg} \sim N(\mu_{\lambda}, \sigma^2_{\lambda}),$$

$$\tau_{jg} \sim N(\mu_{\tau}, \sigma^2_{\tau}),$$

and (9.45)

$$\psi_{jjg} \sim \text{Inv-Gamma}(\nu_{\psi_g}/2, \nu_{\psi_g}\psi_{0_g}/2),$$

which allows the measurement model parameters to vary by groups, as may be suggested by the data, but reflects that a priori we do not believe anything different about them for the different groups. Still other possibilities involve not only different hyperparameters for groups, but different parametric forms as well.

9.8 Flexible Bayesian Modeling

It has been argued that Bayesian approaches to modeling are more flexible than frequentist approaches, and may therefore offer a greater potential to better represent substantive theories (Levy, 2009). This is evidenced in part by researchers turning to Bayesian approaches when they run up against the limits of what is possible in conventional frequentist approaches to modeling and estimation (for examples in factor analysis, see Ansari & Jedidi, 2000; Ansari, Jedidi, & Dube, 2002; Muthén & Asparouhov, 2012; Segawa, Emery, & Curry, 2008; Zhang et al., 2013). This section discusses some examples that highlight some of the flexibility of Bayesian approaches above and beyond those associated with conventional frequentist approaches.

In CFA, this flexibility tis most naturally seen in terms of the loadings when there are multiple latent variables. A conventional approach to CFA specifies observables as either loading on a latent variable or not. A hallmark of conventional CFA is that at least some observables do not load on at least some latent variables. Drawing from Thurstone's (1947) terminology, observables that load on multiple latent variables are referred to as factorially complex, and those that load on only one latent variable are referred to as factorially simple. When the observable is factorially complex, it is often the case that it loads more strongly on a latent variable that is primarily associated with; the (usually) weaker loadings on the other latent variables are sometimes termed *cross-loadings*. Perhaps the most common specification is for all observables to be factorially simple in which all the cross-loadings are 0 (e.g., Figure 9.1b), a situation that is commonly called simple structure, and has also been termed independent clusters (McDonald, 1999), between-item multidimensionality (Adams, Wilson, & Wang, 1997), and heterogeneous test structure (Lucke, 2005).

In conventional CFA, when an observable is specified as loading on the latent variable, the loading is included in the model and estimated from the information in the data. When an observable is specified as not loading on a latent variable, the loading is not included in the model. These loadings may be viewed as indeed being part of the model but constrained to be 0, a perspective that underlies the common use of likelihood ratio tests for models with and without additional loadings (e.g., Bollen, 1989).

Each of these options may be accommodated in a Bayesian approach. The use of a prior distribution that is sufficiently diffuse (e.g., a uniform distribution, a normal distribution with a large variance) yields a posterior that is driven predominantly by the likelihood, and is akin to the frequentist approach in the situation where the loading is included. The frequentist situation where the loading is constrained to 0 may be modeled in a Bayesian approach by specifying a prior for this parameter with all of its mass at 0. Viewing a prior distribution as an expression of a priori beliefs and modeled assumptions, this suggests that the use of prior distribution with a large variance reflects a very weak assumption or belief about the parameter. At the other end of the spectrum, fixing a parameter by concentrating all the mass of the prior at that value represents a particularly strong assumption or belief about the parameter.

Importantly, and in contrast to a frequentist approach, a Bayesian approach easily affords specifications in between these extremes. By choosing a prior distribution of a particular form we can encode various beliefs and at various levels of certainty. This is particularly natural if a prior can be constructed where one of its parameters captures this strength. Using a normal prior for a loading allows for the specification of the certainty via the prior variance. A prior variance of 0 indicates the loading is constrained to be equal to the value of the prior mean; increasing the prior variance represents a weakening of this belief.

Working from this perspective, Muthén and Asparouhov (2012) drew upon the added flexibility of a Bayesian approach over conventional frequentist approaches to construct a model that better reflected their substantive theory. They argued that the constraint that a cross-loading is 0 is in many cases overly stringent. Substantive theory typically does not imply that the cross-loadings are *exactly* 0, just that they are low in magnitude, meaning an observable's dependence on that latent variable is weak relative to its dependence on the latent variable on which it (primarily) loads. Note however that the alternative in a conventional frequentist tradition is to include the parameter and estimate it from the data, which places the parameter on par with the primary, substantive loading. Seeking an alternative to these extremes, Muthén and Asparouhov (2012) advocated the use of prior distributions for cross-loadings that are normal with mean 0 and a small variance. The specification of the variance encodes how willing the analyst is to constrain the cross-loading to be 0 versus allowing it to be dictated by the data. Smaller values of the variance reflect a firmer belief that the cross-loading is exactly 0; larger values reflect an increasing willingness to allow the cross-loading to be farther from 0.

Another example concerns the associations between error terms. In our present development, $\boldsymbol{\Psi}$ was assumed to be diagonal, containing error variances on the diagonal and 0s off the diagonal for the error covariances. This implies that the only source of association for the observables is expressed by the loadings and the latent variables. This can be relaxed in a number of ways. One natural approach is to model $\boldsymbol{\Psi}$ as a complete covariance matrix, like $\boldsymbol{\Phi}$, assigning a prior distribution such as an inverse-Wishart for $\boldsymbol{\Psi}$.

Muthén and Asparouhov (2012) studied this and two other approaches. In one, the error covariance matrix $\boldsymbol{\Psi}$ is defined as

$$\boldsymbol{\Psi} = \boldsymbol{\Omega} + \boldsymbol{\Psi}^*, \tag{9.46}$$

where $\boldsymbol{\Psi}^*$ is a diagonal matrix and $\boldsymbol{\Omega}$ is a covariance matrix that is *not* assumed to be diagonal. $\boldsymbol{\Psi}^*$ may be assigned a prior using the same strategies we had previously developed for $\boldsymbol{\Psi}$, say, using inverse-gamma priors for each of the diagonal elements. $\boldsymbol{\Omega}$ is assigned an inverse-Wishart prior specified such that all the elements are relatively small. As a result $\boldsymbol{\Psi}$ is a covariance matrix with possibly nonzero off-diagonal elements, but these elements are fairly small. In the other approach, the diagonal elements of $\boldsymbol{\Psi}$ are assigned inverse-gamma priors (as in our development) and each of the off-diagonal elements are modeled with a univariate normal prior with mean 0 and a small variance. In this approach, conditional conjugacy is not preserved as the full conditional distributions in the Gibbs sampling scheme are not of known form; a more complex Metropolis-type sampler is utilized.

For our current discussion, the salient point is that these approaches allow for nonzero covariances among all the errors. Such a model is not identified in a frequentist approach; however, this does not necessarily pose the same sort of problems for a Bayesian analysis as it would for a frequentist analysis. More generally, CFA models that are not identified in a frequentist analysis can often by specified by using a Bayesian approach (Scheines et al., 1999).

It is worth taking a step back and considering just what these sorts of developments reveal. First, adopting a Bayesian perspective allows the analyst to specify models akin to the two choices available in a conventional frequentist analysis: (a) let the parameter be estimated by the data by way of using a sufficiently diffuse prior or (b) constrain the parameter to a particular value by way of a prior with all its mass at that point. Second, a Bayesian perspective allows for the analyst to specify models that lie in between these two extremes, by using a prior that is in between a maximally diffuse and degenerate distribution. In the situations described here, this frees the analyst from needing to adhere to the strictness of an equality constraint, without conceding the substantive theory that the parameters in question should be weak in magnitude. We note that the flexibility in using distributional representations to encode a continuum of possibilities has been exploited to considerable advantage in other modeling contexts (e.g., to facilitate partial pooling among groups in multilevel modeling, Gelman & Hill, 2007; Novick et al., 1972). Third, on a technical level, model structures that are unidentified in a frequentist approach may be specified and fitted using a Bayesian approach.

The added flexibility of a Bayesian approach allows for a closer representation of the real-world situation in the modeled space, which allows for a potentially closer correspondence between the model and the substantive theory about the real-world situation. Substantive theory typically does not dictate that the cross-loadings are *exactly* 0, or that *all* of the reasons that observables are associated are captured by the latent variables (Muthén & Asparouhov, 2012). A Bayesian approach allows for some flexibility here while still preserving the core of the substantive theory, which is that the associations can be mainly accounted for by the hypothesized factor structure.

More generally, a Bayesian approach allows for the easy imposition of "hard" constraints (e.g., a parameter equals 0 or is equal to another parameter) and "soft" constraints (e.g., a parameter is likely near 0, or slightly different from another parameter) as warranted by various "hard" and "soft" aspects of the analyst's substantive theory and thinking the model is constructed to represent. The salient point is that the Bayesian approach allows for the choices that are typically made in conventional frequentist modeling, and additionally allows for choices in between these options, in ways that more closely align the model with substantive theory and the desired inferences.

Although these examples were studied in the context of CFA, they have implications for psychometrics more broadly. The possibility of cross-loadings depends on the presence of multiple latent variables, which has been a feature of applications of CFA since its origin. Historically, applications of item response and latent class models (Chapters 11 and 13) have tended to employ one latent variable, though multiple latent variable item response and latent class models are on the rise (Reckase, 2009; Rupp et al., 2010; see Section 11.5 and Chapter 14), and these concepts apply directly to those situations as well.

The example surrounding the residual correlations strikes at the heart of all psychometric models where observables are modeled as depending on latent variables, and is therefore germane to the models covered in the rest of the book. The notion of 0 off-diagonal elements in Ψ is a manifestation of the assumption that the observables are conditionally (locally) independent given the latent variables and the measurement model parameters. Recall that these conditional independence assumptions are aligned with notions of exchangeability (Section 7.2).

Tying these themes together, an assumption of exchangeability allows us to simplify the construction of the joint distribution for the observables. In latent variable psychometric models, this exchangeability is conditional on the latent variables and the measurement model parameters; the observables are modeled as conditionally independent given these entities. However, exchangeability assumptions and implied conditional independence structures in the model reflect the analyst's thinking about the complexities of the real world in terms of modeled entities to facilitate reasoning. This thinking is an inexact account of the more complicated real-world scenario of interest. Allowing error covariances and/or loadings to depart slightly from 0 may be seen as a recognition that the dependence and conditional independence relationships in the model, and the analyst's thinking it represents, is possibly (if not certainly) not exactly correct. Including these parameters with small-variance priors represents a softening of the almost-sure-to-be-wrong-about-the-world independence structures warranted by exchangeability assumptions. This serves to better align the modeled scenario with the real-world scenario it is intended to represent.

This also serves as something of an internal check on the viability of the substantive components in the model and serves as an opportunity to learn about things that were not expected. If an analysis with a small-variance prior around 0 for a parameter in fact yields a posterior where the parameter is rather *different* from 0, this suggests that the original thinking that the weakness of the parameter may need to be revised. In this situation, Muthén and Asparouhov (2012) advocated re-specifying the model so that the parameter has a more diffuse prior. The broader topic of model checking to alert the analyst to weaknesses in the model is taken up in more detail in Chapter 10.

These examples also highlight that the distinction between the prior and the likelihood is inexact. Let us return to the case of a loading being included or excluded from the model. Readers steeped in frequentist traditions may naturally consider this feature of the model to be part of the likelihood, and whether a loading is specified is of course a concern of frequentist analyses, which eschew notions of priors for parameters. On the other hand, we can conceive of a parameter being included or excluded as being a prior distribution specification. In the case of a loading, excluding it may be seen as specifying a prior distribution concentrating all its mass at 0.

Resuming a theme introduced in Section 3.5.2, we might say much the same about other "model specifications" that typically are the focus of conventional modeling approaches, like specifying which observables load on which latent variables, or whether the latent variables or certain errors are correlated. These are just some of the issues analysts address when specifying a CFA model in the conventional frequentist approach to CFA modeling, estimation, and inference.*

But what are these if not prior specifications? As we have seen, they can be expressed as prior distributional specifications. The CFA structure, as typically developed in frequentist traditions, is itself a prior specification. More generally, all of the "model specifications," including when they are framed in frequentist light, may be seen as prior probability specifications.

We close this section with a word of caution. We have argued that a Bayesian approach opens up possibilities for employing models, including those that are not identified in a conventional sense (for conventional treatments of identification in CFA, see, e.g., Bollen, 1989; Mulaik, 2009). The use of a model small-variance prior is an example of this (Muthén & Asparouhov, 2012). However, with these new possibilities come potential pitfalls. We suspect that the rules and guidelines that have been developed in conventional approaches to modeling may not apply as forcefully when employing Bayesian approaches. But just how different things are when operating in a Bayesian framework

* And this is to say nothing about other specifications that often go unstated, like the choice that is almost always made to use normal distributions for the observables and latent variables in CFA and SEM.

has not been worked out in sufficient detail. Analysts would do well to heed the lessons learned in conventional approaches to modeling, and proceed cautiously when exploring unchartered territories. These themes are echoed in the discussion of resolving indeterminacies, to which we now turn.

9.9 Latent Variable Indeterminacies from a Bayesian Modeling Perspective

9.9.1 Illustrating an Indeterminacy via MCMC

The example in Section 9.3 resolved the indeterminacies in the single latent variable model by fixing the mean of the latent variable to 0 and the loading for the first observable to 1. An alternative approach common in CFA is to fix the variance of the latent variable, usually to 1, and free the restriction on the loading. In this section, we illustrate that approach, and potential pitfalls that await the analyst. The model differs from that expressed in (9.38) in terms of the variance of the latent variable and the loadings. Here we have

$$\phi = 1$$

and

$$\lambda_j \sim N(1,10) \quad \text{for } j = 1,\ldots,5.$$

This specification fixes the variance of the latent variable to 1 and releases the constraint on the first loading, specifying it to have the prior as the remaining loadings. All the other expressions remain the same. The DAG for the model is given in Figure 9.5.

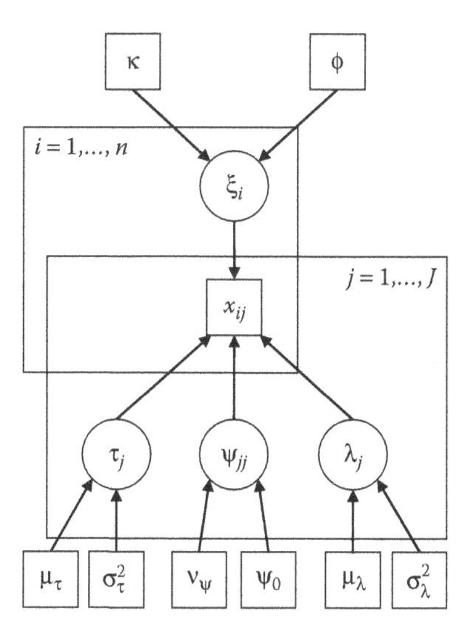

FIGURE 9.5
Directed acyclic graph for the model with one latent variable with fixed latent variable mean and variance (which is denoted by the boxes for κ and ϕ).

WinBUGS code for the model is given as follows.

```
-------------------------------------------------------------------------
##########################################################################
# Model Syntax
##########################################################################
model{

##########################################################################
# Specify the factor analysis measurement model for the observables
##########################################################################
for (i in 1:n){
  for(j in 1:J){
    mu[i,j] <- tau[j] + ksi[i]*lambda[j]
    x[i,j] ~ dnorm(mu[i,j], inv.psi[j])
  }
}


##########################################################################
# Specify the prior distribution for the latent variables
##########################################################################
for (i in 1:n){
  ksi[i] ~ dnorm(kappa, inv.phi)
}


##########################################################################
# Specify the prior distribution for the parameters that govern
# the latent variables
##########################################################################
kappa <- 0
inv.phi <-1
phi <- 1/inv.phi


##########################################################################
# Specify the prior distribution for the measurement model parameters
##########################################################################
for(j in 1:J){
  tau[j] ~ dnorm(3, .1)
    inv.psi[j] ~ dgamma(5, 10)
    psi[j] <- 1/inv.psi[j]
}

for (j in 1:J){
  lambda[j] ~ dnorm(1, .1)
}

} # closes the model
```

```
##########################################################################
# Initial values for two different chains
##########################################################################
list(tau=c(3, 3, 3, 3, 3), lambda=c(3, 3, 3, 3, 3), inv.psi=c(2, 2, 2, 2,
2))

list(tau=c(3, 3, 3, 3, 3), lambda=c(-3, -3, -3, -3, -3), inv.psi=c(2,
2, 2, 2, 2))
--------------------------------------------------------------------------
```

Two chains were run in this analysis using starting values listed above. History plots for the first 1000 iterations for the measurement model parameters for the first observable and for examinee 8's latent variable are given in Figure 9.6. They are representative of the other results for parameters of the same type. By the standards of diagnosing convergence by way of inspection of multiple chains, the behavior of the chains' values for the intercepts and error variances look pretty good. However, the behavior for the loadings (represented by the top plot for λ_1) and the examinees' latent variables (represented by the bottom plot for ξ_8) would seem to be a clear-cut case of a lack of convergence, or at best that one chain has converged and the other has not (but which one?). In fact, the chains—both of them—*have* converged. They are just exploring different parts of the posterior distribution. Figure 9.7 contains density plots of several of the parameters. The multimodality in the posterior for the loadings and the latent variable for examinee 8 can be connected to the separate chains. The chain on the top in the history plots provides the values that constitute the density for positive values; the chain on the bottom in the history plots provides the values that constitute the density for negative values.

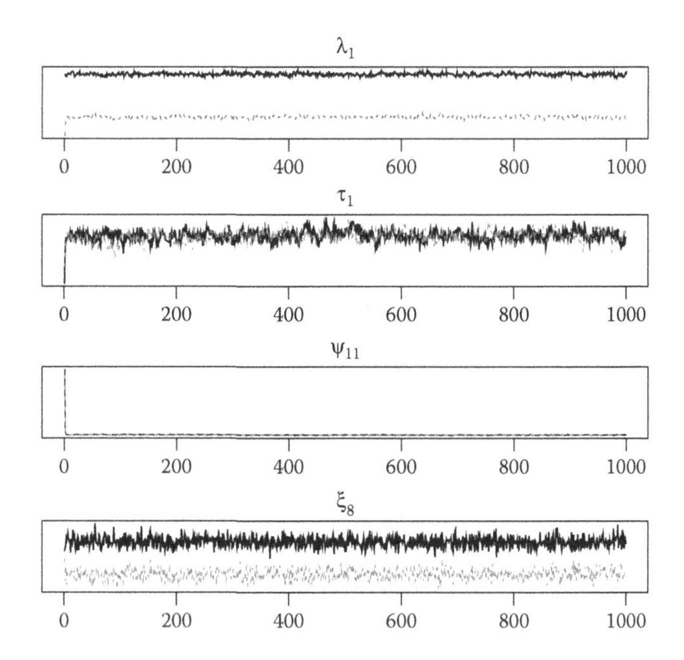

FIGURE 9.6
History plots for parameters for the model with fixed variance and estimated loadings, including the loading (λ_1), intercept (τ_1), and error variance (ψ_{11}) for the first observable, and the latent variable for the eighth examinee (ξ_8).

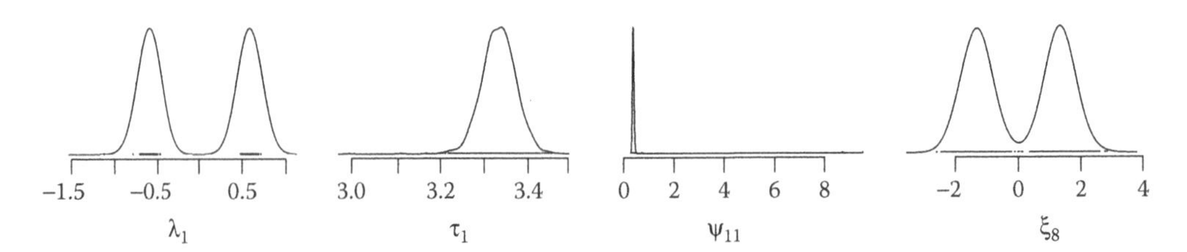

FIGURE 9.7
Density plots for parameters for the model with fixed variance and estimated loadings, including the loading (λ_1), intercept (τ_1), and error variance (ψ_{11}) for the first observable, and the latent variable for the eighth examinee (ξ_8).

The densities in Figure 9.7 are indeed representative of the posterior distribution. The multimodality is a manifestation of the orientation indeterminacy associated with the factor model, namely that the loadings can be reflected by multiplying them by –1 and still reproduce the data equally well. With one orientation, the loadings are positive; with another orientation, the loadings are negative. A similar reflection of positive negative values takes place in the values of the latent variables. Note that there is no reflection in the intercept or error variance.

The multimodality exists in the posterior because of the multimodality in the likelihood—as discussed in Section 9.1.2 the indeterminacy in orientation issue pertains to frequentist estimation as well—and the information in the prior is not strong enough to steer the posterior towards one orientation as opposed to the other. As the information in the prior diminishes, the posterior more closely resembles the likelihood. Thus, MCMC may be a useful tool for exploring multimodality and more generally the shape of the likelihood (see Scheines et al., 1999, for a discussion in CFA; see also Jackman, 2001, for a discussion in item response theory).

This example has illustrated that resolving the indeterminacy in the scale of the latent variable by fixing the variance does not resolve the indeterminacy in orientation. As illustrated in Section 9.3, fixing the first loading to 1 was sufficient to resolve this issue.

9.9.2 Resolving Indeterminacies from a Bayesian Modeling Perspective

The indeterminacies are features of the model for the conditional probability of the data, or the likelihood. In a frequentist framework, they are commonly viewed in terms of an issue of identification, as the data alone are not sufficient to adjudicate among the multiple, equally viable solutions. This needs to be resolved in order for there to be a unique point estimate, such as the MLE.

In a Bayesian analysis, the situation is a little more nuanced. On one hand, this is not an issue of *identification* in the sense of uniquely identifying a solution. In a Bayesian analysis, the solution is the posterior distribution. Thus, concerns about identification are not about whether we have arrived at a unique point, but whether the posterior density is proper. For example, Figure 9.7 depicts the bimodality in the posterior that resulted when the indeterminacy in orientation was not addressed. In a sense, there is nothing "underidentified" about this result, as the posterior is well defined. In another sense, however, we recognize that the multimodality in the posterior density in this situation does not reflect a substantive issue (which can occur), rather it emerges simply from the arbitrary nature of the metric of latent variables. We may wish to resolve the indeterminacy to better support inferences.

Again, we may view this as an issue of the prior specification. When no prior information is supplied about the orientation, the posterior reflects the indeterminacy as the data are insufficient for resolving it. As discussed in Section 9.8, our adopted approach of fixing loadings to take on particular values may be seen as a prior specification. This approach is standard in frequentist analyses. It is also common in Bayesian analysis, particularly if, as was the case here, rather diffuse priors are used for the remaining parameters. However, other possibilities exist, such as fixing values or imposing ordinal constraints on examinee latent variables, which is most viable when there is prior knowledge that two examinees are quite distinct in terms of what the latent variable is intended to capture (Bafumi, Gelman, Park, & Kaplan, 2005; Jackman, 2001), or fitting the more "open" model, without constraints, and then postprocessing the resulting draws from MCMC to rescale to a desired metric (Bafumi et al., 2005, Erosheva & Curtis, 2011; Fox 2010).

Most of the mechanisms used to resolve the indeterminacies may be seen as prior information. Fixing the loading to a particular value (e.g., 1, as was done in the examples in Sections 9.3 and 9.4) may be seen as modeling highly specific prior information about that loading. We may frame this issue in terms of the variance of a normal prior distribution for the loading. A bit more formally, suppose that this loading has a $N(1, \sigma^2_{\lambda_j})$ prior distribution. If $\sigma^2_{\lambda_j} = 0$ the loading is set equal to 1, which effectively resolved the issue. If $\sigma^2_{\lambda_j} = \infty$, we are in effect supplying no prior information, and we will have the multimodality resulting from the indeterminacy.

These two situations are akin to the two choices in a conventional frequentist analysis: either fix the parameter so that the data have no bearing on its value ($\sigma^2_{\lambda_j} = 0$), or let it be estimated solely from the information in the data ($\sigma^2_{\lambda_j} = \infty$). But as was argued in Section 9.8, in a Bayesian analysis there is room in between these extremes.

When $\sigma^2_{\lambda_j} = 0$ the issue was effectively resolved. As we saw in Figures 9.6 and 9.7, with a large but finite value of $\sigma^2_{\lambda_j}$ the indeterminacy rears its head. How small $\sigma^2_{\lambda_j}$ needs to be in order to resolve the indeterminacy depends on the features of the situation, in particular the separation between the modes in the likelihood, and other prior information that may be modeled with an informative prior. What we can say is that increasing the variance $\sigma^2_{\lambda_j}$ moves us further along the undesirable path to the indeterminacy rearing its head and complicating the posterior. See Jackman (2001) for examples of dwindling differences between the modes, which increases the chances of a chain flipping to another mode. See also Loken (2004, 2005) for discussions of the role of seemingly arbitrary constraints in mode separation.

A few additional comments on the use of MCMC in light of the indeterminacy are warranted. The multimodality in the posterior in Figures 9.6 and 9.7 was immediately seen in part because the two chains were run from dispersed starting points. It is tempting to say that if we had only run one of these chains, we would not have seen it. And this is true, for the finite iterations considered. If a chain was run sufficiently long enough it would explore the full posterior and would reveal the multimodality. That is, each chain would eventually "flip" or "jump" and start exploring the space associated with the reflected values. Of course, there is no guarantee when that will happen. In this and many other applications of CFA and item response models that employ continuous latent variables (Chapter 11), the data are informative enough to yield a separation between the modes sufficient to greatly reduce the chance of an individual chain flipping to another mode, such that it might occur only once every thousand years. That would still be in accordance with theorems of the behavior for infinitely long chains, as in the long run the chain would hop back and forth, covering the areas in proportion to the right posterior densities.

Finally, though MCMC may be effective in revealing instances of indeterminacies, it is no substitute for understandings of the model in place. Conceptual, analytical, and empirical analyses, including those in the frequentist traditions, have taught psychometricians a variety of nuances of latent variable models. These lessons and others sure to come should not be dismissed in light of powerful computational tools like MCMC. These issues are subtle, and negotiating the challenges they pose to inference requires more than just running MCMC and uncritically reporting the results.

9.10 Summary and Bibliographic Note

This chapter has focused on Bayesian approaches to CFA. For Bayesian treatments of EFA, see Ghosh and Dunson (2009), Koopman (1978), Lopes and West (2004), and Martin and McDonald (1975). See also Mavridis and Ntzoufras (2014) on exploratory approaches to identifying observables for use in factor analysis. Early Bayesian approaches to CFA may be found in Lee (1981) and Press and Shigemasu (1997) and the references therein. The use of MCMC for Bayesian CFA dates at least to Geweke and Zhou (1996), and gained prominence in psychometrics with publications by Arminger and Muthén (1998) and Scheines et al. (1999).

We have focused on CFA as expressed from an SEM perspective. Accordingly, related Bayesian treatments of CFA may be found in or derived from treatments of Bayesian SEM (Dunson, Palomo, & Bollen, 2005; Kaplan & Depaoli, 2012; Lee, 2007; Levy & Choi, 2013). In fact, viewing SEM as an extension of CFA that structures the distribution of latent variables via regression models, we have covered all the ingredients in Bayesian SEM here. Rowe (2003) and Aitkin and Aitkin (2005) provided recent treatments of CFA not strongly connected to SEM; see also Press and Shigemasu (1997).

We have confined ourselves to CFA assuming conditional normality of the observables and using conditionally conjugate priors for models. A number of variations are possible. Lee (2007) presents an alternative approach using recursive conditional distributions, preserving conjugacy; see also Press and Shigemasu (1997), and Lee and Press (1998) for a sensitivity analysis to prior specifications. Hayashi and Yuan (2003) provide an alternative based on robust estimation and the approach to Bayesian factor analysis proposed by Press and Shigemasu (1997). Ghosh and Dunson (2009) developed the use of thicker-tailed t distributions as priors for loadings.

A number of other extensions have been developed in CFA or in SEM, often in light of a desire to estimate nonstandard models that pose challenges to ML and least-squares estimation. These including models for multiple groups, be they observed (Song & Lee, 2001, 2002b) or latent mixtures (Depaoli, 2013; Lee & Song, 2003; Zhu & Lee, 2001), models with other covariates (Kim, Suh, Kim, Albanese, & Langer, 2013; Lee, Song, & Tang, 2007), latent growth curve and dynamic factor models for longitudinal data (Song & Ferrer, 2012; Song, Lee, & Hser, 2009; Zhang et al., 2013), multilevel models (Ansari & Jedidi, 2000; Ansari et al., 2002; Goldstein & Browne, 2002, 2005; Kim et al., 2013; Song & Lee, 2004), and models that employ thicker-tailed distributions of the data (Lee & Xia, 2008; Zhang et al., 2013). An important class of Bayesian factor analysis treats discrete observables (e.g., Dunson, 2000; Edwards, 2010, Kim et al, 2013), a topic we discuss in the context of item response theory in Chapter 11.

Additional developments have included models with stochastic constraints (Lee, 1992), quadratic, interaction, or other nonlinear relationships among the latent variables (Arminger & Muthén, 1998; Harring, Weiss, & Hsu, 2012; Lee, 2006; Lee & Zhu, 2000; Song & Lee, 2002b, 2004; Lee et al., 2007; Song et al., 2009), models with varying parametric and semiparametric distributional assumptions for the latent variables (Chow, Tang, Yuan, Song, & Zhu, 2011; Fahrmeir & Raach, 2007; Lee, Lu, & Song, 2008; Song et al., 2009; Song, Lu, Cai, & Ip, 2013; Yang, Dunson, & Baird, 2010), procedures for modeling ignorable and nonignorable missingness (Cai & Song, 2010; Lee, 2006; Song & Lee, 2002a), as well as models that integrate many of these features. A number of these are discussed and illustrated in the texts by Lee (2007) and Song and Lee (2012).

Exercises

9.1 Factor analysis models may be seen as simultaneously regressing the observables on the factors.

a. In what ways are the specifications of the factor analysis models analogous to those in regression models of Chapter 6?
b. In what ways are they different?
c. How could regression models of Chapter 6 be converted to factor analysis models, and what changes would need to be made in the inputs to WinBUGS for such a model?

9.2 The CTT model of Chapter 8 may be seen as an instance or special case of factor analysis.

a. Formulate the CTT model in Section 8.3 as a CFA model with one latent variable (factor) in terms of mathematical expressions. Alternatively, what restrictions can be placed on a CFA model with one latent variable to obtain the CTT model?
b. Create a DAG for the model created in (a).
c. Conduct a Bayesian analysis of the model formulated in (a) and (b) in WinBUGS using the data in Table 8.2. Compare your results in terms of convergence and the resulting posterior distribution to that reported in Section 8.3.3.

9.3 An alternative CFA model for the Institutional Integration Scale posits two latent variables, where *Peer Interaction* and *Faculty Interaction* load on one latent variable interpreted as pertaining specifically to *Social Integration*, and the other three observables, *Academic and Intellectual Development*, *Institutional Goal Commitment*, and *Faculty Concern* load on a second latent variable interpreted as pertaining to *Academic Integration*.

a. Formulate the model in terms of mathematical expressions.
b. Specify a path diagram and DAG for the model. How do they compare to those for the model with two latent variables in Section 9.4?
c. Conduct a Bayesian analysis of the model formulated in (a) and (b) in WinBUGS. Interpret the results in terms of convergence and the resulting posterior distribution.

9.4 Figure 9.4 contains an inexact DAG for the model with one latent variable with fixed latent variable mean and fixed loading. Create an exact DAG for the model. How does it compare to that in Figure 9.4 in terms of representation and ease of understanding?

9.5 Reconsider the model in Section 9.9 where the rotational indeterminacy is present. Explain how a chain sampling values near one mode could jump to sample values of another mode in the context of a Gibbs sampler, and in the context of a Metropolis sampler.

10

Model Evaluation

Researchers often use models to account for real-world phenomena in parsimonious and theoretically meaningful ways to guide inference. In this chapter, we take up the task of evaluating models. Model criticism and comparison are core activities in statistical modeling generally, and Bayesian psychometric modeling is no exception. What's more, these are active areas of research and debate within the Bayesian community on appropriate goals, procedures, and conclusions. In this chapter, we aim to survey and discuss methods that have gained popularity in Bayesian psychometrics, and point the reader to existing applications and research on these and other methods. The CFA model of the previous chapter provides us with the first opportunity to illustrate the methods with a rich psychometric model, and bring out ways that more general techniques interact with particularly psychometric concerns.

We describe approaches to the criticism of an individual model in Sections 10.1 and 10.2. In Section 10.3, we turn to the situation in which there are multiple models under consideration and we wish to compare them. In this chapter, we return to the use of θ to generically denote all the unknowns in the model (i.e., latent variables, measurement model parameters, and parameters that govern their priors). When discussing how the methods apply to the CFA examples, we will employ CFA-specific notation, which does not utilize θ.

10.1 Interpretability of the Results

As a first step in critiquing a model, we may consider the interpretation of the results. Does the posterior distribution make sense? Are the substantive conclusions based on model-based reasoning using this model, be they about examinees or tasks, reasonable? If not, what features of the model and the data may be responsible for the unreasonable results? Gelman et al. (2013) argued that when analysts deem that the posterior distribution and inferences are unreasonable, what they are really expressing is that there is additional information available that was not included in the analysis. This underlies the common practice, found in frequentist as well as Bayesian analyses, of revising a model in light of non-sensical results or implications. One place for such additional information is in the prior distribution. This aligns with viewing a prior distribution as a mechanism to rein in or regularize parameter estimates, which is particularly important for parameters that are difficult to estimate based on the data alone. Examples in conventional approaches to psychometrics include negative variances (Heywood cases) and correlations greater than 1 in factor analysis, and likelihood-based estimates of discrimination parameters in item response models (Chapter 11) that are far beyond the range of plausible parameter values. If a Bayesian analysis with a diffuse prior (or a frequentist analysis) yields a posterior distribution (point estimate) for a parameter that is far beyond any credible value, a sensible thing to do is to re-analyze the data, using a more informative prior distribution that indicates that such results are highly unlikely or impossible (Martin & McDonald, 1975; Mislevy, 1986). In conventional approaches, it is often recommended that this should be addressed by the analyst specifying appropriate bounds for such parameters or using software that automatically does this (Gorsuch, 1983; Kline, 2010), which may be seen as incorporating this prior information into the analysis.

In the next chapter on item response theory, we discuss a related situation where the parameters are poorly determined from the data such that there may be many sets of measurement model parameters that are almost equally good at fitting the data, some of which are reasonable and others of which are not. Setting aside research uses that aim for "true" estimates, production uses of item response theory need to use the reasonable ones for very practical reasons. Here again, even a mild prior accomplishes this. Knowing what item parameter estimates *should* look like is kind of prior knowledge about the parameters, but is also about knowing control limits on the factory floor.

10.2 Model Checking

10.2.1 The Goal of Model Checking

A model-based reasoning perspective recognizes that a model is an analyst-created, purpose-driven construction used to facilitate reasoning about a complex real-world situation. Such a perspective readily acknowledges that all models are wrong; the central issue is not whether the model is correct—it is not!—but whether it is useful in the sense that it facilitates adequate inferences for the job at hand. This point has long been recognized in the statistical community (e.g., Box, 1976, 1979; Box & Draper, 1987; Freedman, 1987; Gelman & Shalizi, 2013) as well as in the psychometric community spanning various modeling traditions (e.g., Edwards, 2013; MacCallum, 2003; McDonald, 2010; Preacher, 2006; Thissen, 2001).

Models are convenient fictions—fictions because they are necessarily incorrect, convenient because they are handy tools for representing, processing, and communicating information. Our hope is that they are *useful* convenient fictions, in that they facilitate our arriving at adequate inferences or conclusions. Thissen (2001) noted that useful models live up to the principle Tukey expressed in his famed dictum (1962, pp. 13–14, emphasis in the original): "Far better an approximate answer to the *right* question, which is often vague, than an *exact* answer to the wrong question, which can always be made precise."

Any model, if subjected to sufficiently rigorous examination, can be shown to be false. The goal of model checking is therefore not to judge whether the model is correct, but whether it is useful in this sense. A model is more likely to be useful if it captures salient features of the real-world situation that are important to reasoning in the situation, which are in part defined by the purposes and associated desired inferences. It is less likely to be useful if it ignores aspects of the situation that have a bearing on the desired inferences. We examine whether the model captures or ignores key features by examining the ways in which the model does and does not fit the data. We are less concerned when the misfit is confined to aspects of the model that are peripheral to the desired inferences, but are troubled when the model exhibits considerable misfit to the data in ways that indicate that the inferences may be undermined. Just which aspects, magnitudes, and patterns of misfit are consequential, and which ones are not, is likely to vary based on the analyst's purposes and the model being used. For example, the majority applications of CFA tend not to focus on the estimation of examinees' latent variables, whereas many applications of item response theory (discussed in Chapter 11) do. Accordingly, we may be willing to tolerate certain types or amounts of error in estimating examinees' latent variables in the former case, but less so in the latter case. The onus is on the analyst to know what can be safely ignored, and what requires careful checking and attention. As Box succinctly put it (1976, p. 192), "it is inappropriate to be concerned about mice when there are tigers abroad."

We advocate that Tukey's dictum quoted above applies to model *checking*. In our view, asking whether the model is true is almost always the wrong question, and using sophisticated statistical machinery in pursuit of this question, while an interesting exercise, may be something of a misallocation of resources. In applied work, it is better to devote those resources to addressing the right question, which is whether the inferences emerging from using the model are adequate. As Tukey's dictum suggests, this question is indeed vague, and requires an understanding of what constitutes serious threats to the viability of the inferences and how those threats may reveal themselves in the data or the discrepancy between the model and the data. Just what those threats are and how they manifest themselves, let alone what is to be done about them, is likely to vary based on the uses of the models, the consequences of misfit, and potentially other considerations such as whether revising the model is a viable option in light of the available resources.

10.2.2 The Logic of Model Checking

Model checking aims to evaluate the fit of the data and the model, and proceeds according to a particular logic (Gelman & Shalizi, 2013). A statistical model is a story about how the data could have come about. The story is not correct in detail or in its mechanism, but it does not have to be; if it accords satisfactorily with the patterns in the data, we are warranted in reasoning through it as if it were (bearing in mind the alternative explanations that this logic entails). To address whether a story is plausible, model checking examines the consistency or alignment between the data and the model. The model has implications for what we *ought* to see going on in the data if it were in fact true. Model checking amounts to working out what those implications are, examining whether we *do* see those features in the data, and whether we see *other* features in the data, not present in the model's implications, as well. To the extent that the data do exhibit and conform to the model's implications, we have evidence of model-data fit with respect to that feature of the model and the data. To the extent that the data do not exhibit or conform to those implications, we have evidence of model-data misfit with respect to that feature of the model and the data.

In the following subsections, we cover three approaches to model checking: residual analysis, posterior predictive model checking (PPMC) using test statistics, and PPMC using discrepancy measures. We organize each approach in terms of three steps that roughly correspond to addressing questions concerning: (1) what is going on in our data, possibly in relation to the model, (2) what the model has to say about what should be going on, and then (3) how what is going on in the data compares to what the model says should be going on.

10.2.3 Residual Analysis

Residual analyses have the basic form of the difference between an observed quantity and a model-based implication for that quantity, possibly with adjustments to aid interpretability. Residuals may be framed in terms of the three-step approach as addressing the following questions:

1. What is the feature of the data that is of interest?
2. What does the model imply about this feature?
3. How does the feature of the data compare to what the model implies for that feature?

We will look at a number of examples shortly, to see how the strategy can be applied to a variety of features, how the features are suggested by the way the model is being used, and how evaluating the impact of model-data differences can be evaluated in terms of its effect on targeted inferences.

The above characterization speaks to the *computation* of residuals but belies the key role residual analysis plays in the larger modeling enterprise. The crucial *conceptual* role of residuals is perhaps best expressed in exploratory data analysis (Mosteller & Tukey, 1977; Tukey, 1977; see also Behrens, DiCerbo, Yel, & Levy, 2013), which stresses the foundational relationship DATA = MODEL + RESIDUALS. This reminds us that a model is a simplified representation of the complex real-world situation and is necessarily inexact, implying that residuals will always be present. Exploratory data analysis manipulates this equation to focus on the residuals as what remains when one subtracts the model away from the data. The rationale is that a model being fit embodies the analyst's story—his/her beliefs and understandings in light of purposes—of the real-world situation. Residual analyses proceed by subtracting away the model (i.e., the analyst's story) from the data. What remains is therefore what goes beyond that story. With that story removed, it is easier to see just what is left over; that is, it is easier to see and seek patterns that were previously obscured. Interactive exploration with tools such as those pioneered by Data Desk (Data Description, 2011) make this a real-time detective mission (Tukey, 1969). This kind of interaction is strongly aligned with a view that modeling is a cyclical process of specification, criticism in light of data, and revision (Box, 1976) as opposed to the traditional, less cyclical "fitting a model to data" of confirmatory data analysis (Tukey, 1986).

As a first example, we describe a foundational residual analysis that may be applied across a wide variety of psychometric models. To address the first question, we identify the value of the observables from the examinees as the feature of interest. This is easily obtained from our data as x_{ij}, the value of observable j (such as a test score under classical test theory or an item response under item response theory) for examinee i. To address the second question, we "work through" the model to arrive at its implications. We denote the model-implied value for the value for observable j from examinee i by $E(x_{ij} \mid \theta)$. In the context of CFA, the relevant model parameters are ξ_i for examinee i and τ_j and λ_j for observable j, so $E(x_{ij} \mid \theta) = E(x_{ij} \mid \tau_j, \lambda_j, \xi_i) = \tau_j + \lambda_j \xi_i$. To address the third question, a natural comparison is to form the difference; the residual for the value from examinee i on observable j is

$$\text{Residual}_{ij} = x_{ij} - E(x_{ij} \mid \theta), \tag{10.1}$$

with larger values indicating a larger departure of the observed data from the story of the model.

This residual is applicable in a variety of psychometric modeling scenarios and may be of primary interest, especially in early stages of analysis either to catch data errors or identify instances where the particular situations that led to certain observations are really coming from different mechanisms than most of the data. The overall story that a model seeks to tell, despite variations it allows in its structure, just does not apply very well to these instances, which may need their own special stories—persons who misunderstood directions or followed atypical solution strategies, for example, or coding errors in the data.

There is a value of Residual_{ij} for every examinee on every observable. With a large sample size or number of observables, this may be too many to efficiently investigate one at a time. In addition, the individual values of the observables may not be of primary interest, so we ought to employ residuals that target what is valued or is of chief concern. CFA models are frequently employed to reflect theories regarding the associations among variables. Accordingly, we might prefer a residual that more directly targets associations among the variables to characterize the adequacy of the model. To address the first question, the covariances among the observables are a natural first choice for features of the data. Let $s_{jj'}$ be the element in row j and column j' of the observed covariance matrix $\mathbf{S}$, usually calculated in the CFA literature* by

$$s_{jj'} = \frac{\sum_{i=1}^{n}(x_{ij} - \bar{x}_j)(x_{ij'} - \bar{x}_{j'})}{n-1}. \tag{10.2}$$

To address the second question, we again work through the CFA model to arrive at the model's implications. The model-implied covariance for observables j and j' is denoted by $\sigma(\theta)_{jj'}$, and may be obtained as the element in row j and column j' of the model-implied covariance matrix:

$$\Sigma(\theta) = \Lambda\Phi\Lambda' + \Psi. \tag{10.3}$$

Turning to the third question, a natural comparison is to form the difference; the residual covariance for observables j and j' is then $s_{jj'} - \sigma(\theta)_{jj'}$, which corresponds to the element in row j and column j' of the residual covariance matrix

$$\mathbf{E}_{\text{cov}} = \mathbf{S} - \Sigma(\theta). \tag{10.4}$$

In applications of CFA, we might be primarily interested in modeling the associations among the variables, rather than their univariate features such as their variances. In this case, we might restrict attention to the off-diagonal elements in (10.4). We might also employ the correlation as a basis for model-data fit. Doing so might be warranted if we are less interested in the particular metrics of the variables, as the covariance is metric dependent. Let $\mathbf{R}$ be the sample correlation matrix, where each element $r_{jj'}$ is given by

$$r_{jj'} = \frac{s_{jj'}}{\sqrt{s_{jj}}\sqrt{s_{j'j'}}}. \tag{10.5}$$

Let $\mathbf{P}(\theta)$ denote the model-implied correlation matrix, where each element $\rho(\theta)_{jj'}$ is given by

$$\rho(\theta)_{jj'} = \frac{\sigma(\theta)_{jj'}}{\sqrt{\sigma(\theta)_{jj}}\sqrt{\sigma(\theta)_{j'j'}}}, \tag{10.6}$$

* The covariance could be formulated using n in the denominator, and that may be better justified in the context of model-data fit analyses. Nevertheless, the following form is common in CFA and the difference becomes irrelevant as n increases.

where $\sigma(\theta)_{jj'}$ is the element in row j and column j' of $\Sigma(\theta)$. The residual correlation matrix is then

$$\mathbf{E}_{\text{cor}} = \mathbf{R} - \mathbf{P}(\theta). \tag{10.7}$$

A non-zero residual correlation occurs when the observed correlation differs from the model-implied value, and speaks to the presence of a conditional dependence. In the context of the CFA model for the Institutional Integration Scale introduced in Section 9.3, the *Faculty Interaction* and *Faculty Concern* observables both concern the role of faculty in the student's experience, and may be more correlated than can be accounted for by a single latent variable modeled as underlying all the observables. Similarly, scores derived from tests taken at the same time, or with shared format, often result in higher correlations if these contributing factors are not taken into account. It is not only the size of residuals that matters, but perhaps patterns among them.

Other residuals may be of central importance to applications of CFA. In the context of mean-structure modeling, we may focus on residuals for means of the observables,

$$\mathbf{E}_{\text{mean}} = \bar{\mathbf{x}} - \boldsymbol{\mu}(\theta), \tag{10.8}$$

where $\boldsymbol{\mu}(\theta) = \tau + \Lambda\kappa$ is the vector of model-implied means.

In frequentist analyses, residual analyses are conducted by using a particular set of values for the model parameters obtained from fitting the model. These values are obtained either from a priori specifications (e.g., setting a loading equal to 1) or from point estimates from a frequentist solution. In frequentist approaches, residuals may be used descriptively or in formal inferential settings, the latter of which may rely on asymptotic theory. That is, the magnitude of a residual is judged in comparison with the distribution of values that would obtain if the model were correct.

In a Bayesian analysis, we may compute residuals in a similar manner, using a point summary of the posterior distribution. An analysis more in line with fully Bayesian principles utilizes the complete posterior distribution, rather than a point summary, for any unknowns. This yields a posterior distribution for the model-implied entities as well as the residual values. In a simulation environment such as MCMC, we can obtain an empirical estimate for the posterior distribution for the residual by calculating the value of the residual in each iteration of the chain.

Applications of Bayesian approaches to residual analyses can be found in a variety of psychometric paradigms (Albert & Chib, 1993, 1995; Lee, 2007; Sinharay & Almond, 2007). Albert and Chib (1993) noted that for dichotomous data the frequentist conception of the residual in (10.1) can only take on two values, as the model-implied expected value is a single probability value calculated with a point estimate for all the unknown parameters. In contrast, the Bayesian formulation can take on a continuous distribution, and may therefore be more informative.

Residual analyses suffer from three key limitations. First, though they are well aligned with investigating features for which the model-based implications are simple functions of the parameters, if the feature of interest is not a simple function of the parameters, it may be difficult to proceed. Second, it is not always obvious how to gauge whether the magnitude is large. Third, though the use of the posterior distribution formally incorporates uncertainty in the parameters, there is no formal recognition of the uncertainty associated with the conditional distribution of the data given the parameters. The next sections present alternative model-checking approaches that may address these drawbacks.

10.2.4 Posterior Predictive Model Checking Using Test Statistics

This and the next section describe PPMC (Gelman, Meng, & Stern, 1996; Guttman, 1967; Rubin, 1984). We begin by describing its application using *test statistics*, which we use to refer to quantities that are functions of the data only.* Our three key questions are

1. What is the feature of the data that is of interest?
2. What does the model imply about this feature?
3. How does the feature of the data compare to what the model implies for that feature?

As in residual analyses, we address the first question by defining summaries based on the data. That is, we declare what are the test statistics that represent the features of the data that are of interest. To make this a bit more formal, let T denote a test statistic of interest. Again, prime examples for applications in CFA are the covariances and correlations. Calculating the test statistic on the observed data yields the *realized* value of the test statistic, denoted by $T(\mathbf{x})$.

Where we depart from residual analyses is in addressing the second question concerning the model's implications for those entities. In probabilistic modeling, the story of the model is expressed in probabilistic assumptions, and probabilistic implications follow accordingly (Gelman & Shalizi, 2013). PPMC expresses those implications by way of the posterior predictive distribution for the data, given by

$$p(\mathbf{x}^{postpred} \mid \mathbf{x}) = \int p(\mathbf{x}^{postpred} \mid \theta) p(\theta \mid \mathbf{x}) d\theta, \tag{10.9}$$

where $\mathbf{x}^{postpred}$ denotes *posterior predictive data* of the same size as the observed data, which in the current case of CFA is an $(n \times J)$ matrix. The first term on the right-hand side is the conditional distribution of the posterior predicted data given the model parameters. The second term on the right-hand side is the posterior distribution for the parameters given the observed data. The integration reflects a marginalization. If we knew the values for the model parameters, we would predict data based on those (using the first term). But we do not know the values for the model parameters, so we marginalize over the posterior distribution for them. The result is the posterior predictive distribution on the left-hand side. It is *posterior* because it is conditional on the observed data. It is *predictive* because it refers to data that were not observed. Indeed, $\mathbf{x}^{postpred}$ may be interpreted as data that *could* have been observed, or data that *may* be observed in the future, if the same real-world processes as captured by the model and values for $\boldsymbol{\theta}$ are replicated in the future.

It is worth noting that, in general, the first term on the right-hand side of (10.9) should be $p(\mathbf{x}^{postpred} \mid \theta, \mathbf{x})$. The use of the simpler $p(\mathbf{x}^{postpred} \mid \theta)$ reflects an assumption that the posterior predictive data are conditionally independent of the observed data given the model parameters. This reflects the broader notion that the model captures the salient features of the real-world processes that give rise to the data, and so future or potential (i.e., posterior predictive) data are independent of past (observed) data, given the model.

In practice, the posterior predictive distribution is often empirically approximated using simulation techniques in concert with those used to empirically approximate the posterior

* This specialized use of the phrase "test statistic" differentiates it from the fit measures called "discrepancy measures," which are functions of both the observable data and model entities (see Section 10.2.5).

distribution of the parameters. For each draw from the posterior distribution, we take the additional step of generating a posterior predicted dataset via the conditional probability distribution for the data. Letting $\theta^{(1)},\ldots,\theta^{(R)}$ denote the R simulations of the model parameters from the posterior distribution, we obtain R replicate posterior predicted datasets

$$\begin{aligned} \mathbf{x}^{\text{postpred}(1)} &\sim p(\mathbf{x} \mid \theta = \theta^{(1)}) \\ \mathbf{x}^{\text{postpred}(2)} &\sim p(\mathbf{x} \mid \theta = \theta^{(2)}) \\ &\vdots \\ \mathbf{x}^{\text{postpred}(R)} &\sim p(\mathbf{x} \mid \theta = \theta^{(R)}). \end{aligned} \tag{10.10}$$

The posterior predictive distribution serves as a *reference distribution*, against which the observed data are ultimately compared in terms of the test statistic. In particular, we obtain the posterior predictive distribution for the test statistic by computing the test statistic in each of the posterior predicted datasets, yielding $T(\mathbf{x}^{\text{postpred}(1)}),\ldots,T(\mathbf{x}^{\text{postpred}(R)})$. The posterior predictive distribution constitutes the answer to our second question concerning what the model implies about this feature of data.

Turning to the third question, we now compare the realized value of the test statistic, $T(\mathbf{x})$ to the posterior predictive distribution $T(\mathbf{x}^{\text{postpred}(1)}),\ldots,T(\mathbf{x}^{\text{postpred}(R)})$. We recommend the use of graphical procedures to facilitate the comparison. A simple example in current context involves a histogram or smoothed density for the posterior predictive distribution, with a marker or line indicating the location of the realized value.

The results may be numerically summarized via the posterior predictive p-value,

$$p_{\text{post}} = p(T(\mathbf{x}^{\text{postpred}}) \geq T(\mathbf{x}) \mid \mathbf{x}), \tag{10.11}$$

where the probability is taken over both posterior distribution of θ and the posterior predictive distribution of $\mathbf{x}^{\text{postpred}}$,

$$p_{\text{post}} = \iint I[T(\mathbf{x}^{\text{postpred}}) \geq T(\mathbf{x})] p(\mathbf{x}^{\text{postpred}} \mid \theta) p(\theta \mid \mathbf{x}) d\mathbf{x}^{\text{postpred}} d\theta, \tag{10.12}$$

where I is the indicator function taking on the value of 1 when its argument is true and 0 when its argument is false. An estimate of p_{post} may be obtained in a simulation environment as the proportion of draws in which the posterior predicted value $T(\mathbf{x}^{\text{postpred}})$ exceeds the realized value $T(\mathbf{x})$.

10.2.4.1 Example from CFA

To illustrate, we return to the example introduced in Section 9.3 where we specified a CFA model with one latent variable for five observables derived from the Institutional Integration Scale and examine the correlations. The realized values for the correlations were presented in Table 9.1. In Section 9.3, we fit the model using MCMC and retained 15,000 iterations to empirically approximate the posterior distribution. For each of the 15,000 iterations, we simulated a posterior predicted dataset and computed the correlations from the dataset. It is possible to perform these computations in WinBUGS, and in the current example, the computational burden is fairly light. However, it is often the case that considerable efficiencies may be gained by doing these sorts of computations outside of WinBUGS. This is particularly so for more involved computations such as those involved in the model-checking methods discussed in the following sections. For this reason, we used a modified version of the code from Levy (2011) to import the draws into R and perform the computations associated with PPMC.*

Figure 10.1 displays the densities for the posterior predicted correlations; the vertical line in each corresponds to the realized value (i.e., the correlation in the observed data). Consider first the results for the pairing of *Peer Interaction* and *Faculty Concern*. The realized (observed) correlation is .54. As the vertical line in the graph in the last row and first column of Figure 10.1 reveals, this falls near the middle of the posterior predicted distribution. The position of the realized value in the posterior predictive distribution is summarized numerically via the p_{post} value, which is .65, indicating that 65% of the 15,000 posterior predicted values were greater than the realized value. Substantively, the correlation that was found between these variables is generally consistent with the model's implications for their correlation, which is evidence of adequate model-data fit.

On the other hand, consider the situation for the correlation between *Faculty Interaction* and *Faculty Concern*, depicted in the last row and fourth column of Figure 10.1. The posterior predictive distribution is unimodal and symmetric, centered at .56. The realized value for the correlation is .67, which falls way out in the upper tail of the posterior predictive distribution. The p_{post} value is estimated as the proportion of the 15,000 posterior predicted values that exceeded .67, which is .0004. The correlation that was found between these variables is considerably larger than the model's implications, which is evidence of

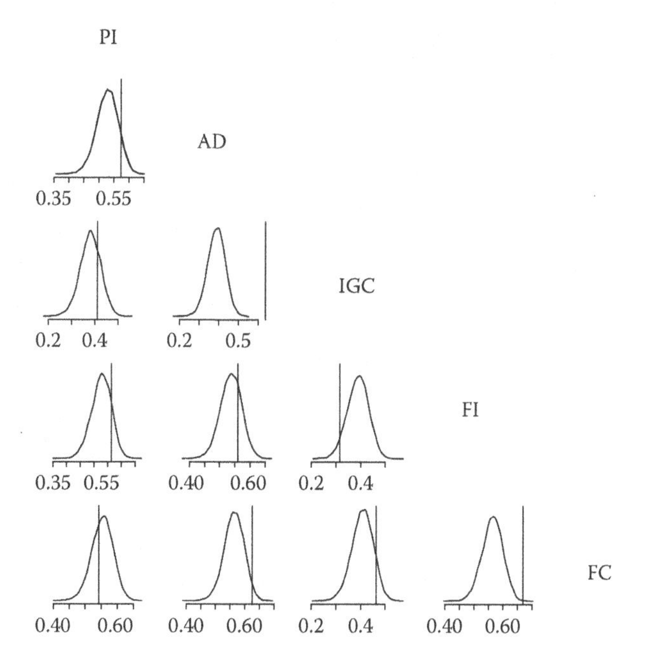

FIGURE 10.1
Posterior predicted densities, with vertical lines for realized values, for the correlations between the latent variables based on the confirmatory factor analysis model with one latent variable (factor) for the observables from the Institutional Integration Scale. PI = peer interaction, FI = faculty interaction, AD = academic and intellectual development, FC = faculty concern, IGC = institutional goal commitment.

* The R code for doing this is available on the website for the book.

model-data misfit. Whether the misfit is substantively meaningful depends on the purpose of the analysis. In this case, where we seek to account for the associations among the variables, the size of the discrepancy between what the model implies (as represented by the posterior predicted distribution) and what was observed is concerning. This is particularly so when we consider the other results for the correlations. A similar result occurred for the pairing of *Academic and Intellectual Development* and *Institutional Goal Commitment* (Figure 10.1, second row, second column), where the misfit is even more pronounced. Here, the realized correlation of .64 is much larger than the model's implications as expressed in the posterior predicted distribution (mean of posterior predicted distribution = .39; estimated $p_{\text{post}} = 0$). In both of those situations, the realized value falls in the upper tail of the posterior predictive distributions, indicating that the model *underpredicts* the associations. We see the opposite for the pairing of *Institutional Goal Commitment* and *Faculty Interaction* (third row, third column). Here, the realized value of .31 falls out the lower tail (mean of posterior predictive distribution = .39, estimated $p_{\text{post}} = .95$), indicating the model *overpredicts* this association, which is also evidence of model-data misfit.

10.2.5 Posterior Predictive Model Checking Using Discrepancy Measures

In the current section, we frame our three questions slightly differently:

1. How discrepant are the data and the model in ways that are of interest?
2. What does the model imply about how discrepant they ought to be?
3. How does the discrepancy based on the observed data compare to what the model implies for this discrepancy?

The key difference from PPMC using test statistics is that here we are focused on a feature of the data *and* the model, rather than a feature of the data alone. Instead of focusing just on characteristics or features of the data, we explicitly build in the comparison between the observed data and the model's implications at the outset by focusing on the *discrepancy* between the data and the model, as in the formulation of a residual. Formally, let $D(\mathbf{x};\theta)$ denote a *discrepancy measure* of interest, where the notation reflects that it is a function of both the data and the model parameters. Casually, if the function of interest is a function of the model parameters as well as the data, it is a discrepancy measure; if it is just a function of the data, it is a test statistic. Many functions that are referred to as "test statistics," "fit statistics," and "fit indices" in the psychometric literature are considered discrepancy measures in the current treatment.

To address the first question, we define a discrepancy measure and evaluate it using the observed data. In a simulation-based environment, this is empirically approximated by calculating the discrepancy measure using each simulated value from the posterior distribution for the parameters. This yields the collection $D(\mathbf{x};\theta^{(1)}),\ldots,D(\mathbf{x};\theta^{(R)})$, which are referred to as the *realized values* of the discrepancy measure. These summarize the discrepancy between the (observed) data and the model. It is a distribution because we have a posterior distribution, rather than just a point estimate, for the model parameters.

To address the second question, we calculate the discrepancy measure using the posterior predicted data from (10.10), yielding an empirical approximation to the distribution of *posterior predicted values* $D(\mathbf{x}^{\text{postpred}(1)};\theta^{(1)}),\ldots,D(\mathbf{x}^{\text{postpred}(R)};\theta^{(R)})$. This distribution expresses what the model implies the discrepancy ought to be for observable data if the model were correct.

Turning to the third question, we now compare the distribution of the realized values, $D(\mathbf{x};\theta^{(1)}),\ldots,D(\mathbf{x};\theta^{(R)})$, to the distribution of posterior predictive values, $D(\mathbf{x}^{\text{postpred}(1)};\theta^{(1)}),\ldots,$

$D(\mathbf{x}^{\text{postpred}(R)};\theta^{(R)})$. In the current context where there is now a distribution of realized values, a simple example involves a scatterplot of the pairs of corresponding realized and posterior predicted values, where for each value r, a point is plotted in the coordinate plane as $(D(\mathbf{x};\theta^{(r)}), D(\mathbf{x}^{\text{postpred}(r)};\theta^{(r)}))$. The results may also be numerically summarized via the posterior predictive p-value,

$$p_{\text{post}} = p(D(\mathbf{x}^{\text{postpred}};\theta) \geq D(\mathbf{x};\theta) \mid \mathbf{x}), \tag{10.13}$$

where the probability is taken over both the posterior distribution of θ and the posterior predictive distribution of $\mathbf{x}^{\text{postpred}}$,

$$p_{\text{post}} = \iint I[D(\mathbf{x}^{\text{postpred}},\theta) \geq D(\mathbf{x},\theta)] p(\mathbf{x}^{\text{postpred}} \mid \theta) p(\theta \mid \mathbf{x}) d\mathbf{x}^{\text{postpred}} d\theta. \tag{10.14}$$

p_{post} may be empirically approximated as the proportion of times that the posterior predicted value exceeds the corresponding realized value.

10.2.5.1 Example from CFA

To illustrate, we return to the CFA example with one latent variable and examine two discrepancy measures that target the overall fit of the model in terms of the covariance structure. The first is the popular likelihood ratio (LR),

$$\text{LR} = (n-1)\ln|\Sigma(\theta)| + \text{tr}(\mathbf{S}\Sigma(\theta)^{-1}) - \log|\mathbf{S}| - J. \tag{10.15}$$

This is commonly referred to as the "χ^2" or the "χ^2 statistic" for the model, as the sampling distribution of LR computed using the sample-based normal-theory ML estimates of θ asymptotically follows a χ^2 distribution under mild regularity conditions (West, Taylor, & Wu, 2012). The second discrepancy measure is the standardized root mean square residual (SRMR; Bentler, 2006)

$$\text{SRMR} = \sqrt{\frac{2\sum_{j=1}^{J}\sum_{j'=1}^{j}[(s_{jj'} - \sigma(\theta)_{jj'}) / (s_{jj}s_{j'j'})]^2}{J(J+1)}}. \tag{10.16}$$

Figure 10.2 contains a scatterplot of the realized and posterior predicted values of LR, with the unit line added as a reference. Observing points that appear to be randomly scattered about the line would have indicated that the realized and posterior predicted values were of similar magnitudes. But in Figure 10.2, it is clearly seen that the realized values are considerably larger than the posterior predictive values. The estimated p_{post} value is the proportion of draws above the line, which in this case was 0. This indicates that the discrepancy between the observed and model-implied covariance structure is larger than the discrepancy implied by the model.

The results for SRMR support this conclusion. The scatterplot for the realized and posterior predicted values is given in Figure 10.3 (panel a), where it is seen that the realized values are consistently larger than their posterior predicted counterparts. The estimated p_{post} value is the proportion of draws above the line, which in this case was .004.

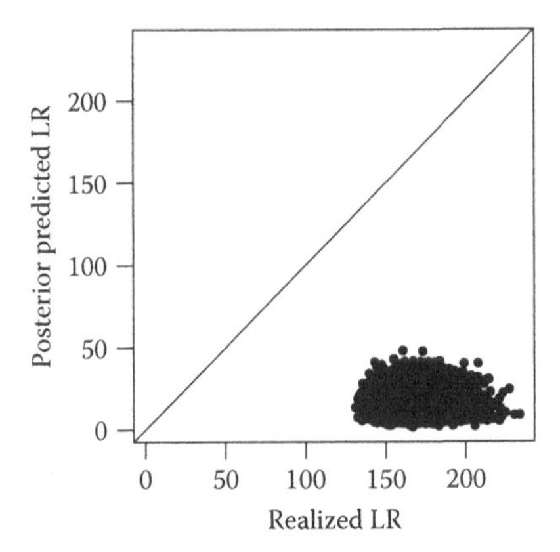

FIGURE 10.2
Scatterplot of the realized and posterior predicted values of the LR, with the unit line, based on the confirmatory factor analysis model with one latent variable (factor) for the observables from the Institutional Integration Scale.

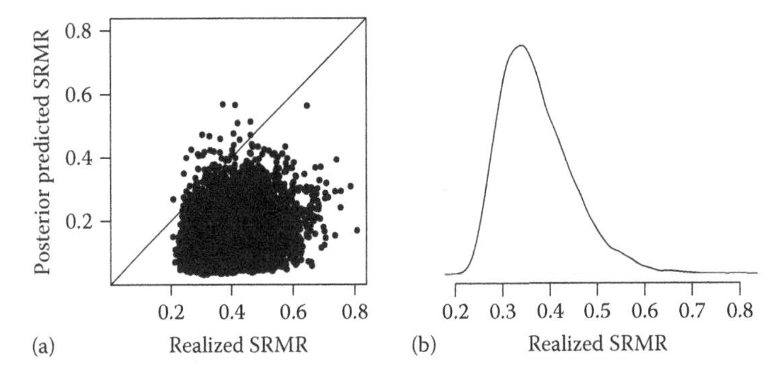

FIGURE 10.3
(a) Scatterplot of the realized and posterior predicted values of the SRMR, with the unit line and (b) density of the realized values of SRMR, based on the confirmatory factor analysis model with one latent variable (factor) for the observables from the Institutional Integration Scale.

The standardization involved renders SRMR to be in a correlation metric, making it a more easily interpretable characterization of fit as opposed to LR (Bentler, 2006; Maydeu-Olivares, 2013). In frequentist approaches, the SRMR is typically evaluated by comparing its realized value computed using the frequentist point estimate to a cutoff based on a value that either (a) represents meaningful misfit as interpreted by the analyst (Maydeu-Olivares, 2013) or (b) prior research suggests is indicative of discriminating between well- and poor-fitting models (e.g., .08; Hu & Bentler, 1999; cf. Fan & Sivo, 2005, 2007; Marsh, Hau, & Wen, 2004; Nye & Drasgow, 2011). In a Bayesian analysis, the use of the posterior distribution for the parameters yields a posterior distribution for realized values of SRMR. In a simulation environment, this posterior distribution is approximated by the realized values obtained based on the simulations. This supports a characterization of its value, possibly with respect to a cutoff. Figure 10.3 (panel b) presents the density based on the 15,000 realized values. The use of a distribution supports a probabilistic evaluation of SRMR. Here, the posterior mean is 0.37, the 95% HPD interval is (.25, .52), and $p(\text{SRMR} \leq .08) = 0$. That is, the typical difference between what the correlations would be if the model is correct and what the actual data produce is likely between .25 and .52, and averages .37—rather high discrepancies, it would seem, for fitted correlations.

Synthesizing the various PPMC analyses, the results from the analysis of LR and SRMR constitute evidence that the model suffers in explaining the overall covariance structure of the observables. The analysis of the individual correlations drills down to a finer grained conclusion, which is that the model suffers chiefly in terms of severely underpredicting the association between *Academic and Intellectual Development* and *Institutional Goal Commitment* and between *Faculty Interaction* and *Faculty Concern* while overpredicting the association between *Institutional Goal Commitment* and *Faculty Interaction*. These results suggest that a model with two latent variables in which the subscales are organized around one latent variable associated with *Faculty* measures and one associated with *Student* measures, as was examined in Section 9.4, might provide better fit (see Exercise 10.2).

10.2.6 Discussion of Posterior Predictive Model Checking

PPMC deserves further discussion, as it has been the most widely applied and studied model criticism method in Bayesian psychometrics, with applications and methodological research in factor analysis (Depaoli, 2012; Levy, 2011; Muthén & Asparouhov, 2012; Scheines et al., 1999), item response theory (Béguin & Glas, 2001; Glas & Meijer, 2003; Karabatsos & Sheu, 2004; Levy, Mislevy, & Sinharay, 2009; Li, Bolt, & Fu, 2006; Sheng & Wikle, 2007; Sinharay, 2005, 2006a; Sinharay, Johnson, & Stern, 2006; Zhu & Stone, 2011, 2012), latent class models (Berkhof, van Mechelen, & Gelman, 2003; Depaoli, 2012; Emons, Glas, Meijer, & Sijtsma, 2003; Hoijtink, 1998; Hoijtink & Molenaar, 1997; van Onna, 2002), and Bayesian networks (Almond et al., 2015; Crawford, 2014; Levy, 2006; Sinharay, 2006b; Yan, Mislevy, & Almond, 2003).

The choice of the test statistics and discrepancy measures in PPMC is crucial. To illustrate this point, let us consider the CFA example using the sample mean for each observable as test statistics. As the reader may verify in Exercise 10.1, the results using the sample mean indicate adequate model-data fit in terms of this feature of the data. If we had *only* used the sample mean in PPMC, we might conclude that the model-data fit is adequate overall. However, to do so would be a serious overstatement, as we have seen that the model suffers in accounting for the associations among the observables.

The takeaway message is that the conclusion of adequate fit extends only as far as the discrepancy measures used. It is always *possible* that we could construct a discrepancy measure that would indicate model-data misfit. More than that, recognizing that our model is necessarily an inexact representation, it is *inevitable* that such a discrepancy measure could be constructed. Model-data misfit in terms of discrepancy measures that capture tangential aspects of a model are not of concern, but model-data misfit in terms of discrepancy measures that capture key aspects of the model are central. It is incumbent on the analyst to use those that are sensitive to particular aspects of misfit that are both possible and a threat to the successful use of the model.

Ideally, an analyst would proceed by identifying aspects of the model that are the most important for the purposes at hand and then employ discrepancy measures that would reveal when those aspects are working well or working poorly. For example, local independence structures are central to the formulation of psychometric models. Incorrectly assuming that observables are locally (conditionally) independent may lead to incorrect estimates of parameters and other variables as well as incorrect estimates of the precision of the estimates. Accordingly, assessing whether the dimensional structure has been sufficiently specified to render the observables close to conditionally independent typically is of great importance when investigating model-data fit of psychometric models (Hambleton, Swaminathan, Cook, Eignor, & Gifford, 1978; Swaminathan, Hambleton, & Rogers, 2007), and a number of discrepancy measures have been proposed to evaluate these features (e.g., Chen & Thissen, 1997; Levy et al., 2009; Levy, Xu, Yel, & Svetina, 2015; Stout et al., 1996).

There is no shortage of residuals, test statistics, and discrepancy measures that have been proposed for evaluating psychometric models, with justifications for broad or targeted use. Research on the performance of these in psychometric models is an active area of research. Note that in this chapter, we have described machinery associated with model checking, but exactly which residuals, statistics, and discrepancy measures should be pursued depends on the model and the purpose as well. This chapter used examples pertinent to certain applications of CFA. We will further illustrate the use of PPMC in the context of item response theory (Section 11.2.5), latent class analysis (Section 13.4.4), and Bayesian networks (Section 14.4.8), and give an indication about how the nature of the models suggests particular features to examine.

The machinery in PPMC is in some ways similar to that used in frequentist approaches. In its current form, hypothesis testing from a frequentist perspective is an amalgamation of different traditions (Gigerenzer, 1993), but may be casually summarized as follows. The null hypothesis implies how data would vary randomly from sample to sample even though the same (fixed) values of the parameters govern the data generating process. This is represented by the sampling distribution of the data, which is equivalent to the first term in the integral on the right-hand side of (10.9), $p(\mathbf{x}^{postpred} \mid \theta)$. In frequentist approaches, the sampling distribution is constructed by using point estimates for unknown parameters. This serves as the reference distribution, against which the observed data are compared. Note that relying on point estimates for defining the sampling distribution suffers in that it ignores the uncertainty in these estimates. This has been recognized as a limitation of frequentist approaches, and one that is not easily circumvented (Molenaar & Hoijtink, 1990; Snijders, 2001; Stone & Hansen, 2000). In PPMC, we employ the posterior distribution for the parameters, $p(\theta \mid \mathbf{x})$, as it appears as the second term inside the integral on the right-hand side of (10.9), which incorporates our uncertainty about the parameters.

The connections and departures from frequentist hypothesis testing approaches raises the question of whether p_{post} values afford the same interpretations as p values in frequentist hypothesis tests. There has been much research and discussion about when they have frequentist interpretations as in hypothesis testing, and whether we should desire them to behave as such, and we will not review these issues here (see Bayarri & Berger 2000a, 2000b; Dahl, 2006; Gelman, 2003, 2007; Johnson, 2007; Meng; 1994b; Robins, van der Vaart, & Ventura, 2000; Rubin 1996a; Stern, 2000). In most situations, we advocate an interpretative lens that views the results of PPMC as diagnostic pieces of evidence regarding, rather than a hypothesis test of, (mis)fit of a model that is known a priori to be incorrect (Box, 1976; Gelman, 2003, 2007; Gelman et al., 1996; Gelman & Shalizi, 2013; MacCallum, 2003; Stern, 2000). From this perspective, p_{post} values have direct interpretations as expressions of our expectation of the extremity in future replications, conditional on the model (Gelman, 2003, 2007). Graphical representations are often employed as summaries of the results of PPMC (e.g., Gelman, 2004), and p_{post} is simply a way to summarize the results numerically. From this perspective, model-data fit analyses using PPMC plays much the same role as residual analyses in exploratory data analysis. PPMC, graphical summaries, and p_{post} have less to do with the probability of rejecting a model in the already-known-to-be-false situation where the model is correct, and more to do with aiding the analyst in characterizing the ways in which the necessarily incorrect but hopefully useful model is (and is not) working well, and the implications of using such a model (Box, 1979; Box & Draper, 1987; Gelman, 2007; McDonald, 2010).

Several other Bayesian approaches to model checking that have been proposed may be viewed as differing from PPMC in how it constructs a distribution for the parameters (the second term on the right-hand side of (10.9)). These include prior predictive model checking (Box, 1980), partial PPMC (Bayarri & Berger, 2000a), and cross-validated PPMC (Evans, 1997, 2000). Still other strategies work to calibrate p_{post} emerging from PPMC to ensure it has frequentist properties (Hjort, Dahl, & Steinbak, 2006). The reader is referred to Levy (2011) for descriptions and illustrations of these procedures in the context of the CFA.

10.3 Model Comparison

When multiple models are under consideration, we may wish to critically compare them and possibly select one for use. In this section, we review several popular strategies commonly used in psychometric modeling.

10.3.1 Comparison via Comparative Model Criticism

One natural approach to model comparison is to critically evaluate each model under consideration using one or more of the techniques just described and compare the results. For example, if PPMC evidences adequate model-data fit for one model but not another, that would support the former over the latter. This approach also allows for the analyst to compare the models in terms of different features of the model-data fit. We might find that one model evidences superior fit in terms of some discrepancy measures, but not others. Though such an approach does not come with built-in criteria for what counts as sufficiently different fit for competing models, such comparisons are often useful at least descriptively.

Such strategies have been gainfully employed in psychometrics, and have compared favorably to alternative methods (Li, Bolt, & Fu 2006; Li, Cohen, Kim, & Cho 2009; Sahu, 2002; Song, Xia, Pan, & Lee, 2011; van Onna, 2002). Rubin and Stern (1994) proposed a posterior predictive checking for model comparisons for determining the number of latent classes in latent class models (Chapter 13), which in principle is applicable to other comparisons of nested models.

10.3.2 Bayes Factor

A classic Bayesian procedure for comparing competing models uses the Bayes factor. To develop this idea, let M_1 and M_2 be two competing models. The posterior odds of M_2 to M_1 is the ratio of posterior probabilities,

$$\begin{aligned}\frac{p(M_2 \mid \mathbf{x})}{p(M_1 \mid \mathbf{x})} &= \frac{p(\mathbf{x} \mid M_2)}{p(\mathbf{x} \mid M_1)} \times \frac{p(M_2)}{p(M_1)} \\ &= \frac{\int_{\theta_{(2)}} p(\mathbf{x} \mid \theta_{(2)}) p(\theta_{(2)} \mid M_2) d\theta_{(2)}}{\int_{\theta_{(1)}} p(\mathbf{x} \mid \theta_{(1)}) p(\theta_{(1)} \mid M_1) d\theta_{(1)}} \times \frac{p(M_2)}{p(M_1)},\end{aligned} \tag{10.17}$$

where $\theta_{(1)}$ is the collection of parameters for M_1, $p(M_1)$ is the prior probability for M_1, and analogous definitions hold for $\theta_{(2)}$ and M_2. The second term on the right-hand side of (10.17) is the ratio of prior probabilities (i.e., the prior odds). The first term on the right-hand side is the ratio of the marginal likelihoods under each model. This term is the *Bayes factor* (e.g., Kass & Raftery, 1995), in particular the Bayes factor for M_2 relative to M_1, which we denote as BF_{21}. Note that the Bayes factor has the form of an LR (and in certain cases specializes to the familiar LR; Kass & Raftery, 1995). The Bayes factor effectively transforms the prior odds into the posterior odds via the likelihoods of the data and as such is sometimes interpreted as the weight of evidence in favor of one model over the other. Importantly, the Bayes factor is not limited to applications in which the models are hierarchically related (nested). Jeffreys (1961) provided a classic set of recommendations for interpreting the magnitudes of Bayes factors; Kass and Raftery (1995) provided a slightly revised set of recommendations, summarized in Table 10.1. Included there are recommendations for working in (twice) the log metric. Note that since the Bayes factor is the ratio of marginal likelihoods, the log of the Bayes factor is equal to the subtraction of the logs of the marginal likelihoods and

$$2\log(\mathrm{BF}_{21}) = 2\log\left(\frac{p(\mathbf{x} \mid M_2)}{p(\mathbf{x} \mid M_1)}\right) = 2(\log[p(\mathbf{x} \mid M_2)] - \log[p(\mathbf{x} \mid M_1)]). \quad (10.18)$$

Bayes factors and its variants have been among the most widely used and studied model comparison approaches in Bayesian psychometrics (Berkhof et al., 2003; Bolt, Cohen, & Wollack, 2001; Fox, 2005b; Hoijtink, Béland, & Vermeulen, 2014; Kang & Cohen, 2007; Klein Entink, Fox, & van der Linden, 2009; Lee & Song, 2003; Li et al., 2006; Li et al., 2009; Raftery, 1993; Sahu, 2002; Song, Lee, & Zhu, 2001; van Onna, 2002; Verhagen & Fox, 2013; Zhu & Stone, 2012).

Bayes factors attempt to capture the evidence in favor of one model as compared to another. When used to compare models representing null and alternative hypotheses, this affords the possibility of concluding that the data provide evidence against the null, much like frequentist hypothesis testing. Importantly, the use of Bayes factors also affords possibly concluding that the data provide evidence in *favor of the null*, or that the data are inconclusive about which model is to be preferred. As the construction in (10.17) suggests, implicit in the use of Bayes factors is the perspective that one or the other model could be a good description of the data. See Gelman et al. (2013) and Gill (2007) for broader discussions and criticisms of Bayes factors, both generally and in specific use contexts.

One set of criticisms involves the challenges in computing Bayes factors, though approximations make Bayes factors more accessible (Bollen, Ray, Zavisca, & Harden, 2012; Carlin & Chib, 1995; Chib, 1995; Chib & Jeliakov, 2001; Han & Carlin, 2001; Kass & Raftery, 1995).

TABLE 10.1

Guidelines for Interpreting the Magnitude of Bayes Factors

$\mathbf{BF_{21}}$	$\mathbf{2log(BF_{21})}$	Evidence in Favor of M_2 and against M_1
1–3	0–2	Not worth more than a bare mention
3–20	2–6	Positive
20–150	6–10	Strong
>150	>10	Very strong

Source: Kass, R. E., & Raftery, A. E. (1995). Bayes factors. *Journal of the American Statistical Association, 90, 773–795.*

In what follows, we adopt an approach that employs approximations to the model's marginal likelihood based on predictive distributions and cross-validation strategies. For the moment, we suppress the conditioning on M as the following applies to computations for individual models. To begin, a cross-validation strategy involves the examination of predictions of a subset $\mathbf{x}_i$ of the data $\mathbf{x}$, when the complement of $\mathbf{x}_i$ is employed to fit the model. Following the notation in Chapter 5, let $\mathbf{x}_{-i}$ denote this complement, such that $\mathbf{x} = (\mathbf{x}_i, \mathbf{x}_{-i})$. The conditional predictive ordinate (CPO) is the value of $\mathbf{x}_i$ in the conditional predictive density, which, assuming conditional independence of $\mathbf{x}_i$ and $\mathbf{x}_{-i}$ given θ is*

$$\text{CPO}_i = p(\mathbf{x}_i \mid \mathbf{x}_{-i}) = \int p(\mathbf{x}_i \mid \theta) p(\theta \mid \mathbf{x}_{-i}) d\theta. \tag{10.19}$$

The collection of $p(\mathbf{x}_i \mid \mathbf{x}_{-i})$ fully determine the marginal likelihood $p(\mathbf{x})$ (Besag, 1974). An estimator of this likelihood, referred as a pseudomarginal likelihood (PsML) is given by (Geisser & Eddy, 1979)

$$\text{PsML} = \hat{p}(\mathbf{x}) = \prod_{i=1}^{n} p(\mathbf{x}_i \mid \mathbf{x}_{-i}). \tag{10.20}$$

The ratio of the PsMLs for two models yields a pseudo Bayes factor (Gelfand, 1996). Equation (10.19) suggests that the CPOs may be estimated by running n separate model-fitting analyses, each one holding out a single examinee and using the remaining $n - 1$ examinees to yield the posterior distribution and the CPO for the held-out examinee. We employ an alternative approach that estimates the CPOs based on a single analysis using MCMC. For a set of R iterations, we compute a Monte Carlo estimate of the CPO (Congdon, 2007),

$$\text{CPO}_i \approx \frac{1}{\frac{1}{R}\sum_{r=1}^{R} [p(\mathbf{x}_i \mid \theta^{(r)})]^{-1}}, \tag{10.21}$$

where $\theta^{(r)}$ contain the values for the unknown parameters from iteration r. We then need to aggregate these over examinees. For numerical stability, we work with logarithms. An estimate of the log of PsML is given by

$$\log(\text{PsML}) \approx \sum_{i=1}^{n} \log(\text{CPO}_i). \tag{10.22}$$

10.3.2.1 Example from CFA

We illustrate these computations in the context of comparing CFA models for the five observables derived from the Institutional Integration Scale. Let M_1 refer to single-latent variable (factor) model from Section 9.3 and let M_2 refer to two-latent variable (factor) model from Section 9.4.

* Note that this has a similar form as the posterior predictive distribution in (10.9), and may be seen as the posterior predictive distribution based on obtaining the posterior distribution for the parameters using $\mathbf{x}_{-i}$. This perspective supports traditional cross-validation approaches, where an observed dataset is split into a cross-validation sample ($\mathbf{x}_i$) and a calibration sample ($\mathbf{x}_{-i}$) (see Bolt et al., 2001 for an application in psychometrics).

For each model, we can compute the core ingredient of the CPO, $[p(\mathbf{x}_i \mid \boldsymbol{\theta}^{(r)})]^{-1}$, in WinBUGS by adding the following lines of code to that given in Sections 9.3.3 and 9.4.3, respectively:

```
------------------------------------------------------------------------
#########################################################################
# Model Syntax to Compute Ingredients for CPO
#########################################################################
for(i in 1:n){
  for(j in 1:J){
    p.x[i,j] <- (1/sqrt(2*3.141593))*sqrt(inv.psi[j])*exp(-
.5*inv.psi[j]*(x[i,j]-mu[i,j])*(x[i,j]-mu[i,j]))
  }

  inv.p.x[i] <- 1/(p.x[i,1]*p.x[i,2]*p.x[i,3]*p.x[i,4]*p.x[i,5])
}
------------------------------------------------------------------------
```

The posterior mean of the node `inv.p.x[i]` is the denominator in (10.21). For each model, we obtained these posterior means for the n examinees, computed the estimate of CPO for each examinee using (10.21), and then the log(PsML) using (10.22). The value of log(PsML) for the single-latent variable (factor) model was −2053.55 and the value of the log(PsML) for the two-latent variable (factor) model was −2030.38. Twice the difference yields an estimate of $2\log(\text{BF}_{21}) = 46.34$. Following the guidelines in Table 10.1, this is interpreted as strong evidence in favor of the model with two latent variables (factors) over the model with one latent variable (factor).

10.3.3 Information Criteria

An alternative method for approximating the Bayes factor is based on transformations of the Bayesian information criterion (BIC) which is constructed as twice the Schwarz (1978) criterion (see also Kass & Raftery, 1995),

$$\text{BIC} = -2\ell(\hat{\boldsymbol{\theta}} \mid \mathbf{x}) + p\ln(n), \tag{10.23}$$

where $\ell(\hat{\boldsymbol{\theta}} \mid \mathbf{x})$ is the maximized log-likelihood and p is the number of parameters. More commonly, analysts use BIC and related information criteria (Burnham & Anderson, 2002) to conduct model selection via examining the values for each model and selecting the model with the smallest value.

Working in this tradition of comparing values of information criteria, Spiegelhalter, Best, Carlin, and van der Linde (2002) introduced the deviance information criterion (DIC),

$$\text{DIC} = \overline{D(\boldsymbol{\theta})} + p_{\text{D}} = 2\overline{D(\boldsymbol{\theta})} - D(\bar{\boldsymbol{\theta}}) + 2p_{\text{D}}, \tag{10.24}$$

where p_{D} is a complexity measure defined as the difference between the posterior mean of the deviance, $\overline{D(\boldsymbol{\theta})}$, and the deviance evaluated at the posterior mean, $D(\bar{\boldsymbol{\theta}})$. Model selection here involves choosing the model with the smallest DIC value, a strategy that is gaining attention in Bayesian psychometrics (Kang & Cohen, 2007; Kang, Cohen, & Sung, 2009; Klein Entink et al., 2009; Lee, Song, & Cai, 2010; Lee, Song, & Tang, 2007; Li et al., 2006, 2009; Zhu & Stone, 2012).

WinBUGS provides values for the DIC and its constituent elements for models. However, WinBUGS does not automatically guard against the use of DIC in situations where it may not be appropriate. Note the use of the deviance evaluated at the posterior mean for the parameters, $D(\bar{\theta})$. This construction relies on the posterior mean being a reasonable point summary of the parameters, as may be relevant in CFA models analyzed here, and item response models of Chapter 11 (see Exercises 11.1 and 11.2). If the posterior mean is not a reasonable summary of the posterior, the DIC may not be appropriate. For example, the latent variables in latent class and Bayesian network models (Chapters 13 and 14) are discrete and often nominal in nature, and a mean is not a reasonable point summary for them. The DIC is not recommended for use in such cases.

The addition of p_D in the DIC is intended to capture the model complexity and penalize the model accordingly. The rationale for this sort of construction comes from Occam's razor, which suggests that, all else equal, we should prefer models that are more parsimonious in that they posit fewer entities or are otherwise simpler. In the case of model comparison, we recognize that models that are more complex (less parsimonious) will provide better fit by virtue of their increased flexibility. When it comes to model selection, the more complex model's capability to provide better fit should therefore be penalized. This sort of rationale is present in the BIC and other information criteria in the form of adjusting the absolute fit as captured by the likelihood by adding a penalty term based on the number of estimated parameters, p. As we have seen in Section 9.8, a maximally diffuse prior is akin to a frequentist approach that has the parameter "in" the model to be estimated, and a prior with its mass concentrated at a single point is akin to having the parameter fixed and not estimated. Whether or not to count the parameter as part of the penalty in these cases is fairly clear. With a prior distribution for a parameter in between these two extremes, it is not as obvious. What is more, not all parameters have the same impact in terms of the capability of the model to fit data (Fan & Sivo, 2005; Preacher, 2006). Model complexity is not just about how many parameters, but which ones. p_D is one approach to capturing model complexity and folding it in to a model comparison process. More advanced discussions of DIC appear in Gelman et al. (2013), Gill (2007), and Plummer (2008).

10.3.3.1 Example from CFA

Returning to the factor analysis models of the Institutional Integration subscales, the DIC for the model with one latent variable (factor) was 3818.59 and the DIC for the model with two latent variables (factors) was 3761.46, agreeing with the Bayes factor in lending support for the latter over the former.

10.3.4 Predictive Criteria

An alternative approach to model comparison focuses on the predictive quality of the models. Working in the context of discrete data, Gilula and Haberman (2001) proposed model comparison procedures based on the expected improvement in prediction for a new observation, and which may be seen as addressing the question, given the information in the data $\mathbf{x}$, how much better would M_2 be expected to predict a new observation than M_1? We can begin to address this in terms of the reduction in entropy. The entropy for a model M_r is

$$\text{Ent}(M_r) = -\sum_{i=1}^{n} p_r(\mathbf{x}_i)\log(p_r(\mathbf{x}_i)), \tag{10.25}$$

where $p_r(\mathbf{x}_i)$ is the modeled probability of $\mathbf{x}_i$ under M_r. The difference in entropy values for two models,

$$\text{Ent}(M_1) - \text{Ent}(M_2), \tag{10.26}$$

captures how much better one model predicts a new observation. The difference is positive if M_2 yields a better prediction, and negative if M_1 yields a better prediction. If M_1 is nested within M_2, the difference is non-negative; however, the construction of the difference does not require that one model be nested within another. Additional extensions or versions are possible. When counting the number of parameters is straightforward, dividing (10.26) by the difference in the number of parameters supports an evaluation of the difference in terms of improvement per parameter. Similarly, dividing (10.26) by n supports an interpretation in terms of the improved prediction for a single observation.

Gilula and Haberman (2001) suggested a criterion for model comparison that is analogous to the proportion of variance accounted for. Letting M_0 denote a baseline model, the proportional improvement obtained by using M_r is

$$\frac{\text{Ent}(M_0) - \text{Ent}(M_r)}{\text{Ent}(M_0)}. \tag{10.27}$$

The above description of entropy-based comparisons applies naturally to discrete data. However, the same properties of entropy do not necessarily hold for continuous data. We illustrate these approaches in the context of latent class analysis for discrete data in Section 13.4.5.

10.3.5 Additional Approaches to Model Comparison

A number of other model comparison procedures have also been proposed, including those that aim to circumvent computational difficulties (Carlin & Chib, 1995) as well as those that do not aim to select the models. Raftery (1993) and Hoeting, Madigan, Raftery, and Volinksy (1999) discussed model averaging, in which the results of different models are pooled. Gelman et al. (2013) discussed model expansion, in which models are extended and model comparison is enacted by analyzing the parameters related to those expansions. See Karabatsos and Walker (in press) for a related approach to model selection for psychometric models. See Gill (2007) and Gelman et al. (2013) for overviews, critical discussions, and additional references on many of the methods discussed here.

Exercises

10.1 Reconsider the CFA model with one latent variable for the data from the Institutional Integration Scale described in Section 9.3.

a. Recreate the Bayesian analysis reported there using WinBUGS (see Section 9.3.3) and:

1. Obtain the DIC for the model.
2. Conduct PPMC using correlations among the observables, LR, and SRMR. Compare your results to those reported here.
3. Conduct PPMC using the means of the observables. Interpret these results in terms of the adequacy of model-data fit. What feature of the model is responsible for this (in)adequacy of fit?

b. Fit the model using ML estimation, and compute LR and SRMR using ML estimates.

c. Compare the results from the frequentist approach in (b) to the Bayesian approach in (a). How do LR and SRMR differ across the two approaches? Why do they differ in these ways?

10.2 Reconsider the CFA model with two latent variables introduced in Section 9.4, in which the subscales are organized around one latent variable associated with *Faculty* measures and one associated with *Student* measures.

a. Recreate the Bayesian analysis reported there using WinBUGS (see Section 9.4.3) and:

1. Obtain the DIC for the model.
2. Conduct PPMC using correlations among the observables, LR, and SRMR. Interpret these results in terms of the adequacy of model-data fit.

b. Compare your results for this model to those for the model with one latent variable in Exercise 10.1.

1. What does model comparison based on DIC suggest?
2. What does the comparison of the PPMC results from the correlations, LR, and SRMR suggest?

11

Item Response Theory

Item response theory (IRT) models employ continuous latent variables to model dichotomous or polytomous observables, as occur frequently in assessment and social and life science settings (e.g., item responses given by students on an achievement test and scored as correct or incorrect, personal preferences rated on a Likert scale by subjects on a survey, votes in favor or against a bill by politicians, or presence or absence of a patient's symptoms). IRT models specify the probability for an observable taking on a particular value as a function of the latent variable for the examinee (subject, politician, and patient) and the measurement model parameters for that observable. The latter are often referred to as *item parameters,* a term that reflects developments that occurred in the context of modeling response data to items in educational assessment (as does the name "item response theory"). In this context, IRT was initially developed for and applied to dichotomous test items. Such items are a special case of assessment tasks; examinees' responses are a special case of work products; the evaluations of responses that become the dependent variables in IRT models are special cases of observable variables. Tasks, work products, and observable variables correspond one-to-one here, and the word "item" is commonly associated with all three of these distinct entities. Ambiguities arise when the correspondence breaks down, as occurs when questions are clustered within the same task or when different evaluation rules of the same multiple-choice response produce different observables (e.g., classifying responses as right/wrong, versus partial credit, versus partially ordered categories of conceptions and misconceptions). In this chapter, we consider only situations where the one-to-one correspondence among task, work product, and observable holds.* Accordingly, we predominantly use the terms "items" and "responses" from educational assessment, though the model applies to situations where this terminology does not apply (e.g., vote for or against, symptom present or absent).

Treatments and overviews of IRT from conventional perspectives can be found in De Ayala (2009), Embretson and Reise (2000), Hambleton and Swaminathan (1985), Lord (1980), McDonald (1999), and van der Linden and Hambleton (1997). These and other conventional sources include Bayesian elements, reflecting that Bayesian perspectives and procedures are more prevalent in IRT than in other psychometric modeling paradigms. This is no doubt in part due to some key difficulties associated with frequentist approaches to IRT that do not arise as forcefully in other psychometric modeling paradigms. The complexities of modeling discrete data with continuous latent variables have led to a variety of Bayesian approaches, and we endeavor to cover the main themes and procedures.

The majority of work in IRT has concerned the use of a single latent variable to model dichotomous observables, and it is in this context that we begin in Section 11.1, reviewing conventional approaches. In Section 11.2, we give a Bayesian approach to modeling dichotomous observables. We extend this foundation to the case of polytomous observables, reviewing conventional approaches in Section 11.3 and turning to Bayesian approaches

* We will see examples where this correspondence breaks down: in Chapter 13, we consider an example where one observable is defined based on work products from two tasks, and in Chapter 14, we consider an example where multiple observables are defined based on work products from a single task.

in Section 11.4. In Section 11.5, we consider models with multiple latent variables. IRT is a workhorse of modern operational assessment, and in Section 11.6 we cover conceptual aspects of several applications of IRT in assessment that are strongly aligned with Bayesian perspectives. In Sections 11.7 and 11.8, we survey and discuss alternative specifications for aspects of the model; the former gives an overview of a number of alternative prior distributions that have been proposed and the latter treats an alternative formulation of IRT. We conclude the chapter with a summary and a bibliographic note in Section 11.9.

11.1 Conventional IRT Models for Dichotomous Observables

11.1.1 Model Specification

Let x_{ij} be the value from examinee i on observable j, and without loss of generality let the possible values for each x_{ij} be 1 (indicating a correct response to an item, success on a task, endorsement of a statement, voting in favor of a bill, having a symptom, etc.) and 0 (indicating an incorrect response to an item or failure on a task, disagreeing with a statement, voting against a bill, not having a symptom etc.). Let $\mathbf{x}$ denote the full collection of the x_{ij}s, which may be arranged in an $(n \times J)$ matrix as in CFA. The most widely used models differ primarily in the specification of the form of the model for the dependence of the observables on the latent variable, and the number of parameters.

Many of the most widely used IRT models in educational measurement may be expressed as a version of

$$P(x_{ij} = 1 \mid \theta_i, d_j, a_j, c_j) = c_j + (1 - c_j)F(a_j\theta_i + d_j), \tag{11.1}$$

with elements described next. θ_i is the latent variable for examinee i. In typical educational measurement contexts, θ_i represents the examinee's proficiency such that higher values indicate more proficiency. d_j, a_j, and c_j are the measurement model (item) parameters, and IRT models are often referred to by how many of these are included per observable. d_j (or a transformation of it described below and denoted b_j) is a location parameter for observable j, often referred to as a difficulty parameter for observable (item) j, as it is strongly related to the overall mean of the observable (which is the proportion correct on an item scored correct or incorrect). a_j is a coefficient for θ_i capturing its relationship to observable j. a_j is often referred to as a discrimination parameter for the observable (item), as it is related to the capability of the observable (item) to discriminate between examinees with lower and higher values along the latent variable. c_j is a lower asymptote parameter for the observable, bounded by 0 and 1, and is the probability that an examinee will have a 1 on the observable as $\theta \to -\infty$. c_j is often referred to as a pseudo-guessing parameter, reflecting that examinees with low proficiency may have a nonzero probability of a correct response due to guessing, as may be prevalent on a multiple-choice achievement test. In practice, F is almost always selected to be the logistic cumulative distribution function or the normal cumulative distribution function, which yields a result for (11.1) that is in the unit interval.

The model may be written equivalently as

$$P(x_{ij} = 1 \mid \theta_i, b_j, a_j, c_j) = c_j + (1 - c_j)F[a_j(\theta_i - b_j)], \tag{11.2}$$

differing from (11.1) by the relationship $d_j = -a_j b_j$. The form in (11.2) is useful for viewing IRT as a scaling model that locates examinees (θs) and item difficulties (bs) on the same scale, particularly when $c_j = 0$ in which case the probability in (11.2) is .5 when $\theta_i = b_j$. The version in (11.1) allows for easier extensions to certain models with multiple latent variables (Section 11.5) as well as a recognition of the connections between IRT and CFA (Section 11.8).

Versions of (11.1) and (11.2) are referred to as the *item response function*, and named in terms of how many of the measurement model parameters they contain. For example, choosing F in (11.1) to be the normal distribution function yields a three-parameter normal ogive (3-PNO) model,

$$P(x_{ij} = 1 \mid \theta_i, d_j, a_j, c_j) = c_j + (1 - c_j)\Phi(a_j\theta_i + d_j), \tag{11.3}$$

where $\Phi(\cdot)$ is the cumulative normal distribution function. Choosing F to be a logistic distribution function would yield the three-parameter logistic (3-PL) model. Figure 11.1 illustrates the 3-PNO item response function for two items. Comparing the two, we see that the item plotted as a dashed line is easier ($d = 0$) than that plotted as a solid line ($d = -4$), in that the probability of observing a 1 is higher for the former than the latter. The slope of the solid item is steeper than the dashed line, as captured by their discrimination parameters ($a = 2$ and $a = 1$, respectively). Finally, the dashed item has a lower asymptote at $c = .25$, whereas the solid item has a lower asymptote at $c = 0$.

Setting $c_j = 0$ yields a two-parameter (2-P) model. Adopting the logistic form, the 2-P logistic model is

$$P(x_{ij} = 1 \mid \theta_i, d_j, a_j) = \Psi(a_j\theta_i + d_j) = \frac{\exp(a_j\theta_i + d_j)}{1 + \exp(a_j\theta_i + d_j)}, \tag{11.4}$$

where $\Psi(\cdot)$ is the cumulative logistic distribution function. Further setting all the a_j equal to one another yields a one-parameter model. One mechanism for resolving an indeterminacy in the latent variable (Section 11.1.2) involves fixing a discrimination parameter to a particular value. In the case of a one-parameter model, this fixes all the discrimination parameters to that value. Choosing that value to be 1, and continuing with the logistic form, we have the one-parameter logistic (1-PL) or the Rasch (1960) model,

$$P(x_{ij} = 1 \mid \theta_i, d_j) = \Psi(\theta_i + d_j) = \frac{\exp(\theta_i + d_j)}{1 + \exp(\theta_i + d_j)}. \tag{11.5}$$

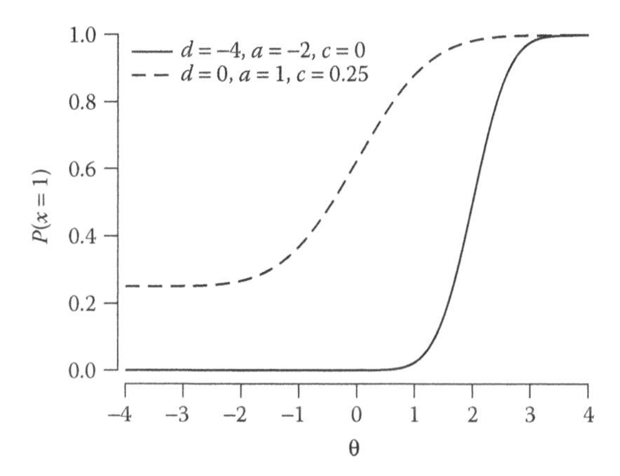

FIGURE 11.1
Item response functions for two items following a three-parameter normal ogive (3-PNO) model.

The 2-PNO and 1-PNO models may be specified by using the normal distribution function rather than the logistic distribution function in (11.4) and (11.5), respectively. The choice between using a normal or logistic distribution function is in some sense arbitrary as the logistic model may be rescaled by multiplying the exponent by 1.701 to make the resulting function nearly indistinguishable from the normal ogive version (e.g., McDonald, 1999), though one may prove more tractable for certain purposes in assessment, or under different estimation paradigms.

11.1.2 Indeterminacies

The 2-P and 3-P models are subject to the same indeterminacies as CFA models (Sections 9.1.2 and 9.9; see also Hambleton & Swaminathan, 1985, and Lord, 1980, for treatments in unidimensional IRT). In the 1-P models, the lone indeterminacy is the location indeterminacy. In IRT the metric of each latent variable is often specified by fixing the mean and the variance of the latent variable, and the orientation of each latent variable is often specified by constraining the discriminations to be positive. As was the case with CFA, options abound, in some cases constraints made on one or more measurement models parameters may be employed to resolve the indeterminacies.

11.1.3 Model Fitting

A variety of approaches have been proposed and are applicable for calibrating models with continuous latent and discrete observable variables from IRT and CFA traditions (Embretson & Reise, 2000; Hambleton & Swaminathan, 1985; Wirth & Edwards, 2007). We briefly characterize marginal maximum likelihood (MML) approaches and variants of it, in line with the general discussion in Section 7.4. Let ω_j denote the collection of measurement model parameters for observable j. For example, for the 3-PNO model in (11.3), $\omega_j = (d_j, a_j, c_j)$. We may write the conditional probability for any particular observed value, whether it is 1 or 0, as

$$P(x_{ij} \mid \theta_i, \omega_j) = P(x_{ij} = 1 \mid \theta_i, \omega_j)^{x_{ij}} (1 - P(x_{ij} = 1 \mid \theta_i, \omega_j))^{1-x_{ij}}. \quad (11.6)$$

As was the case in CTT and CFA, IRT models typically invoke assumptions of independence among examinees and conditional (local) independence among observables. Letting $\theta = (\theta_1, \ldots, \theta_n)$ denote the collection of latent variables for n examinees and letting $\omega = (\omega_1, \ldots, \omega_J)$ denote the full collection of measurement model parameters, the joint probability of the data is then

$$P(\mathbf{x} \mid \theta, \omega) = \prod_{i=1}^{n} P(\mathbf{x}_i \mid \theta_i, \omega) = \prod_{i=1}^{n} \prod_{j=1}^{J} P(x_{ij} \mid \theta_i, \omega_j), \quad (11.7)$$

where $\mathbf{x}_i = (x_{i1}, \ldots, x_{iJ})$ is the collection of J observables for examinee i.

The marginal probability of the observables is obtained via the marginalization as in (7.8),

$$P(\mathbf{x} \mid \omega, \theta_P) = \int_{\theta} P(\mathbf{x} \mid \theta, \omega) p(\theta \mid \theta_P) d\theta, \quad (11.8)$$

where in the current case $P(\mathbf{x} \mid \theta, \omega)$ is given by (11.7). The distribution of the latent variables is typically assumed normal, in which case $\theta_P = (\mu_0, \sigma_0^2)$ are the mean and variance, though other specifications are possible and in general it may be empirically defined.

Once values of the data are observed, (11.8) may be treated as a likelihood function for ω and θ_P. Maximizing this with respect to ω and θ_P yields MML estimates of these parameters. Note that elements of ω and θ_P will drop out of the estimation if they are specified in advance to resolve the indeterminacies. In general, there is no closed form solution, and a variety of methods have been proposed for evaluating the integrals and maximizing the resulting marginal likelihood (Bock & Aitkin, 1981; Bock & Moustaki, 2007; Harwell, Baker, & Zwarts, 1988).

The likelihood may be sufficiently complex that the parameters may be poorly determined from the data. In particular, for 2-P and even more so for 3-P models, discrepant sets of measurement parameter values might yield quite similar model-implied probabilities, at least over the region(s) of the latent continuum where examinees are located (Hulin, Lissak, & Drasgow, 1982). As a result, the likelihood surface is nearly flat in certain regions along one or more dimensions of the parameter space, which may give rise to problems associated with unstable MML estimates. An implausible set of parameter values may produce nearly as high a value for the likelihood, or perhaps even a slightly higher value, than that resulting from a set of plausible parameter values. Moreover, MML estimates are likely to vary considerably from sample to sample, as the sampling variability induces ever-so-slight differences in the likelihood surface. In these sorts of cases, it may be difficult to distinguish between competing sets of values for the measurement model parameters based on the data alone. Conventional practice often turns to the use of a prior distribution for some or all of the measurement model parameters that places more of its density at plausible values to mitigate these problems (Lord, 1986). In a sense, the prior steps in where the data are equivocal to adjudicate in favor of a priori plausible sets of parameters values over a priori implausible ones. With priors specified, application of the same sort of estimation methods for maximizing likelihoods yields Bayes modal estimates (Mislevy, 1986; see also Harwell & Baker, 1991). This has become fairly common if not standard practice, and represents the area where Bayesian approaches to calibration have penetrated operational psychometric practice the most.

Turning to scoring, a variety of approaches have been proposed for estimating examinees' latent variables, including maximum likelihood, weighted likelihood, and Bayesian approaches (Bock & Mislevy, 1982; Embretson & Reise, 2000; Warm, 1989). Of the latter, one option is to obtain the posterior distribution for an examinee's latent variable using point estimates for parameters as in (7.10). Two popular Bayesian methods involve obtaining either the mode (MAP) or the mean (EAP) of the posterior distribution for the examinee's latent variable. These offer advantages in that they yield estimates in situations that pose challenges for frequentist methods, such as situations where a unique maximum of the likelihood does not exist (Samejima, 1973; Yen, Burket, & Sykes, 1991), or for examinees with so-called perfect response patterns of all 1s or 0s. The example in Section 11.2.5 illustrates this point.

11.2 Bayesian Modeling of IRT Models for Dichotomous Observables

11.2.1 Conditional Probability of the Observables

The preceding developments amount to the specification of the conditional probability of the data in an equation-oriented fashion. We restate them here in a compact but precise manner in a more distribution-oriented way, drawing on the equations as needed. The conditional probability of the data is

$$p(\mathbf{x} \mid \theta, \omega) = \prod_{i=1}^{n} p(\mathbf{x}_i \mid \theta_i, \omega) = \prod_{i=1}^{n} \prod_{j=1}^{J} p(x_{ij} \mid \theta_i, \omega_j), \tag{11.9}$$

where

$$x_{ij} \mid \theta_i, \omega_j \sim \text{Bernoulli}[P(x_{ij} = 1 \mid \theta_i, \omega_j)], \tag{11.10}$$

$P(x_{ij} = 1 \mid \theta_i, \omega_j)$ is the item response function, and the ω_j are the associated measurement model (item) parameters for the model.

11.2.2 A Prior Distribution

Equation (11.9) contains unknown latent variables (θ) and unknown measurement model parameters (ω). A Bayesian analysis requires a specification of a prior distribution for these unknowns. As a first step in constructing the prior distribution, we will express the joint prior $p(\theta, \omega)$ for the unknowns introduced so far as the product of priors for $p(\theta)$ and $p(\omega)$; that is,

$$p(\theta, \omega) = p(\theta)p(\omega), \tag{11.11}$$

which reflects an independence assumption between the examinees' latent variables and the measurement model parameters. Specifying the prior distribution comes to specifying distributions for the terms on the right hand side of (11.11), and any hyperprior distributions for any unknown hyperparameters introduced in such specifications.

11.2.2.1 Prior Distribution for the Latent Variables

The first term on the right-hand side of (11.11) is the joint prior distribution of the latent variables, where there is one latent variable for each of n examinees. An exchangeability assumption allows for this joint prior distribution to be factored into the product of common prior distributions,

$$p(\theta) = \prod_{i=1}^{n} p(\theta_i \mid \theta_P), \tag{11.12}$$

where θ_P denotes the hyperparameters for specifying the prior for the θ_i. Commonly, the latent variable is assumed to be normally distributed, in which case $\theta_P = (\mu_\theta, \sigma_\theta^2)$, and the prior for each examinee's latent variable is

$$\theta_i \mid \mu_\theta, \sigma_\theta^2 \sim N(\mu_\theta, \sigma_\theta^2), \tag{11.13}$$

If these hyperparameters are unknown they require a hyperprior distribution, $p(\theta_P)$. A convenient choice is to use the now familiar conditionally conjugate prior structures, namely a normal prior on the mean and an inverse-gamma prior on the variance. In practice, many applications of IRT specify values for the hyperparameters to resolve the location and scale indeterminacies. The choices of the values are arbitrary, with the most popular being $\mu_\theta = 0$ and $\sigma_\theta^2 = 1$.

11.2.2.2 Normal, Truncated-Normal, and Beta Prior Distributions for the Measurement Model Parameters

To complete the specification of the prior distribution, we need to specify the prior distribution for the measurement model (item) parameters, $p(\omega)$. An exchangeability assumption with respect to the observables allows for the factorization of the joint prior distribution for the observables' measurement model parameters into the product of common prior distributions,

$$p(\omega) = \prod_{j=1}^{J} p(\omega_j \mid \omega_P), \tag{11.14}$$

where ω_P denotes the hyperparameters for specifying the prior for the ω_j. If these hyperparameters are unknown they require a hyperprior distribution, $p(\omega_P)$. We consider this case in Section 11.7.4, but for the moment we focus on issues surrounding the specification of the prior for the measurement model parameters themselves.

In CFA, the two dominant approaches to specifying the prior distribution for the measurement model parameters are the conditionally conjugate priors detailed in Section 9.2 or a slightly more complex prior that preserves full conjugacy. Though other specifications are possible, these two are popular in part because they yield easily manageable computational strategies for obtaining the posterior distribution when the data are assumed to be conditionally normally distributed, as has long been popular in CFA. In IRT, the situation is more complex. For many formulations of the popular models, there is no conjugate or conditionally conjugate prior. In addition, a number of developments in Bayesian approaches to IRT, including recommendations for prior distributions, occurred prior to the advent of MCMC when the focus was on employing optimization routines to estimate posterior modes and the curvature of the posterior. The model specifications that simplify the computational burden in one of these contexts may not simplify the computational burdens in others.

In this section, we develop and illustrate a particular set of common choices for prior distributions, introducing an MCMC estimation routine that is flexible enough to handle the situation without conditionally conjugate priors. We return to this issue and discuss alternative prior distributions and associated MCMC routines aligned with them in Sections 11.7 and 11.8.

Assuming a priori independence between the different types of measurement model parameters, we factor the joint prior for each observable's measurement model parameters $\omega_j = (d_j, a_j, c_j)$ into the product of prior distributions for each parameter individually:

$$p(\omega_j) = p(d_j)p(a_j)p(c_j). \tag{11.15}$$

The assumption of exchangeability with respect to the observables supports the specification of a common prior for each instance of the different types of measurement model parameters. The location parameters are continuous and unbounded, and are therefore typically assigned a normal prior distribution,

$$d_j \sim N(\mu_d, \sigma_d^2), \tag{11.16}$$

where μ_d and σ_d^2 are hyperparameters that in the current development are specified by the analyst. Normal prior distributions are similarly common for b parameters in the formulations of IRT that employ the "−b" parameterization.

Turning to the discrimination parameters, recognize that a_j governs the association between the latent variable and the probability of a correct response to the item. It is common to specify a prior distribution that is restricted to be over the positive real line, which restricts the association between the latent variable and the probability of a correct item response to be positive. Doing so resolves the indeterminacy in the orientation of the latent variable. More substantively, in educational assessment it renders an interpretation of the latent variable as one where higher values of the latent variable correspond to higher levels of proficiency as it yields higher probabilities of a correct response. This can be enacted via a normal distribution truncated below at 0,

$$a_j \sim N^+(\mu_a, \sigma_a^2), \tag{11.17}$$

where N^+ denotes the normal distribution truncated to the positive real line and μ_a and σ_a^2 are hyperparameters that in the current development are specified by the analyst. Checks on the restriction implied by the positivity constraint may be conducted using methods discussed in Chapter 10 or via basic descriptive statistics such as the correlation between the item response variable and the total score on a set of items. Furthermore, if an analyst does not wish to impose the restriction that higher values of the latent variable yield higher probabilities of a correct response for some of the items, the positivity restriction in (11.17) can be removed.

The lower asymptote parameters reflect the probability of success on the item when proficiency is low and is bounded by 0 and 1. A natural choice is therefore the beta distribution,

$$c_j \sim \text{Beta}(\alpha_c, \beta_c). \tag{11.18}$$

Note that the hyperparameters in (11.16)–(11.18) are not indexed by j, indicating that each instance of each type of parameter is assigned the same prior in line with the exchangeability assumption. This can be relaxed to specify group-specific hyperparameters reflecting conditional exchangeability of items, possibly given covariates, or even a unique prior for each parameter. We discuss alternative prior structures and parametric forms in Section 11.7.

11.2.3 Posterior Distribution and Graphical Model

The directed acyclic graph (DAG) for the model with fixed values for the mean (μ_θ) and the variance (σ_θ^2) of the latent variables is given in Figure 11.2. Letting $\mathbf{d} = (d_1, \ldots, d_J)$, $\mathbf{a} = (a_1, \ldots, a_J)$, and $\mathbf{c} = (c_1, \ldots, c_J)$, the posterior distribution is

$$p(\boldsymbol{\theta}, \mathbf{d}, \mathbf{a}, \mathbf{c} \mid \mathbf{x}) \propto \prod_{i=1}^{n} \prod_{j=1}^{J} p(x_{ij} \mid \theta_i, d_j, a_j, c_j) p(\theta_i) p(d_j) p(a_j) p(c_j), \tag{11.19}$$

where

$$x_{ij} \mid \theta_i, d_j, a_j, c_j \sim \text{Bernoulli}[P(x_{ij} = 1 \mid \theta_i, d_j, a_j, c_j)] \quad \text{for} \quad i = 1, \ldots, n, j = 1, \ldots, J,$$

$$P(x_{ij} = 1 \mid \theta_i, d_j, a_j, c_j) = c_j + (1 - c_j)\Phi(a_j\theta_i + d_j) \quad \text{for} \quad i = 1, \ldots, n, j = 1, \ldots, J,$$

$$\theta_i \sim N(\mu_\theta, \sigma_\theta^2) \quad \text{for} \quad i = 1, \ldots, n,$$

$$d_j \sim N(\mu_d, \sigma_d^2) \quad \text{for} \quad j = 1, \ldots, J,$$

$$a_j \sim N^+(\mu_a, \sigma_a^2) \quad \text{for} \quad j = 1, \ldots, J,$$

and

$$c_j \sim \text{Beta}(\alpha_c, \beta_c) \quad \text{for} \quad j = 1, \ldots, J.$$

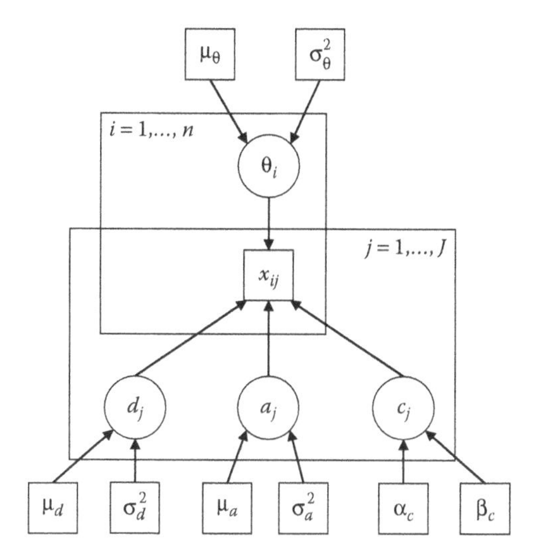

FIGURE 11.2
Directed acyclic graph for a three-parameter item response theory model.

11.2.4 MCMC Estimation

As we do not have the benefit of conditional conjugacy under this specification, we turn to the more flexible but more computationally demanding strategy of a Metropolis–Hastings-within-Gibbs sampler (also known as a single-component Metropolis–Hastings sampler). In this routine, we characterize the posterior in terms of the full conditional distributions and then apply a Metropolis–Hastings step for each full conditional. Here, we lay out the sampling scheme generically (for iteration $t + 1$), and then discuss particular choices of symmetric proposal distributions that enact a Metropolis rather than Metropolis–Hastings sampler.

11.2.4.1 Metropolis–Hastings-within-Gibbs Routine for a 3-P Model for Dichotomous Observables

1. *Sample the latent variables for examinees.* For each examinee, $i = 1, \ldots, n$, sample a uniform variable $U_i \sim \text{Uniform}(0,1)$ and sample a candidate value for the latent variable θ_i^* from a proposal distribution possibly dependent on the current value $\theta_i^{(t)}$, $\theta_i^* \sim q_{\theta_i}(\theta_i^* \mid \theta_i^{(t)})$, and set $\theta_i^{(t+1)} = \theta_i^*$ if $\alpha_i \geq U_i$, where

$$\begin{aligned}\alpha_i &= \min\left[1, \frac{p(\theta_i^* \mid \mu_\theta, \sigma_\theta^2, \omega^{(t)}, \mathbf{x}_i)q_{\theta_i}(\theta_i^{(t)} \mid \theta_i^*)}{p(\theta_i^{(t)} \mid \mu_\theta, \sigma_\theta^2, \omega^{(t)}, \mathbf{x}_i)q_{\theta_i}(\theta_i^* \mid \theta_i^{(t)})}\right] \\ &= \min\left[1, \frac{p(\mathbf{x}_i \mid \theta_i^*, \omega^{(t)})p(\theta_i^* \mid \mu_\theta, \sigma_\theta^2)q_{\theta_i}(\theta_i^{(t)} \mid \theta_i^*)}{p(\mathbf{x}_i \mid \theta_i^{(t)}, \omega^{(t)})p(\theta_i^{(t)} \mid \mu_\theta, \sigma_\theta^2)q_{\theta_i}(\theta_i^* \mid \theta_i^{(t)})}\right]\end{aligned} \tag{11.20}$$

   and $\mathbf{x}_i = (x_{i1}, \ldots, x_{iJ})$ is the collection of J observed values from examinee i; and set $\theta_i^{(t+1)} = \theta_i^{(t)}$ otherwise. Note the use of the values of the measurement model parameters ω from the previous iteration (t).

2. *Sample the measurement model parameters.* For each observable, $j = 1, \ldots, J$, sample the measurement model parameters in a univariate fashion.

a. For each observable, sample a uniform variable $U_{d_j} \sim \text{Uniform}(0,1)$ and sample a candidate value for the observable's location parameter d_j^* from a proposal distribution possibly dependent on the current value $d_j^{(t)}$, $d_j^* \sim q_{d_j}(d_j^* \mid d_j^{(t)})$, and set $d_j^{(t+1)} = d_j^{(*)}$ if $\alpha_{d_j} \geq U_{d_j}$, where

$$\begin{aligned}\alpha_{d_j} &= \min\left[1, \frac{p(d_j^* \mid \mathbf{d}_P, a_j^{(t)}, c_j^{(t)}, \theta^{(t+1)}, \mathbf{x}_j) q_{d_j}(d_j^{(t)} \mid d_j^*)}{p(d_j^{(t)} \mid \mathbf{d}_P, a_j^{(t)}, c_j^{(t)}, \theta^{(t+1)}, \mathbf{x}_j) q_{d_j}(d_j^* \mid d_j^{(t)})}\right] \\ &= \min\left[1, \frac{p(\mathbf{x}_j \mid \theta^{(t+1)}, d_j^*, a_j^{(t)}, c_j^{(t)}) p(d_j^* \mid \mathbf{d}_P) q_{d_j}(d_j^{(t)} \mid d_j^*)}{p(\mathbf{x}_j \mid \theta^{(t+1)}, d_j^{(t)}, a_j^{(t)}, c_j^{(t)}) p(d_j^{(t)} \mid \mathbf{d}_P) q_{d_j}(d_j^* \mid d_j^{(t)})}\right]\end{aligned} \tag{11.21}$$

and $\mathbf{x}_j = (x_{1j}, \ldots, x_{nj})$ is the collection of n observed values for observable j. Note that the current values for the other unknowns include the just-sampled values for θ from step 1 (from iteration $t + 1$), and the values of a_j and c_j from the previous iteration (t).

b. For each observable, sample a uniform variable $U_{a_j} \sim \text{Uniform}(0,1)$ and sample a candidate value for the observable's discrimination parameter a_j^* from a proposal distribution possibly dependent on the current value $a_j^{(t)}$, $a_j^* \sim q_{a_j}(a_j^* \mid a_j^{(t)})$, and set $a_j^{(t+1)} = a_j^{(*)}$ if $\alpha_{a_j} \geq U_{a_j}$, where

$$\begin{aligned}\alpha_{a_j} &= \min\left[1, \frac{p(a_j^* \mid \mathbf{a}_P, d_j^{(t+1)}, c_j^{(t)}, \theta^{(t+1)}, \mathbf{x}_j) q_{a_j}(a_j^{(t)} \mid a_j^*)}{p(a_j^{(t)} \mid \mathbf{a}_P, d_j^{(t+1)}, c_j^{(t)}, \theta^{(t+1)}, \mathbf{x}_j) q_{a_j}(a_j^* \mid a_j^{(t)})}\right] \\ &= \min\left[1, \frac{p(\mathbf{x}_j \mid \theta^{(t+1)}, d_j^{(t+1)}, a_j^*, c_j^{(t)}) p(a_j^* \mid \mathbf{a}_P) q_{a_j}(a_j^{(t)} \mid a_j^*)}{p(\mathbf{x}_j \mid \theta^{(t+1)}, d_j^{(t+1)}, a_j^{(t)}, c_j^{(t)}) p(a_j^{(t)} \mid \mathbf{a}_P) q_{a_j}(a_j^* \mid a_j^{(t)})}\right]\end{aligned} \tag{11.22}$$

and $\mathbf{x}_j = (x_{1j}, \ldots, x_{nj})$ is the collection of n observed values for observable j. Note that the current values for the other unknowns include the just-sampled values for θ and d_j (from iteration $t + 1$), and the value of c_j from the previous iteration (t).

c. For each observable, sample a uniform variable $U_{c_j} \sim \text{Uniform}(0,1)$ and sample a candidate value for the observable's lower asymptote parameter c_j^* from a proposal distribution possibly dependent on the current value $c_j^{(t)}$, $c_j^* \sim q_{c_j}(c_j^* \mid c_j^{(t)})$, and set $c_j^{(t+1)} = c_j^{(*)}$ if $\alpha_{c_j} \geq U_{c_j}$, where

$$\begin{aligned}\alpha_{c_j} &= \min\left[1, \frac{p(c_j^* \mid \mathbf{c}_P, d_j^{(t+1)}, a_j^{(t+1)}, \theta^{(t+1)}, \mathbf{x}_j) q_{c_j}(c_j^{(t)} \mid c_j^*)}{p(c_j^{(t)} \mid \mathbf{c}_P, d_j^{(t+1)}, a_j^{(t+1)}, \theta^{(t+1)}, \mathbf{x}_j) q_{c_j}(c_j^* \mid c_j^{(t)})}\right] \\ &= \min\left[1, \frac{p(\mathbf{x}_j \mid \theta^{(t+1)}, d_j^{(t+1)}, a_j^{(t+1)}, c_j^*) p(c_j^* \mid \mathbf{c}_P) q_{c_j}(c_j^{(t)} \mid c_j^*)}{p(\mathbf{x}_j \mid \theta^{(t+1)}, d_j^{(t+1)}, a_j^{(t+1)}, c_j^{(t)}) p(c_j^{(t)} \mid \mathbf{c}_P) q_{c_j}(c_j^* \mid c_j^{(t)})}\right]\end{aligned} \tag{11.23}$$

and $\mathbf{x}_j = (x_{1j}, \ldots, x_{nj})$ is the collection of n observed values for observable j. Note that the current values for the other unknowns include the just-sampled values for θ, d_j, and a_j (from iteration $t + 1$).

A convenient set of choices for the proposal distributions utilize the following forms:

$$q_{d_j}(d_j^* \mid d_j^{(t)}) = N(b_j^{(t)}, \sigma_{q_{b_j}}^2), \tag{11.24}$$

$$q_{a_j}(a_j^* \mid a_j^{(t)}) = N^+(a_j^{(t)}, \sigma_{q_{a_j}}^2), \tag{11.25}$$

and

$$q_{c_j}(c_j^* \mid c_j^{(t)}) = \text{Uniform}(c_j^{(t)} - \delta, c_j^{(t)} + \delta), \tag{11.26}$$

where $\sigma_{q_{b_j}}^2$, $\sigma_{q_{a_j}}^2$, and δ are parameters that govern the variability of the proposal distributions. These proposal distributions are symmetric with respect to their arguments and the Metropolis–Hastings steps in the routine above reduce to be Metropolis steps, in which case the q_{d_j}, q_{a_j}, and q_{c_j} terms drop out of the calculations of in (11.21)–(11.23).

11.2.5 Example: Law School Admissions Test

To illustrate these procedures, we conduct an analysis of item responses from 1000 examinees to 5 dichotomously scored items on the Law School Admissions Test (LSAT). The data were originally given by Bock and Lieberman (1970) and are reported here in Table 11.1. All of the items were multiple-choice items with five response options.

TABLE 11.1

Frequency of the 32 Item Response Vectors for the Five-Item LSAT Data Example

	Item							Item					
Vector ID	1	2	3	4	5	Frequency	Vector ID	1	2	3	4	5	Frequency
1	0	0	0	0	0	3	17	1	0	0	0	0	10
2	0	0	0	0	1	6	18	1	0	0	0	1	29
3	0	0	0	1	0	2	19	1	0	0	1	0	14
4	0	0	0	1	1	11	20	1	0	0	1	1	81
5	0	0	1	0	0	1	21	1	0	1	0	0	3
6	0	0	1	0	1	1	22	1	0	1	0	1	28
7	0	0	1	1	0	3	23	1	0	1	1	0	15
8	0	0	1	1	1	4	24	1	0	1	1	1	80
9	0	1	0	0	0	1	25	1	1	0	0	0	16
10	0	1	0	0	1	8	26	1	1	0	0	1	56
11	0	1	0	1	0	0	27	1	1	0	1	0	21
12	0	1	0	1	1	16	28	1	1	0	1	1	173
13	0	1	1	0	0	0	29	1	1	1	0	0	11
14	0	1	1	0	1	3	30	1	1	1	0	1	61
15	0	1	1	1	0	2	31	1	1	1	1	0	28
16	0	1	1	1	1	15	32	1	1	1	1	1	298

Source: Bock, R. D., & Lieberman, M. (1970). Fitting a response model for *n* dichotomously scored items. *Psychometrika, 35*, 179–197. Used with kind permission from Springer.

11.2.5.1 Model Specification and Fitting in WinBUGS

We fit a 3-PNO model in (11.3), employing the prior distributions specified in (11.13) and (11.16)–(11.18). For completeness, we write out the model with the chosen values for the hyperparameters as

$$p(\theta,\mathbf{d},\mathbf{a},\mathbf{c}\mid\mathbf{x})\propto\prod_{i=1}^{n}\prod_{j=1}^{J}p(x_{ij}\mid\theta_i,d_j,a_j,c_j)p(\theta_i)p(d_j)p(a_j)p(c_j),$$

where

$$x_{ij}\mid\theta_i,d_j,a_j,c_j\sim\text{Bernoulli}[P(x_{ij}=1\mid\theta_i,d_j,a_j,c_j)]\quad\text{for}\quad i=1,\ldots,1000, j=1,\ldots,5,$$

$$P(x_{ij}=1\mid\theta_i,d_j,a_j,c_j)=c_j+(1-c_j)\Phi(a_j\theta_i+d_j)\quad\text{for}\quad i=1,\ldots,1000, j=1,\ldots,5,$$

$$\theta_i\sim N(0,1)\quad\text{for}\quad i=1,\ldots,1000,$$

$$d_j\sim N(0,2)\quad\text{for}\quad j=1,\ldots,5,$$

$$a_j\sim N^{+}(1,2)\quad\text{for}\quad j=1,\ldots,5,$$

and

$$c_j\sim\text{Beta}(5,17)\quad\text{for}\quad j=1,\ldots,5.$$

As noted in Section 11.2.2, the prior distribution for the latent variables is sufficient to resolve the location and metric indeterminacies in the latent variable. Note the restriction in the truncated normal distribution ensures that each $a_j > 0$, embodying the assumption that the probability of a correct response should be monotonically increasing with proficiency. This restriction is sufficient to resolve the rotational indeterminacy in the latent variable. The prior distributions for the d and a parameters are relatively diffuse. We discuss the particular choice of beta prior distribution for the c parameters in more detail in Section 11.2.6. A more general discussion of specifying the forms and parameters of prior distributions is given in Chapter 3, and Section 11.7 discusses several alternatives popular in IRT.

WinBUGS code for the model and list statements for three sets of initial values for the measurement model parameters are given as follows.

```
-------------------------------------------------------------------------
#########################################################################
# Model Syntax
#########################################################################
model{

#########################################################################
# Specify the item response measurement model for the observables
#########################################################################
for (i in 1:n){
  for(j in 1:J){
    P[i,j] <- c[j]+(1-c[j])*phi(a[j]*theta[i]+d[j])
    x[i,j] ~ dbern(P[i,j])
  }
}
```

```
##########################################################################
# Specify the prior distribution for the latent variables
##########################################################################
for (i in 1:n){
  theta[i] ~ dnorm(0, 1)
}
##########################################################################
# Specify the prior distribution for the measurement model parameters
##########################################################################
for(j in 1:J){
  d[j] ~ dnorm(0, .5)
  a[j] ~ dnorm(1, .5) I(0,)
  c[j] ~ dbeta(5,17)
}

} # closes the model

##########################################################################
# Initial values for three chains
##########################################################################
list(d=c(3, 3, 3, 3, 3), a=c(.1, .1, .1, .1, .1), c=c(.05, .05, .05,
.05, .05))

list(d=c(-3, -3, -3, -3, -3), a=c(3, 3, 3, 3, 3), c=c(.5, .5, .5,
.5, .5))

list(d=c(1, 1, 1, 1, 1), a=c(1, 1, 1, 1, 1), c=c(.2, .2, .2, .2, .2))
--------------------------------------------------------------------------
```

Note that WinBUGS uses the precision as the second argument of the normal distribution. Further, the `I(0,)` in the line specifying the prior for the a parameters enacts that the restriction that the a parameters are positive. The initial values for the measurement model parameters were chosen to represent what we anticipate to be fairly dispersed values for the parameters in the posterior distribution. WinBUGS was used to generate initial values for the latent variables. In this model, WinBUGS uses a slice sampler for the as and the cs, and the Metropolis sampler for the θs and the ds. To complete the specification of the normal proposal distributions, WinBUGS uses the first 4000 iterations as an "adaptive" phase to select an appropriate variance which it then uses for the remaining iterations. Iterations from this adaptive phase should be discarded. Upon running the analysis, there was evidence of convergence (see Section 5.7.2) by 4000 iterations. To be conservative, we discarded an additional 2000 iterations as burn-in.

11.2.5.2 Results

The chains exhibited high serial dependence, with nontrivial autocorrelations for several parameters, particularly those associated with observable (item) 3. To mitigate these effects, we ran each chain for 20,000 iterations after burn-in, yielding 60,000 iterations for use in summarizing the posterior distribution. The marginal posterior distributions were unimodal and mostly symmetric, with several of them exhibiting some skewness. Accordingly, we report the median in addition to the mean as summaries of the central tendency of the marginal posterior distributions for the measurement model parameters, and, as an example, one of the examinee's latent variables in Table 11.2. The location (d) parameters indicate

TABLE 11.2

Summary of the Marginal Posterior Distributions for the Measurement Model Parameters and One Examinee's Latent Variable (θ_{1000}) for the Three Parameter Normal Ogive (3-PNO) Model of the Law School Admissions Test Data

Parameter	Mean	Median	Standard Deviation	95% Highest Posterior Density Interval
d_1	1.41	1.40	0.14	(1.13, 1.68)
d_2	0.28	0.33	0.24	(–0.20, 0.60)
d_3	–0.50	–0.38	0.46	(–1.44, 0.09)
d_4	0.51	0.53	0.16	(0.21, 0.77)
d_5	1.03	1.03	0.13	(0.77, 1.28)
a_1	0.46	0.44	0.18	(0.13, 0.81)
a_2	0.65	0.57	0.35	(0.22, 1.26)
a_3	1.11	0.89	0.65	(0.30, 2.51)
a_4	0.50	0.48	0.18	(0.20, 0.85)
a_5	0.44	0.42	0.17	(0.13, 0.76)
c_1	0.24	0.23	0.09	(0.08, 0.42)
c_2	0.26	0.25	0.10	(0.08, 0.47)
c_3	0.26	0.26	0.08	(0.10, 0.41)
c_4	0.25	0.24	0.10	(0.08, 0.44)
c_5	0.25	0.24	0.09	(0.07, 0.43)
θ_{1000}	0.70	0.69	0.87	(–1.08, 2.35)

that items are fairly easy for these examinees. The hardest item is item 3, which is the only item with most of the HPD interval for the location parameter that is mostly negative. It is also the most discriminating item. The posterior standard deviations and HPD intervals indicate that there is considerable uncertainty about many of the parameters. A related point is that the marginal posteriors for the c parameters are not too different from the Beta(5, 17) prior distribution (rounding to two decimal places: mean = .23, standard deviation = .09, 95% highest density interval of (.07, .40)), suggesting that the data are not very informative about the lower asymptotes, or at least not strongly contradictory to the information in the prior. This is consistent with the items being relatively easy for these examinees.

The last row in Table 11.2 and Figure 11.3 summarizes the marginal posterior distribution for the latent variable for an examinee who correctly answered all the questions. For sake of comparison, Figure 11.3 also depicts the prior distribution and the likelihood function evaluated using the posterior medians of the measurement model parameters. The likelihood function does not have a finite maximum; it increases as $\theta \to \infty$. Conceptually, ever higher values of θ represent ever better accounts of this examinee, where "better" is interpreted in terms of the information in the data as expressed by the likelihood function. However, as Figure 11.3 depicts, the marginal posterior distribution does have a maximum, right around a value of .7. Conceptually, ever higher values of θ *do not* represent ever better accounts of this examinee, where "better" is now interpreted in light of the information in the data as expressed by the likelihood function *and* the information expressed in the prior distribution. On the basis of the posterior distribution, we conclude that the value of θ for this examinee is likely near .7, and almost certainly not greater than 3. Despite their responding correctly to all five of the items, we are pretty sure that their proficiency is not infinite. We suspect analysts using ML would have the same beliefs, and then would face

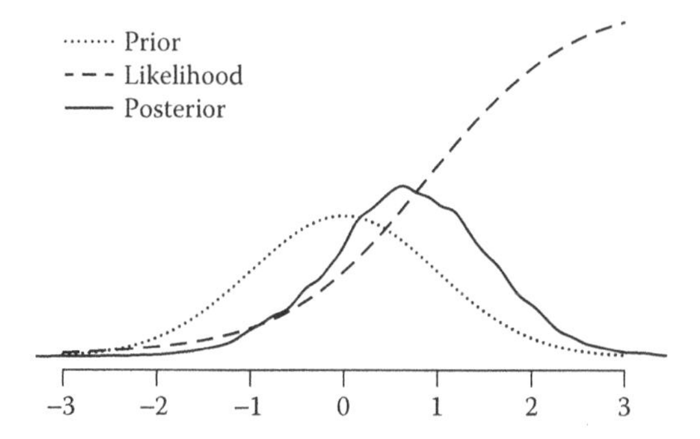

FIGURE 11.3
Marginal posterior distribution (solid line), prior distribution (dotted line), and likelihood function evaluated using the posterior medians for the measurement model parameters (dashed line) for the latent variable for an examinee with all correct responses to the items in the Law School Admissions Test example.

the prospect of reconciling the disparity between their beliefs and the MLE through some post hoc adjustment. An advantage of a Bayesian approach is that the fully probabilistic framework affords a tighter integration among our beliefs, evidence, and statistical model of them. We gain by implementing our beliefs in a statistical framework that includes formal, theory-based, model-checking, and sensitivity tools.

11.2.5.3 Model-Data Fit

The adequacy of the model-data fit may be assessed in a number of ways. Here we focus on a few aspects of model-data fit common to applications of IRT in educational assessment. In the following examples, we use 15,000 iterations from MCMC (5,000 from each of three chains) to conduct PPMC.

We begin by checking the distribution of raw or total scores obtained by summing the scores on the items for each examinee (Hambleton & Han, 2005; Sinharay, 2005). Figure 11.4 compares the frequencies of raw scores in the observed data to the frequencies obtained in the posterior predicted data. To compactly present the latter, a boxplot depicting the interquartile range (the box), median (notch in the middle of the box), and whiskers extending to the 2.5th and 97.5th percentiles graphically represents the posterior predictive distribution of the number of

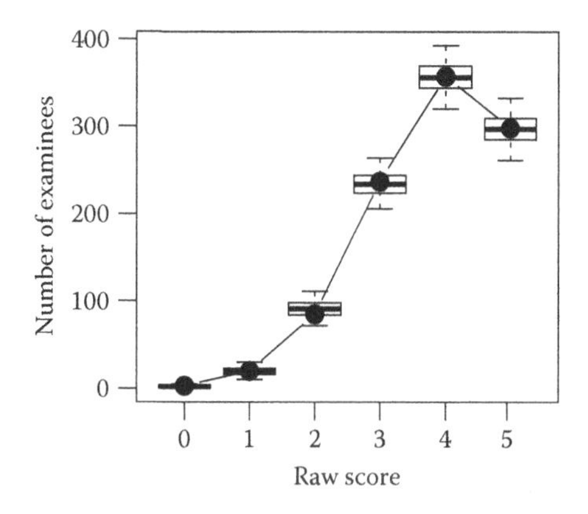

FIGURE 11.4
Observed and posterior predicted raw score distributions for the Law School Admissions Test example. At each raw score, the box depicts the interquartile range, the notch in the middle depicts the median, and the whiskers depict the 2.5th and 97.5th percentiles. The points indicate the frequency in the observed data.

examinees with that particular raw score. The points indicate the number of examinees with that particular raw score in the observed data; line segments connect the points as a visual aid.

We employ a similar representation in Figure 11.5 to investigate the adequacy of item fit (Sinharay, 2006a). Here, a boxplot depicts the posterior predictive distribution of the proportion of examinees that respond correctly to each item at each raw score. Observed proportions are plotted as points, with line segments connecting them. For both the distribution of raw scores and the proportion correct given the raw score, the observed values are well within the posterior predictive distributions, supporting the notion that the model fits adequately in terms of accounting for the distributions of individual items and raw scores.

Two important and related aspects of model-data fit in IRT concern the assumptions of local independence and dimensionality, the latter of which is typically framed as whether the assumed number of latent variables is adequate. To pursue possibilities of local dependence, we employ the standardized model-based covariance (SMBC; Levy et al., 2015); for the pairings of observables (items) j and j',

$$\text{SMBC}_{jj'} = \frac{\dfrac{\sum_{i=1}^{n}(x_{ij} - E(x_{ij} \mid \theta_i, \omega_j))(x_{ij'} - E(x_{ij'} \mid \theta_i, \omega_{j'}))}{n}}{\sqrt{\dfrac{\sum_{i=1}^{n}(x_{ij} - E(x_{ij} \mid \theta_i, \omega_j))^2}{n}}\sqrt{\dfrac{\sum_{i=1}^{n}(x_{ij'} - E(x_{ij'} \mid \theta_i, \omega_{j'}))^2}{n}}},$$

where $E(x_{ij} \mid \theta_i, \omega_j)$ is the conditional expectation of the value of the observable, given by the item response function. In the current context, this is the 3-PNO, and correspondingly $\theta_i = \theta_i$ and $\omega_j = (d_j, a_j, c_j)$. SMBC may be interpreted as a conditional correlation among the observables. To evaluate the adequacy of the assumed unidimensionality of the model, we employ the standardized generalized discrepancy measure (SGDDM, Levy et al., 2015) that is the mean of the absolute values of SMBC over unique observable pairs,

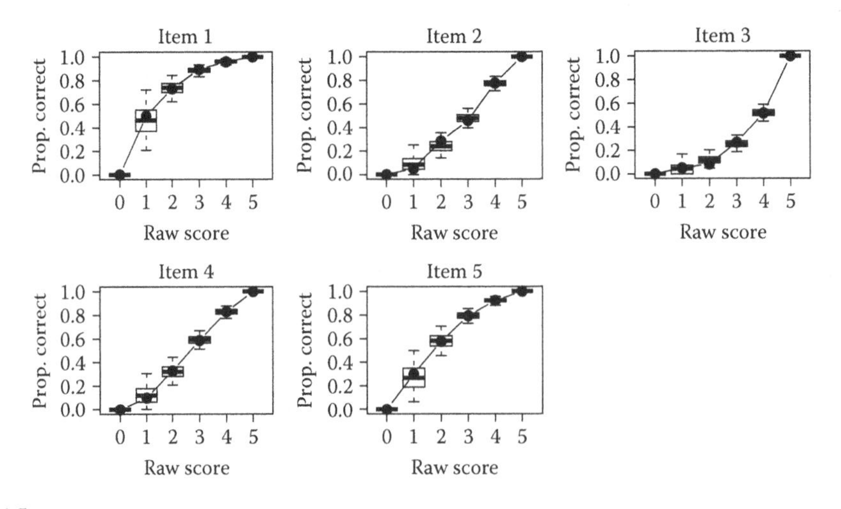

FIGURE 11.5
Item fit plots for the Law School Admissions Test example with observed and posterior predicted proportion correct by raw score. At each raw score, the box depicts the interquartile range, the notch in the middle depicts the median, and the whiskers depict the 2.5th and 97.5th percentiles. The points indicate the frequency in the observed data.

$$\text{SGDDM} = \frac{\sum_{j>j'} |\text{SMBC}_{jj'}|}{J(J-1)/2},$$

and is interpreted as the average magnitude of the conditional correlations among the observables. Scatterplots of the realized and posterior predicted values for SGDDM and SMBC are given in Figures 11.6 and 11.7. The results for SGDDM indicate that, overall, the associations among the observables are well accounted for by the model. The realized values are small and commensurate with the posterior predicted values from the model. The scatterplots for SMBC reveal much the same at the level of the observable-pairs, with realized conditional associations varying around 0 in ways consistent with the posterior predicted values.

The overall finding that the model fits well with respect to the aspects investigated here is unsurprising given the complexity of the 3-PNO and the small number of items.

11.2.6 Rationale for Prior for Lower Asymptote

This section is a case study illustrating the principles analysts may use to specify prior distributions, and how they can be advantageous, illuminated in the context of the specification of a Beta(5,17) prior distribution for the lower asymptote (c) parameters in the analysis of the LSAT data in previous section. Consider first the choice to use of the family of beta distributions. A c parameter is interpreted as a probability, namely the probability that an examinee with a low value of proficiency (as captured by θ) will correctly respond to the item. Accordingly, c parameters are bounded below by 0 and above 1. The beta distribution has its density over [0,1], and is flexible to take on any of several shapes as may be desired (see Figure 2.3). The parameters of the beta distribution may also be naturally interpreted in terms of the prior successes and failures for a corresponding probability. Furthermore, recall that the beta distribution is a conjugate prior for an unknown probability in simple Bernoulli and binomial models. Although not conjugate in the current context, the use of a beta prior does offer conditional conjugacy for data-augmented samplers (see Section 11.8).

A number of rationales may justify the use of a particular beta distribution, in our case the Beta(5,17). One is theoretical consideration. The parameter in question represents the probability of an examinee of extremely low proficiency responding correctly to a multiple-choice question, say, by guessing. A theoretical analysis might reason that an examinee

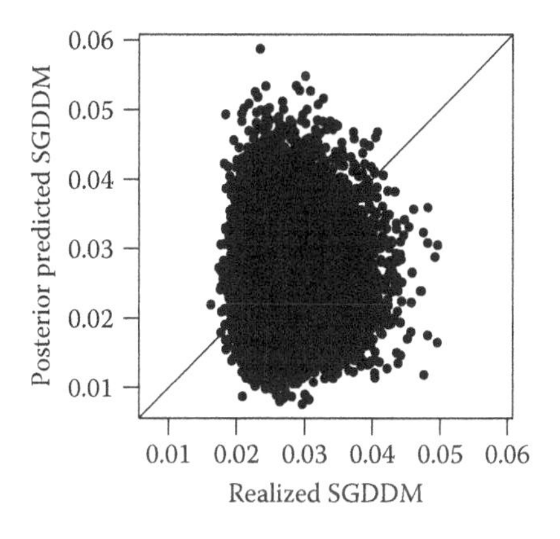

FIGURE 11.6
Scatterplot of the realized and posterior predicted values of the standardized generalized dimensionality discrepancy measure (SGDDM) for the Law School Admissions Test example with the unit line.

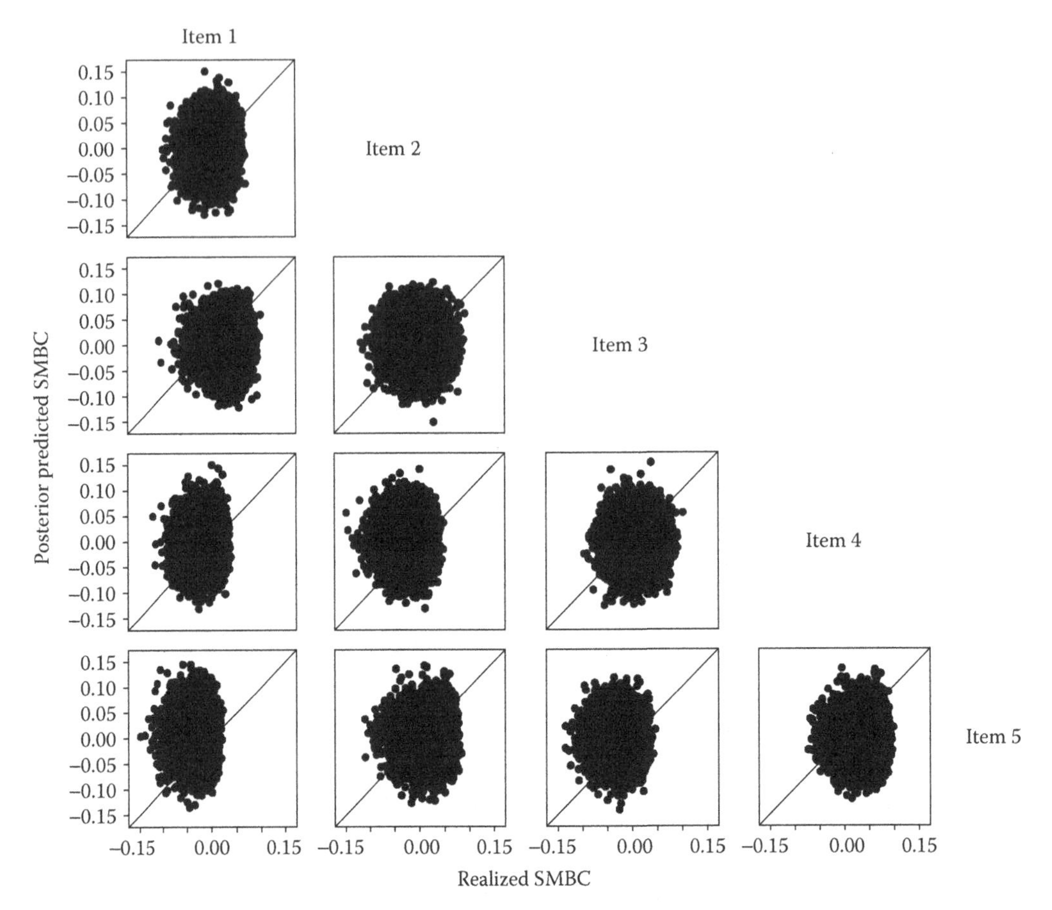

FIGURE 11.7
Scatterplots of realized and posterior predicted values of the standardized model-based covariance (SMBC) for item-pairs from the Law School Admissions Test example, with the unit line.

guessing on a question that has five options would be correct with probability .2, and then build a beta distribution around that anchoring idea. A Beta(5,17) may be interpreted as akin to having seen $5 - 1 = 4$ correct guesses in $5 + 17 - 2 = 20$ attempts, expressing a belief that the probability should be about .2, and weighting that belief as if it was based on 20 observations. It is fair to say that such reasoning could be accomplished even without high levels of familiarity with the features of the assessment situation such as examinee capabilities, item properties, or history of the assessment.

A related strategy *does* capitalize on understanding these issues, usually in form of quantifying beliefs of subject matter experts who have knowledge about the situation that is being modeled. Strategies for eliciting and quantifying such expertise are an emerging issue; references to examples in the broader Bayesian literature were given in Section 3.7. Almond (2010); Almond, Mislevy, Steinberg, Yan, and Williamson (2015); Jiang et al. (2014); Novick and Jackson (1974); and Tsutakawa (1992) presented conceptual and computational strategies for eliciting and encoding information to aid in specifying prior distributions for measurement model parameters in psychometric models.

A different strategy involves past research. If items like the ones in question that have been calibrated have tended to yield certain c values, then the items in question now will likely yield similar c values. In the context of our current example, if analyses of response data of previous items from section 6 of the LSAT had yielded c values around .2, that would be justification for using a Beta(5,17) prior. Note that there is an exchangeability assumption lurking in saying that the current items are "like" other items. Indeed, there are several exchangeability assumptions lurking, for instance about the "other" examinees that took those "other" items, and the context of those interactions such as the stakes associated. It is an open question just how similar other situations should be to be relevant for the current one. Or to put it another way, it is not always obvious when differences make a difference. Should we consider the results for these examinees taking other items that have appeared on section 6 of the LSAT? Other examinees taking the items of current interest? Other examinees taking other items? What about other sections of the LSAT, or other assessments? The answer to these questions is that there is no right answer. These are judgments and decisions that need to be made by the analyst, just like judgments and decisions that are made in specifying other aspects of the model.

These strategies may be combined. If there are five response options for an item, reason and subject matter expertise usually suggest that the chances of guessing the correct one are probably around .2. If other items for which we have estimates of the lower asymptote yielded values between .1 and .4, then the lower asymptote for these items is probably between .1 and .4. These considerations suggest specifying a prior that places higher probabilities for relatively lower values. A Beta(5,17) distribution captures the thrust of these rationales, giving them weight akin to having seen four correct guesses in 20 attempts.

11.3 Conventional IRT Models for Polytomous Observables

A number of IRT models have been proposed for modeling polytomous observables, including those that assume the categories of the observable are nominal (Bock, 1972) or ordered, including the rating scale model (Andrich, 1978), partial credit model (Masters, 1982), generalized partial credit model (Muraki, 1992), and the graded response model (GRM) (Samejima, 1969, 1997). For ordered data, models with comparable parameters that use different parametric forms typically exhibit comparable fit and are equally appropriate (Maydeu-Olivares, Drasgow, & Mead, 1994). Accordingly, in the current development we focus on one such model, namely a version of the GRM for ordered categories (Samejima, 1969, 1997).

To state the model, let x_{ij} refer to the observable variable corresponding to the response from examinee i to item j that can take on integer values that range from 1 to K_j, where K_j denotes the number of possible response categories for observable j. The choice to encode the response categories as integers from $1, \ldots, K_j$ is arbitrary, but reflects how polytomously scored items are often framed (e.g., "rate the following from 1 to 5") and also simplifies the coding of the model in WinBUGS. The GRM specifies the conditional probability for the value from examinee i on observable j being in category k as the difference between the conditional probability of the value being at or above a particular category k and the conditional probability of the value being at or above the next category $(k + 1)$,

$$P(x_{ij} = k \mid \theta_i, \mathbf{d}_j, a_j) = P(x_{ij} \geq k \mid \theta_i, d_{jk}, a_j) - P(x_{ij} \geq k+1 \mid \theta_i, d_{j(k+1)}, a_j), \tag{11.27}$$

where $\mathbf{d}_j = (d_{j1}, \ldots, d_{jK_j})$ is a collection of location parameters and $P(x_{ij} \geq k \mid \theta_i, d_{jk}, a_j)$ is the conditional probability of the value being at or above a particular category k. This is given by a 2-P structure,

$$P(x_{ij} \geq k \mid \theta_i, d_{jk}, a_j) = F(a_j\theta_i + d_{jk}). \tag{11.28}$$

For each observable, the model consists of expressions in (11.27) for $k = 1, \ldots, K_j - 1$. For the highest response category, $P(x_{ij} = K_j \mid \theta_i, \mathbf{d}_j, a_j) = P(x_{ij} \geq K_j \mid \theta_i, d_{jK_j}, a_j)$. Conceptually, the subtraction in (11.27) is not needed, as the probability of a value being at or above a (hypothetical, and in some sense counterfactual) category *above* the highest category is tautologically equal to 0. Similarly, the probability of a value being at or above the lowest category is tautologically equal to 1, which may be seen as setting $d_{j1} = \infty$. Non-negativity in the response probabilities is preserved by restricting $d_{jk} \geq d_{j(k+1)}$.

Much of what was described for conventional modeling of dichotomous observables holds for polytomous observables. With these specifications, no additional indeterminacies in the latent variable are introduced* and the conventional approaches to calibration and scoring are similarly applicable save that the conditional probability for the value of an observable in (11.6) is

$$P(x_{ij} \mid \theta_i, \omega_j) = \prod_{k=1}^{K_j} P(x_{ij} = k \mid \theta_i, \omega_j)^{I(x_{ij}=k)}, \tag{11.29}$$

where I is the indicator function that here takes on a value of 1 when the response from examinee i for observable j is k, and 0 otherwise.

11.4 Bayesian Modeling of IRT Models for Polytomous Observables

11.4.1 Conditional Probability of the Observables

The preceding descriptions are sufficient to describe the conditional probability of the data, albeit in a somewhat equation-oriented way. We formulate the conditional probability of the data in a compact manner here in a more distribution-oriented way, drawing on the equations as needed. The conditional distribution of the data is

$$p(\mathbf{x} \mid \boldsymbol{\theta}, \boldsymbol{\omega}) = \prod_{i=1}^{n} p(\mathbf{x}_i \mid \theta_i, \boldsymbol{\omega}) = \prod_{i=1}^{n} \prod_{j=1}^{J} p(x_{ij} \mid \theta_i, \omega_j), \tag{11.30}$$

where each x_{ij} is specified as a categorical random variable (a generalization of the Bernoulli),

$$x_{ij} \mid \theta_i, \omega_j \sim \text{Categorical}(\mathbf{P}(x_{ij} \mid \theta_i, \omega_j)), \tag{11.31}$$

$\mathbf{P}(x_{ij} \mid \theta_i, \omega_j) = (P(x_{ij} = 1 \mid \theta_i, \omega_j), \ldots, P(x_{ij} = K_j \mid \theta_i, \omega_j))$ are the collection of probabilities for the K_j possible values for observable j, and ω_j are the associated measurement model (item) parameters. In the GRM, the $P(x_{ij} = k \mid \theta_i, \omega_j)$ are given by (11.27) which involve (11.28), and $\omega_j = (\mathbf{d}_j, a_j)$.

* For certain polytomous IRT models, additional indeterminacies are present when considering models with multiple latent variables (Section 11.5), though we do not discuss them here (see Reckase, 2009).

11.4.2 A Prior Distribution

As with the dichotomous IRT models, we specify the prior distribution first by assuming a priori independence between the examinees' latent variables and the measurement model parameters. Once again assuming exchangeability and normality yields the prior structure for examinees latent variables expressed in (11.12) and (11.13).

Turning to the measurement model parameters, we again assume (a) exchangeability and the specification of a common prior for all observables, as expressed in (11.14), (b) a priori independence between the discrimination and location parameters, which in the current context is

$$p(\omega_j) = p(\mathbf{d}_j)p(a_j), \tag{11.32}$$

and (c) a truncated normal prior distribution for the discriminations, as expressed in (11.17).

Thus, the lone difference from the case of dichotomous observables is the specification of the now multiple location parameters for each observable. Again, we will employ normal distributions for the location parameters, though the key issue here is imposing the restriction that $d_{jk} \geq d_{j(k+1)}$.

Recalling that the probability of a value being at or above the lowest category is equal to 1 may be seen as setting $d_{j1} = \infty$, we begin with the prior for d_{j2} and adopt a normal prior as in the case of dichotomous data,

$$d_{j2} \sim N(\mu_{d_2}, \sigma^2_{d_2}), \tag{11.33}$$

where μ_{d_2} and $\sigma^2_{d_2}$ are hyperparameters specified by the analyst.

Several strategies for the remaining location parameters are possible. Here, we specify the remaining location parameters via truncated normal prior distributions, where the truncation bounds the location parameter to be less than the location parameter for the preceding category. A bit more formally,

$$d_{jk} \sim N^{<d_{j(k-1)}}(\mu_{d_k}, \sigma^2_{d_k}) \quad \text{for } k = 3, \ldots, K_j, \tag{11.34}$$

where $N^{<d_{j(k-1)}}$ denotes the normal distribution truncated to be less than $d_{j(k-1)}$. Note that if the "–b" parameterization is used, the model would restrict each location parameter to be greater than the location parameter for the preceding category (e.g., Zhu & Stone, 2011).

11.4.3 Posterior Distribution and Graphical Model

The DAG for the model adopting this prior is given in Figure 11.8 for the case where all observables have $K_j = 5$ categories. Note the directed edges among the d parameters, reflecting that d_{j3} depends on d_{j2}, d_{j4} depends on d_{j3}, and d_{j5} depends on d_{j4}, as implied by the truncation in (11.34).

The posterior distribution for all the unknowns is

$$p(\theta, \mathbf{d}, \mathbf{a} \mid \mathbf{x}) \propto \prod_{i=1}^{n} \prod_{j=1}^{J} p(x_{ij} \mid \theta_i, \mathbf{d}_j, a_j) p(\theta_i) p(a_j) \prod_{k=2}^{K_j} p(d_{jk}), \tag{11.35}$$

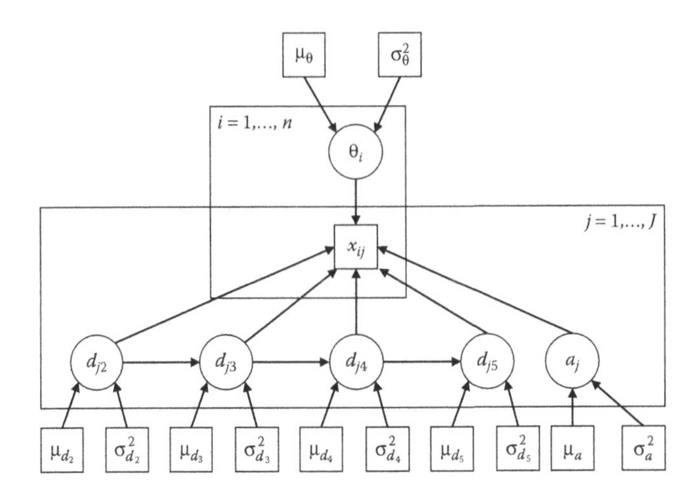

FIGURE 11.8
Directed acyclic graph for the graded response model.

where

$$x_{ij} \mid \theta_i, \mathbf{d}_j, a_j \sim \text{Categorical}(\mathbf{P}(x_{ij} \mid \theta_i, \mathbf{d}_j, a_j)) \quad \text{for } i = 1,\ldots,n, j = 1,\ldots,J,$$

$$\mathbf{P}(x_{ij} \mid \theta_i, \mathbf{d}_j, a_j) = (P(x_{ij} = 1 \mid \theta_i, \mathbf{d}_j, a_j),\ldots,P(x_{ij} = K_j \mid \theta_i, \mathbf{d}_j, a_j)) \quad \text{for } i = 1,\ldots,n, j = 1,\ldots,J,$$

$$P(x_{ij} = k \mid \theta_i, \mathbf{d}_j, a_j) = P(x_{ij} \geq k \mid \theta_i, d_{jk}, a_j) - P(x_{ij} \geq k+1 \mid \theta_i, d_{j(k+1)}, a_j),$$

$$\text{for } i = 1,\ldots,n, j = 1,\ldots,J, k = 1,\ldots,K_j - 1,$$

$$P(x_{ij} = K_j \mid \theta_i, \mathbf{d}_j, a_j) = P(x_{ij} \geq K_j \mid \theta_i, d_{jK_j}, a_j) \quad \text{for } i = 1,\ldots,n, j = 1,\ldots,J,$$

$$P(x_{ij} \geq k \mid \theta_i, d_{jk}, a_j) = F(a_j\theta_i + d_{jk}) \quad \text{for } i = 1,\ldots,n, j = 1,\ldots,J, k = 2,\ldots,K_j,$$

$$P(x_{ij} \geq 1) = 1 \quad \text{for } i = 1,\ldots,n, j = 1,\ldots,J,$$

$$\theta_i \mid \mu_\theta, \sigma^2_\theta \sim N(\mu_\theta, \sigma^2_\theta) \quad \text{for } i = 1,\ldots,n,$$

$$a_j \mid \mu_a, \sigma^2_a \sim N^+(\mu_a, \sigma^2_a) \quad \text{for } j = 1,\ldots,J,$$

$$d_{j2} \sim N(\mu_{d_2}, \sigma^2_{d_2}) \quad \text{for } j = 1,\ldots,J,$$

and

$$d_{jk} \sim N^{<d_{j(k-1)}}(\mu_{d_k}, \sigma^2_{d_k}) \quad \textit{for } j = 1,\ldots,J, k = 3,\ldots,K_j.$$

To conduct MCMC estimation of the posterior distribution the general structure of the Metropolis–Hastings-within-Gibbs approach outlined in Section 11.2.4 may be applied. The key differences are that here we have multiple location parameters per observable, and no lower asymptote. Alternatively, we might employ a sampler that capitalizes on latent response variable formulations outlined in Section 11.8 (see Fox, 2010, who discusses such an approach using bounded uniform priors for the location parameters). Finally, we note that WinBUGS uses an adaptive rejection sampler for this model (Gilks & Wild, 1992).

11.4.4 Example: Peer Interaction Items

To illustrate these procedures, we conduct an analysis based on response data from 500 examinees to the seven items in the *Peer Interaction* subscale of the Institutional Integration Scale described in Section 9.3. All of the responses were scored as integers from 1 to 5, corresponding to increasing agreement with the statement, interpreted as indicative of more positive interactions with peers.

11.4.4.1 Model Specification and Fitting in WinBUGS

In our experience, WinBUGS tends to run into more computational problems with normal-ogive IRT models than logistic IRT models, particularly as model complexity increases. We employ a logistic version of the GRM with prior specifications developed in the previous sections. For completeness, we write out the posterior distribution for the model with the chosen values for the hyperparameters:

$$p(\theta,\mathbf{d},\mathbf{a}\mid\mathbf{x})\propto\prod_{i=1}^{n}\prod_{j=1}^{J}p(x_{ij}\mid\theta_i,\mathbf{d}_j,a_j)p(\theta_i)p(a_j)\prod_{k=2}^{K_j}p(d_{jk}),$$

where

$$x_{ij}\mid\theta_i,\mathbf{d}_j,a_j\sim\text{Categorical}(\mathbf{P}(x_{ij}\mid\theta_i,\mathbf{d}_j,a_j))\quad\text{for } i=1,\ldots,500, j=1,\ldots,7,$$

$$\mathbf{P}(x_{ij}\mid\theta_i,\mathbf{d}_j,a_j)=(P(x_{ij}=1\mid\theta_i,\mathbf{d}_j,a_j),\ldots,P(x_{ij}=5\mid\theta_i,\mathbf{d}_j,a_j))\quad\text{for } i=1,\ldots,500, j=1,\ldots,7,$$

$$P(x_{ij}=k\mid\theta_i,\mathbf{d}_j,a_j)=P(x_{ij}\geq k\mid\theta_i,d_{jk},a_j)-P(x_{ij}\geq k+1\mid\theta_i,d_{j(k+1)},a_j)\quad\text{for } i=1,\ldots,500, j=1,\ldots,7, k=1,\ldots,4,$$

$$P(x_{ij}=5\mid\theta_i,\mathbf{d}_j,a_j)=P(x_{ij}\geq 5\mid\theta_i,d_{j5},a_j)\quad\text{for } i=1,\ldots,500, j=1,\ldots,7,$$

$$P(x_{ij}\geq k\mid\theta_i,d_{jk},a_j)=\frac{\exp(a_j\theta_i+d_{jk})}{1+\exp(a_j\theta_i+d_{jk})}\quad\text{for } i=1,\ldots,500, j=1,\ldots,7, k=2,\ldots,5,$$

$$P(x_{ij}\geq 1)=1\quad\text{for } i=1,\ldots,500, j=1,\ldots,7,$$

$$\theta_i\mid\mu_\theta,\sigma_\theta^2\sim N(0,1)\quad\text{for } i=1,\ldots,500,$$

$$a_j\mid\mu_a,\sigma_a^2\sim N^{+}(0,2)\quad\text{for } j=1,\ldots,7,$$

$$d_{j2}\sim N(2,2)\quad\text{for } j=1,\ldots,7,$$

$$d_{j3}\sim N^{<d_{j2}}(1,2)\quad\text{for } j=1,\ldots,7,$$

$$d_{j4}\sim N^{<d_{j3}}(-1,2)\quad\text{for } j=1,\ldots,7,$$

and

$$d_{j5}\sim N^{<d_{j4}}(-2,2)\quad\text{for } j=1,\ldots,7.$$

WinBUGS code for the model and a list statement for three sets of initial values for the measurement model parameters are given as follows.

```
-------------------------------------------------------------------------
#########################################################################
# Model Syntax
#########################################################################
model{

#########################################################################
# Specify the item response measurement model for the observables
#########################################################################
for (i in 1:n){
  for(j in 1:J){
```

```
    ########################################################################
    # Specify the probabilities of a value being greater than or
    # equal to each category
    ########################################################################
    for(k in 2:(K[j])){

      P.gte[i,j,k] <-
      exp(a[j]*theta[i]+d[j,k])/(1+exp(a[j]*theta[i]+d[j,k]))
    }
    P.gte[i,j,1] <- 1

    ########################################################################
    # Specify the probabilities of a value being equal to each
    # category
    ########################################################################
    for(k in 1:(K[j]-1)){
      P[i,j,k] <- P.gte[i,j,k]-p.gte[i,j,k+1]
    }
    P[i,j,K[j]] <- P.gte[i,j,K[j]]

    ########################################################################
    # Specify the distribution for each observable
    ########################################################################
    x[i,j] ~ dcat(P[i,j,1:K[j]])
  }
}


############################################################################
# Specify the prior distribution for the latent variables
############################################################################
for (i in 1:n){
  theta[i] ~ dnorm(0, 1)
}

############################################################################
# Specify the prior distribution for the measurement model parameters
############################################################################
for(j in 1:J){
  d[j,2] ~ dnorm(2, .5)
  d[j,3] ~ dnorm(1, .5) I(d[j,4],d[j,2])
  d[j,4] ~ dnorm(-1, .5) I(d[j,5],d[j,3])
  d[j,5] ~ dnorm(-2, .5) I( ,d[j,4])
  a[j] ~ dnorm(1, .5) I(0,)
}

} # closes the model


############################################################################
# Initial values for three chains
############################################################################
list(d= structure(.Data= c(
NA, 3, 1, 0, -1,
```

```
NA, 3, 1, 0, -1,
NA, 3, 1, 0, -1,
NA, 3, 1, 0, -1,
NA, 3, 1, 0, -1,
NA, 3, 1, 0, -1,
NA, 3, 1, 0, -1), .Dim=c(7, 5)),
a=c(.1, .1, .1, .1, .1, .1, .1))

list(d= structure(.Data= c(
NA, 2, 0, -1, -2,
NA, 2, 0, -1, -2,
NA, 2, 0, -1, -2,
NA, 2, 0, -1, -2,
NA, 2, 0, -1, -2,
NA, 2, 0, -1, -2,
NA, 2, 0, -1, -2), .Dim=c(7, 5)),
a=c(3, 3, 3, 3, 3, 3, 3))

list(d= structure(.Data= c(
NA, 1, -1, -2, -3,
NA, 1, -1, -2, -3,
NA, 1, -1, -2, -3,
NA, 1, -1, -2, -3,
NA, 1, -1, -2, -3,
NA, 1, -1, -2, -3,
NA, 1, -1, -2, -3), .Dim=c(7, 5)),
a=c(1, 1, 1, 1, 1, 1, 1))
-----------------------------------------------------------------------
```

The line for `d[j,5]` expresses the prior distribution for the d_{j5}. The use of the `I( ,d[j,4])` imposes the boundary that $d_{j5} < d_{j4}$. Note that there is no lower bound here. In the line for `d[j,4]`, the `I(d[j,5],d[j,3])` imposes the boundary. In particular, the second argument imposes the boundary that $d_{j4} < d_{j3}$, in accordance with the specification of the model and Figure 11.8. Though conceptually this is sufficient, WinBUGS also requires the first argument as a lower bound to honor to the constraint that $d_{j5} < d_{j4}$, even though that is imposed in the line for `d[j,5]` as just discussed. The line for `d[j,3]` is structured similarly.

The list statements with the initial values also deserve some comment. The d parameters are arranged as a matrix of $J = 7$ rows and $K = 5$ columns. For each observable (row), the first entry is "NA," as the d_{j1} parameters are not really in the model (recall, they may be conceived of as all being infinite). The remaining entries for each observable is numeric, giving the initial value for $d_{j2},\ldots, d_{j5}$.

11.4.4.2 Results

The model was run with three chains as just described using WinBUGS to generate initial values for the latent variables. There is evidence of convergence (see Section 5.7.2) within a few hundred iterations. To be conservative, we discarded 1,000 iterations as burn-in, and ran each chain for 15,000 more iterations after burn-in, yielding 45,000 iterations for use in summarizing the posterior distribution. The marginal posterior densities were unimodal and fairly symmetric. Summary statistics for these marginal posterior distributions are reported in Table 11.3, and may be used to characterize the results. For each item, the location parameters are fairly well spread out over the latent continuum. And for each item,

TABLE 11.3

Summary of the Marginal Posterior Distribution for the Measurement Model Parameters from the Logistic Graded Response Model of the *Peer Interaction* Data

Parameter	Mean	SD[a]	95% HPD[b] Interval	Parameter	Mean	SD[a]	95% HPD[b] Interval
a_1	2.15	0.16	(1.83, 2.48)	d_{42}	6.41	0.45	(5.54, 7.28)
a_2	3.53	0.27	(3.01, 4.06)	d_{43}	2.75	0.25	(2.26, 3.26)
a_3	3.97	0.30	(3.39, 4.57)	d_{44}	−0.74	0.21	(−1.15, −0.34)
a_4	3.57	0.26	(3.07, 4.08)	d_{45}	−4.78	0.34	(−5.47, −4.11)
a_5	2.46	0.18	(2.11, 2.82)	d_{52}	4.82	0.32	(4.19, 5.45)
a_6	1.87	0.15	(1.59, 2.16)	d_{53}	2.73	0.21	(2.32, 3.15)
a_6	1.50	0.13	(1.26, 1.75)	d_{54}	−0.31	0.16	(−0.62, −0.00)
d_{12}	4.97	0.34	(4.32, 5.64)	d_{55}	−3.58	0.25	(−4.07, −3.10)
d_{13}	2.98	0.21	(2.57, 3.39)	d_{62}	3.86	0.25	(3.38, 4.37)
d_{14}	0.68	0.15	(0.39, 0.97)	d_{63}	1.98	0.17	(1.66, 2.31)
d_{15}	−2.49	0.19	(−2.87, −2.12)	d_{64}	−0.46	0.13	(−0.72, −0.19)
d_{22}	5.24	0.38	(4.52, 5.98)	d_{65}	−3.19	0.21	(−3.61, −2.78)
d_{23}	1.23	0.21	(0.82, 1.64)	d_{72}	4.30	0.29	(0.06, 4.88)
d_{24}	−1.72	0.22	(−2.16, −1.29)	d_{73}	2.47	0.17	(2.14, 2.80)
d_{25}	−5.09	0.36	(−5.81, −4.39)	d_{74}	0.29	0.12	(0.05, 0.52)
d_{32}	6.57	0.46	(5.66, 7.47)	d_{75}	−2.72	0.19	(−3.10, −2.36)
d_{33}	3.39	0.30	(2.83, 4.00)				
d_{34}	−1.25	0.23	(−1.71, −0.80)				
d_{35}	−5.52	0.42	(−6.33, −4.71)				

[a] SD = Standard Deviation.
[b] HPD = Highest Posterior Density.

the extreme location parameters (d_{j2} and d_{j5}) have higher posterior standard deviations than those in the middle, reflecting relatively less information in accordance with fewer examinees using the extreme response categories.

11.5 Multidimensional IRT Models

The models presented so far may be extended to include multiple latent variables for each examinee. The most popular of the multivariate or multidimensional IRT (MIRT) models formulates a linear combination of the latent variables akin to that in CFA. A 3P-MIRT model is given by

$$P(x_{ij} = 1 \mid \boldsymbol{\theta}_i, d_j, \mathbf{a}_j, c_j) = c_j + (1 - c_j)F(\mathbf{a}_j'\boldsymbol{\theta}_i + d_j), \tag{11.36}$$

where, in addition to the terms defined previously, $\boldsymbol{\theta}_i = (\theta_{i1}, \ldots, \theta_{iM})'$ is a vector of M latent variables that characterize examinee i and $\mathbf{a}_j = (a_{j1}, \ldots, a_{jM})'$ is a vector of M coefficients or discrimination parameters for observable j that capture the discriminating power of the associated examinee variables. As with unidimensional IRT models, popular choices for F include the logistic and normal distributions. As in was the case in CFA, confirmatory approaches to MIRT typically specify some observables as not being dependent on some of the latent variables. Commonly, many observables are modeled as being factorially simple in that they depend on only on one of the latent variables.

The expansion to multiple latent variables and multiple discrimination parameters per observable necessitates an expansion of the prior specifications for these unknowns. For the latent variables, the situation is akin to that of CFA with multiple latent variables, and we may specify a multivariate normal prior distribution,

$$p(\boldsymbol{\theta}) = \prod_{i=1}^{n} p(\boldsymbol{\theta}_i)$$

and (11.37)

$$\boldsymbol{\theta}_i \sim N(\boldsymbol{\mu}_\theta, \boldsymbol{\Sigma}_\theta),$$

where $\boldsymbol{\mu}_\theta$ and $\boldsymbol{\Sigma}_\theta$ are hyperparameters to be specified.

Conventional approaches to model estimation mimic those for unidimensional models (Bock, Gibbons, & Muraki, 1988; McDonald, 1999; Reckase, 2009; Wirth & Edwards, 2007) The MIRT model here is subject to the same indeterminacies as those discussed for CFA in Section 9.1.2 (see also Davey, Oshima, & Lee, 1996, and Jackman, 2001, for discussions in the context of MIRT). These may be addressed in the same ways, which may involve specifying values for $\boldsymbol{\mu}_\theta$ and $\boldsymbol{\Sigma}_\theta$. Alternatively, if $\boldsymbol{\mu}_\theta$ and $\boldsymbol{\Sigma}_\theta$ are treated as unknown they are assigned prior distributions, such as the conditionally conjugate forms that were introduced in the context of CFA,

$$p(\boldsymbol{\mu}_\theta) = \prod_{m=1}^{M} p(\mu_{\theta_m}),$$
$$\mu_{\theta_m} \sim N(\mu_{\mu_\theta}, \sigma^2_{\mu_\theta}), \tag{11.38}$$

and

$$\boldsymbol{\Sigma}_\theta \sim \text{Inv-Wishart}(\boldsymbol{\Sigma}_{\theta_0}, d). \tag{11.39}$$

For the multiple discrimination parameters, we may again draw upon the connection to the specifications introduced for the multiple loadings in CFA and specify normal distributions,

$$p(\mathbf{a}_j) = \prod_{m=1}^{M} p(a_{jm}),$$
$$a_{jm} \sim N^{+}(\mu_a, \sigma^2_a), \tag{11.40}$$

now with the positivity restriction as in the unidimensional IRT models. The 2P versions of this type of MIRT model have been extended to modeling polytomous observables. A multidimensional GRM model is specified by replacing (11.28) with

$$P(x_{ij} \geq k \mid \boldsymbol{\theta}_i, d_{jk}, \mathbf{a}_j) = F(\mathbf{a}_j' \boldsymbol{\theta}_i + d_{jk}). \tag{11.41}$$

The preceding MIRT models may be referred to as *compensatory* models, which reflects that high values on one latent variable may compensate for low values on other latent variables to yield a high probability of a correct response. *Conjunctive* MIRT models specify the response probability in such a way that low values on one latent variable cannot be compensated by high values of another; rather, high values on all latent variables are needed to yield a high probability of a correct response. This is typically operationalized via specifying the response function via a product of probabilities. Although we could employ the normal ogive form and/or the "+d" parameterizations, this is typically done using logistic versions of one-parameter structures with the "–b" parameterization (Embretson, 1984, 1997; Whitely, 1980),

$$P(x_{ij}=1 \mid \boldsymbol{\theta}_i, \mathbf{b}_j) = \prod_{m=1}^{M} \frac{\exp(\theta_{im} - b_{jm})}{1+\exp(\theta_{im} - b_{jm})}, \tag{11.42}$$

where $\mathbf{b}_j = (b_{j1}, \ldots, b_{jM})$ is the collection of M location parameters for observable j, and b_{jm} is interpreted as the difficulty parameter for component m of the item.

The expansion to multiple location parameters per observable calls for an expansion of the prior specifications for these unknowns. Following an assumption of exchangeability with respect to how difficult each item is along each dimension, we may assign a common prior to each of the location parameters. As in previous situations, normal distributions are a natural choice for these parameters,

$$b_{jm} \sim N(\mu_b, \sigma_b^2), \tag{11.43}$$

where μ_b and σ_b^2 are hyperparameters specified by the analyst.

11.6 Illustrative Applications

IRT has grown to become a dominant approach to psychometric modeling in educational assessment, and accordingly has been leveraged to accomplish a number of aims involved in this enterprise. In this section, we briefly review several assessment activities that are strongly aligned with the Bayesian perspective on reasoning and inference.

11.6.1 Computerized Adaptive Testing

In computerized adaptive testing (CAT; van der Linden & Glas, 2010; Wainer, Dorans, Flaugher, Green, & Mislevy, 2000), examinees are presented items that are tailored to them. When an examinee completes an item, the next item administered to them is selected or constructed on the fly. The idea is that we can take into account how the examinee is performing to select an item that is best for them, or at least better than having chosen the items before having observed their performance. CAT may be seen as an instance of the four process delivery architecture reviewed in Section 1.4.3. Somewhat casually, the processes in a CAT typically correspond to moving clockwise around Figure 1.7 beginning with the upper-left node as follows:

1. Select the best item to administer to the examinee based on beliefs about the examinee's proficiency.
2. Administer the item and capture the examinee's response/behavior (via computer).

3. Evaluate the examinee's response/behavior (say, for correctness of their response).
4. On the basis of their performance, update beliefs about the examinee's proficiency.
5. Repeat steps 1–5 until some stopping rule has been reached (e.g., beliefs about the examinee are sufficiently precise).

The notion of what constitutes the best item in step 1 is usually based on optimizing some criterion related to the amount of information we expect to gain from observing the examinee's performance on the item, possibly subject to various constraints. We do not focus on this aspect beyond noting that our choice for the next item to administer obviously depends on what we believe about the examinee. Loosely speaking, if we believe an examinee to be very proficient, we ought to select a more challenging item, while if we believe an examinee to not be very proficient, we ought to select a less challenging item. This reflects an intuition that not much is learned by observing a proficient examinee successfully complete an easy item or observing a low-proficiency examinee struggle with a difficult item, and this is indeed how things turn out when formalized.

We focus on how we revise beliefs about examinee proficiency in step 4. Here, we are reasoning from the evidence provided by the examinee's response. How are we to capture the evidentiary contribution of a response to our reasoning process of revising our beliefs about the examinee's proficiency? By expressing the problem as a probability model, such as IRT, we express the reasoning problem using the associated components of the model. Once the response to the item is observed, the item response function for the item plays the role of the likelihood function. If the response is correct, the likelihood is the item response function; suppressing the notation for the measurement model (item) parameters, $L(\theta; x=1) = P(x=1 \mid \theta)$. If the response is incorrect, the likelihood is one minus the item response function, $L(\theta; x=0) = 1 - P(x=1 \mid \theta)$. The item response function serves to express the evidentiary contribution of the response to that particular item. A Bayesian approach offers a natural solution here; the likelihood may be combined with a prior distribution for θ to yield an updated or posterior distribution for θ via Bayes' theorem. This distribution then serves as the prior distribution for the next cycle through steps 1–4. By repeatedly cycling through steps 1–4, we keep on updating our beliefs in light of the accruing evidence from the increasing number of observations. At any point in time, our distribution for θ is a posterior distribution relative to the responses we have observed, and is a prior distribution for the response we are about to observe.

Several other comments are worth making. First, this example highlights an important benefit of adopting a probability-based approach to evidentiary reasoning in assessment. We see that our psychometric models, be they IRT or some other variety, express the evidentiary contribution of the observation for our reasoning about the examinee. We formulate these reasoning structures in one direction—specifying observable behavior as conditional on unobservable parameters—only to reverse that direction when doing assessment and making inferences about the examinee. We may view our psychometric models as the avenues we take when conducting evidentiary reasoning from observations to beliefs about the examinee. For dichotomous models, the item response function indicates the neighborhood of θ values where the observation is most informative, which is surrounding b in the 1-P and 2-P models and slightly higher than b in the 3-P model. Sharply increasing item response functions characterized by relatively high values of a and low values of c offer more evidence than slowly increasing item response functions for making inferences about where the examinee is with respect to that neighborhood.

Second, Bayesian procedures have been developed for many of the processes involved in CAT (Lewis & Sheehan, 1990; Owen, 1969, 1975, Segall, 1996; van der Linden, 1998; van der

Linden & Pashley 2010) as well as supporting or follow-up investigations (Bradlow, Weiss, & Cho, 1998). As noted above, we can update the distribution for θ in step 4 via Bayes' theorem. We may select an item in step 1 based on Bayesian arguments such as minimizing the expected posterior variance of θ or optimizing other criteria. And we can define a stopping rule based on when the posterior variance or standard deviation is sufficiently small.

Third, though the Bayesian machinery just mentioned is a fruitful way to enact these steps in CAT, one may adopt frequentist machinery to these ends. Instead of using Bayes' theorem in step 4, we might use ML estimation of θ once at least one correct and one incorrect response is observed. We may select the next item based on maximizing the item information function, and define a stopping rule based on the standard error being sufficiently small. And yet our CAT process will still have undertones that strongly align with Bayesian inference, namely the logic of updating what we believe about the examinee in light of what was just observed, and preparing to update that further in light of new information that is to come.

11.6.2 Collateral Information about Examinees and Items

Modeling collateral information refers to the formal inclusion of additional sources of information to facilitate inference. Such collateral information is often expressed via covariates in a model for examinees and/or for items (De Boeck & Wilson, 2004). For examinees, this involves variables related to their proficiencies captured by the latent variables. The multiple-group CFA model presented in Section 9.7 is one example, where group membership was a covariate, and each examinee's latent variable was specified as following a group-specific prior distribution. More generally, we may embed discrete covariates (e.g., gender, whether the examinee has received instruction on the material being assessed) and continuous covariates (e.g., age, performance on other assessments of the material) in regression-like structures for examinees.* In the current context of IRT, we expand the prior distribution for examinees' latent variables in (11.13) to be an augmented regression structure,

$$\theta_i \mid \boldsymbol{\beta}_\theta, \sigma^2_{\varepsilon\theta}, \mathbf{y}_i \sim N(\mathbf{y}_i \boldsymbol{\beta}_\theta, \sigma^2_{\varepsilon\theta}), \tag{11.44}$$

where $\mathbf{y}_i$ is a $(1 \times [Q_\theta + 1])$ augmented predictor matrix obtained by combining a scalar of 1 with a vector of values for examinee i on Q_θ predictors, $\boldsymbol{\beta}_\theta$ is a $([Q_\theta + 1] \times 1)$ augmented vector of coefficients, and $\sigma^2_{\varepsilon_\theta}$ is the error variance in the regression structure.

For items, collateral information comes in the form of features of the item that are known that relate to their psychometric properties. Structured IRT models extend the foundational IRT setup by modeling the measurement model parameters as dependent on relevant features of the item. Research to date has focused principally on features that influence item difficulty and therefore involve models for location parameters. We expand the prior distribution for items' location parameters in (11.16) to be

$$d_j \mid \boldsymbol{\beta}_d, \sigma^2_{\varepsilon d}, \mathbf{w}_j \sim N(\mathbf{w}_j \boldsymbol{\beta}_d, \sigma^2_{\varepsilon d}), \tag{11.45}$$

* In this and the following section, our focus is on the mechanics of incorporating collateral information about examinees into the model. The propriety of doing so is a situation-specific question involves purposes and values associated with the model and the desired inferences, points we take up in more detail in Section 15.2.

where $\mathbf{w}_j$ is a $(1 \times [Q_d + 1])$ augmented predictor matrix obtained by combining a scalar of 1 with a vector of values for item j on Q_d predictors, $\boldsymbol{\beta}_d$ is a $([Q_d + 1] \times 1)$ augmented vector of coefficients, and $\sigma^2_{\varepsilon_d}$ is the error variance in the regression structure.

Basic IRT and other psychometric models formally model differences among examinees and items, but offer no explanatory account of those differences. Expanding the model to incorporate covariates for examinee and item parameters allows for the modeling of hypotheses for the underlying reasons for the differences. On the examinee side, we can model features of the individual that drive differences in proficiency, for example via the latent regression in (11.44). On the item side, we can model how the psychometric properties depend on features of the item, for example via the regression in (11.45). This is particularly powerful when those item features are part of a larger substantive theory about the knowledge or cognitive processes they require or evoke from examinees engaging with them (Rupp & Mislevy, 2007).

Models of this type align with several aspects of Bayesian thinking. Exchangeability is assumed conditional on the covariates, and these models are another example of the perspective on model construction that invokes hierarchical structures in pursuit of conditional exchangeability. At a more conceptual level, the situation here is one in which our beliefs about the unknown parameters are shaped in part by things other than just the examinees' responses. This collateral information constitutes evidence used in our reasoning, and can be naturally built into the model using a Bayesian approach.

11.6.3 Plausible Values

Large-scale surveys are often used to facilitate inference about groups. For example, the National Assessment of Educational Progress (NAEP) aims to support inferences about the students in the United States as a whole, or in meaningful subgroups, rather than individuals (Mazzeo, Lazer, & Zieky, 2006; von Davier, Sinharay, Oranje, & Beaton, 2007). This is accomplished by administering items to individuals and aggregating up to the desired group level. A simple approach to performing this aggregation would estimate the distribution of the latent variables for a group by the distribution of the individual estimates. Practical constraints prevent the administration of sufficient numbers of items to support accurate estimates at the individual level. Happily though, we may obtain accurate group-level estimates without first obtaining accurate estimates for the individuals that comprise the group. This counterintuitive claim is aligned with the general statistical principle that when there is uncertainty in lower levels of hierarchical designs, different procedures are optimal for inferences at different levels rather than one single procedure being optimal for all, a point famously illustrated by Stein's (1956; James & Stein, 1961) work on the estimation of the mean vector of a multivariate normal distribution. In our context, what might be needed for accurate estimation at the lower level of individuals' latent variables, namely administering many informative items to each individual, is not needed for accurate estimation at the higher level, namely estimation of the group's distribution of latent variables (Mislevy, Beaton, Kaplan, & Sheehan, 1992).

Importantly, absent precise estimates at the individual level, naïve aggregations to the group level are insufficient owing to the failure to acknowledge and properly propagate uncertainty. With only a few items per examinee, the uncertainty associated with each individual's latent variable is considerable; ignoring that via simple aggregations is untenable (Mislevy et al., 1992). In summary, precise estimates at the individual are not available, but they are not needed—what are needed are characterizations of the information about the distributions obtained from individuals that (1) properly acknowledge the uncertainty and (2) lead to correct estimates of population-level parameters.

To accomplish this, NAEP and other large scale surveys employ *plausible value* methodology (Mislevy, 1991; Mislevy et al, 1992; Mislevy, Johnson, & Muraki, 1992). For each examinee, a set of plausible values is obtained via draws from the posterior distribution for the examinee's latent variable(s). For simplicity, we consider the case of a single latent variable. For each examinee, we construct the posterior distribution for the latent variable conditional on the observables for the examinee $\mathbf{x}_i$ and collateral information $\mathbf{y}_i$ if the latter are available,

$$\begin{aligned} p(\theta_i \mid \mathbf{x}_i, \mathbf{y}_i) &\propto p(\mathbf{x}_i \mid \theta_i, \mathbf{y}_i) p(\theta_i \mid \mathbf{y}_i) \\ &= p(\mathbf{x}_i \mid \theta_i) p(\theta_i \mid \mathbf{y}_i), \end{aligned} \tag{11.46}$$

where $p(\theta_i \mid \mathbf{y}_i)$ is the prior distribution for θ_i, here formulated as conditional on the collateral variables $\mathbf{y}_i$, and the second line follows from the conditional independence assumption. Plausible value methodology involves taking draws from such posterior distributions.

Analyzing the draws supports inferences at the group level; details can be found in the cited sources. For our purposes, we call out a few key points. On the computational front, the draws for the values of the θs in any one iteration of an MCMC sampler represent a set of plausible values for the examinees' θs. Multiple draws, from multiple iterations of the MCMC sampler, constitute multiple plausible values for the examinees' θs. Conceptually, the draws (plausible values) are taken from the posterior distributions for individuals' θs building in everything we know, believe, and have estimated from data about their relationships with all the collateral variables in the model, in addition to the observables. Finally, we note that, with respect to the group-level distribution of latent variables, the individuals' latent variables may be conceived as missing data. Cast in this light, plausible value methodology is an instance of using Rubin's (1987) multiple imputation approach to missing data analysis (Mislevy et al., 1992), a topic we take up in more detail in Chapter 12.

Note that in (11.46), $p(\mathbf{x}_i \mid \theta_i)$ is the likelihood function for the examinee's latent variable induced by the observed values in $\mathbf{x}_i$ and treating the measurement model parameters as known. Similarly $p(\theta_i \mid \mathbf{y}_i)$ is the prior distribution treating the parameters that govern the dependence of the latent variables on the covariates (e.g., the parameters $\boldsymbol{\beta}_\theta$ and $\sigma^2_{\varepsilon\theta}$ in the latent regression model in (11.44)) as known. This reflects that operational demands necessitate practical solutions to addressing the complexities of large-scale assessment surveys, such as drawing on (a) multiple imputation analyses that account for the sampling design, but use point estimates of the parameters of the IRT model and the latent regression model (Beaton, 1987), (b) estimates of the IRT parameters that ignore the complex sampling (Scott & Ip, 2002), and (c) superpopulation-based analyses for educational surveys with hierarchical structures assuming error-free dependent variables (Longford, 1995; Raudenbush, Fotiu, & Cheong, 1999). As noted in Section 7.4, Johnson and Jenkins (2005) developed a fully Bayesian model that allows for the estimation of a joint model addressing all of these design features. Using simulated and real data from operational NAEP assessments, they found that both the standard analysis with its piece-wise approximations and their unified model provided consistent estimates of subpopulation features, but the unified model more appropriately captured the variance of those estimates. By treating IRT item parameters and population variances as known the standard analysis tended to underestimate posterior uncertainty by about 10%. What's more, the unified model provided more stable estimates of sampling variance than the standard procedures.

11.7 Alternative Prior Distributions for Measurement Model Parameters

The lack of a general class of conjugate or conditionally conjugate priors for IRT models and the rise of Bayesian approaches to IRT that leverage optimization routines for point estimation has led to a number of prior specifications that have been proposed for measurement model parameters. In this section, we present several alternative choices.

11.7.1 Alternative Univariate Prior Specifications for Models for Dichotomous Observables

Several prior distributions are commonly employed for measurement model parameters in IRT. For discrimination parameters, researchers have used a normal distribution (without truncation). Setting the prior such that the mean is sufficiently larger than 0 and the variance is sufficiently small so that most of the density is over values larger than 0 effectively resolves the rotational indeterminacy. Another choice is a lognormal prior distribution,

$$a_j \sim \log N(\mu_a, \sigma_a^2), \tag{11.47}$$

where the lognormal distribution implies that $\ln(a_j) \sim N(\mu_a, \sigma_a^2)$.

For lower asymptote parameters, alternatives to the beta distribution include a uniform distribution over some portion between 0 and 1, or the specification of a normal prior distribution for a transformation of c_j that places it over the real line. For example, we may employ the logit transformation,

$$\text{logit}(c_j) = \Psi^{-1}(c_j) = \log\left(\frac{c_j}{1-c_j}\right),$$

and then specify a normal prior (Mislevy, 1986),

$$\Psi^{-1}(c_j) \sim N(\mu_c, \sigma_c^2). \tag{11.48}$$

Similarly, the probit transformation $\Phi^{-1}(c_j)$ may be employed, with the result modeled as a normal distribution (Fox, 2010).

11.7.2 Alternative Univariate Prior Specifications for Models for Polytomous Observables

The lognormal or unrestricted normal distributions for discriminations just described in the context of dichotomous models may be employed in polytomous models as well. Polytomous models bring with them the additional location parameters, which we focus on here.

In the analysis of the GRM in Section 11.4, we employed a strategy of specifying a normal prior for the first location parameter for each observable and then specifying the remaining location parameters as via truncated normal distributions, where the truncation bounds the location parameter to be less than the location parameter for the preceding category.

A different strategy specifies the remaining location parameters as successive deviations from the location parameter for the preceding category in such a way that the order restrictions are preserved. A bit more formally,

$$d_{jk} = d_{j(k-1)} - s_{jk} \quad \text{for } k = 3, \ldots, K_j. \tag{11.49}$$

The deviation terms are then specified as following a distribution, such that they are nonnegative. A natural choice would be a uniform distribution bounded below by 0. Setting the maximum to be, say, 10 or larger, enacts a diffuse prior over the latent continuum in most applications of IRT. An alternative strategy that does not require the analyst to specify a maximum of the deviation capitalizes on the log transformation of a normal distribution over the real line:

$$s_{jk} = \exp(l_{jk}),$$

where (11.50)

$$l_{jk} \sim N(\mu_l, \sigma_l^2) \quad \text{for} \quad k = 3, \ldots, K_j.$$

A generalization of this model specifies the distribution for l_{jk} to vary over j or k. A third strategy specifies unrestricted versions of all the location parameters included in the model,

$$d_{jk}^* \sim N(\mu_d, \sigma_d^2) \quad \text{for} \quad k = 2, \ldots, K_j, \tag{11.51}$$

and then sets the location parameters in terms of order statistics of these unrestricted versions (Curtis, 2010; Plummer, 2010). For observable j,

$$\begin{aligned} d_{j2} &= d_{j,[K_j]}^*, \\ d_{j3} &= d_{j,[K_j-1]}^*, \\ &\vdots \end{aligned} \tag{11.52}$$

and

$$d_{jK} = d_{j,[2]}^*.$$

The reverse ordering of the order statistics with the location parameters (i.e., as we go *up* from d_{j2} to d_{jK}, we go *down* from the K_jth to the second-order statistic) is due to the construction of the GRM in (11.27) in terms of probabilities of being at or above a category. This strategy of using order statistics is a bit more straightforward in versions of the GRM model that work with cumulative probabilities, in which we model the probabilities of being at or below a certain category. In that case the kth location parameter corresponds to the kth-order statistic from the unrestricted versions (Curtis, 2010).

11.7.3 Multivariate Prior Specifications for Dependence among Parameters

The use of univariate priors for the measurement model parameters reflects an assumption of a priori independence. Although this does not force the parameters to be independent in the posterior, it does not formally model their possible dependence. We take up this topic here, working in the context of models for dichotomous data; the same principles apply to models of polytomous data.

We might expect location and discrimination parameters to be dependent, as they are primarily reflective of first- and second-order moments, respectively, which are dependent in dichotomous data (Carroll, 1945; see Fox, 2010, pp. 36–37 for a related argument). Similarly, we might expect location and lower asymptote parameters to be dependent as higher d and c parameters represent competing explanations for higher rates of success on the item.

To formally model the dependence among the measurement model parameters, we may employ a multivariate prior for the ω_j (Mislevy, 1986). For the 2-P models, a natural extension of the univariate approach involves a truncated multivariate normal prior, where the truncation applies to a_j. In this case, $\omega_j = (d_j, a_j)'$ and

$$\omega_j \sim N(\mu_\omega, \Sigma_\omega)\, I_{R^+}(a_j) \tag{11.53}$$

where μ_ω and Σ_ω are hyperparameters and $I_{R^+}(a_j)$ is the indicator function that is equal to 1 when a_j is in the region defined by the positive real line, and 0 otherwise.*

In software environments where the truncated multivariate normal is difficult to implement (e.g., WinBUGS), or if a 3-P model is of interest, an alternative strategy involves using some of the alternative forms presented in this section. The multivariate normal prior may be specified without the truncation; a judicious choice for the prior mean and variance for the a parameters may help to resolve the rotational indeterminacy. Alternatively, to preserve the positivity restriction on the a parameters, we may employ a log transformation. The idea here is to take the a parameters, which we wish to bound below by 0, and express them on the real line. Along the same lines, we can define a function that maps from the restricted space of the c parameters, which we bound by 0 and 1, to the real line via the logit or probit transformation. That is, we work with a reparameterization that uses the identity transformation for the locations, the log transformation for the discriminations, and the logit transformation for the lower asymptotes:

$$\delta_j = d_j,$$

$$\alpha_j = \log(a_j),$$

and

$$\gamma_j = \Psi^{-1}(c_j) = \log\left(\frac{c_j}{1-c_j}\right).$$

* To point out a connection with the simpler, more common notation used earlier in the univariate case, note that (11.17) can be written in indicator notation as $a_j \sim N^+(\mu_a, \sigma_a^2) I_{R^+}$.

We can then specify an (unrestricted) multivariate normal prior on the transformed set of parameters $\eta_j = (\delta_j, \alpha_j, \gamma_j)'$,

$$\eta_j \sim N(\mu_\eta, \Sigma_\eta). \tag{11.54}$$

The original parameters can be recovered by taking the inverses of the transformations involved in the reparameterizations,

$$d_j = \delta_j,$$

$$a_j = \exp(\alpha_j),$$

and

$$c_j = \Psi(\gamma_j) = \frac{\exp(\gamma_j)}{1 + \exp(\gamma_j)}.$$

The hyperparameters μ_η and Σ_η may be specified by the user. As "thinking" in the log and inverse-logit parameterizations is not natural to most, it may be useful to use the original parameterization as a guide for defining the hyperparameter values and, roughly speaking, apply the corresponding transformations to these values. In the example in Section 11.2.5, we employed an $N(0, 2)$ prior for the d parameters. As the δs are obtained by the identity transformation, a comparable specification using the multivariate approach would simply specify $\mu_\delta = 0$ and $\sigma_\delta^2 = 2$. For the a parameters, we employed an $N^+(1,2)$ prior. Taking the log of the mode of this distribution (here, 1) yields a suggested value of $\mu_\alpha = 0$. If the geometric average of the slopes were 1, then choosing $\sigma_\alpha^2 = .25$ yields a prior where the a parameters are highly likely to be between about 1/3 and 3 (Mislevy, 1986), which is in line with the $N^+(1,2)$ prior. For the c parameters, we employed a Beta(5,17) prior, which has a mean of $5/22 \approx .227$. Applying the logit function to that value yields a suggested value of $\mu_\gamma = -1.225$, and using $\sigma_\gamma^2 = .25$ yields a distribution of the cs that closely matches the Beta(5, 17). These means may be collected into μ_η and these variances may be collected into the diagonal of Σ_η.

The off-diagonal elements of Σ_η capture the dependency among the items' measurement model parameters. One source of information about their dependence comes in the form of analyses of the items with other examinees or from analyses of similar items. The example in Section 11.2.5 specified a prior distribution reflecting a priori independence among the measurement model parameters; however, they were correlated in the posterior. These correlations, suitably transformed to reflect the reparameterization, could be the basis for a future analysis of either (a) these items with different examinees, (b) different items with these examinees, or (c) different items with different examinees. (We would *not* use these posterior correlations as the prior in a reanalysis of the same items and same examinees; see Exercise 11.9). Note the implicit exchangeability assumptions inherent in using the results of this analysis in any of scenarios (a)–(c): The results from the current analysis would be relevant for an analysis of (a) different examinees if the two sets of examinees were exchangeable with respect to the current items. They would be relevant for an analysis of (b) different items if the sets of items were exchangeable with respect to the current examinees. They would be relevant for an analysis of (c) different items with different examinees if both exchangeability assumptions are warranted. To the extent that these exchangeability assumptions are not warranted, the use of the current strategies may be suspect. Next, we describe a mechanism that potentially accommodates these situations.

11.7.4 A Hierarchical Prior Specification

As we have seen, one perspective on Bayesian modeling views the role of the prior distribution as augmenting the information from the data. From this viewpoint, the posterior distribution will be shrunk toward the prior distribution, relative to the likelihood. This most clearly manifests itself in the result common to Bayesian analyses where the posterior mean for a parameter is a precision-weighted average of the prior mean and the mean suggested by the data. This may be seen as the prior distribution regularizing or reining in the estimates that would be obtained by consideration of the data alone. This can be particularly advantageous in situations where the parameters may be ill-defined by the data, such as in IRT. The analysis of the latent variable for an examinee with a "perfect" response vector of all correct answers in Section 11.2.5 was an example of this, as is the situation mentioned in Section 11.1.3 where multiple sets of values of the item parameters produce nearly the same model-implied probabilities yielding complicated likelihoods.

However, if the prior distribution is poorly specified, the effect will shrink the posterior *away* from the actual center of the parameter values, which may have deleterious effects on interpretations of those parameters and other inferences (e.g., claims about examinees). To address this, a hierarchical prior structure can be specified, where a second-level prior structure is placed on the hyperparameters for the measurement model.

We develop these ideas in the context of the univariate approach to prior specifications. For the location parameters, we again specify a normal prior,

$$d_j \sim N(\mu_d, \sigma_d^2). \tag{11.55}$$

However, instead of specifying values for these hyperparameters, we now specify a hyperprior for these parameters. As these are parameters of a normal distribution, a conditionally conjugate specification is given by

$$\mu_d \sim N(\mu_{\mu_d}, \sigma_{\mu_d}^2) \tag{11.56}$$

and

$$\sigma_d^2 \sim \text{Inv-Gamma}(\nu_d/2, \nu_d \sigma_{0_d}^2/2), \tag{11.57}$$

where μ_{μ_d}, $\sigma_{\mu_d}^2$, ν_d, and $\sigma_{0_d}^2$ are hyperparameters specified by the analyst. Viewed as a hierarchy, (11.55) represents the first level of the hierarchy, and (11.56) and (11.57) represent the second level of the hierarchy.

First consider the mean of the location parameters, μ_d, which is a parameter in the first level of the hierarchy. μ_d captures the central tendency of the prior distribution for the ds. The value of μ_d may be thought of as an answer to the question "What should we shrink the estimate of each d_j towards?" In the single-level prior specification for the ds, the value of μ_d is chosen by the analyst, which may be problematic if chosen poorly. Our current focus is to examine how the hierarchical prior structure offers a mechanism to, in a sense, relieve the pressure placed on the analyst to make such a choice. To delve into how, consider the full conditional for μ_d given by

$$\mu_d \mid \mathbf{d} \sim N(\mu_{\mu_d \mid \mu_{\mu_d}, \sigma_{\mu_d}^2, \mathbf{d}}, \sigma^2_{\mu_d \mid \mu_{\mu_d}, \sigma_{\mu_d}^2, \mathbf{d}}), \tag{11.58}$$

where

$$\mu_{\mu_d|\mu_{\mu_d},\sigma^2_{\mu_d},\mathbf{d}} = \frac{\dfrac{\mu_{\mu_d}}{\sigma^2_{\mu_d}} + \dfrac{J\bar{d}}{\sigma^2_d}}{\dfrac{1}{\sigma^2_{\mu_d}} + \dfrac{J}{\sigma^2_d}},$$

$$\sigma^2_{\mu_d|\mu_{\mu_d},\sigma^2_{\mu_d},\mathbf{d}} = \frac{1}{\dfrac{1}{\sigma^2_{\mu_d}} + \dfrac{J}{\sigma^2_d}},$$

$$\mathbf{d} = (d_1,\ldots,d_J),$$

and

$$\bar{d} = 1/j\sum_{J=1}^{J} d_j.$$

The prior variance $\sigma^2_{\mu_d}$ may be seen as controlling the degree to which the hierarchical prior structure departs from the single-level prior with known hyperparameters. For small values of $\sigma^2_{\mu_d}$, the mean of the location parameters (i.e., μ_d, which is a parameter at level-1 in the hierarchy) is more strongly influenced by the prior mean at the second level of the hierarchy. In the limit, as $\sigma^2_{\mu_d} \to 0$, $\mu_{\mu_d|\mu_{\mu_d},\sigma^2_{\mu_d},\mathbf{d}} \to \mu_{\mu_d}$ and the two-level hierarchical prior degenerates into the single-level prior with known hyperparameters. For large values of $\sigma^2_{\mu_d}$, the mean of the location parameters (i.e., μ_d, which is a parameter at level-1 in the hierarchy) is more strongly influenced by the values of the location parameters suggested by the data. In the limit, $\mu_{\mu_d|\mu_{\mu_d},\sigma^2_{\mu_d},\mathbf{d}} \to \bar{d}$ as $\sigma^2_{\mu_d} \to \infty$. Casually, this corresponds to answering the question "What should we shrink the estimate of each d_j towards?" with "the mean of the ds as suggested by the data." This same limit for the posterior mean, $\mu_{\mu_d|\mu_{\mu_d},\sigma^2_{\mu_d},\mathbf{d}} \to \bar{d}$, obtains as $J \to \infty$.

More generally, the hierarchical prior structure has the attractive feature of retaining the core goal of specifying a prior distribution for the posterior distribution to be shrunk towards (relative to the possibly unstable likelihood), but allows *where* that prior distribution is located to depend to some degree on the information in the data. How much it depends on the data is governed principally by relative amounts of information in the prior, captured by $\sigma^2_{\mu_d}$, and in the data, captured by J.

μ_d governs *where* we should shrink the data-based estimate of d_j towards. The other parameter in the first level of the prior distribution for the ds in (11.55), σ^2_d, governs *how much* we should shrink the data-based estimate. If σ^2_d is large, there will be relatively less shrinkage, and the posterior for each d_j will be more strongly influenced by the data. If σ^2_d is small, there will be relatively more shrinkage. The limiting case of this has $\sigma^2_d \to 0$, in which case $d_j \to \mu_d$ and the data become irrelevant. In the single-level model, the analyst's choice of σ^2_d expresses the variability of the d parameters and constitutes an answer to the question: "How much should we shrink each data-based estimate?" By building a hierarchical prior structure as in (11.57), we again have a mechanism for relieving the pressure on the analyst to specify a value, and enact a balancing of the analyst's prior beliefs, expressed in (11.57), with the information in the data. Generally speaking, the information in the data becomes more important for σ^2_d—and therefore our answer to the question "How much should we shrink each data-based estimate?"—as ν_d decreases or J increases.

A hierarchical prior structure provides an elegant mechanism to address any uneasy feelings the analyst may have about specifying values for the hyperparameters of prior distributions. It allows the influence of the analyst's prior beliefs to be "softened" a bit. Instead of being represented in the first (and only) level of a prior distribution which more directly influence the posterior distribution for the parameters of interest, they are bumped up to a second level and are "further away" from the parameters of interest.

It is also very much a Bayesian mechanism. To the extent that the analyst is unsure about what to specify as the hyperparameters, we do what we always do in Bayesian modeling when encountered with things we are not sure of: assign a distribution! After conditioning on the observed data, the posterior distributions for these parameters (the hyperparameters of the first level of the prior distribution) will themselves be a mix of the analyst's prior beliefs (expressed in the second level of the hierarchy) and the data.

A similar logic applies for a hierarchical prior structure on the other measurement model parameters. For the discrimination parameters, the first level of the prior is

$$a_j \sim N^+(\mu_a, \sigma_a^2), \tag{11.59}$$

and instead of specifying values for the hyperparameters, we again specify normal and inverse-gamma hyperprior distributions as the second level in the hierarchy,

$$\mu_a \sim N(\mu_{\mu_a}, \sigma_{\mu_a}^2) \tag{11.60}$$

and

$$\sigma_a^2 \sim \text{Inv-Gamma}(\nu_a / 2, \nu_a \sigma_{0_a}^2 / 2), \tag{11.61}$$

where μ_{μ_a}, $\sigma_{\mu_a}^2$, ν_a, and $\sigma_{0_a}^2$ are hyperparameters specified by the analyst. Normal and inverse-gamma hyperprior distributions may be similarly used if the lognormal prior for discriminations in (11.47) is employed.

For the lower asymptote parameters, the first level of the prior is

$$c_j \sim \text{Beta}(\alpha_c, \beta_c), \tag{11.62}$$

and instead of specifying values for the hyperparameters, we specify a second level of the hierarchy. An interpretation of the parameters of the Beta distribution in terms of counts of successes and failures suggests the use of a Poisson distribution at the second level of the hierarchy (Levy, 2014):

$$\alpha_c \sim \text{Poisson}(\lambda_{\alpha_c}) \tag{11.63}$$

and

$$\beta_c \sim \text{Poisson}(\lambda_{\beta_c}), \tag{11.64}$$

where λ_{α_c} and λ_{β_c} are hyperparameters specified by the analyst. If a logit-normal (see Equation 11.48) or probit-normal prior distribution is specified for lower asymptotes, we are again in a situation where normal and inverse-gamma distributions may be convenient choices for the second level of the hierarchy.

A hierarchical construction on a multivariate prior for the measurement model parameters supports modeling the dependence among them. Recall our multivariate normal prior for the set of transformed parameters

$$\eta_j \sim N(\mu_\eta, \Sigma_\eta).$$

Rather than specify values for these hyperparameters, they may be assigned hyperprior distributions. Desires for conditional conjugacy imply a multivariate normal hyperprior for μ_η and an inverse-Wishart hyperprior for Σ_η. In addition to allowing where and by how much the data-suggested estimates should be shrunk, this approach allows for learning about the dependence among the measurement model parameters, as expressed by the posterior distribution for the off-diagonal elements of Σ_η (see Exercises 11.7 and 11.8).

11.8 Latent Response Variable Formulation and Data-Augmented Gibbs Sampling

Albert (1992; see also Albert & Chib, 1993) introduced an alternative strategy that formulates the model via a *latent response variable* (e.g., Wirth & Edwards, 2007) and enacts a data-augmented Gibbs sampler (Tanner & Wong, 1987). Thinking in terms of the examinee variables, a latent response variable resides "in between" the usual latent variable and an observable. A latent response variable is a continuous variable posited to underlie the discrete observable variable; the discrete observable results from a process of discretizing the continuous latent response variable. This is a reasonable interpretation in some circumstances (e.g., in surveys, there is continuum of age that underlies the examinee's response to a question of whether they are at least 18 years old). In other circumstances we may be hard pressed to conceive of a continuous variable as underlying a discrete observable. This does not necessarily undermine formulating the model in this way, as it may be viewed as a modeling device or piece of machinery. Importantly, the use of latent response variables does not change the model substantively; it is simply a reparameterization (really, an extra-parameterization) that yields a number of advantages. For the 2-PNO model, these include illuminating connections between IRT models and normal-theory CFA and CTT models (Kamata & Bauer, 2008; Lord & Novick, 1968; McDonald, 1999; Takane & de Leeuw, 1987), and closed form expressions for the full conditional distributions resulting from conjugacy, which we focus on here. Only slight differences are needed to implement a data-augmented Gibbs sampler for the logistic models (Maris & Maris, 2002; see also Fox, 2010).

Various treatments of the normal-ogive model in terms of an underlying latent response variable can be found in the IRT and CFA literatures, many of which are formally equivalent (Kamata & Bauer, 2008). This notion of a continuous latent response variable is common and is particularly natural in CFA traditions, as it may be seen as a way to move from modeling discrete variables to the more-natural-to-CFA situation of continuous variables modeled as dependent on continuous latent variables. Casually, a CFA perspective might see the latent response variable formulation as a way to "undo" the discretization, putting us squarely back in the camp of linear, normal-theory CFA. As we will indeed see, introducing the latent response variable results in the same or nearly the same full conditional distributions as those in CFA.

We focus on the case of dichotomous observables; extensions to polytomous observables contain the same key ideas. For each observable, the 2-PNO model is essentially a probit model where the predictor is a latent variable (θ). The probit model may be thought of as regression model with a latent outcome which is censored and observed only in terms of its sign (Jackman, 2009). Let x_{ij}^* denote the latent response variable for examinee i and observable j, where

$$x_{ij}^* = a_j\theta_i + d_j + \varepsilon_{ij} \tag{11.65}$$

and, for each j, $\varepsilon_{ij} \sim N(0, \sigma_{\varepsilon_j}^2)$. Distributionally, this may be expressed as

$$x_{ij}^* \sim N(a_j\theta_i + d_j, \sigma_{\varepsilon_j}^2). \tag{11.66}$$

The probability for an observable taking on a value of 1 is then modeled as the probability that the latent response variable is greater than or equal to a threshold for the observable, γ_j,

$$\begin{aligned} P(x_{ij} = 1 \mid \theta_i, d_j, a_j, \varepsilon_{ij}) &= P(x_{ij}^* \geq \gamma_j \mid \theta_i, d_j, a_j, \varepsilon_{ij}) \\ &= P(a_j\theta_i + d_j + \varepsilon_{ij} \geq \gamma_j \mid \theta_i, d_j, a_j, \varepsilon_{ij}). \end{aligned} \tag{11.67}$$

It is clear from the expression on the right-hand side of (11.67) that a location indeterminacy is present, as a constant can be added to both d_j and γ_j and the probability will be preserved. An (arbitrary) constraint is needed to resolve this. Interestingly, CFA strategies rooted in covariance structure analysis typically assume that the location parameters in the latent response variable equation (here, the ds) are 0, and the thresholds (γs) remain in the model. In IRT traditions, it is more common to assume that the thresholds are 0 and the location parameters remain in the model.

It is less obvious that an indeterminacy in scale exists as well. This should not be surprising—latent variables do not have inherent metrics, so the introduction of a latent response variable brings with it the need to resolve these indeterminacies. For each observable, this can be resolved by setting the $\sigma_{\varepsilon_j}^2$ equal to a constant; the value of 1 is often chosen for convenience (see Kamata & Bauer, 2008, for details on this and other alternatives). This renders $\varepsilon_{ij} \sim N(0,1)$, which implies that (11.67) may be expressed as

$$\begin{aligned} P(x_{ij} = 1 \mid \theta_i, d_j, a_j) &= P(x_{ij}^* \geq 0 \mid \theta_i, d_j, a_j) = P(a_j\theta_i + d_j + \varepsilon_{ij} \geq 0 \mid \theta_i, d_j, a_j) \\ &= P(\varepsilon_{ij} \geq -(a_j\theta_i + d_j) \mid \theta_i, d_j, a_j) = P(\varepsilon_{ij} < a_j\theta_i + d_j \mid \theta_i, d_j, a_j) \\ &= \Phi(a_j\theta_i + d_j). \end{aligned} \tag{11.68}$$

Making the now familiar assumptions of a priori independence between the latent variables and the measurement model parameters, exchangeability of examinees, and exchangeability of the observables, the posterior distribution is then

$$\begin{aligned} p(\boldsymbol{\theta}, \mathbf{d}, \mathbf{a}, \mathbf{x}^* \mid \mathbf{x}) &\propto p(\mathbf{x} \mid \boldsymbol{\theta}, \mathbf{d}, \mathbf{a}, \mathbf{x}^*) p(\boldsymbol{\theta}, \mathbf{d}, \mathbf{a}, \mathbf{x}^*) \\ &= p(\mathbf{x} \mid \mathbf{x}^*) p(\mathbf{x}^* \mid \boldsymbol{\theta}, \mathbf{d}, \mathbf{a}) p(\boldsymbol{\theta}) p(\mathbf{d}) p(\mathbf{a}) \\ &= \prod_{i=1}^{n} \prod_{j=1}^{J} p(x_{ij} \mid x_{ij}^*) p(x_{ij}^* \mid \theta_i, d_j, a_j) p(\theta_i) p(d_j) p(a_j), \end{aligned} \tag{11.69}$$

where

$$x_{ij} \mid x_{ij}^* = \begin{cases} 1 \text{ if } x_{ij}^* \geq 0 \\ 0 \text{ if } x_{ij}^* < 0 \end{cases} \quad \text{for } i = 1, \ldots, n, j = 1, \ldots, J$$

and

$$x_{ij}^* \mid \theta_i, d_j, a_j \sim N(a_j\theta_i + d_j, 1) \quad \text{for } i = 1, \ldots, n, j = 1, \ldots, J.$$

Employing the normal prior for examinee latent variables in (11.13) and the truncated multivariate normal prior for the measurement model parameters in (11.53), the full conditional distributions are expressed in the following equations (see Appendix A for derivations).

On the left-hand side, the parameter in question is written as conditional on all the other relevant parameters and data.* On the right-hand of each of the following equations, we give the parametric form for the full conditional distribution. In several places we denote the arguments of the distribution (e.g., mean and variance for a normal distribution) with subscripts denoting that it refers to the full conditional distribution; the subscripts are then just the conditioning notation of the left-hand side.

For the examinee latent variables, introducing the latent response variables make the model akin to a CFA model, with the continuous latent response variables playing the role of the observables in CFA. The full conditional distribution for examinee i is

$$\theta_i \mid \mathbf{d},\mathbf{a},\mathbf{x}_i^* \sim N(\mu_{\theta_i|\mathbf{d},\mathbf{a},x_i^*},\sigma^2_{\theta_i|\mathbf{d},\mathbf{a},x_i^*}), \tag{11.70}$$

where

$$\mu_{\theta_i|\mathbf{d},\mathbf{a},x_i^*} = \frac{\dfrac{\mu_\theta}{\sigma^2_\theta}+\sum_j a_j(x_{ij}^*-d_j)}{\dfrac{1}{\sigma^2_\theta}+\sum_j a_j^2},$$

$$\sigma^2_{\theta_i|\mathbf{d},\mathbf{a},x_i^*} = \frac{1}{\dfrac{1}{\sigma^2_\theta}+\sum_j a_j^2},$$

$\mathbf{d}=(d_1,\ldots,d_J)$, $\mathbf{a}=(a_1,\ldots,a_J)$, and $\mathbf{x}_i^*=(x_{i1}^*,\ldots,x_{iJ}^*)$. The same structure applies to all the examinees.

For each examinee i and observable j, the associated latent response variable has a normal distribution given by its prior, now truncated to be positive or negative depending on whether the corresponding observable was 1 or 0,

$$x_{ij}^* \mid \theta_i,d_j,a_j,x_{ij} \sim \begin{cases} N(a_j\theta_i+d_j,1)\, I(x_{ij}^*\geq 0) & \text{if } x_{ij}=1 \\ N(a_j\theta_i+d_j,1)\, I(x_{ij}^*<0) & \text{if } x_{ij}=0, \end{cases} \tag{11.71}$$

where I is the indicator function.

Turning to the measurement model parameters, we present the full conditionals for any observable j. The same structure applies to all the observables. The full conditional distribution for the measurement model parameters $\omega_j=(d_j,a_j)$ is

$$\omega_j \mid \boldsymbol{\theta},\mathbf{x}_j^* \sim \mathbf{N}(\mu_{\omega_j|\theta,x_j^*},\Sigma_{\omega_j|\theta,x_j^*})\, \boldsymbol{I}(\mathbf{a}_j>0), \tag{11.72}$$

where

$$\mu_{\omega_j|\theta,x_j^*} = (\Sigma_\omega^{-1}+\theta_A'\theta_A)^{-1}(\Sigma_\omega^{-1}\mu_\omega+\theta_A'\mathbf{x}_j^*),$$

$$\Sigma_{\omega_j|\theta,x_j^*} = (\Sigma_\omega^{-1}+\theta_A'\theta_A)^{-1},$$

* We suppress the role of specified hyperparameters in this notation; see Appendix A for a presentation that formally includes the hyperparameters.

and θ_A is the $(n \times 2)$ *augmented* matrix of latent variables obtained by combining an $(n \times 1)$ column vector of 1s to the $(n \times 1)$ column vector of latent variables θ.

There are a number of variations on the latent response variable model that have been proposed or implemented in software, differing primarily in terms of how they resolve the indeterminacies associated with the latent response variables and formulate the truncation (Kamata & Bauer, 2008). The model may also be extended to the 3-PNO, in which case a data-augmentation scheme may be enacted by introducing an additional set of auxiliary variables (Béguin & Glas, 2001; Sahu, 2002; see Fu, Tao, & Shi, 2009 for 3-PL models).

As noted above, the latent response variable formulation is natural and quite common in a factor analytic perspective for modeling discrete variables (Wirth & Edwards, 2007). The equivalence of FA and IRT versions of the models (for the latter, 2-P versions that exclude lower asymptotes) has been demonstrated and repeatedly recognized (Takane & de Leeuw, 1987; see also Kamata & Bauer, 2008). However, as noted by Wirth and Edwards (2007), conflicting misconceptions in each of these paradigms may be traced to historical estimation traditions. In IRT, traditional estimation proceeds by integrating over the prior distribution of the latent variables to produce a marginal likelihood for the measurement model parameters. This becomes increasingly difficult as the number of latent variables increases. As a consequence, IRT analysts sometimes believe their models are suited to situations with few latent variables, and the number of observables is for the most part inconsequential. On the other hand, traditional FA approaches to estimation in the presence of discrete data involve the computation of polychoric correlations. These are then then submitted to traditional CFA routines, which easily accommodate many latent variables. The difficulties arise with the computation of the polychoric correlations, which involves integration over the distribution of the latent response variables. As there is a latent response variable for each observable, this becomes increasingly difficult as the number of observables increases. As a consequence, FA analysts sometimes believe their models are suited to situations with relatively few observable variables, and the number of latent variables is for the most part inconsequential. Thus, despite the recognition of the equivalence of IRT and FA perspectives on the model with continuous latent variables for discrete data, analysts steeped in the traditions of either of these approaches might have a view of their models that is unnecessarily limited by the estimation machinery associated with them. A fully Bayesian approach, with any of the MCMC sampling algorithms described above, may be a unifying and illuminating framework, freeing analysts from certain misconceptions.

11.9 Summary and Bibliographic Note

There is a wide world of IRT, and we have restricted our focus to some of the more popular IRT models. Similarly, there is considerable variation in Bayesian model specifications in IRT, owing in part to the variation in applications of IRT, computational challenges of IRT models, and the rise of Bayesian approaches prior to the advent of MCMC. Early work on fully Bayesian IRT models was conducted by Swaminathan and Gifford (1982, 1985, 1986), Tsutakawa and Lin (1986), Mislevy (1986), and Tsutakawa (1992). Albert (1992) conducted the first analysis of Bayesian IRT using MCMC by way of a data-augmentation Gibbs sampling solution for the 2-PNO model. The algorithm was extended to handle polytomous data by Albert and Chib (1993). The growth in Bayesian IRT via MCMC was precipitated by Patz and Junker (1999a, 1999b), who introduced a more flexible Metropolis–Hastings-within-Gibbs approach demonstrating how logistic IRT models for dichotomous data, polytomous data, missing data, and rater effects could be handled without requiring particular forms of prior distributions. The flexibility of MCMC estimation has allowed the expansion to more complex IRT models.

We have confined our treatment to commonly used parametric models where the latent variable is monotonically related to the probabilities of dichotomous or ordered polytomous responses. Fox (2010) provides a textbook length treatment of such Bayesian IRT models emphasizing hierarchical formulations. Various Bayesian approaches to compensatory and conjunctive MIRT for dichotomous and polytomous variables have been discussed by Babcock (2011), Béguin and Glas (2001), Bolt and Lall (2003), Clinton, Jackman, and Rivers (2004), Edwards (2010), Jackman (2001), Kim et al. (2013), Maris and Maris (2002), Reckase (2009), Sheng and Wikle (2008), Yao and Boughton (2007), and Yao and Schwarz (2006).

Bayesian approaches to other models for dichotomous and polytomous data include other specifications for ordinal and nominal data (Dunson, 2000; Wollack, Bolt, Cohen, & Lee, 2002; Yao & Schwarz, 2006) as well as for dichotomous data representing alternative response processes (Bolfarine & Bazan, 2010; Bolt, Deng, & Lee, 2014; Bolt, Wollack, & Suh, 2012; Jin & Wang, 2014; Loken & Rulison, 2010), including randomized response administrations (Fox, 2005a; Fox, Klein Entink, & Avetisyan, 2014). Bayesian models have also been developed for semiparametric (Miyazaki & Hoshino, 2009) and nonparametric models (Duncan & MacEachern, 2008; Fujimoto & Karabatsos, 2014; Karabatsos & Sheu, 2004; Karabatsos & Walker, in press) as well as unfolding IRT models that posit a nonmonotonic relationship between an item response and the latent variable as may be prevalent in psychological assessment (de la Torre, Stark, & Chernyshenko, 2006; Johnson & Junker, 2003; Roberts & Thompson, 2011; Wang, Liu, & Wu, 2013).

Procedures for accommodating ignorable and nonignorable missingness have also been developed (Fu, Tao, & Shi, 2010; Maier 2002; Patz & Junker, 1999a), and we discuss these topics in greater detail in the next chapter. Chang, Tsai, and Hsu (2014) described a Bayesian approach to modeling a related situation in which the time limit of an assessment affects the response process.

Section 11.7.4 described basic ideas of modeling latent variables and measurement model parameters with a hierarchical structure. Applications and extensions of this include models for rater effects (Patz, Junker, Johnson, & Mariano, 2002), testlet effects (Wainer, Bradlow, & Wang, 2007), differential functioning of items in subpopulations (Frederickx, Tuerlinckx, De Boeck, & Magis, 2010; Fukuhara & Kamata, 2011; Soares, Gonçalves, & Gamerman, 2009; Verhagen & Fox, 2013), items organized around areas of the domain (Janssen, Tuerlinckx, Meulders, & De Boeck, 2000), and families of items with members designed to reflect a common pattern (Geerlings, Glas, & van der Linden, 2011; Johnson & Sinharay, 2005; Sinharay, Johnson, & Williamson, 2003). Similarly, we may gainfully employ multilevel structures for the organization of examinees based on groupings or covariates (Fox, 2005b; Fox & Glas, 2001; Huang & Wang, 2014a; Jiao & Zhang, 2015; Maier, 2002; Natesan, Limbers, & Varni, 2010) or modeling change over time (Segawa, 2005). See Fox (2010) for a treatment of a number of these models.

Many of these models may be seen as instances of incorporating collateral information in the form of covariates or conditioning variables for model specifications. Finite mixture IRT models extend this idea to where the conditioning variables are discrete and latent (Bolt, Cohen, & Wollack, 2001; Cho, Cohen, & Kim, 2014; Finch & French, 2012; Meyer, 2010) and possibly dependent on covariates (Dai, 2013), which has been gainfully employed in the analysis of differential functioning (Cohen & Bolt, 2005; Samuelsen, 2008) including in situations recognizing the hierarchical organization of examinees (Cho & Cohen, 2010). Collateral information about items may be used to model the measurement model parameters, and shed interpretative light on differences among latent groups (Choi & Wilson, 2015).

An alternative source of collateral information may come in the form of other aspects of the examinee–item interaction, such as performance on items measuring other latent variables in MIRT (de la Torre, 2009; de la Torre & Song, 2009), the selection of distractors (Bolt et al., 2012), and response times (Klein Entink et al., 2009; Meyer, 2010; van der Linden; 2007; Wang, Fan, Chang, & Douglas, 2013).

The majority of adaptive testing has focused on adapting the items presented to the examinee in support of scoring. The same principles may be applied in support of optimizing calibration of measurement model parameters (van der Linden & Ren, 2014). The principles of adaptive testing are more general than the particular forms of IRT models and apply equally to other models (Vos, 1999) including those that employ discrete latent variables as in latent class and Bayesian network models discussed in Chapters 13 and 14 (for focused treatments of adaptive testing using models with discrete latent variables, see Cheng, 2009; Collins, Greer, & Huang, 1996; Jones, 2014; Macready & Dayton, 1992; Marshall, 1981; Rudner, 2009; Welch & Frick, 1993). See Almond and Mislevy (1999) for a general account of adaptive testing via graphical models, including the incorporation of collateral information.

Exercises

11.1 Recreate the 3-PNO analysis of the LSAT data in Section 11.2.5. Verify your results with those reported here and obtain the DIC.

11.2 Compare the results from the 3-PNO to those from 1-PNO and 2-PNO models of the LSAT data.

a. For each of the 1-PNO and 2-PNO models, obtain the posterior distribution for the measurement model parameters, conduct PPMC using the statistics and discrepancy measures reported in Section 11.2.5, and obtain the DIC.

b. Compare the results for the measurement model parameters to those in Table 11.2. How are the similar, how are they different, and why?

c. Compare the models in terms of their fit.

11.3 Consider what occurs if there are fewer items. Fit the 3-PNO model to the first four items of the LSAT data.

a. Compare the results for the measurement model parameters to those in Table 11.2. How are they similar, how are they different, and why?

b. Compare the result for examinee 1,000 to that in Table 11.2. How are they similar, how are they different, and why?

11.4 Fit a multidimensional GRM-L model for the item response data from the *Peer Interaction* and *Academic and Intellectual Development* subscales of the Institutional Integration Scale.

a. Specify and write out the DAG for a model with two latent variables, where the items associated with each subscale load on one latent variable. Be sure to consider the resolution of the indeterminacies in the model.

b. Interpret the results for the measurement model parameters and the parameters that govern the distribution of the latent variables. Compare the results for the correlation between the latent variables to the correlation between the observed subscale scores of .57.

11.5 In the description of CAT, it was suggested that ML could be used to estimate θ after each response if the examinee has at least one correct and at least one incorrect response. Why is this needed for ML estimation? Why is it not needed for Bayesian estimation?

11.6 Consider a 1-PL model that specifies collateral information for examinees and the items in the forms of (11.44) and (11.45).

a. Write out the DAG for the model.

b. For each parameter, write out what other entities need to be conditioned on in the full conditional distribution.

11.7 Analyze a 3-PNO model for the LSAT data in Table 11.1 using a prior distribution that formally models the dependence among the measurement model parameters.

a. Write out the DAG for the model.

b. How do the results for the measurement model parameters compare to those for the model that specifies a priori independence among the parameters reported in Table 11.2?

11.8 Consider what occurs if data arrive sequentially.

a. Using the same prior as you specified in Exercise 11.7, analyze a 3-PNO model for a random subset of 500 examinees from the LSAT data. How do the results differ from those in Exercise 11.6?

b. Use the information in the posterior from part (a) to specify a new prior distribution for the measurement model parameters.

c. Obtain the posterior distribution now using the remaining 500 examinees. How do the results differ from those in Exercise 11.6 and part (a)?

11.9 In Section 11.7.3, we described how the results for an analysis of data from examinee responses to a set of items could be used as the basis for the prior distribution of an analysis of (a) the same items with different examinees, (b) different items with the same examinees, or (c) different items with different examinees, but *not* (d) a reanalysis of the same items and the same examinees. Why would this latter analysis be dubious?

11.10 Show that, under the stated assumptions and model specifications, the full conditional distribution for μ_d in the hierarchical prior is given by (11.58).

12

Missing Data Modeling

Missing values for potentially observable variables are ubiquitous in psychometric and related social science scenarios for a variety of reasons, including those that may or may not be planned, anticipated, or controlled by the assessor. This chapter covers Bayesian perspectives on missing data as they play out in psychometrics through the lens of Rubin's (1976) missing-data framework. The key insight is this: when there is a "hole in the data," a Bayesian perspective tells us to build a predictive distribution for what we might have seen, given all the data we did see, our model for the data had it all been observed, and a model that expresses our belief about the mechanisms that caused part of the data to be missing (Rubin, 1977).

In Section 12.1, we review foundational concepts pertaining to missing data and introduce a running example that we will revisit in several of the following sections. In Section 12.2, we focus on conducting inference when the missingness is ignorable, and in Section 12.3, we focus on inference when the missingness is nonignorable. In Section 12.4, we review multiple imputation, drawing connections to the fully Bayesian analyses of missing data. In Section 12.5, we discuss connections and departures among missing data, latent variables, and model parameters. Section 12.6 concludes the chapter with a summary and bibliographic note.

12.1 Core Concepts in Missing Data Theory

When some values for the data are missing, we may partition the full set of *potentially* observable variables $\mathbf{x}$ into two subsets, $\mathbf{x} = (\mathbf{x}_{obs}, \mathbf{x}_{mis})$, where $\mathbf{x}_{obs}$ are the *observed data* and $\mathbf{x}_{mis}$ are the *missing data*. Though we do not know the values of $\mathbf{x}_{mis}$, we do know *that* they are missing. That is, we know *which* values are observed and which are missing. If we again let $\mathbf{x}$ be the collection of potentially observable values from n examinees to J variables, let I_{ij} be an indicator of the presence or absence of the value from examinee i to observable j, where $I_{ij} = 1$ if the value of x_{ij} is observed (known) and $I_{ij} = 0$ if the value x_{ij} is missing (unknown). Let $\mathbf{I}$ denote the full collection of the I_{ij}. We may view the missingness indicator as inducing the partition $\mathbf{x} = (\mathbf{x}_{obs}, \mathbf{x}_{mis})$. In this light, $\mathbf{x}$ represents *all* possible values for the observables, some of which are known ($\mathbf{x}_{obs}$), others of which are obscured ($\mathbf{x}_{mis}$). Which ones are known and which are obscured are reflected by $\mathbf{I}$, so a model for $\mathbf{I}$ is therefore a model for the mechanism(s) at play that result in the presence or absence of values for $\mathbf{x}$.

Consider the joint distribution of $\mathbf{x}$ and $\mathbf{I}$ conditional on parameters that govern their distribution, denoted as $p(\mathbf{x}, \mathbf{I} \mid \theta, \phi)$ where, in addition to the terms previously introduced, ϕ are parameters relevant for characterizing the missingness mechanism and θ are the parameters that govern the distribution of the data. Here, we return to the use of θ to denote all the parameters that govern the distribution of the data—latent variables, measurement model parameters, and parameters that govern the distributions of those entities. In our examples below, we primarily focus on scoring in unidimensional IRT, and so in most of the cases we discuss, θ just refers to the latent variable in IRT. When we consider examples that expand the situation such that θ includes measurement model parameters, we will note it as such.

The joint distribution for $\mathbf{x}$ and $\mathbf{I}$ may be constructed as

$$p(\mathbf{x},\mathbf{I} \mid \theta,\phi) = p(\mathbf{x} \mid \theta)p(\mathbf{I} \mid \mathbf{x},\phi). \tag{12.1}$$

The first term on the right-hand side of (12.1) is the familiar conditional distribution of $\mathbf{x}$ given parameters θ. What is new is the second term, the conditional distribution of the missingness given ϕ and $\mathbf{x}$, which may be interpreted as a model for the missingness. In most cases, the joint parameter space of θ and ϕ can be factored into a θ-space and a ϕ-space. In some cases, such as the intentional-omitting example we will be considering, missingness can depend directly on θ. We will then either consider θ an element of ϕ, or denote the dependence explicitly as $p(\mathbf{I} \mid \mathbf{x},\theta,\phi)$.

The role of $\mathbf{x}$ in the conditioning here is crucial. Equation (12.1) states that the missingness mechanism may depend on $\mathbf{x}$, including potentially both the observed values $\mathbf{x}_{\text{obs}}$ and missing values $\mathbf{x}_{\text{mis}}$. We briefly cover the various ways this plays out.

Missing data are *missing completely at random* (MCAR) if the probability of missingness does not depend on either the observed or unobserved values:

$$p(\mathbf{I} \mid \mathbf{x},\phi) = p(\mathbf{I} \mid \mathbf{x}_{\text{obs}},\mathbf{x}_{\text{mis}},\phi) = p(\mathbf{I} \mid \phi). \tag{12.2}$$

Missing data are *missing at random* (MAR) if the probability of missingness depends on the observed values $\mathbf{x}_{\text{obs}}$, but once they are conditioned on, the probability of missingness is independent of the actual unobserved values $\mathbf{x}_{\text{mis}}$:

$$p(\mathbf{I} \mid \mathbf{x},\phi) = p(\mathbf{I} \mid \mathbf{x}_{\text{obs}},\mathbf{x}_{\text{mis}},\phi) = p(\mathbf{I} \mid \mathbf{x}_{\text{obs}},\phi). \tag{12.3}$$

Missing data are *missing not at random* (MNAR) if the probability of missingness depends on the unobserved values $\mathbf{x}_{\text{mis}}$, even when the observed values $\mathbf{x}_{\text{obs}}$ are taken into account:

$$p(\mathbf{I} \mid \mathbf{x},\phi) = p(\mathbf{I} \mid \mathbf{x}_{\text{obs}},\mathbf{x}_{\text{mis}},\phi). \tag{12.4}$$

As $\mathbf{x} = (\mathbf{x}_{\text{obs}}, \mathbf{x}_{\text{mis}})$, (12.4) is tautological. Nevertheless, it is useful for communicating that $\mathbf{x}_{\text{mis}}$ is relevant to the missingness mechanisms, and as a foil for the previous two characterizations of missingness mechanisms. Note that MAR is a special case of MNAR, and MCAR is a special case of MAR.

Example

We will use a simple running example from IRT to illustrate the key ideas of modeling missing data. We will focus our attention on inference about θ for a single examinee, and can therefore omit the subscripting by i to simplify notation. There are three test items, with dichotomous responses x_j. The 1-PL (Rasch) model is assumed to hold for the item responses, and the item difficulty parameters expressed as b_j are known to be 0, 1, and –1, respectively. In the examinee population, $\theta \sim f(\theta)$. The missingness indicator I_j is 1 if a response is observed and 0 if it is missing. We will consider inference about θ for a single examinee, Sally, for whom $\mathbf{I} = (1,1,0)$ and $\mathbf{x} = (1,0,*)$. That is, we have observed her responses to only the first two items, so $\mathbf{x}_{\text{obs}} = (x_1, x_2) = (1,0)$ and $\mathbf{x}_{\text{mis}} = (x_3)$.

We will see that even when the Rasch model is assumed to be wholly sufficient for the process of how item responses are produced, our inference about θ depends materially on the process of how the response to x_3 came to be missing. We will consider several situations: using multiple test forms, adaptive testing, Sally running out of time, and Sally intentionally omitting a response.

12.2 Inference under Ignorability

12.2.1 Likelihood-Based Inference

Given values for $\mathbf{I}$ and $\mathbf{x}$, including the hypothetically observable, but not actually observed values in $\mathbf{x}_{mis}$, (12.1) may be viewed as a *complete-data* likelihood function of the (unknown) parameters θ and ϕ. Of course, not all the elements in $\mathbf{x}$ are known. Assuming that the joint parameters space of θ and ϕ can be factored into a θ-space and a ϕ-space, the *observed-data* likelihood is given by

$$\begin{aligned} p(\mathbf{x}_{obs}, \mathbf{I} \mid \theta, \phi) &= \int p(\mathbf{x}, \mathbf{I} \mid \theta, \phi) d\mathbf{x}_{mis} \\ &= \int p(\mathbf{x} \mid \theta) p(\mathbf{I} \mid \mathbf{x}, \phi) d\mathbf{x}_{mis} \\ &= \int p(\mathbf{x}_{obs}, \mathbf{x}_{mis} \mid \theta) p(\mathbf{I} \mid \mathbf{x}_{obs}, \mathbf{x}_{mis}, \phi) d\mathbf{x}_{mis}. \end{aligned} \tag{12.5}$$

The observed-data likelihood may be recognized as an average over all of the possible complete-data likelihoods $p(\mathbf{x}_{obs}, \mathbf{x}_{mis} \mid \theta)$ that accord with the observed data $\mathbf{x}_{obs}$, each weighted by the probability of the (known) missingness pattern ($\mathbf{I}$) given the observed data ($\mathbf{x}_{obs}$) and the possible missing data ($\mathbf{x}_{mis}$).

A likelihood analysis that *ignores* the missing data mechanisms works with a likelihood based only on $\mathbf{x}_{obs}$, namely

$$p(\mathbf{x}_{obs} \mid \theta) = \int p(\mathbf{x} \mid \theta) d\mathbf{x}_{mis} = \int p(\mathbf{x}_{obs}, \mathbf{x}_{mis} \mid \theta) d\mathbf{x}_{mis}. \tag{12.6}$$

This forms the basis of full-information ML approaches to modeling missing data (Enders, 2010).

When is it safe to work with the likelihood in (12.6) rather than the one in (12.5)? Rubin (1976, Theorem 7.1) showed that in addition to the assumption of the factorable parameter space, when the missing data are MAR (or its special case, MCAR), the missingness mechanisms are considered *ignorable* for likelihood-based inference. This means that an analysis can safely ignore the reasons for missingness when conducting inference about θ through the likelihood. That is, the likelihood based on $\mathbf{x}_{obs}$ in (12.6) yields the same inferences about θ as does the likelihood based on ($\mathbf{x}_{obs}$, $\mathbf{I}$) in (12.5). Ignoring the missingness mechanism essentially treats the realized values of $\mathbf{I}$ as fixed, ignoring its stochastic process. If the data are MAR (or MCAR), nothing is lost here for inferences based on just likelihoods. In contrast, mechanisms that are MNAR are nonignorable; the pattern of missingness, $\mathbf{I}$, contains information about θ beyond the information contained in $\mathbf{x}_{obs}$. Inference for θ depends on the model for the missingness, captured in part by the model for ϕ.

Example (continued)

Multiple test forms. Suppose that there are two forms of the example test, the first consisting of Items 1 and 2, the second consisting of Items 1 and 3. Sally was assigned to Test Form 1 by a coin flip with outcomes denoted as H and T for the two sides of the coin. Letting ϕ represent the outcome, the assignment rule was to administer Form 1 if ϕ = H and Form 2 if ϕ = T.* The missingness process $p(\mathbf{I} \mid \mathbf{x}_{obs}, \mathbf{x}_{mis}, \phi)$ can be written as

$$p(\mathbf{I} \mid \mathbf{x}_{obs}, \mathbf{x}_{mis}, \phi) = \begin{cases} p(\mathbf{I} = (1,1,0) \mid \mathbf{x}, \phi) = 1 & \text{if} \quad \phi = \text{H} \\ p(\mathbf{I} = (1,0,1) \mid \mathbf{x}, \phi) = 1 & \text{if} \quad \phi = \text{T} \\ p(\mathbf{I} = (i_1, i_2, i_3) \mid \mathbf{x}, \phi) = 0 & \text{if} \quad (i_1, i_2, i_3) \notin \{(1,1,0),(1,0,1)\}. \end{cases}$$

Because the coin flip does not depend on what Sally's responses were, or would have been had she also responded to Item 3, it follows that $p(\mathbf{I} = (1,1,0) \mid \mathbf{x}, \phi) = p(\mathbf{I} = (1,1,0) \mid \phi)$ and $p(\mathbf{I} = (1,0,1) \mid \mathbf{x}, \phi) = p(\mathbf{I} = (1,0,1) \mid \phi)$, so the missingness is MCAR. The complete-data likelihood function simplifies to $p(\mathbf{x}_{obs} \mid \theta)$, which is the likelihood ignoring the missingness process.

Adaptive testing. Suppose Sally took a computerized adaptive test (CAT), where the middle-difficulty item, Item 1, is presented first. If the examinee's response is right, the more difficult Item 2 is administered, and if it is wrong, the easier Item 3 is administered. In this case,

$$\begin{cases} p(\mathbf{I} = (1,1,0) \mid \mathbf{x}) = 1 & \text{if} \quad x_1 = 1 \\ p(\mathbf{I} = (1,0,1) \mid \mathbf{x}) = 1 & \text{if} \quad x_1 = 0 \\ p(\mathbf{I} = (i_1, i_2, i_3) \mid \mathbf{x}) = 0 & \text{if} \quad (i_1, i_2, i_3) \notin \{(1,1,0),(1,0,1)\}. \end{cases}$$

The missingness process now depends on Sally's response to Item 1, which is part of $\mathbf{x}_{obs}$, but not on her response to Item 3, which is part of $\mathbf{x}_{mis}$. This missingness is not MCAR, but it is still MAR, so again the likelihood ignoring the missingness is appropriate under likelihood inference.

Running out of time. Suppose Sally was administered a test form with all three items, and worked sequentially from the beginning. Sally answered the first two, producing $\mathbf{x}_{obs}$ = (1,0), but ran out of time before looking at Item 3. We have assumed that the Rasch model accounts for response probabilities, but there is a speed factor ϕ at work as well. It may well be the case that ϕ is correlated to θ—say, fast examinees are generally more proficient than slow examinees—so that $p(\theta,\phi) = p(\theta \mid \phi)p(\phi)$. Here the missingness process depends on ϕ and not $\mathbf{x}$, so MAR holds; and ϕ is distinct from θ in the sense of separate parameter spaces. The missingness is again ignorable under likelihood inference, and the MLE of Sally's θ depends only on her first two responses.

12.2.2 Bayesian Inference

The situation for Bayesian inference is similar, and it should be unsurprising that the conditions for ignorability in Bayesian inference are similar to those for likelihood-based inference, only requiring an additional requirement on the prior distribution. To see this, consider the joint posterior distribution for θ and ϕ

* This is a simplified form administration process, but relates to more complicated procedures in which the order of forms are predetermined and administered in a spiraled manner in a classroom (e.g., Mazzeo et al., 2006).

$$p(\theta, \phi \mid \mathbf{I}, \mathbf{x}_{\text{obs}}) = \int p(\theta, \phi, \mathbf{x}_{\text{mis}} \mid \mathbf{I}, \mathbf{x}_{\text{obs}}) d\mathbf{x}_{\text{mis}}$$
$$\propto \int p(\mathbf{x}_{\text{obs}}, \mathbf{x}_{\text{mis}}, \mathbf{I} \mid \theta, \phi) p(\theta, \phi) d\mathbf{x}_{\text{mis}} \tag{12.7}$$
$$= \int p(\mathbf{x}_{\text{obs}}, \mathbf{x}_{\text{mis}} \mid \theta) p(\mathbf{I} \mid \mathbf{x}_{\text{obs}}, \mathbf{x}_{\text{mis}}, \phi) d\mathbf{x}_{\text{mis}} p(\theta, \phi).$$

The first line in (12.7) marginalizes with respect to $\mathbf{x}_{\text{mis}}$. The second line follows from Bayes' theorem, and the third line substitutes in the observed-data likelihood from (12.5). This suggests that the observed-data posterior distribution may be recognized as an average over all of the complete-data posterior distributions that accord with the observed data $\mathbf{x}_{\text{obs}}$, with weights given by the probability of the (known) missingness pattern (**I**) given the observed data ($\mathbf{x}_{\text{obs}}$) and the possible missing data ($\mathbf{x}_{\text{mis}}$).

A Bayesian analysis that ignores the missing data mechanisms works with a posterior distribution for θ based only on $\mathbf{x}_{\text{obs}}$, namely

$$p(\theta \mid \mathbf{x}_{\text{obs}}) \propto p(\mathbf{x}_{\text{obs}} \mid \theta) p(\theta)$$
$$= \int p(\mathbf{x}_{\text{obs}}, \mathbf{x}_{\text{mis}} \mid \theta) d\mathbf{x}_{\text{mis}} p(\theta). \tag{12.8}$$

When is it safe to work with the posterior distribution in (12.8) rather than the one in (12.7)? Rubin (1976, theorem 8.1) showed that when the missing data are MAR (or its special case, MCAR), the parameter space can be factored into a θ-space and a ϕ-space, and these parameters are independent in the prior,

$$p(\theta, \phi) = p(\theta) p(\phi), \tag{12.9}$$

then the missingness mechanisms are ignorable for Bayesian inference about θ. This can be understood by simplifying the general characterization of the joint posterior for θ and ϕ in (12.7) in accordance with these assumptions,

$$p(\theta, \phi \mid \mathbf{I}, \mathbf{x}_{\text{obs}}) \propto \int p(\mathbf{x}_{\text{obs}}, \mathbf{x}_{\text{mis}} \mid \theta) p(\mathbf{I} \mid \mathbf{x}_{\text{obs}}, \phi)\, d\mathbf{x}_{\text{mis}} p(\theta, \phi)$$
$$= \{p(\theta) \int p(\mathbf{x}_{\text{obs}}, \mathbf{x}_{\text{mis}} \mid \theta)\, d\mathbf{x}_{\text{mis}}\} \{p(\phi) p(\mathbf{I} \mid \mathbf{x}_{\text{obs}}, \phi)\}. \tag{12.10}$$

The first line in (12.10) follows from the general characterization of joint posterior for θ and ϕ in (12.7) and the MAR assumption, in that the distribution of **I** may depend on $\mathbf{x}_{\text{obs}}$, but not $\mathbf{x}_{\text{mis}}$. The second line follows from the a priori independence of the parameters in (12.9). This last line reveals that the joint posterior may be factored into two terms in braces: one for θ and one for ϕ. Focusing on the former, we have the simpler posterior for θ based only on $\mathbf{x}_{\text{obs}}$ in (12.8). This means that, under the stated assumptions, an analysis can safely ignore the reasons for missingness when conducting Bayesian inference about θ. That is, the posterior based on $\mathbf{x}_{\text{obs}}$ in (12.8) yields the same inferences about θ as does the posterior based on ($\mathbf{x}_{\text{obs}}$, **I**) in (12.10). This essentially treats the realized values of **I** as fixed, ignoring its stochastic process. If the data are MAR (or MCAR), nothing is lost here.

In contrast, there are two cases when missingness mechanisms are nonignorable under Bayesian inference:

1. *Nonfactorable parameter space.* When the missingness process is MAR and θ and ϕ are distinct parameter spaces, ignorability holds for likelihood inference but not for Bayesian inference if $p(\theta, \phi)$ does not factor as in (12.9). (For example, in an application of IRT where θ is proficiency and ϕ is tendency to omit items, suppose we believe that low-proficiency examinees are more likely to omit items; then $p(\theta, \phi) \neq p(\theta)p(\phi)$.) Here, inference for θ depends not only on $\mathbf{x}_{obs}$ but on indirect information about θ through $\mathbf{I}$, in the form of

$$p(\theta \mid \mathbf{I}) = \int p(\theta \mid \phi) p(\phi \mid \mathbf{I})\, d\phi.$$

2. *MNAR mechanisms.* When the missingness process is MNAR and nonignorable under likelihood inference, it is usually nonignorable under Bayesian inference as well.* Again inference for θ depends on the model for the missingness, as both $\mathbf{I}$ and $\mathbf{x}_{mis}$ hold information about θ.

Methods for conducting inference under nonignorability will be discussed in Section 12.3.

Under ignorability, a Bayesian analysis seeks the posterior distribution based only on the $\mathbf{x}_{obs}$ given in (12.8). Recognizing that $p(\mathbf{x}_{obs}, \mathbf{x}_{mis}, \theta) = p(\mathbf{x}_{obs}, \mathbf{x}_{mis} \mid \theta)p(\theta)$, the last line in (12.8) may be interpreted as saying that the posterior distribution for the model parameters given the observed data may be obtained by first constructing the joint distribution of the model parameters, observed data, and the missing data, and then integrating out the latter.

When an analytical evaluation of the integral is not feasible, we may once again turn to simulation strategies. A Gibbs sampling approach may be enacted by iteratively sampling from the full conditionals

$$\theta \sim p(\theta \mid \mathbf{x}_{obs}, \mathbf{x}_{mis})$$

and (12.11)

$$\mathbf{x}_{mis} \sim p(\mathbf{x}_{mis} \mid \mathbf{x}_{obs}, \theta).$$

This level of generality is sufficient for our purposes, which is to highlight that there is a full conditional for $\mathbf{x}_{mis}$. In practice, we may employ variations on this theme. For example, a univariate approach to Gibbs sampling would work with the univariate full conditionals, where each parameter in θ is conditioned on the remaining parameters in θ in addition to $\mathbf{x}_{obs}$ and $\mathbf{x}_{mis}$, and each missing value in $\mathbf{x}_{mis}$ is conditioned on the remaining missing values in $\mathbf{x}_{mis}$ in addition to θ and $\mathbf{x}_{obs}$. Of course, the full conditional distributions may simplify in light of the conditional independence assumptions in the model. In particular,

* MAR and distinctness are sufficient but not necessary conditions for ignorability. One can construct cases in which a missingness process is MNAR and missingness is nonignorable under likelihood inference, yet a specially built $p(\boldsymbol{\theta}, \boldsymbol{\phi})$ "cancels out" the confounding and ignorability obtains under Bayesian inference.

the local independence assumption in psychometric models implies that, given the model parameters, the observables are conditionally independent. Under local independence, the full conditional for the missing data simplifies to

$$\mathbf{x}_{\text{mis}} \sim p(\mathbf{x}_{\text{mis}} \mid \theta). \tag{12.12}$$

Gibbs sampling proceeds by iteratively drawing from the full conditionals. The marginal distribution for θ from a chain of draws approximates the posterior distribution for θ in (12.8).

Example (continued)

The previous block of examples showed that under IRT assumptions, the cases of multiple test forms, adaptive testing, and running out of time were ignorable under likelihood inference. We revisit them here, and see that Bayesian ignorability still holds if (12.9) holds. If (12.9) does not hold, we have Bayesian nonignorable inference of the *Nonfactorable parameter space* type.

Multiple test forms. The missingness process for multiple test forms is still MAR, because it does not depend on an examinee's responses to the items not presented. We saw that the missingness was ignorable under likelihood inference. It is ignorable under Bayesian inference as well, *as long as form assignment does not depend on* θ. Such is the case if the form is assigned by a coin flip, the result of which has nothing to do with the examinee's θ. Instead, suppose eighth-grade students are assigned to the harder Test Form 1 with probability .67 and the easier Test Form 2 with probability .33; fourth-grade students are assigned randomly to the forms with probabilities .25 and .75, respectively. Again ϕ is a coin flip, but with differently biased coins at the two grades. If $p(\theta \mid \text{Grade} = 4) \neq p(\theta \mid \text{Grade} = 8)$, then $p(\theta, \phi \mid \text{Grade} = k) = p(\theta \mid \text{Grade} = k)p(\phi \mid \text{Grade} = k)$ but $p(\theta, \phi) \neq p(\theta)p(\phi)$. Ignorability holds under Bayesian inference only given grade, and the correct "conditionally ignorability" posterior is $p(\theta \mid \mathbf{x}_{\text{obs}}, \text{Grade} = k) \propto p(\mathbf{x}_{\text{obs}} \mid \theta)p(\theta \mid \text{Grade} = k)$. If we know that Sally is a fourth grader, we should calculate her posterior distribution using the Grade 4 prior $p(\theta \mid \text{Grade} = 4)$. (If we know she was assigned Form 1 under the grade-dependent administration scheme but we do not know her grade, the prior we should use is a mixture of the Grade 4 and 8 distributions with weights proportional to .25 and .67, the probability of assignment to Form 1 to students in the two grades respectively.)

Adaptive testing. Similar results hold for ignorability under Bayesian inference, with the exception that if the initial item or the item selection probabilities depend on some examinee covariate Z, it must be conditioned on in both the prior and posterior: $p(\theta \mid \mathbf{x}_{\text{obs}}, Z{=}z) \propto p(\mathbf{x}_{\text{obs}} \mid \theta)p(\theta \mid Z{=}z)$.

Running out of time. Recall that $\mathbf{x}_{\text{mis}}$ is MAR if the responses are missing because the examinee ran out of time before getting to them, and speed ϕ is assumed to be a separate parameter from θ. Ignorability holds under likelihood inference whether or not (12.9) holds. If (12.9) does not hold, θ and ϕ are not distinct in the required Bayesian sense and the missingness is not ignorable under Bayesian inference. Inference must proceed with a marginal posterior for θ that is proportional to the IRT likelihood times a prior that takes into account how many items the examinee has reached, or $p(\mathbf{x}_{\text{obs}} \mid \theta)p(\theta \mid \mathbf{I})$ where $p(\theta \mid \mathbf{I}) = \int p(\theta \mid \phi)p(\mathbf{I} \mid \phi)p(\phi)d\phi$.

What this means for Sally is that her likelihood function based on Items 1 and 2 remains the same, but a prior for θ has also been incorporated that reflects the distribution of examinees who, like Sally, did not reach Item 3. If this distribution is lower than the undifferentiated population, for example, Sally's posterior distribution will be shifted downward and her Bayes modal (MAP) estimate will be lower than her MLE would have been.

12.3 Inference under Nonignorability

When the missingness is nonignorable, the situation is more complex as we cannot conduct inference for θ without consideration of the model for missingness. We will not be able to factor the posterior distribution as before, and thus must consider the full joint posterior distribution for all unknown entities (θ, ϕ, $\mathbf{x}_{mis}$) given the known data ($\mathbf{x}_{obs}$, $\mathbf{I}$),

$$p(\theta,\phi,\mathbf{x}_{mis} \mid \mathbf{I},\mathbf{x}_{obs}) \propto p(\mathbf{I} \mid \mathbf{x}_{obs},\mathbf{x}_{mis},\theta,\phi)p(\mathbf{x}_{obs},\mathbf{x}_{mis} \mid \theta)p(\phi \mid \theta)p(\theta). \tag{12.13}$$

The key component is the first term on the right-hand side, which is the model for the missingness pattern given the observed data, missing data, parameters that govern the distribution of the data, and additional parameters that govern the missingness mechanism.

Given a structure for the missingness process, we may employ MCMC procedures to empirically approximate the posterior distribution. A Gibbs sampler would iteratively draw from the full conditionals for the unknowns θ, ϕ, and $\mathbf{x}_{mis}$. What needs to be conditioned on and the actual form of each full conditional will depend on the dependence and conditional independence structures in the model (e.g., does $\mathbf{I}$ depend on some or all of the other entities in the model?).

The marginal distribution for θ from the chain of draws approximates the marginal posterior distribution for θ,

$$p(\theta \mid \mathbf{I},\mathbf{x}_{obs}) = \int\int p(\theta,\phi,\mathbf{x}_{mis} \mid \mathbf{I},\mathbf{x}_{obs})d\phi d\mathbf{x}_{mis}. \tag{12.14}$$

Similarly, MCMC may be used to approximate the marginal distribution of ϕ or its dependence on θ if it is of inferential interest.

Example (continued)

Intentional Omits. Suppose Sally chooses not to respond to a dichotomously scored item on the three-item test if her probability of getting the item correct is less than .5. Note that her probability depends on both her proficiency variable θ and the item's parameters, both included in θ in the following equations for convenience. The missingness model is then

$$\begin{aligned} p(\mathbf{I} \mid \mathbf{x}_{obs},\mathbf{x}_{mis},\theta,\phi) &= p(\mathbf{I} \mid \theta) \\ &= \begin{cases} I_j = 1 \text{ if } p(\text{correct} \mid \theta) \geq .5 \\ I_j = 0 \text{ if } p(\text{correct} \mid \theta) < .5. \end{cases} \end{aligned} \tag{12.15}$$

and the term for ϕ drops out of the model.

A more complex model for the missingness expands on this to specify that the value will be missing if Sally *thinks* that her probability of getting the item correct is less than .5. In this case, ϕ may contain person-specific parameters that capture the accuracy of that person's assessment of her or his probabilities of correctly answering the items. Let ϕ denote the parameter for Sally. In this case, the missingness model is then

$$\begin{aligned} p(\mathbf{I} \mid \mathbf{x}_{obs},\mathbf{x}_{mis},\theta,\phi) &= p(\mathbf{I} \mid \theta,\phi) \\ &= \begin{cases} I_j = 1 \text{ if } p(\text{believe correct} \mid \theta,\phi) \geq .5 \\ I_j = 0 \text{ if } p(\text{believe correct} \mid \theta,\phi) < .5. \end{cases} \end{aligned} \tag{12.16}$$

Furthermore, the examinees' judgment parameters may be related to their proficiency, task features, or their true probability of correctly responding. For example, examinees with higher proficiency may be better judges of whether they know the correct answer. The dependence of examinee judgment parameters on the model parameters is captured by $p(\phi \mid \theta)$. At the low end, examinees' beliefs about their chances of correctness might be totally unrelated to the actual correctness. This situation approaches MAR. At the high end, the examinees' beliefs might be so related to the actual correctness that they know which ones they would get correct, and selectively answer them accordingly. In an analog of this situation, Sally video-records herself attempting 100 basket free throws. She makes only 20, but she knows exactly which ones they are. She edits the video down to an impressive string of 20 successful attempts in a row. We are wrong to take 20-out-of-20 to estimate her free-throw percentage.

As long as intentional omitting depends on $\mathbf{x}_{mis}$, the missingness is not MAR, and it is not ignorable under likelihood inference. We can estimate the missingness function $p(\mathbf{I} \mid \mathbf{x}_{obs}, \mathbf{x}_{mis}, \theta, \phi)$ empirically from a study like that of Sheriffs and Boomer (1954). Their subjects answered or omitted items under standard conditions, then later provided responses to the items they omitted. It is then an empirical question as to whether the results would support acceptable inference from a scoring function based on only $\mathbf{x}_{obs}$ and $\mathbf{I}$ (e.g., Moustaki & Knott, 2000). As a computational expedient for operational work, Lord (1974) suggested treating omitted responses in multiple-choice tests as fractionally correct. This is consistent with positing that $p(x_j = 1 \mid I_j = 0) = c$, where c is the reciprocal of the number of response alternatives. Note that neither θ nor any ϕ appears in this formulation. The justification is that this is the optimal strategy for omitting when an examinee's beliefs are accurate.

12.4 Multiple Imputation

In this section, we review a closely related strategy for conducting inference in light of missing data, highlighting its connections to the previously described procedures.

A natural approach to dealing with missing data is to impute a value for each missing data point. That is, for each missing data point, we fill in the empty spot with a value. The result is a complete dataset that is then subjected to whatever analysis would have been conducted had the dataset been fully observed from the outset. Enders (2010) reviewed a number of these procedures, as well as their limitations. A problem common to many of these procedures is that they understate our uncertainty, which manifests itself in underestimating posterior standard deviations or standard errors.

Multiple imputation tackles the problems posed by missing data for inference by imputing *multiple* values for each missing data point and guiding the subsequent analyses in a way that properly accounts for our uncertainty (Rubin 1978, 1987, 1996b). As we will see, it has a distinctly Bayesian grounding, but it supports frequentist as well as Bayesian approaches to inference, making it popular among analysts of various persuasions (Enders, 2010; Gelman et al., 2013). In this section, we briefly give an overview of multiple imputation, highlighting its Bayesian nature and connections to the procedures in Section 12.2.2.

Multiple imputation approaches to inference in light of missing data may be seen as proceeding in three phases (Enders, 2010). The first is the imputation phase, in which multiple complete datasets are constructed based on augmenting the observed data with values for the missing data. The second is an analysis phase, in which the desired analysis is conducted for each of the just-constructed complete datasets. The third phase is a pooling phase in which the results from the analyses in the second phase are synthesized. The following sections detail these phases, and then revisit the plausible values approach to estimating distributions of latent variables introduced in the previous chapter. It is a straightforward application of multiple imputation for missing data.

12.4.1 Imputation Phase

In the first phase, multiple values are imputed for the missing data. The imputations are draws from the posterior predictive distribution

$$p(\mathbf{x}_{\text{mis}} \mid \mathbf{I}, \mathbf{x}_{\text{obs}}) = \int\int p(\boldsymbol{\theta}, \boldsymbol{\phi}, \mathbf{x}_{\text{mis}} \mid \mathbf{I}, \mathbf{x}_{\text{obs}}) d\boldsymbol{\theta} d\boldsymbol{\phi}. \tag{12.17}$$

Under the general case of MNAR, this amounts to the marginalization of the joint posterior in (12.13). From a Bayesian perspective, the posterior predictive distribution in (12.17) is the answer to the question of how to impute missing values: what we believe about each missing value is fully represented by its predictive distribution given the assumed model, the observations, and the prior distributions for the unknown parameters (Little & Rubin, 2002; Rubin, 1977).

Multiple imputation is typically employed using the observed-data likelihood reflecting an assumption of MAR, and we focus on this case in the following development. Owing to the factorization of the posterior in (12.10), the joint posterior distribution for the unknowns $\boldsymbol{\theta}$ and $\mathbf{x}_{\text{mis}}$ given $\mathbf{x}_{\text{obs}}$ is

$$\begin{aligned} p(\boldsymbol{\theta}, \mathbf{x}_{\text{mis}} \mid \mathbf{x}_{\text{obs}}) &\propto p(\mathbf{x}_{\text{obs}}, \mathbf{x}_{\text{mis}} \mid \boldsymbol{\theta}) p(\boldsymbol{\theta}) \\ &= p(\mathbf{x}_{\text{mis}} \mid \boldsymbol{\theta}, \mathbf{x}_{\text{obs}}) p(\mathbf{x}_{\text{obs}} \mid \boldsymbol{\theta}) p(\boldsymbol{\theta}) \\ &\propto p(\mathbf{x}_{\text{mis}} \mid \boldsymbol{\theta}, \mathbf{x}_{\text{obs}}) p(\boldsymbol{\theta} \mid \mathbf{x}_{\text{obs}}) \\ &= p(\mathbf{x}_{\text{mis}} \mid \boldsymbol{\theta}) p(\boldsymbol{\theta} \mid \mathbf{x}_{\text{obs}}), \end{aligned} \tag{12.18}$$

where the last line follows from local independence of observables in psychometric models. Marginalizing over the parameters yields the posterior predictive distribution for the missing data

$$\begin{aligned} p(\mathbf{x}_{\text{mis}} \mid \mathbf{x}_{\text{obs}}) &= \int p(\boldsymbol{\theta}, \mathbf{x}_{\text{mis}} \mid \mathbf{x}_{\text{obs}}) d\boldsymbol{\theta} \\ &= \int p(\mathbf{x}_{\text{mis}} \mid \boldsymbol{\theta}) p(\boldsymbol{\theta} \mid \mathbf{x}_{\text{obs}}) d\boldsymbol{\theta}. \end{aligned} \tag{12.19}$$

A value drawn from this posterior predictive distribution constitutes a single imputation.

Multiple imputations are obtained by repeatedly drawing from this distribution. This is typically carried out by cycling through two steps, which correspond to an iteration of the Gibbs sampler that draws from the full conditional distributions for the parameters and the missing data. The full conditional distributions are:

$$\begin{aligned} \boldsymbol{\theta} &\sim p(\boldsymbol{\theta} \mid \mathbf{x}_{\text{obs}}, \mathbf{x}_{\text{mis}}), \\ \mathbf{x}_{\text{mis}} &\sim p(\mathbf{x}_{\text{mis}} \mid \mathbf{x}_{\text{obs}}, \boldsymbol{\theta}). \end{aligned} \tag{12.20}$$

Note that $p(\mathbf{x}_{mis} \mid \mathbf{x}_{obs}, \theta) = p(\mathbf{x}_{mis} \mid \theta)$ under local independence. First, we simulate a value for θ from its full conditional distribution given the observed data $\mathbf{x}_{obs}$ and initial values for the missing data $\mathbf{x}_{mis}$. Next, we simulate a value for $\mathbf{x}_{mis}$ from its full conditional distribution given $\mathbf{x}_{obs}$ and the just-simulated value for θ. To generate the next imputation, we simulate a new value for θ from its full conditional distribution, using the just-simulated value for $\mathbf{x}_{mis}$, and then use that value to simulate $\mathbf{x}_{mis}$ from its full conditional. Upon convergence of the chain, the distribution of the draws for $(\theta, \mathbf{x}_{mis})$ approximates the joint posterior distribution. Accordingly, the marginal distribution for $\mathbf{x}_{mis}$ from a chain of draws approximates the posterior predictive distribution for $\mathbf{x}_{mis}$ in (12.19). Each draw for $\mathbf{x}_{mis}$ constitutes an imputation.

Importantly, the process for obtaining multiple imputations using MCMC is the same as the process for conducting Bayesian inference about θ using MCMC. Both involve obtaining draws from the joint posterior distribution of parameters and the missing data given the observed data. The full conditionals in (12.20) are the same as the those in (12.11). In Section 12.2.2, where we focused on inference, we marginalized over the missing data to yield the marginal posterior distribution for the parameters. In a simulation environment, this amounted to simulating from the joint posterior distribution and then paying attention only to the values for the parameters. In this treatment focusing on imputations, we marginalize over the parameters to yield the posterior predictive distribution for the missing data to yield imputations. In a simulation environment, this amounts to simulating from the joint posterior distribution and then paying attention only to the values for the missing data. The central point is that the procedures are the same. The mechanisms for obtaining multiple imputations are the same mechanism for conducting Bayesian inference. Producing imputations also yields Bayesian inference for the parameters. Conducting Bayesian inference for the parameters also produces imputations. Instead of discarding the draws for the parameters in multiple imputation, they could be retained and used to conduct inference for the parameters. Instead of discarding the draws for the missing data in conducting inference, they could be retained and viewed as imputations. The same holds for Bayesian inference under the more general case of nonignorability. The draws for $\mathbf{x}_{mis}$, when simulating values from the joint posterior distribution in (12.13) amount to multiple imputations.

Under this approach, when we formulate a single model there is no need to distinguish between an imputation phase and an analysis phase. We conduct inference for the parameters at the same time as we obtain imputations. In fact, as noted in Section 12.2.2, we do not even need to pay attention to the imputations to perform inference. The payoff of multiple imputation strategies to missing data analysis, and in particular, the distinction between imputation and analysis phases, comes in situations when a dataset is intended to be used in different contexts (Rubin, 1996b): by different analysts, at different times, with different models; or when auxiliary variables not of inferential interest may be useful in predicting the missing values (Meng, 1994a; Rubin, 1996b; Schafer, 2003). Essentially, the imputation phase handles the missingness and the resulting multiple completed datasets are turned over as a set to other analysts for whatever analyses they have in mind in the analysis phase, with guidelines for how to combine the results obtained from the multiple datasets in the pooling phase.

With this distinction between imputation and analysis phases, the model and parameters used for imputations need not be the same as those used in the analysis phase. To make this explicit, in the imputation phase we specify an *imputation* model with parameters ψ that specifies the conditional distribution for the data given ψ and a prior distribution for ψ. In general, ψ need not equal θ. Once imputations are obtained, we proceed to the analysis phase for inference about θ.

12.4.1.1 Obtaining Independent Imputations from MCMC

It is important that the imputations be independent of one another to use the simple pooling procedures discussed in Section 12.4.3. Independence is achieved when the values for $\mathbf{x}_{mis}$ are drawn independently. In the more general case of dependent draws from MCMC sampling schemes, there are two strategies for generating independent imputations. In one approach, multiple chains are run independently, capitalizing on the fact that though the iterations are dependent *within* chains, they are independent *between* chains. Suppose we wish to obtain K imputations. We run K chains independently and, following an evaluation of convergence, obtain one draw from each chain, that is, a vector with a value for all of the nonobserved variables in the problem drawn in that particular cycle in that chain. In another approach, a single chain is run, and the draws are thinned by an amount needed to render the remaining draws approximately independent.

Example (continued)

Sally's missing response is MAR in the cases of multiple forms, adaptive testing, and running out of time. Obtaining a draw for x_3 is therefore accomplished using the results of the ignorability analysis under Bayesian inference. If no covariate Z is involved in form assignment or in item selection algorithms, then an imputation for x_3 is drawn in these two cases from

$$\begin{aligned} p(\mathbf{x}_{mis} \mid \mathbf{x}_{obs}, \theta) &= p(\mathbf{x}_{mis} \mid \theta)\ p(\theta \mid \mathbf{x}_{obs}) \\ &= p(x_3 \mid \theta)\ p(\theta \mid x_1 = 1, x_2 = 0) \\ &\propto p(x_3 \mid \theta)\ p(x_1 = 1, x_2 = 0 \mid \theta) p(\theta). \end{aligned}$$

That is, first the likelihood for θ that is induced by observing x_1 and x_2, $p(x_1 = 1, x_2 = 0 \mid \theta)$, is combined with the prior $p(\theta)$ to produce a predictive distribution for θ given what we have learned from $\mathbf{x}_{obs}$. Then, we draw a value of θ from it. Finally, we draw a value of x_3 from the IRT function $p(x_3 \mid \theta)$ evaluated with that θ draw.

Because of MAR, the procedure is the same in running out of time except that the information about θ contained in the fact that Sally did not reach Item 3 must be taken into account. Let ϕ again represent a speed parameter which may be correlated with θ. We see the difference in the imputation model in terms of the predictive distribution for θ:

$$\begin{aligned} p(\mathbf{x}_{mis} \mid \mathbf{I}, \mathbf{x}_{obs}, \theta, \phi) &= p(\mathbf{x}_{mis} \mid \theta)\ p(\theta \mid \mathbf{x}_{obs}, \mathbf{I}, \phi) \\ &= p(x_3 \mid \theta)\ p(\theta \mid x_1 = 1, x_2 = 0, \mathbf{I}, \phi) \\ &\propto p(x_3 \mid \theta)\ p(x_1 = 1, x_2 = 0 \mid \theta) p(\theta \mid \mathbf{I}, \phi) \\ &= p(x_3 \mid \theta)\ p(x_1 = 1, x_2 = 0 \mid \theta) p(\theta \mid \phi) p(\phi \mid \mathbf{I} = (1,1,0)) \\ &\propto p(x_3 \mid \theta)\ p(x_1 = 1, x_2 = 0 \mid \theta) p(\theta \mid \phi) p(\mathbf{I} = (1,1,0) \mid \phi) p(\phi). \end{aligned}$$

The key difference is that the distribution for θ is now conditional on ϕ, and information about ϕ is contained in its posterior $p(\phi \mid \mathbf{I} = (1,1,0)) \propto p(\mathbf{I} = (1,1,0) \mid \phi) p(\phi)$. In practice, a value for ϕ is first drawn from its predictive distribution given $\mathbf{I} = (1,1,0)$. Then a value for θ is drawn from $p(\theta \mid \phi)$ evaluated with this ϕ draw. Finally we draw a value of x_3 from the IRT function $p(x_3 \mid \theta)$ evaluated with that θ draw.

12.4.2 Analysis Phase

As a result of the first phase, some number K multiple imputations for the missing data are obtained, and can be used to construct K multiple completed datasets. In the analysis phase, each of these datasets is analyzed as if they were the lone dataset, using either Bayesian or frequentist methods. Here, inference centers on the model parameters. For simplicity of exposition let us consider a single parameter θ. Let $\tilde{\theta}_k$ be some point estimate or summary from the analysis of completed dataset K, and let $\tilde{W}_k$ be the associated variance. For example, in a frequentist context, $\tilde{\theta}_k$ may be the MLE for θ from fitting a model to dataset K, and $\tilde{W}_k$ is then the squared standard error of the MLE. In a Bayesian context, $\tilde{\theta}_k$ may be the posterior mean for θ from fitting a model (including a prior distribution) to dataset K and $\tilde{W}_k$ is then the posterior variance of θ. In a simulation-based approach to fitting Bayesian models, note that we also would have K sets of simulations, each representing draws from the posterior distribution for θ based on one of the K datasets.

12.4.3 Pooling Phase

In the final phase, we pool the results from the analyses of the K completed datasets. If the analysis phase yielded a set of K point estimates or summaries and their associated variances, then they may be pooled in ways derived by Rubin (1987). The multiple imputation point estimate or summary is given by

$$\bar{\theta} = \frac{1}{K}\sum_{k=1}^{K}\tilde{\theta}_k. \tag{12.21}$$

The total variability associated with this estimate is constructed from two components,

$$V = \bar{W} + \frac{K+1}{K}B, \tag{12.22}$$

where the first component captures the variability *within* imputations,

$$\bar{W} = \frac{1}{K}\sum_{k=1}^{K}\tilde{W}_k, \tag{12.23}$$

and the second captures the variability *between* imputations,

$$B = \frac{1}{K-1}\sum_{k=1}^{K}(\tilde{\theta}_k - \bar{\theta})^2. \tag{12.24}$$

If the previous phase involved a Bayesian analysis simulating values from the posterior (for each completed dataset), the pooling phase amounts to combining the K sets of simulations from the analyses of the completed datasets, effectively creating a single set of simulations (Gelman et al., 2013). This combined set of simulations may then be summarized as we would any individual set of simulations in the usual ways (e.g., using graphical summaries or point or interval summaries).

12.4.4 Plausible Values Revisited

The previous chapter described the plausible values methodology used in many large-scale surveys of examinee proficiency to estimate population characteristics, from test forms with too few items to estimate individual examinees' proficiencies. Having developed machinery in this chapter to handle missing data from a Bayesian perspective, we now see the plausible values machinery as a straightforward application of the Bayesian way to view missing data and the multiple imputation approach to deal with it (Mislevy, 1991; Rubin, 1977, 1987).

The key insight here is realizing that in a latent variable model such as IRT, the examinee proficiency variable θ can be viewed as missing, for everyone, regardless of its value. Thus latent variables are always MAR, for everyone. The traditional use of θ in IRT for an individual's proficiency and the generic use of θ in this chapter for variables that are inferential targets makes the connection a little less obvious. However, the imputation model for plausible values for individual students (Equation 11.46) is an instance of (12.8), with $\theta_i \equiv \mathbf{x}_{\text{mis}} \equiv \theta$ and $(\mathbf{x}_i, \mathbf{y}_i) \equiv \mathbf{x}_{\text{obs}}$. Note that the predictive distribution involves the population level parameters, such as β_θ and $\sigma^2_{\varepsilon\theta}$ in the latent regression model used in surveys like NAEP. Features of the population distribution are then calculated using the examinee-level imputations by the three phases described above.

12.5 Latent Variables, Missing Data, Parameters, and Unknowns

The previous discussion highlighted that, in plausible value analyses, the latent variables were treated as missing values. In this section, we set out some connections and departures among perspectives on latent variables associated with missing data, Bayesian computation, and the modeling perspective developed in this book. To begin, Bollen (2002) advocated an interpretation of latent variables in terms of missingness. On this account, a latent variable is a variable in which there are no values for some or all of its observations. The latent variables we employ in psychometric models are on conceptual par with observable variables; they are just observable variables with missing values. This accords nicely with Bayesian modeling and inference in two senses: one that concerns computational strategies, and one that takes a high-level view of Bayesian inference. First, computational strategies such as the expectation-maximization (EM) algorithm, data augmentation, and MCMC algorithms for computation in Bayesian modeling treat latent variables as missing data. In Gibbs sampling, the full conditional distributions that define the posterior include distributions for the examinees' latent variables and distributions for the examinees' missing data. An iteration of the MCMC sampler yields drawn values for the latent variables and drawn values for the missing data. Practically then, what we call a latent variable and what we call a variable with missing data does not matter.

Second, latent variables structure the joint distributions of observables. Structuring joint distributions has been the role of parameters, and in this light latent variables are also referred to as person parameters, particularly when they of inferential interest. If we have latent variables, we have a posterior distribution for them. If we have missing data values, we have a posterior distribution for them. And we have had a posterior distribution for unknown parameters all along. To summarize, on this account *latent variables* may be seen as akin to *missing data*. And they may be seen as *parameters*, in particular, *unknown parameters*. We can achieve a synthesis by recognizing that all three have the following in common: (1) they are entities in a probability model constructed by an analyst to aid in reasoning and (2) they are not known with certainty. From a high-level perspective, in Bayesian inference we place entities into one of two big buckets, one labeled KNOWNS and the other labeled UNKNOWNS. Missing data, latent variables and parameters all get placed in the latter and have the same status: they are unknowns in a probability model, and through the machinery of Bayes' theorem we can condition on the knowns in the model to obtain the posterior distribution for them. This synthesis is most sensible when in the context of both elements (1) and (2) just mentioned. The point is that *once inside* the probability model, we can use the same machinery—Bayes' theorem, setting up and drawing from full conditionals, and so on—to coherently tackle all of them.

Importantly, this does not suggest that we should discontinue the use of separate terms, as they convey potentially crucial differences. We gain by understanding what kinds of distinctions among these various entities are useful and which are meaningless *for various purposes*. We may accrue considerable benefits from capitalizing on the connections among these entities when suitable. The distinctions are somewhat artificial when it comes to estimation, and ignoring them may be quite effective in being able to trick computer programs into doing what we want. In Section 13.5.2, we discuss how treating latent variables as missing data opens up possibilities for resolving estimation difficulties in the context of latent class analyses. Seeing past the distinctions may also facilitate understanding connections among different domains. In this respect, Little and Rubin's (2002) book *Statistical Analysis with Missing Data* exploits this power to great effect; we see that missing data in surveys, experimental design, latent variable modeling, semisupervised learning, and resolving mixtures may all be seen as variations of the same basic ideas.

On the other hand, we may draw conceptual benefits from having distinct terms for different entities. To see this, first consider the various entities in the bucket labeled as KNOWNS, which includes things like data, fixed values of parameters, and specified values of hyperparameters in the highest level of a hierarchical prior specification. Again, the high-level perspective views Bayesian inference as conditioning on the *knowns in the model* to obtain the posterior distribution for unknowns. We flesh out the twin emphases on "knowns" and "in the model." Here, the emphasis on "knowns" calls out that this is a technical point about statistical machinery—the knowns are those things that are conditioned on in Bayes' theorem. However, lumping them all together may obscure key distinctions among them. The emphasis on "in the model" calls out that they are knowns in the context of our reasoning about some situation of interest. Data are things we observe or collect from the world. On the other hand, fixing parameters may be based on the need to resolve indeterminacies. Similarly, hyperparameters in the highest level of hierarchical priors are not known because they are observed, but because we stipulate them as "what we know" from previous experience, expectations, and awareness of how their values affect what happens in estimation—usually in a manner vaguely explicated if at all. Thus, the grounding, conceptual meaning, and role played by the various entities that all get tossed into KNOWNS bucket vary considerably.

Much the same can be said for latent variables, missing data, and parameters that occupy the UNKNOWNS bucket. They may be treated the same by the statistical machinery, but they may reflect differences in grounding, conceptual meaning, and the roles they play. An important distinction between missing data and latent variables may be seen by reconsidering our characterization of the role of latent variables. From a measurement perspective, latent variables are employed in psychometric models where what is ultimately of inferential interest is not directly observable. From a statistical perspective, latent variables are used in structuring the distribution of variables we can observe. De Finetti's representation theorem for exchangeability is a powerful tool for structuring high dimensional joint distributions of observables in terms of simpler conditional distributions given latent variables that induce conditional independence. Pearl (1988) pointed out that conditional independence is such a central tool in building and reasoning through models, when the variables that induce conditional independence are unknown or unavailable, we create them. This reveals a key distinction between latent variables and missing data. Whereas missing data are things that we do not observe but potentially could have under different circumstances or choices, this is not the case for the latent variables, at least how they are employed in psychometrics. We do not choose whether to *observe* them; we choose whether to *introduce* them when constructing a model. It is not that they are entities out there in the world that we happened to not have observed but could have under different choices or circumstances; there are not values out there in the world to observe. The terminology may represent an interpretative shorthand—we often use the term "missing data" to stand for values that in some sense we conceivably could have observed, "latent variable" to stand for things we could not, and "parameter" or "person parameter" to highlight that it is of inferential interest.*

On this account, latent variables are entities introduced for and used in a modeling context to organize our thinking and reason through the messy real-world problem of making sense of examinees' behaviors. The behaviors are situated and our reasoning task is one of interpreting certain behaviors by certain examinees in certain contexts. Our thinking and reasoning are likewise situated, against a backdrop of many things we know or believe with some degree of confidence, and a potentially related backdrop of purposes and constraints. And it is in this context of a complex reasoning task that we employ latent variables to summarize and capture our thinking about the examinees. Adopting the role of the assessor or analyst, the nature and values of latent variables have more to do with what is going on in our heads than what's going on in the examinees' heads. A latent variable's purpose—to organize our thinking in the context of a model constructed to reason from situated data to situated ends—and its role as a parameter on which other things depend—namely observable variables that are rendered conditionally independent—reveal that a latent variable is quite a different beast than what is usually denoted by the term "missing data."

To summarize, the distinctions among these are often not relevant for doing work *within* the probability model, but are indeed typically relevant for *how* they arrive in the model and what we think about what a model and model-based reasoning has to say about them.

12.6 Summary and Bibliographic Note

Missing data are prevalent in assessment. Inference regarding model parameters can proceed using just the observed data when the missingness is ignorable. In a Bayesian analysis, ignorability holds when the missing data are MAR, the parameters that govern the missingness mechanism are distinct from and a priori independent of those that govern the data. In this

* Without advancing a causal account, we suspect that the relative paucity of use of the term "person parameter" in CFA is related to the relative lack of applications in which inferences about the values of the person parameters/latent variables are sought. The relatively more frequent use of "person parameter" in IRT is related to the relative plethora of applications in which inferences about the values of the person parameters/latent variables are sought.

case, correct inferences result from analyses that take the pattern of missingness as given, and simply use $p(\mathbf{x}_{obs} \mid \theta)$ as the likelihood. Most statistical packages now offer this option. However, if the missingness is nonignorable, inference about the parameters requires including a model for missingness. A brief review of multiple imputation was proffered, and it was emphasized that the imputation phase is explicitly Bayesian. A Bayesian perspective on modeling brings to light connections among missing data, latent variables, and parameters.

In addition to the foundational work of Rubin (1978, 1987, 1996b), excellent didactic treatments of multiple imputation can be found in Enders (2010), Schafer and Graham (2002), and Sinharay, Stern, and Russell (2001). In psychometrics, procedures and examples for modeling ignorable and nonignorable missingness have also been developed in CFA (Cai & Song, 2010; Lee, 2006; Song & Lee, 2002a) and IRT (Fu, Tao, & Shi, 2010; Maier 2002; Patz & Junker, 1999a). Additional examples, analyses, and notable discussion can be found in Mislevy (in press) in the context of IRT and Lee (2007) in the context of CFA and SEM.

Exercises

12.1 Beginning with the joint posterior distribution for θ, ϕ, and $\mathbf{x}_{mis}$, show that the posterior distribution for θ and ϕ is separable under the assumptions of factorization of the parameter space into a θ-space and a ϕ-space, MAR, and a priori independence of θ and ϕ.

12.2 You receive six values x from a population in which $x \sim N(\mu, 1)$:

$$0.99, -0.68, -0.23, -0.89, -1.16, -2.27.$$

Use the prior $\mu \sim N(0,1)$. Obtain the posterior for μ, including the posterior mean and variance, assuming the six values are a random sample from the population.

12.3 Use the same population distribution, prior, and six observations from Exercise 12.2, but now assume that the six values are $\mathbf{x}_{obs}$ from a random sample of 10 from the population, where the missing values are MCAR.

a. Use multiple imputation to create five imputed data sets.

b. Obtain the posterior distribution for μ in each completed data set separately.

c. Obtain the posterior distribution for μ under Rubin's formulas for pooling the information across multiple imputations.

d. Compare the results of Exercises 12.2, 12.3(b), and 12.3(c). How do they differ, and why?

12.4 Analysts who are new to multiple imputation are sometimes suspicious that answers are too accurate because "we are making up data." Do you agree or disagree? Explain your answer.

12.5 Build an imputation model for Sally's response for x_3 under random assignment to the two multiple forms.

12.6 Build an imputation model for Sally's response for x_3 under random assignment to the two forms, but with probabilities .67 and .33 to Forms 1 and 2 for eighth graders and .25 and .75 for fourth graders, and $p(\theta \mid \text{Grade} = 4) = N(-1,1)$ and $p(\theta \mid \text{Grade} = 8) = N(1,1)$, under the following states of knowledge:

a. You know Sally is in Grade 4.

b. You know Sally is in Grade 8.

c. You do not know what grade Sally is in.

12.7 Suppose Sally's response for x_3 is missing because she ran out of time. To simplify, suppose only Item 3 is subject to the time limit. Assume $\theta \sim N(0,1)$. Suppose the missingness process is such that the probability an examinee *will* reach Item 3 is $\Phi^{-1}(\phi)$—that is, the cumulative normal probability of the speed parameter ϕ, where $\phi \sim N(0,1)$.

a. Build an imputation model for Sally's response for x_3, additionally assuming the correlation between ϕ and θ is 0.

b. Build an imputation model for Sally's response for x_3, additionally assuming the correlation between ϕ and θ is .5.

c. Build an imputation model for Sally's response for x_3, additionally assuming the correlation between ϕ and θ is –.5.

d. How would estimates of Sally's θ differ under conditions (a)–(c)?

12.8 Suppose Sally intentionally omits x_3. Assume $\theta \sim N(0,1)$. Build an imputation model for Sally's response for x_3, assuming the missingness process is the one described in (12.15).

12.9 Suppose Sally intentionally omits x_3. Assume $\theta \sim N(0,1)$. Build an imputation model for Sally's response for x_3, assuming the missingness process is the one described in (12.16). Note that you will need to provide additional structure for the missing process.

13

Latent Class Analysis

Latent class analysis (LCA) departs from the psychometric models previously presented in that it employs discrete latent variables. We focus on the case of discrete observables as well. Analysts employ discrete latent variables when they want to organize their thinking about examinees in terms of categories or groupings, as opposed to organizing their thinking about examinees along a continuum by using continuous latent variables.* We have in fact already encountered simple latent class models in the subtraction proficiency example in Chapters 1 and 7 and the medical diagnosis example in Chapter 2. In those treatments we confined our analyses to the context of scoring examinees, where the measurement model parameters and the parameters governing the distribution of the latent variables were known. Here, we expand our treatment and conduct inference for these parameters of the model as well as for examinees. We will see a number of the themes present in CTT, CFA, and IRT, including conditional independence assumptions and hierarchical specifications.

Treatments and overviews of LCA from conventional perspectives can be found in Collins and Lanza (2010), Dayton (1999), Dayton and Macready (2007), Lazarsfeld and Henry (1968), and McCutcheon (1987). These and other conventional sources emphasize Bayesian elements for some aspects of inference, but not others. LCA is often employed in an exploratory manner, where the goal is to determine the number of latent classes (i.e., levels of the latent variable) and examine the pattern of dependence of the observables on those classes. This chapter treats more confirmatory flavors of LCA, in which a particular number of latent classes are specified, possibly with constraints on the conditional probabilities that capture the dependence structure of the observables (Croon, 1990; Dayton & Macready, 2007; Goodman, 1974; Hoijtink, 1998; Lindsay, Clogg, & Grego, 1991; van Onna, 2002).

In Section 13.1, we review conventional approaches to LCA. In Section 13.2, we present a Bayesian analysis for the general case of polytomous observable and latent variables. In Section 13.3, we present a Bayesian analysis for the special case of dichotomous observable and latent variables. An example is given in Section 13.4. In Section 13.5, we discuss strategies for resolving indeterminacies associated with the use of discrete latent variables. We conclude this chapter in Section 13.6 with a summary and bibliographic note.

* The distinction between continuous and discrete latent variables is not always sharp, as models that posit differences here may be exactly or approximately equivalent (e.g., Haertel, 1990; Heinen, 1996; von Davier, 2008).

13.1 Conventional LCA

13.1.1 Model Specification

As in IRT model specifications, let $\mathbf{x}$ be the full collection of potentially observable values from n examinees to J observable discrete variables. As before in the general polytomous case, we code each x_{ij} as integers $1,\ldots, K_j$. For ease of exposition, we proceed assuming that all the observables have the same number of response categories, in which case the subscript of "j" can be dropped from K_j. This may be relaxed if the observables have differing numbers of response categories. In Section 13.3, when we examine the special case of dichotomous x_{ij}s, we will once again adopt the coding of the value as either 0 or 1. The data setup here is the same as in IRT models; the difference lies in the specification of the latent variables and the dependence of the observables on the latent variables. Let θ_i stand for a discrete latent variable for examinee i; and let $\boldsymbol{\theta} = (\theta_1,\ldots,\theta_n)$ stand for the collection of latent variables from n examinees. In general, $\theta_i \in \{1,\ldots,C\}$, where C is the total number of latent classes. We refer to the latent classes as class 1, class 2, and so on. In general, this labeling should be interpreted as a nominal feature, not one that necessarily represents any particular ordering or magnitude.

Let π_{cjk} denote the conditional probability of observing a value of k for observable j given that the examinee is a member of latent class c, and let $\boldsymbol{\pi}_j$ denote the collection of the π_{cjk} parameters for observable j. The role of the π_{cjk} parameters may be seen in Table 13.1, which presents the conditional probabilities for the values of an observable (j) given the latent class value for the examinee when there are $C = 2$ latent classes and $K = 3$ possible values of the observable. Accordingly, $\boldsymbol{\pi}_j$ are the measurement model parameters for observable j. The entries in the first column in Table 13.1 reflect that, for each possible value of c, the conditional probabilities are restricted such that for each class c,

$$\sum_{k=1}^{K} \pi_{cjk} = \sum_{k=1}^{K} P(x_{ij} = k \mid \theta_i = c) = 1.$$

A traditional formulation of a C-class model specifies the probability that examinee i has a value of k on observable j as

$$P(x_{ij} = k) = \sum_{c=1}^{C} P(x_{ij} = k \mid \theta_i = c, \boldsymbol{\pi}_j)\gamma_c, \qquad (13.1)$$

TABLE 13.1

Conditional Probability Table for Observable j with $K = 3$ Categories, Given Membership in One of $C = 2$ Classes (θ), Where π_{cjk} Denotes the Probability of Responding to Observable j in Category k Given Membership in Class c

	Category (k)		
Latent Class (θ)	**1**	**2**	**3**
1	$\pi_{1j1} = 1-(\pi_{1j2} + \pi_{1j3})$	π_{1j2}	π_{1j3}
2	$\pi_{2j1} = 1-(\pi_{2j2} + \pi_{2j3})$	π_{2j2}	π_{2j3}

where $\gamma_c = P(\theta_i = c)$ is the proportion of examinees in class c, also referred to as the size of class c, which are subjected to the restriction that

$$\sum_{c=1}^{C} \gamma_c = \sum_{c=1}^{C} P(\theta_i = c) = 1.$$

In other words, (13.1) says that the marginal probability of the response $x_{ij} = k$ is the weighted average of the probability of this response from examinees in each class c, weighted by the class proportions.

The framework can also accommodate multiple discrete latent variables by constructing a single discrete latent variable that captures all the possible patterns generated from the multiple discrete variables. Consider a situation in which we conceive of three dichotomous latent variables, each of which can take a value of 1 or 2. There are $2^3 = 8$ possible *profiles* corresponding to combinations of values on these variables: [1,1,1], [1,1,2], [1,2,1], [1,2,2], [2,1,1], [2,1,2], [2,2,1], and [2,2,2], where the sequence of three numbers refers to the levels of the three variables (e.g., [2,1,2] refers to an examinee being in the second class on latent variable 1, the first class on latent variable 2, and the second class on latent variable 3). We could reconceive the situation as one in which there is a single latent variable with eight classes, each corresponding to one of the profiles. In this way, we can naturally fold the situation of multiple discrete latent variables into one in which there is a single discrete latent variable, and we are squarely back to the model described in (13.1). We therefore restrict our attention in the rest of this chapter to the situation in which there is a single discrete latent variable. In Chapter 14, we revisit this situation and describe approaches to modeling multiple distinct discrete latent variables.

13.1.2 Indeterminacy in LCA

The labeling of the latent variable in LCA as formulated so far is arbitrary. There is nothing inherent in the model that determines which class we call class 1, which class we call class 2, and so on. As a result, LCA is subject to problems associated with label switching. For any LCA model, another model holds that is substantively equivalent, but differs in how the classes are labeled (i.e., what is class 1 in one model is class 2 in another model). This may be seen as multimodality in the likelihood resulting from different labeling schemes (Loken, 2004). This is sometimes viewed as an issue of identification, though we conceive of it as an indeterminacy akin to those associated with the metric of continuous latent variables. Continuous latent variables do not come with an inherent metric, and discrete latent variables do not come with inherent labels. We delve deeper into the indeterminacies from a Bayesian perspective in Section 13.5.

13.1.3 Model Fitting

The dominant paradigm in calibrating LCA models involves maximum likelihood estimation of $\boldsymbol{\pi} = (\boldsymbol{\pi}_1, \ldots, \boldsymbol{\pi}_J)$ and $\boldsymbol{\gamma} = (\gamma_c, \ldots, \gamma_C)$ based on the marginal probability distribution of the observables. The development is akin to that for CFA and IRT, and relies on the same assumptions of independence among examinees and conditional (local) independence among observables. The joint probability of the data conditional on the latent variables and measurement model parameters is

$$P(\mathbf{x} \mid \boldsymbol{\theta}, \boldsymbol{\pi}) = \prod_{i=1}^{n} P(\mathbf{x}_i \mid \theta_i = c, \boldsymbol{\pi}) = \prod_{i=1}^{n} \prod_{j=1}^{J} P(x_{ij} \mid \theta_i = c, \boldsymbol{\pi}_j), \tag{13.2}$$

where $\mathbf{x}_i = (x_{i1},\ldots, x_{iJ})$ is the collection of J observables for examinee i,

$$P(x_{ij} \mid \theta_i = c, \pi_j) = \prod_{k=1}^{K} \pi_{cjk}^{\;I(x_{ij}=k)}, \tag{13.3}$$

and I is the indicator function that here takes a value of 1 when the response from examinee i for observable j is k, and 0 otherwise. The marginal probability of the observables is a version of (7.8), instantiated with a discrete latent variable:

$$P(\mathbf{x} \mid \pi, \gamma) = \prod_{i=1}^{n} \sum_{\theta_i=1}^{C} p(\mathbf{x}_i \mid \theta_i, \pi) p(\theta_i \mid \gamma). \tag{13.4}$$

The distribution of the latent variables is a categorical distribution with category probabilities or proportions contained in $\boldsymbol{\gamma}$. Once values for the data are observed, (13.4) may be treated as a likelihood function for $\boldsymbol{\pi}$ and $\boldsymbol{\gamma}$. In general, there is no closed-form solution, and numerical methods are typically employed (Dempster, Laird, & Rubin, 1977; Goodman, 1974).

Turning to scoring, the consensus approach is to use Bayes' theorem to obtain the posterior distribution for an examinee's latent variable using point estimates for the response-probability and class-proportion parameters, as in (7.10). The resulting posterior distribution may be reported as such or summarized via the MAP, which amounts to estimating the value of the latent variable as the class with the highest posterior probability.

13.2 Bayesian LCA

This section develops a Bayesian approach to LCA for the general case of possibly polytomous latent and observable variables. We then turn to a popular special case of dichotomous latent and observable variables for a fuller discussion in Section 13.3, and a worked-through example in Section 13.4.

13.2.1 Conditional Probability of the Observables

The preceding developments ground the specification of the conditional probability of the data, expressed here from a distributional perspective as

$$p(\mathbf{x} \mid \theta, \pi) = \prod_{i=1}^{n} p(\mathbf{x}_i \mid \theta_i, \pi) = \prod_{i=1}^{n} \prod_{j=1}^{J} p(x_{ij} \mid \theta_i = c, \pi_j), \tag{13.5}$$

where for observables taking values coded $1,\ldots, K$,

$$(x_{ij} \mid \theta_i = c, \pi_j) \sim \text{Categorical}(\pi_{cj}), \tag{13.6}$$

and $\pi_{cj} = (\pi_{cj1},\ldots,\pi_{cjK})$ are the conditional probabilities for observable j associated with class c, captured in row c of tables like that in Table 13.1.

13.2.2 Prior Distribution

As a first step in constructing the prior distribution, we factor the joint prior for the parameters introduced so far as

$$p(\theta, \pi) = p(\theta)p(\pi) \tag{13.7}$$

following an assumption of independence between the examinees' latent variables and the measurement model parameters. Specifying the prior distribution involves specifying distributions for the terms on the right-hand side of (13.7) and any hyperprior distributions for any unknown hyperparameters introduced in these specifications.

13.2.2.1 Prior Distribution for Latent Variables

An exchangeability assumption allows for the factorization of the joint prior distribution for the examinees' latent variables into the product of common prior distributions,

$$p(\theta) = \prod_{i=1}^{n} p(\theta_i \mid \theta_P), \tag{13.8}$$

where θ_P denotes the hyperparameters for specifying the prior for each θ_i. In the current context, the latent variable is assumed to be categorically distributed, in which case $\theta_P = \gamma = (\gamma_1, \ldots, \gamma_C)$ are the latent class proportions and

$$\theta_i \mid \gamma \sim \text{Categorical}(\gamma). \tag{13.9}$$

13.2.2.2 Prior Distribution for Parameters That Govern the Distribution of the Latent Variables

If the elements of γ are unknown, they require a prior distribution. A convenient choice is to use the Dirichlet distribution,

$$\gamma \sim \text{Dirichlet}(\alpha_\gamma), \tag{13.10}$$

where $\boldsymbol{\alpha}_\gamma = (\alpha_{\gamma 1}, \ldots, \alpha_{\gamma C})$ is a collection of hyperparameters. The Dirichlet is a generalization of the beta distribution to the polytomous case. Just as the beta distribution is a conjugate prior for the Bernoulli and binomial likelihoods, the Dirichlet is a conjugate prior for the polytomous extensions of the Bernoulli and binomial, namely the categorical and multinomial, respectively (Spiegelhalter & Lauritzen, 1990). Suppose a collection of probabilities or proportions $\boldsymbol{\gamma}$ is assigned a Dirichlet prior, $\gamma \sim \text{Dirichlet}(\alpha)$, and some variable $\mathbf{y}$ is modeled as $\mathbf{y} \sim \text{Multinomial}(\gamma)$, where $\mathbf{y} = (y_1, \ldots, y_C)$ contains counts of the number of observations in the C categories. The posterior distribution for $\boldsymbol{\gamma}$ is also a Dirichlet, specifically $\gamma \mid \mathbf{y} \sim \text{Dirichlet}(\alpha_1 + y_1, \alpha_2 + y_2, \ldots, \alpha_C + y_C)$. In our context, we do not have the counts of the observations in each of the C categories, as θs are unknown. However, we will make use of this conjugacy relationship in MCMC sampling.

13.2.2.3 Prior Distribution for Measurement Model Parameters

To develop the second term on the right-hand side of (13.7), an assumption of exchangeability allows for the factorization of the joint prior distribution for the measurement model parameters into the product over common prior distributions,

$$p(\pi) = \prod_{j=1}^{J} p(\pi_j \mid \pi_P) = \prod_{j=1}^{J} \prod_{c=1}^{C} p(\pi_{cj} \mid \pi_P), \tag{13.11}$$

where $\boldsymbol{\pi}_P$ denotes the hyperparameters spelled out in more detail below. Each $\boldsymbol{\pi}_j$ is composed of C vectors of length K that are the conditional probabilities for the observable taking a value of k given membership in class c, for $k = 1,\ldots, K$, and $c = 1,\ldots, C$. More formally, $\pi_j = (\pi_{1j},\ldots,\pi_{Cj})$, where each $\pi_{cj} = (\pi_{cj1},\ldots,\pi_{cjK})$. The right-hand side of (13.11) reflects a prior independence assumption among the $\boldsymbol{\pi}_{cj}$. As each $\boldsymbol{\pi}_{cj}$ is a vector of conditional probabilities for a categorically distributed variable, a natural choice is to again employ Dirichlet prior distributions,

$$\pi_{cj} \sim \text{Dirichlet}(\alpha_{\pi_c}), \tag{13.12}$$

where $\alpha_{\pi_c} = (\alpha_{\pi_{c1}},\ldots,\alpha_{\pi_{cK}})$ are hyperparameters that govern the Dirichlet distribution for the conditional probabilities given membership in latent class c. The exchangeability assumption is reflected in the absence of the j in the subscript for $\boldsymbol{\alpha}_{\pi_c}$. This may be relaxed if exchangeability is only partial or conditional on other variables; in the limit, unique prior distributions may be specified for each observable. Continuing the development assuming exchangeability, the full collection of hyperparameters for the measurement model parameters is $\pi_P = (\alpha_{\pi_c},\ldots,\alpha_{\pi_C})$.

13.2.3 Posterior Distribution and Graphical Model

The DAG for the model is given in Figure 13.1. Collecting the preceding developments, the posterior distribution for all the unknowns is

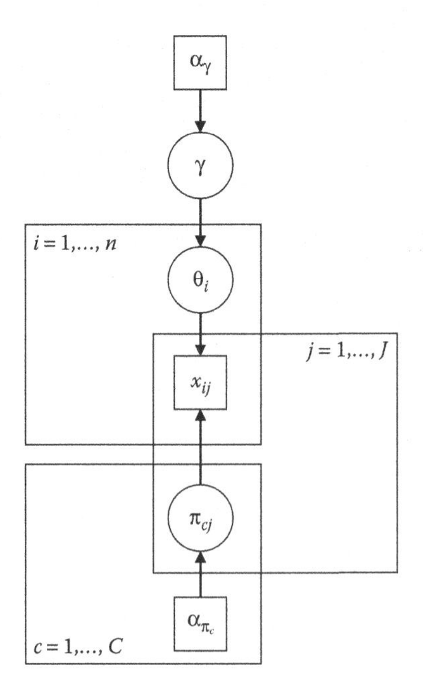

FIGURE 13.1
Directed acyclic graph for a latent class model for J observables from each of n examinees, with a single discrete latent variable with C classes.

$$\begin{aligned}p(\theta,\gamma,\pi \mid \mathbf{x}) &\propto p(\mathbf{x} \mid \theta,\gamma,\pi)p(\theta,\gamma,\pi)\\ &= p(\mathbf{x} \mid \theta,\pi)p(\theta \mid \gamma)p(\gamma)p(\pi)\\ &= \prod_{i=1}^{n}\prod_{j=1}^{J} p(x_{ij} \mid \theta_i,\pi_j)p(\theta_i \mid \gamma)p(\gamma)\prod_{c=1}^{C} p(\pi_{cj}),\end{aligned} \tag{13.13}$$

where

$$\begin{aligned}(x_{ij} \mid \theta_i = c, \pi_j) &\sim \text{Categorical}(\pi_{cj}) \text{ for } i = 1,\ldots,n,\ j = 1,\ldots,J,\\ \theta_i \mid \gamma &\sim \text{Categorical}(\gamma) \text{ for } i = 1,\ldots,n,\\ \gamma &\sim \text{Dirichlet}(\alpha_\gamma),\end{aligned}$$

and

$$\pi_{cj} \sim \text{Dirichlet}(\alpha_{\pi_c}) \text{ for } c = 1,\ldots,C,\ j = 1,\ldots,J.$$

This specification is fairly "open" in the sense that it does not include any constraints that may be imposed to resolve the indeterminacies and prevent label switching or to reflect substantive theory. For ease of exposition, we continue with the current specification when describing the full conditional distributions and then MCMC approaches next. We return to this issue in the context of an example in Section 13.4.

13.2.4 MCMC Estimation

We express each full conditional distribution in the following equations. On the left-hand side, the parameter in question is written as conditional on all the other relevant parameters and data.* Beginning with the examinees' latent variables, we present the full conditional for any examinee i. The same structure applies to all the examinees. The full conditional distribution for the latent variable for examinee i is

$$\theta_i \mid \gamma,\pi,\mathbf{x}_i \sim \text{Categorical}(\mathbf{s}_i), \tag{13.14}$$

where $\mathbf{x}_i = (x_{i1}, \ldots, x_{iJ})$ refers to the collection of J observed values from examinee i, $\mathbf{s}_i = (s_{i1},\ldots,s_{iC})$

$$s_{ic} = \frac{\prod_{j=1}^{J}\prod_{k=1}^{K}(\pi_{cjk})^{I(x_{ij}=k)}\gamma_c}{\sum_{g=1}^{C}\prod_{j=1}^{J}\prod_{k=1}^{K}(\pi_{gjk})^{I(x_{ij}=k)}\gamma_g}, \tag{13.15}$$

and I is the indicator function that here takes a value of 1 when the response from examinee i for observable j is k, and 0 otherwise.

* We suppress the role of specified hyperparameters in this notation; see Appendix A for a presentation that formally includes the hyperparameters.

The full conditional for the latent class proportions is

$$\gamma \mid \theta \sim \text{Dirichlet}(\alpha_{\gamma_1} + r_1, \ldots, \alpha_{\gamma_C} + r_C), \tag{13.16}$$

where $r_1, \ldots, r_C$ are counts of the number of examinees for whom $\theta = c$ (i.e., are members of class c).

Turning to the measurement model parameters, we present the full conditional for $\boldsymbol{\pi}_{cj}$, which is the collection of conditional probabilities for the possible values of observable j given the examinee is a member of latent class c. The same structure applies to all the observables, for all classes. The full conditional distribution for each $\boldsymbol{\pi}_{cj}$ is

$$\pi_{cj} \mid \theta, \mathbf{x}_j \sim \text{Dirichlet}(\alpha_{\pi_{c1}} + r_{cj1}, \ldots, \alpha_{\pi_{cK}} + r_{cjK}), \tag{13.17}$$

where $\mathbf{x}_j = (x_{1j}, \ldots, x_{nj})$ refers to the collection of n observed values for observable j and $r_{cj1}, \ldots, r_{cjK}$ are counts of the of the number of examinees for whom $\theta = c$ (i.e., are members of class c) who have a value of k for observable j. Derivations for these full conditional distributions are given in Appendix A.

A Gibbs sampler is constructed by iteratively drawing from these full conditional distributions using the just-drawn values for the conditioned parameters, generically written below (for iteration $t + 1$).

13.2.4.1 Gibbs Sampling Routine for Polytomous Latent and Observable Variables

1. *Sample the latent variables for examinees.* For each examinee $i = 1, \ldots, n$, use the values of the measurement model parameters and parameters that govern the latent distribution from the previous iteration, $\boldsymbol{\pi}^{(t)}$ and $\boldsymbol{\gamma}^{(t)}$ for $\boldsymbol{\pi}$ and $\boldsymbol{\gamma}$, respectively, to compute $\mathbf{s}_i^{(t)} = (s_{i1}^{(t)}, \ldots, s_{iC}^{(t)})$ where each

$$s_{ic}^{(t)} = \frac{\prod_{j=1}^{J} \prod_{k=1}^{K} (\pi_{cjk}^{(t)})^{I(x_{ij}=k)} \gamma_c^{(t)}}{\sum_{g=1}^{C} \prod_{j=1}^{J} \prod_{k=1}^{K} (\pi_{gjk}^{(t)})^{I(x_{ij}=k)} \gamma_g^{(t)}}.$$

   Sample a value for the latent variable from

$$\theta_i^{(t+1)} \mid \gamma^{(t)}, \pi^{(t)}, \mathbf{x}_i \sim \text{Categorical}(\mathbf{s}_i^{(t)}), \tag{13.18}$$

   where $\mathbf{x}_i = (x_{i1}, \ldots, x_{iJ})$ is the collection of J observed values from examinee i.

2. *Sample the parameters for the latent variable distribution.* Count the number of examinees in each class based on the just-sampled values of the latent variables, $\theta^{(t+1)} = (\theta_1^{(t+1)}, \ldots, \theta_n^{(t+1)})$, yielding the frequencies in each class $(r_1^{(t+1)}, \ldots, r_C^{(t+1)})$. Sample a value for the latent class proportions using these values for $(r_1, \ldots, r_C)$ from

$$\gamma^{(t+1)} \mid \theta^{(t+1)} \sim \text{Dirichlet}(\alpha_{\gamma_1} + r_1^{(t+1)}, \ldots, \alpha_{\gamma_C} + r_C^{(t+1)}). \tag{13.19}$$

3. *Sample the measurement model parameters.* For each observable $j = 1,\ldots, J$, use the values of the latent variables from step (1), $\theta^{(t+1)}$, and count the number of examinees in each class that had a value of k for the observable. These counts are the class- and observable-specific frequencies of values in each category of the observable $(r_{cj1}^{(t+1)},\ldots,r_{cjK}^{(t+1)})$. For each observable j, and each latent class c, use these values for $(r_{cj1},\ldots,r_{cjK})$ and sample a value for the conditional probabilities of observing a value in the categories from

$$\pi_{cj}^{(t+1)} \mid \theta^{(t+1)}, \mathbf{x}_j \sim \text{Dirichlet}(\alpha_{\pi_{c1}} + r_{cj1}^{(t+1)},\ldots,\alpha_{\pi_{cK}} + r_{cjK}^{(t+1)}), \tag{13.20}$$

where $\mathbf{x}_j = (x_{1j}, \ldots, x_{nj})$ is the collection of n observed values for observable j.

13.3 Bayesian Analysis for Dichotomous Latent and Observable Variables

A model with dichotomous latent and observable variables can be specified in terms of the model just developed, namely where $C = 2$ and $K = 2$. This is a popular special case, and affords the possibility of a somewhat simpler specification. We make all of the same conditional independence assumptions, and focus on the distributional forms.

13.3.1 Conditional Probability of the Observables

Coding the dichotomous observables as 0 and 1, we may specify the conditional distribution of each observable as

$$(x_{ij} \mid \theta_i = c, \pi_j) \sim \text{Bernoulli}(\pi_{cj}), \tag{13.21}$$

where π_{cj} is the element in $\boldsymbol{\pi}_j$ that captures the conditional probability of x_{ij} being a value of 1 given an examinee is in class c.

13.3.2 Prior Distribution

As before, we have $\theta = (\theta_1,\ldots,\theta_n)$, now with each θ_i being dichotomous. Accordingly, we may again make use of the Bernoulli specification. For each examinee,

$$\theta_i \mid \gamma \sim \text{Bernoulli}(\gamma). \tag{13.22}$$

We note that under the usual specification of Bernoulli variables, the two classes of the latent variable are coded as 0 and 1, with γ being the proportion of examinees in the class coded as 1. The proportion of examinees in the class coded as 0 is $1-\gamma$. In some situations, this coding may be desirable. In the next chapter, we will use this coding for a latent variable capturing whether an examinee possesses a skill (coded as 1) or does not possess the skill (coded as 0). Alternatively, we may prefer coding the latent variable as 1 and 2. This is usually the case when programming a model in WinBUGS (or other software) in terms of tables of conditional probabilities, such as that in Table 13.1. Here, we need to reference the rows of the table defined by the possible values of the latent variable, and it is sometimes difficult or impossible to reference the 0th row of a table; referring to the rows (latent classes) as 1 and 2 rather than 0 and 1 is more convenient. One option is to move to the categorical specification in (13.9); in the current case of a dichotomous latent variable, the collection of probabilities in the categorical variable specification reduces to $\boldsymbol{\gamma} = (\gamma, 1-\gamma)$. A second option is to recode the result of a Bernoulli specification from its natural 0/1 coding to a 1/2 coding. This could be done formally via the introduction of a new parameter, but we prefer the following expression intended to communicate that θ_i is the result of taking a Bernoulli random variable and adding 1, effectively converting 0/1 to 1/2 coding:

$$\theta_i \mid \gamma \sim \text{Bernoulli}(\gamma) + 1. \tag{13.23}$$

Either way, we are left with a single parameter γ that governs the distribution of the dichotomous latent variable, in effect capturing the probability of being in a particular class, regardless of whether it is coded as 1 in the 0/1 coding or 2 in the 1/2 coding. We therefore specify a beta prior for γ:

$$\gamma \sim \text{Beta}(\alpha_\gamma, \beta_\gamma), \tag{13.24}$$

where α_γ and β_γ are hyperparameters.

Turning to the measurement model parameters, we now have for each observable $\pi_j = (\pi_{1j}, \pi_{2j})$, where each π_{cj} is the conditional probability of the observable (j) taking a value of 1 given membership in class c (coded here as 1/2). As each π_{cj} is the probability for a dichotomous variable, a natural choice is again the beta distribution,

$$\pi_{cj} \sim \text{Beta}(\alpha_{\pi_c}, \beta_{\pi_c}), \tag{13.25}$$

where α_{π_c} and β_{π_c} are hyperparameters.

13.3.3 The Complete Model and Posterior Distribution

Collecting the preceding developments, the posterior distribution for all the unknowns is

$$p(\boldsymbol{\theta}, \gamma, \boldsymbol{\pi} \mid \mathbf{x}) \propto \prod_{i=1}^{n} \prod_{j=1}^{J} p(x_{ij} \mid \theta_i, \boldsymbol{\pi}_j) p(\theta_i \mid \gamma) p(\gamma) \prod_{c=1}^{C} p(\pi_{cj}), \tag{13.26}$$

where $\boldsymbol{\pi} = (\boldsymbol{\pi}_1, \ldots, \boldsymbol{\pi}_J)$, $\boldsymbol{\pi}_j = (\pi_{1j}, \pi_{2j})$, π_{cj} is the probability of a value 1 on observable j given membership in class c, and, adopting the Bernoulli-plus-one construction in (13.23),

$$(x_{ij} \mid \theta_i = c, \boldsymbol{\pi}_j) \sim \text{Bernoulli}(\pi_{cj}) \text{ for } i = 1, \ldots, n, j = 1, \ldots, J,$$

$$\theta_i \mid \gamma \sim \text{Bernoulli}(\gamma) + 1 \text{ for } i = 1, \ldots, n,$$

$$\gamma \sim \text{Beta}(\alpha_\gamma, \beta_\gamma),$$

and

$$\pi_{cj} \sim \text{Beta}(\alpha_{\pi_c}, \beta_{\pi_c}) \text{ for } c = 1, 2, j = 1, \ldots, J.$$

13.3.4 MCMC Estimation

We present the full conditional distributions for the parameters, coding the dichotomous latent variable as taking values of 1 or 2, in which case γ is the proportion of examinees in the class coded as 2. Beginning with the examinees' latent variables, we present the full conditional for any examinee i; the same structure applies to all the examinees. The full conditional distribution for the latent variable for examinee i is

$$\theta_i \mid \gamma, \pi, \mathbf{x}_i \sim \text{Bernoulli}(s_i) + 1, \tag{13.27}$$

where

$$s_i = \frac{p(\mathbf{x}_i \mid \theta_i = 2, \pi) p(\theta_i = 2 \mid \gamma)}{p(\mathbf{x}_i)}$$

$$= \frac{\prod_{j=1}^{J} (\pi_{2j})^{x_{ij}} (1 - \pi_{2j})^{1-x_{ij}} \gamma}{\prod_{j=1}^{J} (\pi_{2j})^{x_{ij}} (1 - \pi_{2j})^{1-x_{ij}} \gamma + \prod_{j=1}^{J} (\pi_{1j})^{x_{ij}} (1 - \pi_{1j})^{1-x_{ij}} (1 - \gamma)}.$$

The full conditional for the proportion of examinees in latent class 2 is

$$\gamma \mid \boldsymbol{\theta} \sim \text{Beta}(\alpha_\gamma + r_2, \beta_\gamma + r_1), \tag{13.28}$$

where r_c is the number of examinees for whom $\theta = c$ (i.e., are members of class c), where $r_1 + r_2 = n$.

Turning to the measurement model parameters, we present the full conditional for π_{cj}, the conditional probability that the value of observable j is 1 given the examinee is a member of latent class c. The same structure applies to all the observables, for all classes. The full conditional for π_{cj} is

$$\pi_{cj} \mid \boldsymbol{\theta}, \mathbf{x}_j \sim \text{Beta}(\alpha_{\pi_c} + r_{cj}, \beta_{\pi_c} + r_c - r_{cj}), \tag{13.29}$$

where r_{cj} is the number of examinees for whom $\theta = c$ (i.e., are members of class c) that have a value of 1 for observable j. Derivations for these full conditional distributions are given in Appendix A.

A Gibbs sampler is constructed by iteratively drawing from the full conditional distributions defined by (13.27)–(13.29) using the just-drawn values for the conditioned parameters, generically written below (for iteration $t + 1$).

13.3.4.1 Gibbs Sampling Routine for Dichotomous Latent and Observable Variables

1. *Sample the latent variables for examinees.* For each examinee $i = 1, \ldots, n$, use the values of the measurement model parameters and parameters that govern the latent distribution from the previous iteration, $\boldsymbol{\pi}^{(t)}$ and $\gamma^{(t)}$ for $\boldsymbol{\pi}$ and γ, respectively, to compute

$$s_i^{(t)} = \frac{p(\mathbf{x}_i \mid \theta_i = 2, \boldsymbol{\pi}^{(t)}) p(\theta_i = 2 \mid \gamma^{(t)})}{p(\mathbf{x}_i)}$$

$$= \frac{\prod_{j=1}^{J} (\pi_{2j}^{(t)})^{x_{ij}} (1 - \pi_{2j}^{(t)})^{1-x_{ij}} \gamma^{(t)}}{\prod_{j=1}^{J} (\pi_{2j}^{(t)})^{x_{ij}} (1 - \pi_{2j}^{(t)})^{1-x_{ij}} \gamma^{(t)} + \prod_{j=1}^{J} (\pi_{1j}^{(t)})^{x_{ij}} (1 - \pi_{1j}^{(t)})^{1-x_{ij}} (1 - \gamma^{(t)})}.$$

Sample a value for the latent variable from

$$\theta_i^{(t+1)} \mid \gamma^{(t)}, \pi^{(t)}, \mathbf{x}_i \sim \text{Bernoulli}(s_i^{(t)}) + 1, \tag{13.30}$$

where $\mathbf{x}_i = (x_{i1}, \ldots, x_{iJ})$ is the collection of J observed values from examinee i.

2. *Sample the parameters for the latent variable distribution.* Count the number of examinees in class 2 based on the just-sampled values of the latent variables, $\boldsymbol{\theta}^{(t+1)} = (\theta_1^{(t+1)}, \ldots, \theta_n^{(t+1)})$, yielding the frequency $r_2^{(t+1)}$. Compute $r_1^{(t+1)} = n - r_2^{(t+1)}$ and use this value for r to sample a value for the proportion of examinees in latent class 2 from

$$\gamma^{(t+1)} \mid \boldsymbol{\theta}^{(t+1)} \sim \text{Beta}(\alpha_\gamma + r_2^{(t+1)}, \beta_\gamma + r_1^{(t+1)}). \tag{13.31}$$

3. *Sample the measurement model parameters.* For each observable $j = 1, \ldots, J$, and each class $c = 1, 2$, count the number of examinees in each class (out of the $r_c^{(t+1)}$) that had a value of 1 for the observable, yielding the class- and observable-specific frequency of values of 1, denoted as $r_{cj}^{(t+1)}$. For each observable j, and each latent class c, using these values for r_{cj}, sample a value for the conditional probability of a value 1 on the observable from

$$\pi_{cj}^{(t+1)} \mid \boldsymbol{\theta}^{(t+1)}, \mathbf{x}_j \sim \text{Beta}(\alpha_{\pi_c} + r_{cj}^{(t+1)}, \beta_{\pi_c} + r_c^{(t+1)} - r_{cj}^{(t+1)}), \tag{13.32}$$

where $\mathbf{x}_j = (x_{1j}, \ldots, x_{nj})$ is the collection of n observed values for observable j.

13.4 Example: Academic Cheating

We illustrate a Bayesian approach to LCA via an analysis of response data from university juniors and seniors on a survey of cheating behavior previously reported by Dayton (1999). Responses from $n = 319$ students were obtained to five questions that asked students whether or not, during their undergraduate years, they had (1) lied to avoid taking an exam, (2) lied to avoid handing a term paper in on time, (3) purchased a term paper to hand in as their own, (4) obtained a copy of an exam prior to taking the exam, and (5) copied answers during an exam from someone sitting near to them. The proportions of students endorsing each of these were .11, .12, .03, .04, and .21, respectively. Following Dayton (1999), as the third and fourth items had such small frequencies of positive responses, they were pooled in a manner that produced a single observable that took on two categories: 1 if a student had given a positive response to either or both items, and 0 if a student had given a negative response to both items. The proportion of 1s on this observable was .07. This leaves us with four observables for analysis, where the third observable is derived from the combination of the third and fourth items and the fourth observable corresponds to the fifth item. Coding positive responses as 1 and negative responses as 0, Table 13.2 contains the observed frequencies for the vectors of observables.

13.4.1 Complete Model and Posterior Distribution

We fit a model that specifies two latent classes, which we endeavor to interpret as cheater and noncheater classes. To illustrate various possibilities, we employ beta-Bernoulli

TABLE 13.2

Frequency of Response Vectors for the Cheating Data

	Observable				
Vector ID	**1**	**2**	**3**	**4**	**Frequency**
1	0	0	0	0	207
2	1	0	0	0	10
3	0	1	0	0	13
4	1	1	0	0	11
5	0	0	1	0	7
6	1	0	1	0	1
7	0	1	1	0	1
8	1	1	1	0	1
9	0	0	0	1	46
10	1	0	0	1	3
11	0	1	0	1	4
12	1	1	0	1	4
13	0	0	1	1	5
14	1	0	1	1	2
15	0	1	1	1	2
16	1	1	1	1	2

Source: Dayton, C. M. (1999). *Latent class scaling analysis*. Thousand Oaks, CA: SAGE Publications. With permission.

specifications for the conditional probabilities of the dichotomous observables and Dirichlet-categorical specifications for the probabilities of the dichotomous latent variable.

For completeness, we write out the posterior distribution for the model with the chosen values for the hyperparameters:

$$p(\theta,\gamma,\pi \mid \mathbf{x}) \propto \prod_{i=1}^{n}\prod_{j=1}^{J} p(x_{ij} \mid \theta_i,\pi_j)p(\theta_i \mid \gamma)p(\gamma)\prod_{c=1}^{C} p(\pi_{cj}), \tag{13.33}$$

where

$$(x_{ij} \mid \theta_i = c,\pi_j) \sim \text{Bernoulli}(\pi_{cj}) \text{ for } i = 1,\ldots,319, j = 1,\ldots,4,$$

$$\theta_i \mid \gamma \sim \text{Categorical}(\gamma) \text{ for } i = 1,\ldots,319,$$

$$\gamma \sim \text{Dirichlet}(1,1),$$

$$\pi_{cj} \sim \text{Beta}(1,1) \text{ for } c = 1,2, j = 1,2,3,$$

$$\pi_{14} \sim \text{Beta}(1,1),$$

and

$$\pi_{24} \sim \text{Beta}(1,1)\, I(\pi_{24} > \pi_{14}).$$

The DAG for the model is given in Figure 13.2.

Under this formulation, π_{cj} is the conditional probability of a positive response for observable j given membership in latent class c. For all but the last observables these are specified as following Beta(1,1) (i.e., uniform) priors. This is also used as the prior for the

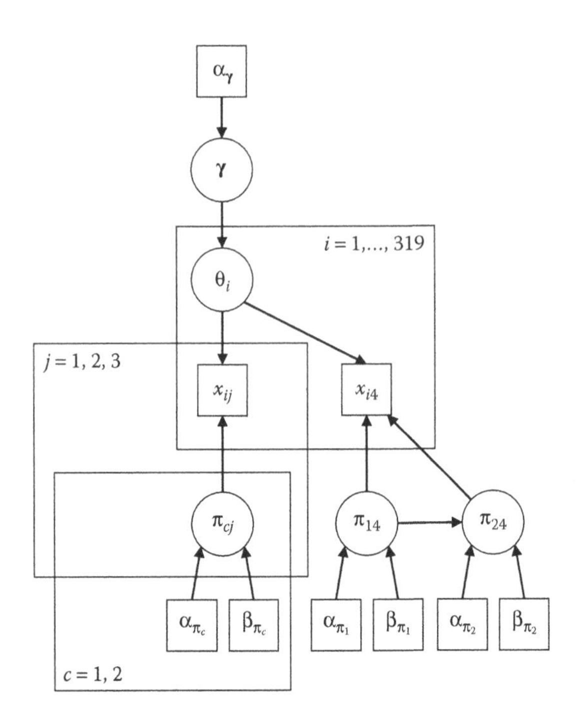

FIGURE 13.2
Directed acyclic graph for a 2-class latent class model for the academic cheating example. The nodes for the values for the fourth observable (x_{i4}), the conditional probabilities of the value for the fourth observable (π_{14} and π_{24}), and the hyperparameters that govern their prior distribution (α_{π_1}, β_{π_1}, α_{π_2}, and β_{π_2}) lie outside the plates over j and c owing to the constraint that $\pi_{24} > \pi_{14}$.

conditional probabilities for the last observable, subject to the constraint that the conditional probability for class 2 is larger than that for class 1. This is sufficient to resolve the indeterminacy in the latent variable. Conceptually, this constraint renders class 2 to be the class that was more likely to have copied answers. The distinct treatment of the last observable manifests itself in terms of the observable and the conditional probabilities (π_{14} and π_{24}) lying outside the plate over the remaining observables. Additionally, the DAG expresses that π_{24} depends on π_{14}, as implied by the last line of specifications in (13.33). The constraint amounts to specifying truncated beta prior distributions for these parameters and implies that the corresponding full conditional distributions are truncated beta distributions. Analogously, the full conditional distribution for the probability vector for a polytomous variable so constrained is a truncated Dirichlet distribution (Gelfand, Smith, & Lee, 1992; Nadarajah & Kotz, 2006; Sedransk, Monahan, & Chiu, 1985).

13.4.2 MCMC Estimation

The Gibbs sampler in the current case draws elements from each of those sketched in Sections 13.2.4 and 13.3.4. By using the Dirichlet-categorical specification for the latent variable, steps 1 and 2 are based on those in (13.18) and (13.19). By using the beta-Bernoulli specification for the observables, step 3 is based on that in (13.32). Importantly, although the notation would differ depending on whether we employed the Dirichlet-categorical or beta-Bernoulli specifications, the computations would be the same.

One salient feature of the model in this example is the constraint that $\pi_{24} > \pi_{14}$. As a result, the full conditionals for these parameters will be truncated beta distributions. For π_{14},

$$\pi_{14} \mid \theta, \pi_{24}, \mathbf{x}_j \sim \text{Beta}^{<\pi_{24}}(\alpha_{\pi_1} + r_{14}, \beta_{\pi_1} + r_1 - r_{14}), \tag{13.34}$$

where $\text{Beta}^{<\pi_{24}}$ denotes the beta distribution truncated to be less than π_{24}, r_1 is the number of examinees for whom $\theta = 1$ (i.e., are members of class 1), and r_{14} is the number of examinees for whom $\theta = 1$ (i.e., are members of class 1) that have a value of 1 for observable 4.

For π_{24},

$$\pi_{24} \mid \theta, \pi_{14}, \mathbf{x}_j \sim \text{Beta}^{>\pi_{14}}(\alpha_{\pi_2} + r_{24}, \beta_{\pi_2} + r_2 - r_{24}), \tag{13.35}$$

where $\text{Beta}^{>\pi_{14}}$ denotes the beta distribution truncated to be greater than π_{14}, r_2 is the number of examinees for whom $\theta = 2$ (i.e., are members of class 2), and r_{24} is the number of examinees for whom $\theta = 2$ (i.e., are members of class 2) that have a value of 1 for observable 4.

The Gibbs sampler is modified accordingly. Suppose in each iteration we draw a value for π_{14} before we draw a value for π_{24}. The instantiations of (13.32) then become

$$\pi_{14}^{(t+1)} \mid \theta^{(t+1)}, \pi_{24}^{(t)}, \mathbf{x}_j \sim \text{Beta}^{<\pi_{24}^{(t)}}(\alpha_{\pi_1} + r_{14}^{(t+1)}, \beta_{\pi_1} + r_1^{(t+1)} - r_{14}^{(t+1)}) \tag{13.36}$$

and

$$\pi_{24}^{(t+1)} \mid \theta^{(t+1)}, \pi_{14}^{(t+1)}, \mathbf{x}_j \sim \text{Beta}^{>\pi_{14}^{(t+1)}}(\alpha_{\pi_2} + r_{24}^{(t+1)}, \beta_{\pi_2} + r_2^{(t+1)} - r_{24}^{(t+1)}). \tag{13.37}$$

To illustrate, we step through the computations of the first iteration of the Gibbs sampler. We begin by specifying initial values. We will say more shortly about choosing start values in Section 13.4.3 where we fit the model in WinBUGS, using multiple sets of dispersed start values. The ones used here were arbitrarily chosen for the sake of illustration. For the conditional probabilities for the observables, we set $\pi_{11}^{(0)} = .4$, $\pi_{21}^{(0)} = .8$, $\pi_{12}^{(0)} = .3$, $\pi_{22}^{(0)} = .7$, $\pi_{13}^{(0)} = .2$, $\pi_{23}^{(0)} = .5$, $\pi_{14}^{(0)} = .1$, and $\pi_{24}^{(0)} = .5$. Note that these last two honor the constraint that $\pi_{24} > \pi_{14}$. For the latent variables, $\theta_1^{(0)}, \ldots, \theta_{219}^{(0)}$ are set to 1 and $\theta_{220}^{(0)}, \ldots, \theta_{319}^{(0)}$ are set to 2. Finally, we set the initial value for the proportion of examinees in latent class 2 to be .2. In the Dirichlet-categorical specification adopted here, this amounts to specifying $\gamma^{(0)} = (\gamma_1^{(0)}, \gamma_2^{(0)}) = (.8, .2)$. The first iteration of the Gibbs sampler is composed of the following steps.

1. *Sample the latent variables for examinees.* For the first examinee, the response vector was (0,0,0,0). We use these values to compute the ingredients of $s_{ic}^{(0)}$ based on each class:

$$\prod_{j=1}^{J} (\pi_{1j}^{(0)})^{x_{ij}} (1 - \pi_{1j}^{(0)})^{1 - x_{ij}} \gamma_1^{(0)} = (.4)^0(.8)^1 \times (.3)^0(.7)^1 \times (.2)^0(.8)^1 \times (.1)^0(.9)^1 \times .8 = .06048,$$

$$\prod_{j=1}^{J} (\pi_{2j}^{(0)})^{x_{ij}} (1 - \pi_{2j}^{(0)})^{1 - x_{ij}} \gamma_2^{(0)} = (.8)^0(.2)^1 \times (.7)^0(.3)^1 \times (.6)^0(.4)^1 \times (.5)^0(.5)^1 \times .2 = .0024,$$

and then

$$s_{i1}^{(0)} = \frac{\prod_{j=1}^{J} (\pi_{1j}^{(0)})^{x_{ij}} (1-\pi_{1j}^{(0)})^{1-x_{ij}} \gamma_1^{(0)}}{\prod_{j=1}^{J} (\pi_{2j}^{(0)})^{X_{ij}} (1-\pi_{2j}^{(0)})^{1-x_{ij}} \gamma_2^{(0)} + \prod_{j=1}^{J} (\pi_{1j}^{(0)})^{x_{ij}} (1-\pi_{1j}^{(0)})^{1-x_{ij}} \gamma_1^{(0)}} = \frac{.06048}{.0024+.06048} \approx .96.$$

and

$$s_{i2}^{(0)} = \frac{\prod_{j} (\pi_{2j}^{(0)})^{x_{ij}} (1-\pi_{2j}^{(0)})^{1-x_{ij}} \gamma_2^{(0)}}{\prod_{j} (\pi_{2j}^{(0)})^{x_{ij}} (1-\pi_{2j}^{(0)})^{1-x_{ij}} \gamma_2^{(0)} + \prod_{j} (\pi_{1j}^{(0)})^{x_{ij}} (1-\pi_{1j}^{(0)})^{1-x_{ij}} \gamma_1^{(0)}} = \frac{.0024}{.0024+.06048} \approx .04.$$

We then sample a value for the latent variable from (13.18) using these values to define the distribution,

$$\theta_i^{(1)} \mid \gamma^{(0)}, \pi^{(0)}, \mathbf{x}_i \sim \text{Categorical}(.96, .04).$$

This process is repeated for each examinee, yielding a value of 1 or 2 for each examinee.

2. *Sample the parameters for the latent variable distribution.* Count the number of examinees in class 2 in the just-sampled values of the latent variables, $\theta^{(1)} = (\theta_1^{(1)}, \ldots, \theta_n^{(1)})$. This frequency is $r_2^{(1)}$, which in the current case was 17. Compute $r_1^{(1)} = n - r_2^{(1)} = 319 - 17 = 302$ and sample a value for the latent class proportions from (13.19),

$$\begin{aligned} \gamma^{(1)} \mid \theta^{(1)} &\sim \text{Dirichlet}(\alpha_{\gamma_1} + r_1^{(1)}, \alpha_{\gamma_2} + r_2^{(1)}) \\ &= \text{Dirichlet}(1+302, 1+17) \\ &= \text{Dirichlet}(303, 18). \end{aligned}$$

The sampled value was $\gamma^{(1)} \approx (.96, .04)$.

3. *Sample the measurement model parameters.* For each observable, $j = 1, \ldots, J$ and class $c = 1, 2$, count the number of examinees in each class (out of the $r_c^{(1)}$) that had a value of 1 for the observable, yielding the class- and observable-specific frequency of values of 1, denoted as $r_{cj}^{(1)}$. The counts are shown in Table 13.3.

For observable 1, sample a value for π_{11} as in (13.29) using $r_1^{(1)} = 302$ and $r_{11}^{(1)} = 27$,

$$\pi_{11}^{(1)} \mid \theta^{(1)}, \mathbf{x}_j \sim \text{Beta}(28, 276).$$

TABLE 13.3

Class- and Observable-Specific Counts from the First Iteration of a Gibbs Sampler for a 2-Class Latent Class Model for the Cheating Data Example

	Observable			
Class	1	2	3	4
1	$r_{11}^{(1)} = 27$	$r_{12}^{(1)} = 30$	$r_{13}^{(1)} = 14$	$r_{14}^{(1)} = 55$
2	$r_{21}^{(1)} = 7$	$r_{22}^{(1)} = 8$	$r_{23}^{(1)} = 7$	$r_{24}^{(1)} = 13$

The drawn value was .08. Continuing with observable 1, we sample a value for π_{21} as in (13.29) using $r_2^{(1)} = 17$ and $r_{21}^{(1)} = 7$,

$$\pi_{21}^{(1)} \mid \boldsymbol{\theta}^{(1)}, \mathbf{x}_j \sim \text{Beta}(8, 11).$$

The drawn value was .29. Likewise for observables 2 and 3, we have

$$\pi_{12}^{(1)} \mid \boldsymbol{\theta}^{(1)}, \mathbf{x}_j \sim \text{Beta}(31, 273),$$

$$\pi_{22}^{(1)} \mid \boldsymbol{\theta}^{(1)}, \mathbf{x}_j \sim \text{Beta}(9, 10),$$

$$\pi_{13}^{(1)} \mid \boldsymbol{\theta}^{(1)}, \mathbf{x}_j \sim \text{Beta}(15, 289),$$

and

$$\pi_{23}^{(1)} \mid \boldsymbol{\theta}^{(1)}, \mathbf{x}_j \sim \text{Beta}(8, 11),$$

yielding drawn values of $\pi_{12}^{(1)} = .09$, $\pi_{22}^{(1)} = .76$, $\pi_{13}^{(1)} = .05$, and $\pi_{23}^{(1)} = .41$. For observable 4, we sample a value for π_{14} as in (13.36), noting that $\pi_{24}^{(0)} = .5$

$$\pi_{14}^{(1)} \mid \boldsymbol{\theta}^{(1)}, \pi_{24}^{(0)}, \mathbf{x}_j \sim \text{Beta}^{<.5}(56, 248),$$

yielding a value $\pi_{14}^{(1)} = .22$. This value is then used in sampling from the full conditional for π_{24} as in (13.37),

$$\pi_{24}^{(1)} \mid \boldsymbol{\theta}^{(1)}, \pi_{14}^{(1)}, \mathbf{x}_j \sim \text{Beta}^{>.22}(14, 5),$$

yielding a value of $\pi_{24}^{(1)} = .68$.

13.4.3 WinBUGS

The WinBUGS code and three sets of initial values are given as follows.

```
------------------------------------------------------------------------
########################################################################
# Model Syntax
########################################################################
model{

########################################################################
# Conditional probability of the observables via a latent class model
########################################################################

for (i in 1:n){
  for(j in 1:J){
           x[i,j] ~ dbern(pi[theta[i],j])
      }
}
```

```
########################################################################
# Prior distribution for the latent variables
########################################################################
for(i in 1:n){
  theta[i] ~ dcat(gamma[])
}
########################################################################
# Prior distribution for the parameters that govern the distribution
# of the latent variables
########################################################################
gamma[1:C] ~ ddirch(alpha_gamma[])
for(c in 1:C){
  alpha_gamma[c] <- 1
}

########################################################################
# Prior distribution for the measurement model parameters
########################################################################
for(c in 1:C){
  for(j in 1:(J-1)){
    pi[c,j] ~ dbeta(1,1)
  }
}

pi[1,J] ~ dbeta(1,1) I( ,pi[2,J])
pi[2,J] ~ dbeta(1,1) I(pi[1,J], )

} # closes the model

########################################################################
# Initial values for three chains
########################################################################
list(gamma=c(.9, .1),
pi= structure(.Data= c(
.37, .20, .06, .04,
.41, .47, .32, .19)
, .Dim=c(2, 4)))

list(gamma=c(.1, .9),
pi= structure(.Data= c(
.58, .62, .69, .77,
.81, .84, .90, .88)
, .Dim=c(2, 4)))

list(gamma=c(.5, .5),
pi= structure(.Data= c(
.32, .49, .29, .61,
.48, .54, .44, .70)
, .Dim=c(2, 4)))
------------------------------------------------------------------------
```

The final lines of the model express the prior distribution for the conditional probabilities of a value of 1 for the last observable. The use of the `I(pi[1,J], )` in the last line ensures that $\pi_{24} > \pi_{14}$. Though conceptually this is sufficient, WinBUGS also requires the construction in the previous line, where `I( ,pi[2,J])` ensures that $\pi_{14} < \pi_{24}$.

The model was fit in WinBUGS using three chains from dispersed starting values for the parameters listed above in the code, using WinBUGS to generate values for the latent variables. There was evidence of fast convergence (see Section 5.7.2). To be conservative, for each chain we discarded the first 1,000 iterations and ran an additional 20,000 iterations for use in inference.

The marginal posterior densities for the conditional probabilities and latent class proportions are plotted in Figure 13.3 and numerically summarized in Table 13.4. The summaries reported here suggest that, for all the observables, the conditional probability of a value 1 given the examinee in latent class 2 is higher than the conditional probability in latent class 1 (i.e., $\pi_{2j} > \pi_{1j}$ for all j). This was imposed via the prior distribution for the last observable, but not the others. This lends credence to the desired interpretation of class 2

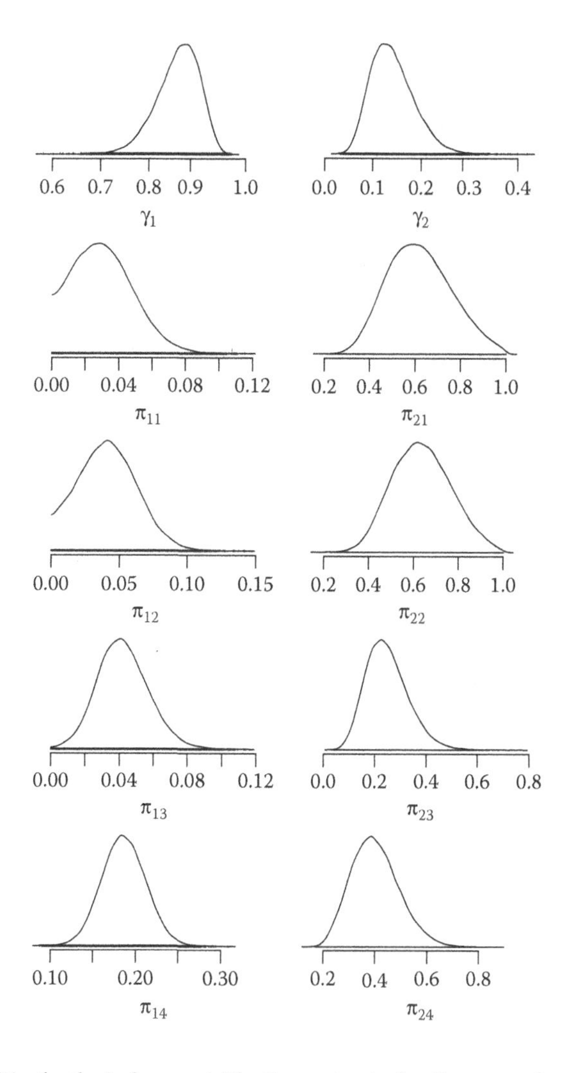

FIGURE 13.3
Marginal posterior densities for the 2-class model for the academic cheating example, where γ_c is the proportion of examinees in class c and π_{cj} is the conditional probability of observing a value of 1 on observable j given the examinee is a member of class c.

TABLE 13.4

Summary of the Posterior Distribution for a 2-Class Latent Class Model for the Academic Cheating Data

Parameter	Median	Standard Deviation	95% Highest Posterior Density Interval
π_{11}	0.03	0.02	(<.01, 0.06)
π_{21}	0.61	0.14	(0.36, 0.90)
π_{12}	0.04	0.02	(<.01, 0.08)
π_{22}	0.63	0.13	(0.40, 0.90)
π_{13}	0.04	0.01	(0.01, 0.07)
π_{23}	0.24	0.08	(0.10, 0.41)
π_{14}	0.19	0.03	(0.14, 0.24)
π_{24}	0.40	0.09	(0.23, 0.58)
γ_1	0.86	0.04	(0.77, 0.94)
γ_2	0.14	0.04	(0.06, 0.23)

as those who are more inclined to cheat. The marginal posterior for $\boldsymbol{\gamma}$ indicates that the proportion of students in the cheater class is around .14, and we are 95% certain that the proportion is between .06 and .23. Table 13.5 reports the marginal posterior probability of membership in class 2 for each response vector under the column labeled "Full Posterior." Positive responses to the first two observables are more strongly indicative of membership

TABLE 13.5

Posterior Probability of Membership in Class 2 (The Nominally Labeled "Cheaters" Class) for Each Response Vector in the Cheating Data Example Computed in Two Ways[a]

	Observable				$p(\theta = 2 \mid \mathbf{x})$	
Vector ID	**1**	**2**	**3**	**4**	**Full Posterior**	**Posterior Medians**
1	0	0	0	0	.02	.02
2	1	0	0	0	.44	.43
3	0	1	0	0	.39	.38
4	1	1	0	0	.96	.97
5	0	0	1	0	.12	.10
6	1	0	1	0	.79	.85
7	0	1	1	0	.75	.82
8	1	1	1	0	.99	.996
9	0	0	0	1	.05	.04
10	1	0	0	1	.64	.68
11	0	1	0	1	.60	.63
12	1	1	0	1	.98	.99
13	0	0	1	1	.25	.24
14	1	0	1	1	.90	.94
15	0	1	1	1	.87	.93
16	1	1	1	1	.998	.998

[a] The results labeled "Full Posterior" were obtained by marginalizing over the posterior distribution for the measurement model parameters and the parameters that govern the distribution of latent class membership. The results labeled "Posterior Medians" were obtained by using the marginal posterior medians as the values for the measurement model parameters and the parameters that govern the distribution of latent class membership. Values rounded to two decimal places, except where doing so they would round to 1.00, in which case they are rounded to three decimal places.

in class 2 (i.e., the cheating class). A positive response to either yields a posterior probability in latent class 2 that exceeds .5.

13.4.4 Model-Data Fit

The adequacy of the model-data fit may be assessed in a number of ways, including variations on the ways employed in the CFA and IRT examples. Here we focus on twin aspects of observable fit (OF) and person fit (PF). In Section 11.2.5, we examined observable (item) fit using graphical methods in PPMC. Here we employ an approach based on using residuals as discrepancy measures in PPMC to examine both OF and PF. We investigate fit using simple discrepancy measures examined by Yan, Mislevy, and Almond (2003; see also Almond et al., 2015). This is constructed by considering the squared Pearson residual for examinee i and observable j, which may be written in the notation of discrepancy measures in the current context of LCA as

$$\mathrm{V}_{ij}(x_{ij},\theta_i,\pi_j)=\frac{(x_{ij}-P_{ij})^2}{P_{ij}(1-P_{ij})}, \tag{13.38}$$

where $P_{ij}=E(x_{ij}\mid\theta_i,\pi_j)$ is the probability of $x_{ij}=1$. In the current context, P_{ij} is equal to π_{cj} for examinees for which $\theta_i=c$ (i.e., are members of class c). A discrepancy measure targeting OF for observable j is then given by

$$\mathrm{OF}_j(\mathbf{x}_j,\theta,\pi_j)=\left(\frac{1}{n}\sum_{i=1}^{n}\mathrm{V}_{ij}(x_{ij},\theta_i,\pi_j)\right)^{\frac{1}{2}}, \tag{13.39}$$

which is a root mean square error taken with respect to the values for the observable, where $\mathbf{x}_j=(x_{1j},\ldots,x_{nj})$ is the collection of observed values for observable j from the n examinees. Taking the root mean square with respect to each examinee yields a discrepancy measure for PF. For examinee i,

$$\mathrm{PF}_i(\mathbf{x}_i,\theta,\pi_j)=\left(\frac{1}{J}\sum_{j=1}^{J}\mathrm{V}_{ij}(x_{ij},\theta_i,\pi_j)\right)^{\frac{1}{2}}, \tag{13.40}$$

where $\mathbf{x}_i=(x_{i1},\ldots,x_{iJ})$ is the collection of J observed values from examinee i.

We conducted PPMC employing (13.39) and (13.40), using 15,000 iterations from MCMC (5,000 from each of the three chains). We computed p_{post} as a numerical summary of the results. The values of p_{post} for the analyses of (13.39) for the four observables were .54, .54, .53, and .49, suggesting that the model performs well in terms of accounting for the univariate distributions of the observables.

Turning to examinees, we confine our discussion here to a select set of results. Examinees with a response pattern of (0,0,0,0) had p_{post} values of .29, which is evidence of adequate fit. This is unsurprising as this was the most common response pattern in the data and therefore the posterior is attuned to this pattern. Examinees with a response pattern of (1,1,0,0) had p_{post} values of .68, which is evidence of adequate fit. This result is sensible in that it conforms to the model's implication that the first two items are endorsed more often than the others. Examinees with a response pattern of (0,0,1,1) had p_{post} values of .03, which were the smallest in the data. This is evidence of a lack of fit for these examinees. This reflects that the model's implications are that the items forming the last two observables are endorsed rarely, and with less frequency than the first two items. The examinees in question do not behave in accordance with these implications, resulting in poorer PF. This lack of PF may undermine the inferences for the examinees in question. Generally, poor PF suggests that the inferential argument expressed in the psychometric model may be dubious, at least for the examinees in question.

Note however that in the current case, the structure of the model does not imply that having a value of 1 on the last two observables must be accompanied by having a value of 1 on the first two, as would be in the case of latent class models for deterministic versions of Guttman scaling (Dayton & Macready, 1976; Guttman, 1947; Proctor, 1970). This model allows for more patterns, though the results of fitting the model suggest that a pattern of (0,0,1,1) should occur rarely. And that is indeed the case, as only 5 of the 319 examinees exhibited this pattern. The results from PPMC for these examinees amount to a red flag indicating that additional considerations may be needed to understand the viability of the model or alternative explanations for these examinees.

13.4.5 Model Comparison

We illustrate the use of the predictive criteria described in Section 10.3.4. More specifically, we compare the 2-class model, denoted as M_2, with a baseline model where the observables from examinees are independent and, for each observable, the probability of a value of 1 is given by proportion of 1s in the sample. This baseline model may be thought of as a 1-class model in which all the examinees are in the same class, and we denote it as M_1. For sake of illustration, we employ the posterior medians as point summaries of the parameters when computing the entropy for the 2-class model. We will have more to say about the use of point summaries of these parameters and the consequences for scoring in Section 13.4.6.

Following (10.25), the entropy for the 2-class model was $\text{Ent}(M_2) = 78.29$, which is lower than the entropy for the 1-class model, $\text{Ent}(M_1) = 85.69$. This yields a difference (Equation 10.26) of

$$\text{Ent}(M_1) - \text{Ent}(M_2) = 85.69 - 78.29 = 7.4.$$

This being positive indicates that the 2-class model is expected to yield better predictions for a new observation than the 1-class model. Following (10.27), the proportional improvement obtained by using the 2-class model is, rounding to two decimal places,

$$\frac{\text{Ent}(M_1) - \text{Ent}(M_2)}{\text{Ent}(M_1)} = \frac{7.4}{85.69} \approx .09.$$

This indicates that the modeling of two separate classes improves prediction a bit, a little under 10%. Under Gilula and Haberman's (2001) suggestion of thinking of this index as analogous to a proportion of variance explained, the improvement seems worthwhile.

13.4.6 Comparing the Use of the Full Posterior with Point Summaries in Scoring

It is instructive to consider how the inferences for examinees would be affected by using point values for the $\boldsymbol{\pi}$ and $\boldsymbol{\gamma}$ parameters rather than the full posterior distribution. In conventional approaches to LCA, (frequentist) point estimates of $\boldsymbol{\pi}$ and $\boldsymbol{\gamma}$ are obtained first and then used to obtain the posterior distribution for latent class membership. In the present example, we compare the use of the full posterior with an approach that uses the posterior medians reported in Table 13.4. The posterior probability of membership in class 2 for each response vector using the posterior medians for $\boldsymbol{\pi}$ and $\boldsymbol{\gamma}$ is reported in Table 13.5 in columns adjacent to those obtained from using the full posterior distribution. The differences are slight, but reveal that the use of the posterior medians yields posterior probabilities for examinees that are more extreme than when the full posterior distribution is used. Looking at it going in the other direction, when we use the full posterior distribution rather than the medians, the posterior probabilities of membership in class 2 are closer to .5. This may be interpreted as having less certainty regarding latent class membership. Echoing the discussion in Section 7.4, the posterior distribution for an examinee using a point for the elements of $\boldsymbol{\pi}$ and $\boldsymbol{\gamma}$ yields the posterior distribution *as if* the $\boldsymbol{\pi}$ and $\boldsymbol{\gamma}$ parameters took on the values used; nowhere is our uncertainty regarding those values incorporated in the analysis. A bit more formally, when using the point values $\pi = \pi_0$ and $\gamma = \gamma_0$, the posterior distribution for an examinee is $p(\theta_i \mid \mathbf{x}_i) = p(\theta_i \mid \mathbf{x}_i, \pi = \pi_0, \gamma = \gamma_0)$. When the full posterior distribution is used, the posterior distribution for an examinee becomes

$$p(\theta_i \mid \mathbf{x}_i) = \int\int p(\theta_i \mid \mathbf{x}_i, \pi, \gamma) p(\pi, \gamma \mid \mathbf{x}) d\gamma d\pi,$$

which will be more diffuse to the extent that there is variability in $\boldsymbol{\pi}$ and $\boldsymbol{\gamma}$. Conceptually, by using the full posterior distribution for $\boldsymbol{\pi}$ and $\boldsymbol{\gamma}$, our uncertainty in those parameters is acknowledged and induces some additional uncertainty about the examinees. To the extent that there is considerable uncertainty about $\boldsymbol{\pi}$ and $\boldsymbol{\gamma}$, ignoring that uncertainty by the use of a point value (e.g., a point summary of the posterior, or an MLE) may be deleterious in inferences regarding examinees.

13.5 Latent Variable Indeterminacies from a Bayesian Modeling Perspective

13.5.1 Illustrating the Indeterminacy

The example in Section 13.4 resolved the indeterminacy in the latent variable via a constraint on the conditional probabilities for the last observable. Consider now a model that differs only in that it does not include this constraint, namely one that specifies $\pi_{cj} \sim \text{Beta}(1,1)$ for all c and all j. The WinBUGS code may be enacted by taking the code given in Section 13.4.3 and removing the `I()` constructions in the last two lines.

Two chains were run in this analysis from starting values generated in WinBUGS. History plots and associated density plots for the first 1,000 iterations for the latent class proportions and the conditional probabilities for the first observable are given in Figure 13.4. The results for the conditional probabilities for the remaining observables followed the same patterns as that shown here.

A cursory look at the history plots would suggest that at most one but not both the chains have converged. In fact, they have converged, but are exploring different regions of the posterior distribution, which results in the multimodality of the density plots. The chain on the top of the history plot for γ_1 and π_{21} provides the values that constitute the density for the larger values of those parameters. This chain on the bottom of the history plot for γ_2 and π_{11} providing the values that constitute the density for the smaller values of those parameters. The other chain in each case provides the complementary values.

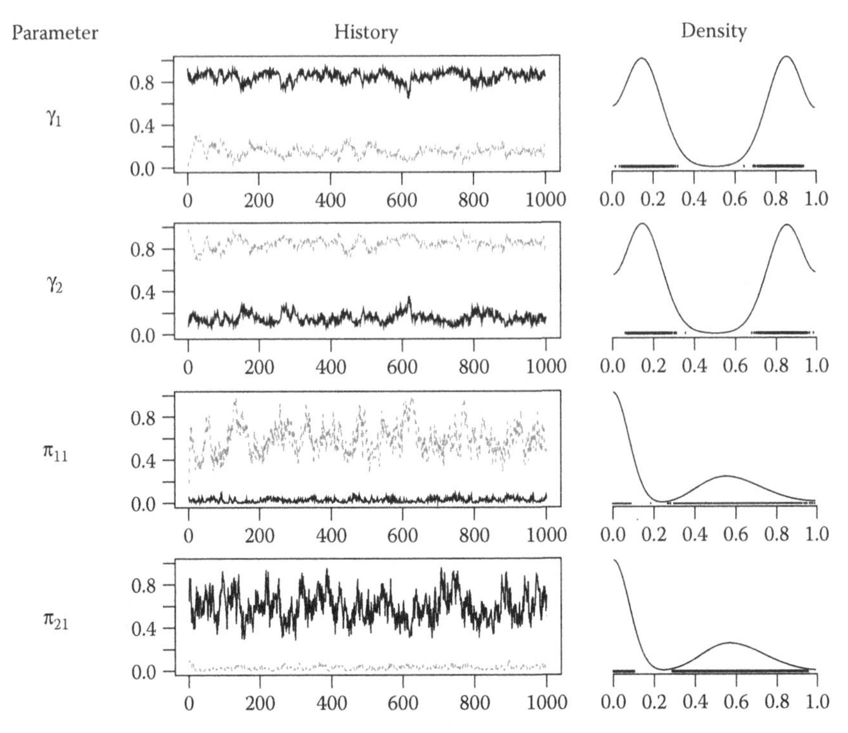

FIGURE 13.4
History plots and density plots for parameters for the 2-class model without constraints to resolve the indeterminacy for the academic cheating example, including the proportion of examinees in class 1 (γ_1), the proportion of examinees in class 2 (γ_2), the conditional probability of observing a value of 1 on the first observable given membership in class 1 (π_{11}), and the conditional probability of observing a value of 1 on the first observable given membership in class 2 (π_{21}).

The densities in Figure 13.4 are indeed representative of the posterior distribution. The multimodality is a manifestation of the indeterminacy associated with the discrete latent variable, namely that the roles of the classes can be interchanged: the latent class proportion and conditional probabilities associated with one class could be associated with another class (and vice versa) and still reproduce the data equally well. The multimodality in the posterior exists because of the multimodality in the likelihood—as discussed in Section 13.1.2 the indeterminacy and associated label switching pertain to frequentist estimation as well—and the information in the prior is not strong enough to steer the posterior toward one orientation as opposed to the other. As the prior is quite diffuse, the posterior closely resembles the likelihood. This example illustrates how MCMC may be a useful tool for exploring multimodality and the shape of the likelihood more broadly.

13.5.2 Resolving the Indeterminacy

There are several strategies that may be adopted to resolving the indeterminacy in the latent variable in Bayesian approaches to LCA (Chung, Loken, & Schafer, 2004; Hoijtink, 1998; Loken, 2004; Richardson & Green, 1997; Rodríguez & Walker, 2014). The analysis in Section 13.4 illustrated the use of a constraint on the conditional probabilities of the observables, $\pi_{24} > \pi_{14}$, labeling the second class as the one with the higher conditional probability of a value of 1 on the last observable. This approach extends naturally to the case of polytomous observables, in which case $\pi_{CjK} > \cdots > \pi_{1jK}$ labels the classes in order corresponding to increases in the conditional probabilities of the highest value of a particular observable (j). Any unique ordering would be sufficient (e.g., ordering on another observable, or on another response category).

Another option involves a specification of the latent class proportions. Setting $\gamma_1 < \cdots < \gamma_C$ renders the latent classes to be labeled according to their size. Again, any unique ordering would be sufficient (e.g., reverse order of their size, $\gamma_1 > \cdots > \gamma_C$).

Still another option involves assigning one or more examinees to each of the classes. For computational and interpretative ease, we may select examinees with the most extreme response patterns. In the case of the cheating data example, we may take one or more examinees with a response pattern of (0,0,0,0) and assign them to a particular class, say, class 1, and take one or more examinees with a response pattern of (1,1,1,1) and assign them to the other particular class, here class 2. This strategy is similar to setting the values of the latent variables for certain examinees to resolve the indeterminacies associated with continuous latent variables in IRT and CFA. This can be accomplished by capitalizing on computational strategies that treat latent variables as missing data. In WinBUGS, we would supply values of the latent variable as part of the data statement, with numerical values for those examinees assigned to classes, and "NA" for the remaining examinees. This approach suffers from certain drawbacks. First, the assigning of examinees to classes is a fairly strong assumption. On the other hand, when working with a 2-class model for the cheating data, it stands to reason that examinees with the response patterns of (0,0,0,0) and (1,1,1,1) ought to be in the different classes; any other account would strain credulity. Still, this approach is limited in that posterior inference and expressions of posterior uncertainty are not available for those examinees assigned to classes; they are treated as belonging to the classes with certainty.

All of the strategies just described may also be adopted in frequentist approaches to LCA modeling and estimation. A Bayesian approach offers an additional variation on each of these in the form of alternative prior distributions. Instead of imposing a constraint such as $\pi_{24} > \pi_{14}$, we may express the desire to have the second class have higher conditional probability of a value of 1 on the last item via the hyperparameters of the prior distribution. For example, specifying $\pi_{14} \sim \text{Beta}(2,5)$ and $\pi_{24} \sim \text{Beta}(5,2)$ would orient the posterior toward a class labeling where the second class has higher probabilities of a positive response to the last item. But it does not constrain the posterior to honor this notion. An example in the next chapter illustrates this approach. Similarly, instead of constraining the latent class proportions to be ordered in size, the prior distribution may be specified using hyperparameters to orient the posterior toward a particular class labeling. Likewise, instead of assigning examinees to particular classes with certainty, we may specify a different prior distribution that orients the solution as such.

Each approach may in fact be fairly weak in supplying information to resolve the indeterminacy if the model does not accord with the imposed constraints. For example, if two classes have comparable sizes, adopting the strategy that orders them by class sizes may not perform well in resolving the indeterminacy when fitting the model. This applies even more so for the use of prior distributions that orients but does not constrain the posterior to just one labeling scheme.

Another strategy involves analyzing a model without constraints needed to resolve the indeterminacy and then relabeling results based on inspection of the results. In our cheating data example, we may examine the trace plots in Figure 13.4, recognize the situation for what it is, and simply relabel the classes in one chain to accord with that of another. This would address the situation here where multiple chains are exploring different regions of the (multimodal) posterior. Things are more complicated when label switching occurs *within* a chain, as is possible and indeed guaranteed to happen for an infinitely long chain. This calls for more complicated algorithms for postprocessing or relabeling each successive iteration from MCMC (Rodríguez & Walker, 2014; Stephens, 2000).

13.5.3 The Blurred Line between the Prior and the Likelihood

The preceding discussion underscores two larger points. First, echoing themes from our treatments of CFA and IRT, we can recognize that imposing constraints on the parameters is itself a type of prior specification. Specifying $\pi_{14} \sim \text{Beta}(2,5)$ and $\pi_{24} \sim \text{Beta}(5,2)$ constitutes a prior where π_{24} is *probably* larger than π_{14}. We can control how likely that is by manipulating the forms and or hyperparameters of these prior distributions. The constraint that $\pi_{24} > \pi_{14}$ may be seen as a limiting case of these prior expectations, where π_{24} is *certainly* larger than π_{14}. Second, though the indeterminacy exists in the likelihood, we have been discussing resolutions in terms of prior distributions. Should a specification such as $\pi_{24} > \pi_{14}$ be viewed as a feature of the likelihood, or of the prior? Echoing the discussions in Sections 3.5, 3.6, and 9.8, one perspective is that such a question is largely irrelevant. Distinctions between the likelihood and the prior may be useful for certain purposes, but may be unnecessary, or worse, inhibitive to understanding.

13.6 Summary and Bibliographic Note

This chapter has focused on Bayesian approaches to LCA in a confirmatory setting. For Bayesian treatments of LCA in more exploratory settings, see Garrett and Zeger (2000), Pan and Huang (2014), Richardson and Green (1997), and Rubin and Stern (1994). Garrett, Eaton, and Zeger (2002) provide a Bayesian account of LCA that fits models with different numbers of classes from a confirmatory model comparison perspective. Additional treatments of Bayesian approaches to confirmatory LCA with emphases on the use of ordinality constraints may be found in Hoijtink (1998) and van Onna (2002). Hojtink and Molenaar (1997) provided a related treatment that invoked connections with nonparametric MIRT.

Extensions include Bayesian approaches for producing point and interval estimates for quantities derived from the basic LCA parameters (Garrett, Eaton, & Zeger 2002; Lanza, Collins, Schafer, & Flaherty, 2005), fitting models that are unidentified in frequentist conceptions (Evans, Gilula, & Guttman, 1989), incorporating covariates (Chung, Flaherty, & Schafer, 2006), or modeling the conditional probabilities via linear logistic models (Garrett & Zeger, 2000) as in Formann (1985, 1992). Examples of the last of these will be developed in the context of Bayesian networks in the next chapter. Similarly, many extensions of the LCA models can be grouped under the heading of involving models with multiple discrete latent variables, including latent transition and profile analyses (Chung & Anthony, 2013; Chung, Lanza, & Loken, 2008; Chung, Walls, & Park, 2007; Lanza, Collins, Schafer, & Flaherty, 2005), diagnostic classification models (Rupp et al., 2010), and Bayesian networks (Almond et al., 2015). Several of these are treated more explicitly in the next chapter.

Exercises

13.1 The first step in the Gibbs sampler described in Section 13.2.4 involves sampling the values of the latent variables for examinees. In Section 13.4.2, we showed the computation for the first iteration for examinees with the response vector of (0,0,0,0) in the cheating data example. Show the computations for the remaining response vectors.

13.2 The third step in the Gibbs sampler described in Section 13.2.4 involves sampling the values for the measurement model parameters for dichotomous observables. In Section 13.4.2, we showed the computation for the first iteration for π_{11} in the cheating data example. Show the computations for the remaining measurement model parameters.

13.3 Replicate the analysis of the 2-class model for the cheating data in Section 13.4 using WinBUGS.

a. Verify your results for the measurement model parameters with those reported here.

b. When fitting the model, compute the probabilities that the conditional probabilities of a positive response are higher in class 2 than class 1 for each observable. (Hint: recall the WinBUGS code in Exercise 6.3).

c. Interpret the probabilities from part (b).

13.4 Conduct an analysis of a 3-class model for the cheating data.

a. Specify a model that yields an ordinal interpretation of the classes in terms of their proclivity for cheating.

b. Obtain and interpret the posterior distribution for the measurement model parameters and the latent class proportions.

13.5 The chains for the model in the example in Section 13.5 explored different parts of the posterior distribution, corresponding to different labels of the classes.

a. Explain why label switching will eventually happen within a chain.

b. Explain how this could occur in the context of a Gibbs sampler, and in the context of a Metropolis sampler.

14

Bayesian Networks

Bayesian networks (BNs) model multivariate distributions of discrete variables. When brought to bear in psychometric contexts they are typically constructed as modeling discrete observables as dependent on discrete latent variables. As such they are similar to the LCA models discussed in Chapter 13. There, we focused on models with a single latent variable. In this chapter, we focus mainly on models with multiple latent variables. In this light, BNs may be seen as generalizations of LCA models as developed in Chapter 13. In another light, BNs may be seen as instances of LCA models more broadly conceived, as models with multiple discrete latent variables may be recast as having a single discrete latent variable (see Section 13.1.1).

However, BNs are not as frequently used in the psychometric community as CFA, IRT, and LCA models. Accordingly, we organize this chapter beginning with a broader treatment of BNs in Section 14.1, discussing them in terms that do not immediately map onto latent variable psychometric models. We then focus on how BNs can be leveraged as psychometric models and briefly describe how the developments of the previous chapter apply to the current case in Sections 14.2 and 14.3. We then discuss three examples of other psychometric models that can be framed as BNs in Sections 14.4 through 14.6. Section 14.7 provides a brief summary and bibliographic note regarding related treatments and extensions. Additional accounts of BNs from a general statistical perspective may be found in Jensen (1996, 2001), Neapolitan (2004), and Pearl (1988). Almond et al., (2015) provided a broad treatment, focusing on their application in educational assessment.

14.1 Overview of BNs

14.1.1 Description of BNs

We begin our general description of BNs with a simple four-variable example. Consider a system of four variables, W, X, Y, and Z, each representing one of four skills a student may possess. A BN is a statistical model that structures the joint distribution for a set of discrete variables by recursively specifying conditional distributions. Any joint distribution can be expressed as a product of a distribution for the first variable, a distribution for the second variable conditional on the first, a distribution for the third variable conditional on the first and second, and so on. The joint distribution for our system of four variables (W, X, Y, Z) can be written as

$$p(W,X,Y,Z) = p(W)p(X \mid W)p(Y \mid W,X)p(Z \mid W,X,Y). \tag{14.1}$$

Any ordering is possible; for example we might specify a reversal of this ordering and specify a distribution for Z, the conditional distribution for Y given Z, and so on. Certain orderings may be preferable according to certain criteria, including interpretability and implications of substantive theory. In particular, we will capitalize on conditional independence relationships to organize our thinking. One such model is

$$p(W,X,Y,Z) = p(W)p(X)p(Y \mid W,X)p(Z \mid Y), \tag{14.2}$$

which departs from (14.1) in expressing X as independent of W, and Z as conditionally independent of W and X given Y. The model in (14.2) expresses a structuring of the joint distribution in accordance with a theory of skill acquisition in which

- Skills W and X are mastered first and independently;
- Mastery of skill Y then depends on whether skills W and X are mastered; and
- Mastery of skill Z depends on whether skill Y is mastered, but not on mastery of skills W or X above and beyond mastery of skill Y.

We can now characterize a BN as a model that structures the joint distribution. For a set of variables generically represented as **v**, a BN structures the joint distribution as

$$p(\mathbf{v}) = \prod_{v \in \mathbf{v}} p(v \mid pa(v)), \tag{14.3}$$

where $pa(v)$ are the parents of variable v, namely, the variables on which v directly depends. If v has no parents $p(v \mid pa(v))$ is taken as the unconditional (marginal) distribution of v.

14.1.2 BNs as Graphical Models

As the characterization of variables as dependent on their parents suggests, BNs are strongly associated with graphical models, owing to the correspondence between the structure of a graph and the conditional independence relationships in the model (Pearl, 1988, 2009). BNs may be characterized as DAG models in which all of the nodes correspond to discrete variables. The DAG for the BN in (14.2) is depicted in Figure 14.1, and expresses that the joint distribution may be factored in accordance with the right-hand side of (14.2). Furthermore, the conditional independence relationships may be read off the graph. Given its parents, children, and other parents of its children, each variable is conditionally independent of all other variables (Pearl, 1988, 2009). For example, the DAG in Figure 14.1 indicates that Z depends on Y (its parent), but once we condition on Y, Z is independent of W and X.

14.1.3 Inference Using BNs

The models are referred to as *Bayesian* networks because, once data have been observed, they support the application of Bayes' theorem across multivariate systems. The DAG

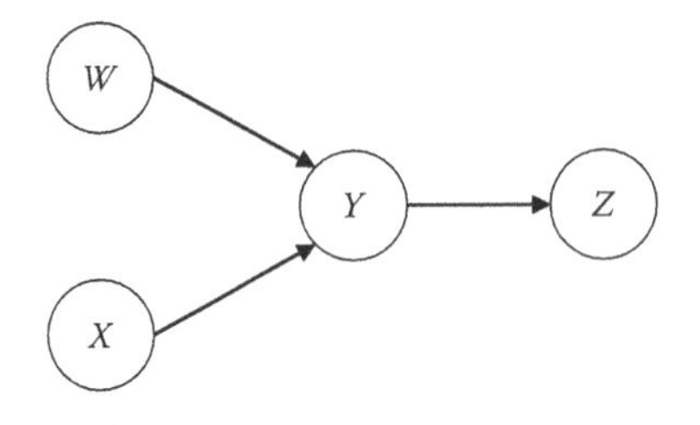

FIGURE 14.1
Directed acyclic graph for four variables in the model in (14.2).

plays a key role here too: in addition to structuring the joint distribution of a system of variables, the DAG structures the computations involved in obtaining the posterior distribution (Lauritzen & Spiegelhalter, 1988; Pearl, 1988).

14.2 BNs as Psychometric Models

BNs can be employed as psychometric models by specifying observable variables as dependent on latent* variables for an examinee. In assessment applications, once values for the observables are known they are entered into the network and the posterior distribution of the remaining variables—primarily that for the latent variables of inferential interest—is obtained. In the parlance of Chapter 7, scoring for examinees is conducted by Bayesian inference.

We have already encountered several examples of models that may be seen as BNs, specifically the subtraction proficiency example in Chapters 1 and 7, the medical diagnosis example in Chapter 2, and the LCA models in Chapter 13. We revisit some of these before turning to other BN psychometric models. As with LCA, the characterization of BNs as involving only discrete variables refers to a just-persons perspective on the model, meaning that the examinee variables are discrete. As we will see, modeling the conditional probabilities may involve parameters that are specified as continuous random variables. For the moment, we confine ourselves to the just-persons perspective before expanding our lens in Section 14.3.

14.2.1 Latent Class Models as BNs

The medical diagnosis example introduced in Chapter 2 may be seen as a BN. The DAG for the model is given in Figure 14.2 where C is the patient's cancer status (Yes or No) and M is the result of the mammography screening (Positive or Negative). This is the simplest of BNs of any substance, containing just two variables, each of which can take on two values, with a directed edge linking them.

Following the structure of the DAG, the joint distribution is given by

$$p(C, M) = p(C)p(M \mid C). \tag{14.4}$$

The elements on the right-hand side of (14.4) are given in Tables 2.1 and 2.2, repeated here as Tables 14.1 and 14.2 for $p(M \mid C)$ and $p(C)$, respectively.

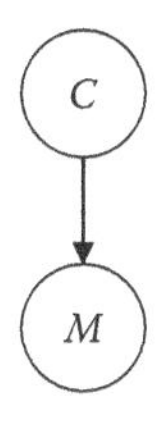

FIGURE 14.2
Directed acyclic graph for the medical diagnosis example, where the mammogram result (M) is modeled as dependent on cancer status (C).

* In the BN literature, latent variables are also referred to as *hidden* variables.

TABLE 14.1
Conditional Probability of the Mammography Result (M) Given Breast Cancer (C)

	Mammography Result	
Breast Cancer	**Positive**	**Negative**
Yes	.90	.10
No	.07	.93

TABLE 14.2
Prior Probability of Breast Cancer (C)

Breast Cancer	Probability
Yes	.008
No	.992

Once the value of the observable variable M is known, the posterior distribution for C is obtained via Bayes' theorem. Revisiting the example in Section 2.2, once it is observed that M = Positive, the posterior probability for a patient's cancer status, rounding to two decimal places, is $p(C = \text{Yes} \mid M = \text{Positive}) \approx .09$ and $p(C = \text{No} \mid M = \text{Positive}) \approx .91$.

The medical diagnosis model is a simple LCA model with one dichotomous observable. The more general LCA model expressed in Chapter 13 may be seen as expanding the situation to one where there are J possibly polytomous observables modeled as dependent on, and rendered conditionally independent by, a possibly polytomous latent variable. As the variables are all discrete, the LCA model may be seen as a BN. The just-persons DAG for the model is given in Figure 14.3, which is just the structure of the canonical psychometric model characterized in Chapter 7.

Following the structure of the DAG, the joint distribution for any examinee (suppressing the subscripting by i) is given by

$$p(x_1,\ldots,x_J,\theta) = \prod_{j=1}^{J} p(x_j \mid \theta)p(\theta). \tag{14.5}$$

14.2.2 BNs with Multiple Latent Variables

We now extend our treatment to consider BNs with multiple latent variables. Figure 14.4 contains a DAG with three latent variables and eight observable variables. We considered multiple continuous latent variables in the context of CFA (Chapter 9) and then in IRT (Section 11.5). In both of these contexts, we specified a multivariate normal distribution

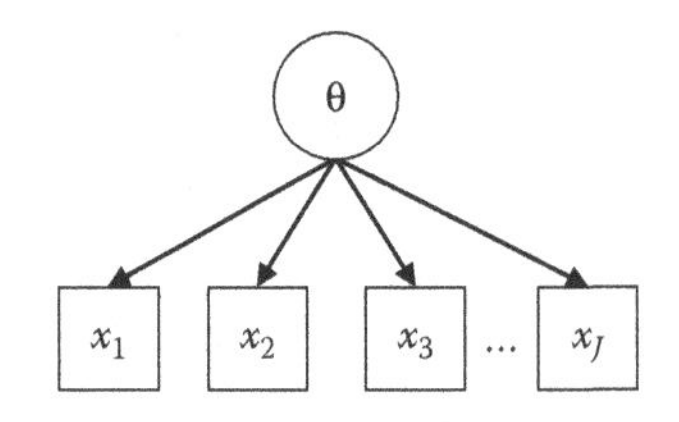

FIGURE 14.3
Directed acyclic graph for the latent class model as a Bayesian network.

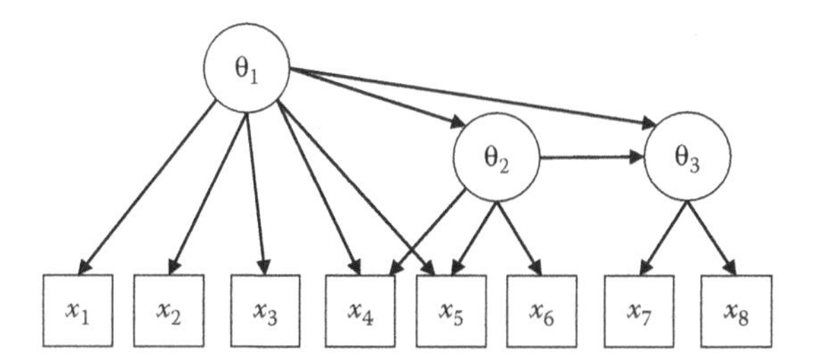

FIGURE 14.4
Directed acyclic graph for a Bayesian network with three latent variables, illustrating factorially simple and factorially complex observables, and dependence among multiple latent variables.

for the multiple latent variables. In the current context with multiple discrete latent variables, we model the associations among the latent variables by recursively specifying distributions for the latent variables in a univariate fashion, conditioning on the previously specified latent variables to reflect the dependence. As expressed by the directed edges in Figure 14.4, we will specify a distribution for θ_1, a distribution for θ_2 conditional on its parent, θ_1, and a distribution for θ_3 conditional on its parents, θ_1 and θ_2. As we will see, the form of the conditional distributions for latent variables mimic those of the conditional distributions for the observables.*

Turning to the observables in Figure 14.4, the first three observables (x_1, x_2, x_3) and the last three observables (x_6, x_7, x_8) are modeled as being factorially simple, each having one latent parent variable and may be modeled like the (unidimensional) LCA models in Chapter 13. Observables x_4 and x_5 are modeled as factorially complex, having both θ_1 and θ_2 as parents. For each combination of the parent variables, there is a conditional distribution for the possible values of the observable. Table 14.3 illustrates this, with a conditional distribution for a dichotomous observable coded as 0 and 1 that depends on two dichotomous latent variables, each coded as taking on values of 1 or 2. The directed edges connecting the latent variables indicate a dependence structure. As all the latent variables are discrete, the dependence may also be expressed in terms of conditional probability tables.

TABLE 14.3
Conditional Probability Table for Observable j That Depends on Two Latent Variables, Where $\pi_{(ab)jk}$ Is the Probability of Observing a Value of k for Observable j When $\theta_1 = a$ and $\theta_2 = b$

		$p(x_j \mid \theta_1, \theta_2)$	
θ_1	θ_2	0	1
1	1	$\pi_{(11)j0} = 1 - \pi_{(11)j1}$	$\pi_{(11)j1}$
1	2	$\pi_{(12)j0} = 1 - \pi_{(12)j1}$	$\pi_{(12)j1}$
2	1	$\pi_{(21)j0} = 1 - \pi_{(21)j1}$	$\pi_{(21)j1}$
2	2	$\pi_{(22)j0} = 1 - \pi_{(22)j1}$	$\pi_{(22)j1}$

* We note that the same approach may be taken with continuous latent variables, where the associations among the latent variables are modeled with direct effects. As is the case here, not every latent variable needs to be associated with every other latent variable. This approach makes the model a member the larger family of structural equation modeling. See Levy and Choi (2013) for a treatment of Bayesian structural equation modeling in line with the perspectives adopted here. For additional treatments, see Dunson, Palomo, and Bollen (2005); Kaplan & Depaoli (2012); and Lee (2007).

Following the structure of the DAG, the joint distribution is given by

$$p(x_1,\ldots,x_8,\theta_1,\theta_2,\theta_3)=\prod_{j=1}^{3}p(x_j\mid\theta_1)\prod_{j=4}^{5}p(x_j\mid\theta_1,\theta_2)p(x_6\mid\theta_2)\prod_{j=7}^{8}p(x_j\mid\theta_3)\times p(\theta_1)p(\theta_2\mid\theta_1)p(\theta_3\mid\theta_1,\theta_2). \quad (14.6)$$

14.3 Fitting BNs

The presentation so far has assumed that the conditional probability tables capturing the dependence among the entities and the marginal probabilities for variables with no parents are known. Inference then concerns scoring, which is conducted by obtaining the posterior distribution for the examinees' latent variables given the values of the observables. This is enacted via recursive application of Bayes' theorem across the system of variables. If the probability tables are not known, the situation expands to consider them as well. As with LCA, conventional approaches employ ML estimation of the conditional probabilities.*

In the fully Bayesian approach presented here, Dirichlet and beta distributions are convenient choices for the prior distribution for unknown (conditional) probabilities for categorically and Bernoulli distributed variables (Almond et al., 2015; Neapolitan, 2004). For BNs with observable variables only and complete data, these distributions are conjugate priors (Spiegelhalter & Lauritzen, 1990). Full conjugacy is lost in models with latent variables or missing data, but conditional conjugacy is preserved and makes for efficient Gibbs sampling.

Once again let π denote the full collection of conditional probabilities for the values of the observables given the latent variables, and now let γ denote the full collection of probabilities that govern the distribution of the latent variables. Returning to the example in (14.6), $\gamma=(\gamma_1,\gamma_2,\gamma_3)$ where γ_1 contains the unconditional probabilities for θ_1 (which define $p(\theta_1)$), γ_2 contains the conditional probabilities for θ_2 given θ_1 (which define $p(\theta_2\mid\theta_1)$), and γ_3 contains the conditional probabilities for θ_3 given θ_1 and θ_2 (which define $p(\theta_3\mid\theta_1,\theta_2)$). Making the familiar assumptions of independence among $\boldsymbol{\pi}$ and θ and among $\boldsymbol{\pi}$ and γ, the posterior distribution for all the unknowns is

$$\begin{aligned} p(\boldsymbol{\theta},\gamma,\pi\mid\mathbf{x}) &\propto p(\mathbf{x}\mid\boldsymbol{\theta},\gamma,\pi)p(\boldsymbol{\theta},\gamma,\pi) \\ &= p(\mathbf{x}\mid\boldsymbol{\theta},\pi)p(\boldsymbol{\theta}\mid\gamma)p(\gamma)p(\pi) \\ &= \prod_{i=1}^{n}\prod_{j=1}^{J}p(x_{ij}\mid\theta_i,\pi_j)p(\theta_i\mid\gamma)p(\gamma)p(\pi_j), \end{aligned} \quad (14.7)$$

where $p(x_{ij}\mid\theta_i,\pi_j)$ is the conditional probability distribution for the value of observable j for examinee i. As in the LCA model, this is given by the measurement model parameters associated with observable j and examinee i's values for the latent variables, and $p(\boldsymbol{\pi}_j)$ is the prior distribution for the measurement model parameters. Next, $p(\boldsymbol{\theta}_i\mid\gamma)=p(\theta_{i1}\mid\gamma_1)p(\theta_{i2}\mid\theta_{i1},\gamma_2)p(\theta_{i3}\mid\theta_{i1},\theta_{i2},\gamma_3)$ is the prior distribution for the multiple latent variables for examinee i, and $p(\gamma)$ is the prior distribution for the parameters that govern the distribution of the latent variables.

* This is the semantically confusing circumstance in which the conditional probabilities in a *Bayesian* network are estimated via *frequentist* methods.

This differs from the posterior distribution for the LCA model in (13.13) in two respects. First, we now have multiple latent variables for each examinee. Second, as a result of the possibly different patterns of dependence of the observables on the latent variables, the prior distribution for the measurement model parameters for each observable $\boldsymbol{\pi}_j$ is not further factored. In any particular application, common dependence structures and exchangeability assumptions may support further simplifications of the model in (14.7). As the forms of the distributions presented here are the same as those in LCA, namely categorical (or Bernoulli) distributions for latent and observable variables and Dirichlet (or beta) prior distributions for the probabilities that govern those distributions, the full conditional distributions may be derived in a straightforward extension of those for LCA in Chapter 13 (see also Appendix A).

The development so far has characterized BNs in a fairly "open" manner. Commonly, BNs are specified in ways that may be seen as constrained versions of these models. Such constraints may be employed for a variety of reasons including resolving indeterminacies in latent variables as discussed in Chapter 13, specifying the model to reflect substantive theory, and reducing the parameterization. The latter issue frequently arises in BNs when variables have many possible categories and/or many parent variables that result in many possible combinations of the values of the parent variables. Letting K_j denote the number of categories for observable j and letting C_m denote the number of categories for latent variable m, then assuming the observables are conditionally (locally) independent given the latent variables, there are

$$\sum_{j=1}^{J}(K_j-1)\prod_{m:\theta_m\in pa(x_j)}(C_m)$$

conditional probabilities altogether for the observables. In these cases, the conditional probability tables become large, increasing the number of entities in need of specification and estimation. This can be difficult if the data are sparse with respect to the parameters. Impositions made to reduce the parameterization should be in line with substantive theory. Of course, constraints motivated by substantive theory often yield a reduced parameterization. In the rest of this chapter, we illustrate these ideas in the discussing three examples of BNs as psychometric models.

14.4 Diagnostic Classification Models

Many cognitive diagnosis models or diagnostic classification models (DCMs; DiBello, Roussos, & Stout, 2007; Rupp et al., 2010) may be characterized as modeling discrete observables as dependent on multiple discrete latent variables. To simplify the presentation, we confine the current discussion of DCMs to those that specify dichotomous observables and dichotomous latent variables. In educational assessment, each latent variable is postulated to represent a skill, attribute, or other aspect of proficiency. The levels of the latent variable correspond to whether the examinee does or does not possesses (alternatively, has mastered or has not mastered) the associated skill. A key goal of most applications of DCMs is to characterize examinees with respect to the constellation of skills they do or do not possess. Central to this goal, DCMs usually specify the dependencies of the observables on the latent variables in ways that reflect substantive theory about how examinees bring together these skills to complete tasks. The models may also specify particular relationships among the latent variables in accordance with substantive theory. DCMs may be seen as related to, and perhaps versions of, a number of different types of models (Rupp et al., 2010). BNs may be seen as enacting diagnostic measurement akin to DCMs (Almond, DiBello, Moulder, & Zapata-Rivera, 2007; Rupp et al., 2010). Further many DCMs may be cast as BNs that model the conditional probabilities in the BN via parameters that both express the theory of cognition and simplify the conditional probabilities that define the BN.

14.4.1 Mixed-Number Fraction Subtraction Example

We further define and illustrate these points in the context of an example in the domain of mixed-number fraction subtraction. Klein, Birnbaum, Standiford, and Tatsuoka (1981) identified two methods that students use to tackle these sorts of problems. Briefly, they are as follows:

- Method A: Convert mixed numbers to improper fractions, subtract, and reduce if necessary.
- Method B: Separate mixed numbers into a whole number part and a fractional part, subtract as two sub-problems, borrowing one from minuend whole number if necessary, and simplify and reduce if necessary.

Tatsuoka (1984) reported on analyses of student responses to test items designed to illuminate which method they were using, and characterized the sorts of errors they were making with the various procedures. Mislevy (1995) described BN models for middle school students' performances on mixed-number subtraction tasks based on Tatsuoka's (1987, 1990) data and cognitive analyses. Model-fitting strategies and model-data fit analyses for portions of this model that we describe here have additionally been discussed by Almond et al. (2015); Mislevy, Almond, Yan, and Steinberg (1999); Sinharay and Almond (2007); Yan, Almond, and Mislevy (2004); and Yan et al., (2003), and we draw from their work in the current presentation. Related DCM and cognitive diagnostic modeling of data from these tasks have been investigated by de la Torre and Douglas (2004, 2008); DeCarlo (2011, 2012); Henson, Templin, and Willse (2009); and Tatsuoka (2002).

We confine ourselves to a model for Method B and observables associated with 15 items for which it is not necessary to find a common denominator; extensions to this model are briefly discussed in Section 14.4.8. Following Mislevy (1995), the BN model involves linking the observables to five latent variables representing the components of executing Method B for these tasks. Following the parlance of DCMs for these sorts of situations, we refer to these latent variables as the attributes or skills:

Skill 1: Basic fraction subtraction;

Skill 2: Simplify/reduce a fraction or mixed number;

Skill 3: Separate a whole number from a fraction;

Skill 4: Borrow one from the whole number part to the fraction part in a given mixed number;

Skill 5: Convert a whole number to a fraction.

We specify a dichotomous latent variable for each skill, intended to capture our beliefs about whether the examinee possesses or does not possess each skill. Inference centers on diagnosing which skills are possessed by which examinees. In principle, there are $2^5 = 32$ possible *skill profiles* defined by combinations of the values of the five dichotomous latent variables.

Picking up on the final point in Section 14.3, even with the assumption that the observables are conditionally (locally) independent given the latent variables, there are 511 (conditional) probabilities in need of specification: 31 are needed to model the joint distribution of the five latent variables;* and 480 are needed to model the dependence of the observables on the latent variables (as for each of 15 observables there is a conditional probability table with 32 conditional probabilities, one for each skill profile). Happily, substantive theory and cognitive analyses of examinee responses to these types of tasks (Klein et al., 1981; Tatsuoka, 1984, 1987, 1990, 2009) implies several simplifications in both portions of the model, as described in the next two subsections.

14.4.2 Structure of the Distribution for the Latent Variables

With five dichotomous latent variables, there are 32 logically possible skill profiles defined as combinations of the values of the latent variables. However, based on cognitive analyses of student responses (Tatsuoka, 1987, 1990, 2009), Mislevy (1995) and Mislevy et al. (1999) proposed a model for the latent variables that simplifies things in that some of the skill profiles are entertained as distinct, some are pooled or collapsed together, and others are ruled out as implausible.

To facilitate the presentation, we will at times employ a just-persons perspective on the model. The DAG for this portion of the model, which comprises the student model in ECD terms, is depicted in Figure 14.5. For each examinee i the nodes $\theta_{i1},\ldots,\theta_{i5}$ are the latent variables for *Skill 1*,..., *Skill 5*; the other node will be explained in the course of our description of the model.

The structure and details of the dependencies reflect particular hypotheses concerning the various aspects of proficiency. θ_1 is modeled as a parent of the others because normally students would gain proficiency on *Skill 1* (basic fraction subtraction) before the others. As before, we begin by specifying a Bernoulli distribution for the value of θ_1; for each examinee

$$\theta_{i1} \mid \gamma_1 \sim \text{Bernoulli}(\gamma_1). \tag{14.8}$$

We proceed by modeling the values of θ_2 as conditional on θ_1; for each examinee

$$(\theta_{i2} \mid \theta_{i1} = z, \gamma_2) \sim \text{Bernoulli}(\gamma_{2z}) \quad \text{for } z = 0,1, \tag{14.9}$$

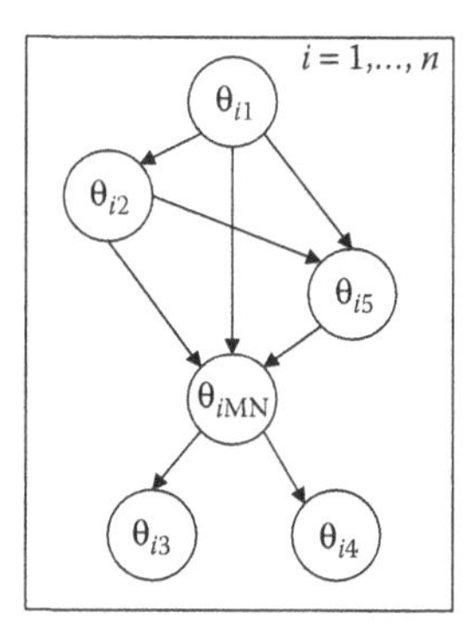

FIGURE 14.5
Directed acyclic graph fragment for the latent variables in the mixed-number subtraction example.

* These may be formulated in a number of ways. One way is to think of the 31 being needed to uniquely define the joint distribution of the 32 skill profiles, the last probability being constrained by the requirement that they sum to 1.

where $\gamma_2 = (\gamma_{20}, \gamma_{21})$, with γ_{2z} denoting the conditional probability of θ_2 taking a value of 1 given $\theta_1 = z$. Table 14.4 demonstrates the probability structure for θ_1 and for θ_2 given θ_1.

Following the flow of the DAG, we now turn to the specification of the values of θ_5. In principle, we could specify a unique distribution for each combination of θ_5 given θ_1 and θ_2, yielding four conditional distributions. Following Mislevy et al. (1999), a simplification is afforded by specifying a unique distribution depending on whether the examinee has mastered neither, one, or both of θ_1 and θ_2. For each examinee,

$$(\theta_{i5} \mid \theta_{i1}, \theta_{i2}, \gamma_5) = (\theta_{i5} \mid \theta_{i1} + \theta_{i2} = z, \gamma_5) \sim \text{Bernoulli}(\gamma_{5z}) \quad \text{for } z = 0,1,2, \tag{14.10}$$

where $\gamma_5 = (\gamma_{50}, \gamma_{51}, \gamma_{52})$, with γ_{5z} denoting the conditional probability of θ_5 taking a value of 1 given $\theta_1 + \theta_2 = z$. Table 14.5 demonstrates the probability structure. Note that there are four rows for the possible conditional distributions, but only three parameters: γ_{50}, γ_{51}, and γ_{52}. The choice to model the conditional distribution of θ_5 given θ_1 and θ_2 in this way reflects the substantive belief that we need not distinguish between the situation in which an examinee possesses *Skill 1* but not *Skill 2* and the situation in which an examinee possesses *Skill 2* but not *Skill 1*. In addition, it effects a small reduction in terms of the number of parameters, modeling the four conditional probabilities in terms of three parameters.

These relationships may be viewed as expressing the dependencies as a type of prerequisite relationship. If $\gamma_{21} > \gamma_{20}$, this implies that students are more likely to be proficient with respect to *Skill 2* if they are proficient on *Skill 1*. Similarly, if $\gamma_{52} > \gamma_{51} > \gamma_{50}$, students are more likely to possess *Skill 5* if they possess *Skill 1* and *Skill 2* than if they possess just one or neither of *Skill 1* and *Skill 2*. These structures are viewed as *soft* prerequisites, because they express that possessing certain skills is less likely, but not impossible, if they are lacking some other skills.

TABLE 14.4

Probability Tables for θ_1 and θ_2 in the Mixed-Number Subtraction Example, Where γ_1 Is the Probability That $\theta_1 = 1$ and γ_{2c} Is the Probability That $\theta_2 = 1$ Given the Value of $\theta_1 = c$

		$p(\theta_2 \mid \theta_1)$	
θ_1	**Probability**	**0**	**1**
0	$1-\gamma_1$	$1-\gamma_{20}$	γ_{20}
1	γ_1	$1-\gamma_{21}$	γ_{21}

TABLE 14.5

Conditional Probability Table for θ_5 in the Mixed-Number Subtraction Example, Where γ_{5z} Is the Probability That $\theta_5 = 1$ Given the Sum of θ_1 and θ_2, Denoted by z

			$p(\theta_5 \mid \theta_1, \theta_2)$	
θ_1	θ_2	z	**0**	**1**
0	0	0	$1-\gamma_{50}$	γ_{50}
0	1	1	$1-\gamma_{51}$	γ_{51}
1	0	1	$1-\gamma_{51}$	γ_{51}
1	1	2	$1-\gamma_{52}$	γ_{52}

In contrast, the model hypothesizes a *hard* prerequisite relationship between *Skill 3* and *Skill 4*. Examinees must possess *Skill 3* to possess *Skill 4*; not possessing *Skill 3* implies not possessing *Skill 4*. The implication is that of the four logically possible combinations of values of θ_3 and θ_4, only three are deemed possible in the model; the situation where $\theta_3 = 0$ and $\theta_4 = 1$ is excluded. We could specify this portion of the model in terms of a conditional distribution for θ_3 given θ_1, θ_2, and θ_5, and then a conditional distribution for θ_4 additionally conditioning on $\theta_3 = 1$ (see Exercise 14.2). Following Mislevy et al. (1999) and Almond et al. (2015), we take an alternative approach and introduce an intermediary polytomous variable that enacts the hard prerequisite relationship. This is expressed in the DAG as θ_{MN}, standing for mixed number skills, and represents the count of *Skill 3* and *Skill 4* possessed by an examinee. This is specified as a categorical variable taking on values of 0, 1, or 2 corresponding to neither, one, or both of *Skill 3* and *Skill 4*. As we did with θ_5, we adopt a simplification based on specifying a conditional distribution given the count of the parent skills the examinee possesses: with the count taking on values of 0, 1, 2, 3 corresponding to whether the examinee possesses none, one, two, or all three of its parents, *Skill 1*, *Skill 2*, and *Skill 5*. Formally, we have

$$\begin{aligned}(\theta_{iMN} \mid \theta_{i1}, \theta_{i2}, \theta_{i5}, \gamma_{MN}) &= (\theta_{iMN} \mid \theta_{i1} + \theta_{i2} + \theta_{i5} = z, \gamma_{MN}) \\ &\sim \text{Categorical}(\gamma_{MN,z,0}, \gamma_{MN,z,1}, \gamma_{MN,z,2}) \quad \text{for } z = 0,1,2,3,\end{aligned} \tag{14.11}$$

subject to the usual restriction that, for each value of z,

$$\gamma_{MN,z,0} + \gamma_{MN,z,1} + \gamma_{MN,z,2} = 1. \tag{14.12}$$

Table 14.6 lays out the conditional probability structure for θ_{MN}. The restriction in (14.12) implies that there are in principle 16 free elements in Table 14.6. The choice to model the conditional distribution of θ_{MN} given the number of preceding skills possessed reduces this to eight parameters, and reflects the substantive belief that possessing *Skill 3* and *Skill 4* depends on how many of *Skill 1*, *Skill 2*, and *Skill 5* the examinee possesses, not which ones.

TABLE 14.6

Conditional Probability Table for θ_{MN} in the Mixed-Number Subtraction Example, Where $\gamma_{MN,z,a}$ Is the Probability That $\theta_{MN} = a$ Given the Sum of θ_1, θ_2, and θ_5, Denoted by z

				$p(\theta_{MN} \mid \theta_1, \theta_2, \theta_5)$		
θ_1	θ_2	θ_5	z	**0**	**1**	**2**
0	0	0	0	$\gamma_{MN,0,0}$	$\gamma_{MN,0,1}$	$\gamma_{MN,0,2}$
0	0	1	1	$\gamma_{MN,1,0}$	$\gamma_{MN,1,1}$	$\gamma_{MN,1,2}$
0	1	0	1	$\gamma_{MN,1,0}$	$\gamma_{MN,1,1}$	$\gamma_{MN,1,2}$
0	1	1	2	$\gamma_{MN,2,0}$	$\gamma_{MN,2,1}$	$\gamma_{MN,2,2}$
1	0	0	1	$\gamma_{MN,1,0}$	$\gamma_{MN,1,1}$	$\gamma_{MN,1,2}$
1	0	1	2	$\gamma_{MN,2,0}$	$\gamma_{MN,2,1}$	$\gamma_{MN,2,2}$
1	1	0	2	$\gamma_{MN,2,0}$	$\gamma_{MN,2,1}$	$\gamma_{MN,2,2}$
1	1	1	3	$\gamma_{MN,3,0}$	$\gamma_{MN,3,1}$	$\gamma_{MN,3,2}$

So far, this specifies that the number of mixed number skills (i.e., *Skill 3* and *Skill 4*) an examinee possesses depends on possessing the other skills, but this does not encode the prerequisite relationship among *Skill 3* and *Skill 4*. To accomplish this, we include the following deterministic dependence structures for θ_3 and θ_4 given θ_{MN}:

$$\theta_{i3} \mid \theta_{iMN} = \begin{cases} 0 \text{ if } \theta_{MN} = 0 \\ 1 \text{ if } \theta_{MN} = 1 \text{ or } 2 \end{cases} \tag{14.13}$$

and

$$\theta_{i4} \mid \theta_{iMN} = \begin{cases} 0 \text{ if } \theta_{MN} = 0 \text{ or } 1 \\ 1 \text{ if } \theta_{MN} = 2. \end{cases} \tag{14.14}$$

Tying this all together, the distribution for the examinees latent variables is

$$\begin{aligned} p(\theta \mid \gamma) &= \prod_{i=1}^{n} p(\theta_i \mid \gamma) \\ &= \prod_{i=1}^{n} p(\theta_{i1} \mid \gamma_1) p(\theta_{i2} \mid \theta_{i1}, \gamma_2) p(\theta_{i5} \mid \theta_{i1}, \theta_{i2}, \gamma_5) \\ &\quad \times p(\theta_{iMN} \mid \theta_{i1}, \theta_{i2}, \theta_{i5}, \gamma_{MN}) p(\theta_{i3} \mid \theta_{iMN}) p(\theta_{i4} \mid \theta_{iMN}), \end{aligned} \tag{14.15}$$

where $\gamma = (\gamma_1, \gamma_2, \gamma_5, \gamma_{MN})$.

14.4.3 Structure of the Conditional Distribution for the Observable Variables

Substantive theory also implies considerable simplifications to the model for the conditional distributions of the observables on the latent variables. To begin, not every task requires each skill. Table 14.7 lists the text of the items and marks the skills they depend on, as well as responses from select examinees discussed later. It is seen that two of the observables are factorially simple with only one latent variable parent. Of the factorially complex observables, nine have two latent variable parents, three have three latent variable parents, and one has four latent variable parents. These patterns of dependence may be depicted by the pattern of directed edges from latent to observable variables that are present in a just-persons DAG as in CFA and IRT models with multiple latent variables (as we will see later in Figure 14.6). In DCMs, this same information is often expressed mathematically in a *Q*-matrix, which is a $(J \times M)$ incidence matrix where $q_{jm} = 1$ if observable j depends on latent variable m, and 0 otherwise. This is just the information in the middle of Table 14.7, where entries of "X" are converted to "1", and empty cells are converted to "0".* The jth row of the *Q*-matrix, $\mathbf{q}_j = (q_{j1}, \ldots, q_{jM})$ contains elements declaring which latent variables the jth observable depends on.

The second simplification due to substantive theory pertains to the factorially complex observables, in particular it concerns *how* these observables depend on the latent variables. The model was specified to reflect the theory that, for each task, an examinee who possesses all of the relevant skills should successfully complete the task, but an examinee who does not

* As the *Q*-matrix captures which observables depend on which latent variables, it is similar to the matrix of loadings in CFA and the matrix of discrimination parameters in compensatory MIRT.

TABLE 14.7

Q-matrix and Examinee Records for Two Examinees for the Mixed-Number Subtraction Example

		Skill					Examinee	
Item	**Text**	**1**	**2**	**3**	**4**	**5**	**527**	**171**
6	$\frac{6}{7}-\frac{4}{7}$	X					1	0
8	$\frac{2}{3}-\frac{2}{3}$	X					1	0
12	$\frac{11}{8}-\frac{1}{8}$	X	X				1	1
9	$3\frac{7}{8}-2$	X		X			1	0
14	$3\frac{4}{5}-3\frac{2}{5}$	X		X			1	0
16	$4\frac{5}{7}-1\frac{4}{7}$	X		X			1	1
4	$3\frac{1}{2}-2\frac{3}{2}$	X		X	X		1	0
11	$4\frac{1}{3}-2\frac{4}{3}$	X		X	X		0	0
17	$7\frac{3}{5}-\frac{4}{5}$	X		X	X		0	1
18	$4\frac{1}{10}-2\frac{8}{10}$	X		X	X		0	1
20	$4\frac{1}{3}-1\frac{5}{3}$	X		X	X		0	1
7	$3-2\frac{1}{5}$	X		X	X	X	0	0
15	$2-\frac{1}{3}$	X		X	X	X	1	0
19	$4-1\frac{4}{3}$	X		X	X	X	0	0
10	$4\frac{4}{12}-2\frac{7}{12}$	X	X	X	X		0	0

Source: A portion of this is based on Table 4 of "Probability-Based Inference in Cognitive Diagnosis" by Robert J. Mislevy, Published in 1994 by ETS as Research Report RR-94-3-ONR, and is Used with Permission from ETS.

possess all of the relevant skills should not successfully complete the task. This is an example of what may be called a conjunctive condensation rule (Klein et al., 1981; Rupp et al., 2010). The result of the conjunction of the relevant latent variables for observable j can be expressed as

$$\delta_{ij} = \prod_{m=1}^{M} \theta_{im}^{q_{jm}} = \begin{cases} 1 \text{ if examinee } i \text{ has mastered all of the skills required for task } j \\ 0 \text{ otherwise.} \end{cases} \tag{14.16}$$

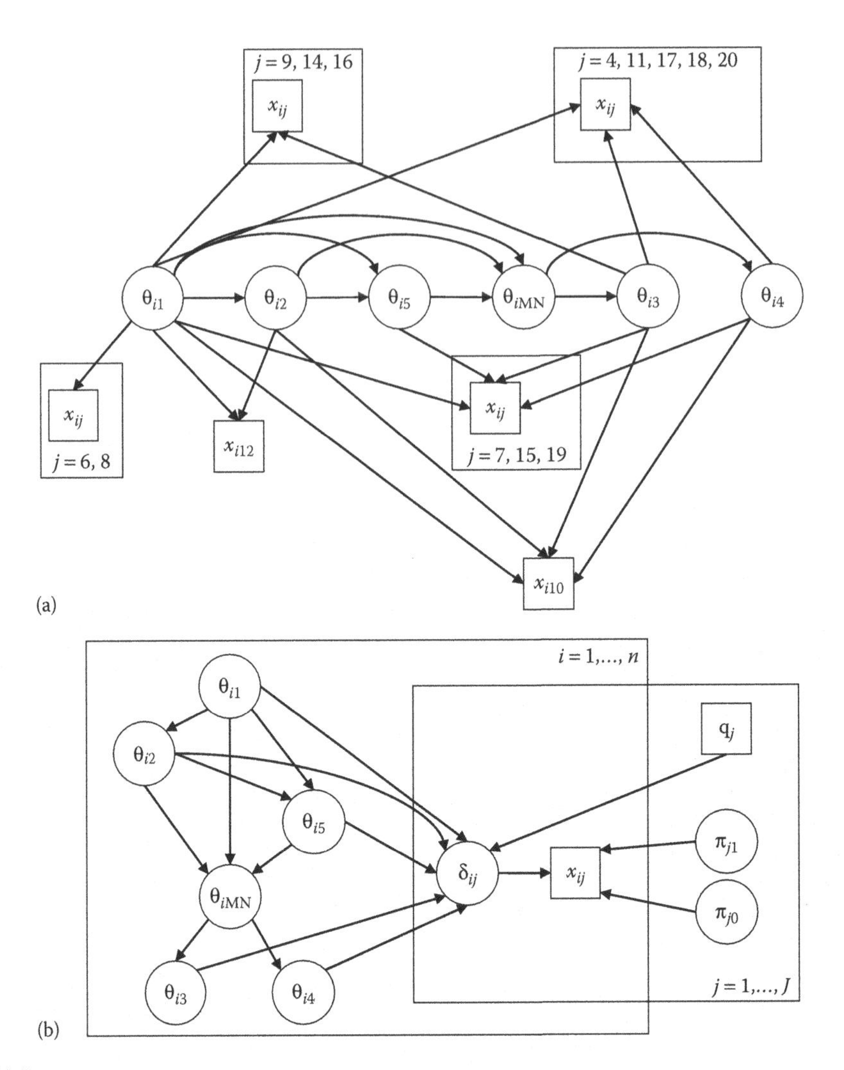

FIGURE 14.6
Two versions of the directed acyclic graph for the mixed-number subtraction example: (a) just-persons version highlighting which observables depend on which latent variables and (b) an expanded version that highlights the role of the Q-matrix and the measurement model parameters.

We then define the conditional probabilities of a value of 1 on observable j as $\pi_j = (\pi_{j1}, \pi_{j0})$, where π_{j1} denotes the conditional probability given the examinee possesses all of the relevant skills for task j (i.e., $\delta_{ij} = 1$) and π_{j0} denotes the conditional probability given the examinee does not possess all of the relevant skills for task j (i.e., $\delta_{ij} = 0$). The conditional distribution for the value from examinee i for observable j is specified as

$$(x_{ij} \mid \theta_i, \pi_j) = (x_{ij} \mid \delta_{ij} = z, \pi_j) \sim \text{Bernoulli}(\pi_{jz}) \quad \text{for } z = 0,1. \tag{14.17}$$

The model then structures the conditional distribution of an observable given the latent variables in terms of just two parameters, π_{j1} and π_{j0}, further effecting a simplification. We note in passing that many DCMs refer to π_{j0} as a *guessing* parameter, as it captures the probability of a correct response given the examinee does not possess the requisite skills. Such DCMs typically use a parameterization that defines $\pi_{j1} = 1 - s_j$, where s_j is a *slip* parameter capturing the probability that an examinee who does have the requisite skills will respond incorrectly.

Tying this all together, the conditional probability of the observables is given by

$$p(\mathbf{x} \mid \boldsymbol{\theta}, \boldsymbol{\pi}) = \prod_{i=1}^{n} p(\mathbf{x}_i \mid \boldsymbol{\theta}_i, \boldsymbol{\pi}) = \prod_{i=1}^{n} \prod_{j=1}^{J} p(x_{ij} \mid \boldsymbol{\theta}_i, \boldsymbol{\pi}_j), \tag{14.18}$$

where $p(x_{ij} \mid \boldsymbol{\theta}_i, \boldsymbol{\pi}_j)$ is given by (14.17). Figure 14.6 contains two versions of the DAG for the model that now includes the conditional distribution of the data. The first is a just-persons version that only contains examinee variables. It is distinguished by the presence of multiple nodes for the observables, with plates indicating replication over sets of observables. It highlights which observables depend on which latent variables, essentially communicating the information contained in the *Q*-matrix. This is masked in the second version, which collapses the multiple nodes for observables and more strongly emphasizes the parent–child relationships in the probability model. This version highlights the role of the *Q*-matrix, which, in combination with the latent student model variables, dictates whether the examinee has mastered all the requisite skills for a particular observable (δ_{ij}). It also includes the measurement model parameters, π_{j1} and π_{j0}, which, in combination with the δs, govern the distribution of the observables.

14.4.4 Completing the Model Specification

The two preceding sections have introduced the prior distribution for the latent variables and the conditional distribution of the data, both of which introduced new parameters in need of specification. We treat them in turn.

14.4.4.1 Prior Distributions for the Parameters That Govern the Distribution of the Latent Variables

So far we have developed the model by characterizing the dependencies among the latent variables via the parameters in $\boldsymbol{\gamma}$. To complete the specification of the prior distribution for the latent variables, the unknown values in $\boldsymbol{\gamma}$ require a prior specification. As in LCA we employ beta and Dirichlet prior distributions. For the parameter that governs the distribution of *Skill 1*, we specify

$$\gamma_1 \sim \text{Beta}(21, 6). \tag{14.19}$$

This reflects a belief that the probability that an examinee possesses *Skill 1* is akin to having seen $21 - 1 = 20$ examinees who have mastered *Skill 1* and $6 - 1 = 5$ examinees who have not. Put another way, it reflects a belief that proportion of examinees who have mastered *Skill 1* is about .8 (i.e., 20/25), and this is given a weight akin to 25 observations.* Similarly, we specify

$$\gamma_{21} \sim \text{Beta}(21, 6), \tag{14.20}$$

* The choice to weight the prior belief in this way may reflect the expression of confidence by subject matter experts, analyses of previous data, or a desire to have our prior beliefs influence the posterior in a certain amount relative to the data at hand.

which reflects a belief that if an examinee possesses *Skill 1*, the probability that she also possesses *Skill 2* is about .8, and this is given a weight akin to 25 observations. Conversely, if a student does *not* possess *Skill 1*, we would not expect them to possess *Skill 2*. This is embodied by specifying

$$\gamma_{20} \sim \text{Beta}(6,21), \tag{14.21}$$

which expresses that the probability is likely around .2, again afforded a weight akin to 25 observations. Turning to the parameters that govern the conditional distribution for *Skill 5*, a similar line of reasoning leads to expressing that the probability of an examinee possessing *Skill 3* is about .8 if she possesses both *Skill 1* and *Skill 2*, about .5 if she possesses one of *Skill 1* and *Skill 2*, and about .2 if she possesses neither *Skill 1* nor *Skill 2*. The following prior encodes these beliefs, again affording a weight of 25 observations to each:

$$\gamma_{50} \sim \text{Beta}(6,21), \tag{14.22}$$

$$\gamma_{51} \sim \text{Beta}(13.5,13.5), \tag{14.23}$$

and

$$\gamma_{52} \sim \text{Beta}(21,6). \tag{14.24}$$

As θ_{MN} has three categories, we employ Dirichlet prior distributions for the conditional probabilities. Letting $\gamma_{MN,z} = (\gamma_{MN,z,0}, \gamma_{MN,z,1}, \gamma_{MN,z,2})$ denote the collection of probabilities for the three categories of θ_{MN} given the number of *Skills 1, 2,* and *5* possessed is *z*,

$$\gamma_{MN,0} \sim \text{Dirichlet}(16,8,6), \tag{14.25}$$

$$\gamma_{MN,1} \sim \text{Dirichlet}(12,10,8), \tag{14.26}$$

$$\gamma_{MN,2} \sim \text{Dirichlet}(8,10,12), \tag{14.27}$$

and

$$\gamma_{MN,3} \sim \text{Dirichlet}(6,8,16). \tag{14.28}$$

These reflect the belief that the more skills among *Skills 1, 2,* and *5* that an examinee possesses, the more likely she is to possess *Skills 3* and *4*. The weight afforded to the prior in each case is akin to having observed these patterns in 27 observations, a slight departure from the weight of 25 afforded to the other parameters, but one that more easily allows for the division into three values.

14.4.4.2 Prior Distributions for the Parameters That Govern the Distribution of the Observables

For each observable, we have two conditional probabilities of correctly completing the task: one conditional on the examinee possessing all the requisite skills (π_{j1}), and one conditional on the examinee not possessing all of the requisite skills (π_{j0}). Assumptions of exchangeability imply the use of a common prior for the π_{j1}s for all the observables, and a common prior for the π_{j0}s for all the observables. We employ the following priors:

$$\pi_{j1} \sim \text{Beta}(23.5,3.5) \tag{14.29}$$

and

$$\pi_{j0} \sim \text{Beta}(3.5, 23.5). \tag{14.30}$$

These choices reflect the beliefs that examinees who possess the requisite skills are highly likely to correctly complete the task (probability of about .90) and that examinees who do not possess all the requisite skills are highly unlikely to correctly complete the task (probability of about .10), with weights on these beliefs akin to having seen 25 observations of each situation.

14.4.5 The Complete Model and the Posterior Distribution

The DAG for the model including the hyperparameters is given in Figure 14.7. Collecting the preceding specifications, the posterior distribution is

$$\begin{aligned} p(\boldsymbol{\theta}, \boldsymbol{\gamma}, \boldsymbol{\pi} \mid \mathbf{x}) &\propto p(\mathbf{x} \mid \boldsymbol{\theta}, \boldsymbol{\gamma}, \boldsymbol{\pi}) p(\boldsymbol{\theta}, \boldsymbol{\gamma}, \boldsymbol{\pi}) \\ &= p(\mathbf{x} \mid \boldsymbol{\theta}, \boldsymbol{\pi}) p(\boldsymbol{\theta} \mid \boldsymbol{\gamma}) p(\boldsymbol{\gamma}) p(\boldsymbol{\pi}) \\ &= \prod_{i=1}^{n} \prod_{j=1}^{J} p(x_{ij} \mid \boldsymbol{\theta}_i, \boldsymbol{\pi}_j) p(\boldsymbol{\theta}_i \mid \boldsymbol{\gamma}) p(\boldsymbol{\gamma}) \prod_{c=0}^{1} p(\pi_{jc}). \end{aligned} \tag{14.31}$$

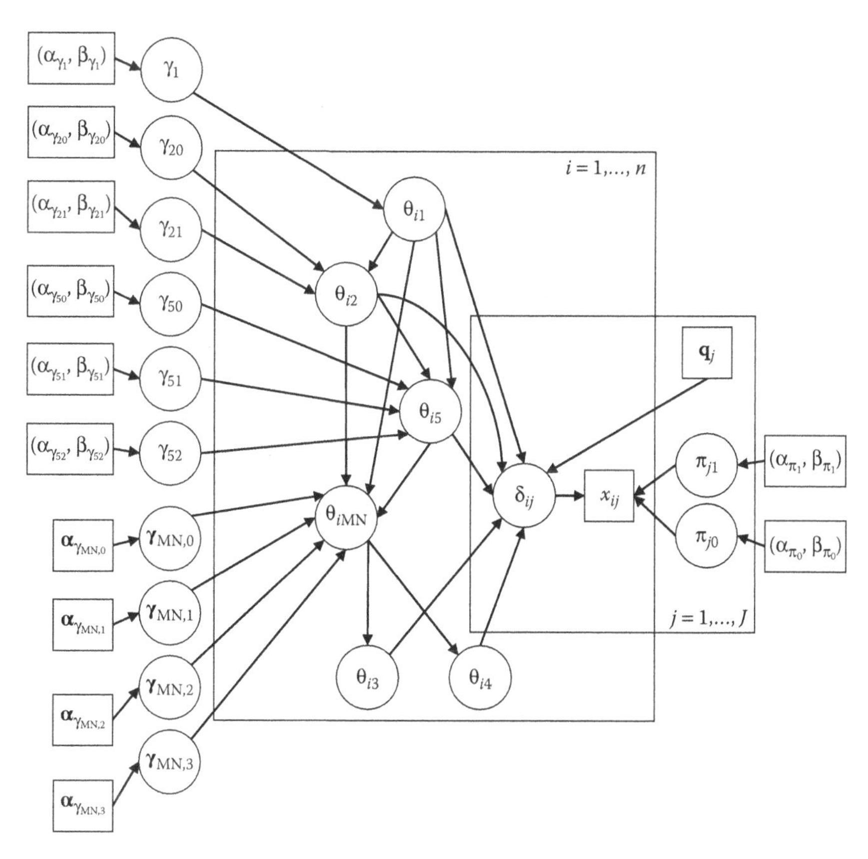

FIGURE 14.7
Directed acyclic graph for the mixed-number subtraction example including hyperparameters.

Note the similarity between the structure of (14.31) and the posterior distribution for the LCA model in (13.13). The key difference is that we now have multiple latent variables for each examinee. This results in a multivariate distribution of the latent variables for each examinee

$$p(\theta_i \mid \gamma) = p(\theta_{i1} \mid \gamma_1)p(\theta_{i2} \mid \theta_{i1}, \gamma_2)p(\theta_{i5} \mid \theta_{i1}, \theta_{i2}, \gamma_5)p(\theta_{iMN} \mid \theta_{i1}, \theta_{i2}, \theta_{i5}, \gamma_{MN})p(\theta_{i3} \mid \theta_{iMN})p(\theta_{i4} \mid \theta_{iMN}).$$

The remaining terms in (14.31) are

$$(x_{ij} \mid \theta_i, \pi_j) = (x_{ij} \mid \delta_{ij} = z, \pi_j) \sim \text{Bernoulli}(\pi_{jz}) \quad \text{for } i = 1, \ldots, n,\ j = 1, \ldots, 15,\ z = 0, 1,$$

$$\delta_{ij} = \prod_{m=1}^{M} \theta_{im}^{q_{jm}} \quad \text{for } i = 1, \ldots, n,\ j = 1, \ldots, 15,$$

$$\theta_{i1} \mid \gamma_1 \sim \text{Bernoulli}(\gamma_1) \quad \text{for } i = 1, \ldots, n,$$

$$(\theta_{i2} \mid \theta_{i1} = z, \gamma_2) \sim \text{Bernoulli}(\gamma_{2z}) \quad \text{for } i = 1, \ldots, n,\ z = 0, 1,$$

$$(\theta_{i5} \mid \theta_{i1} + \theta_{i2} = z, \gamma_5) \sim \text{Bernoulli}(\gamma_{5z}) \quad \text{for } i = 1, \ldots, n,\ z = 0, 1, 2,$$

$$(\theta_{iMN} \mid \theta_{i1} + \theta_{i2} + \theta_{i5} = z, \gamma_{MN}) \sim \text{Categorical}(\gamma_{MN,z,0}, \gamma_{MN,z,1}, \gamma_{MN,z,2}) \quad \text{for } i = 1, \ldots, n,\ z = 0, 1, 2, 3,$$

$$\theta_{i3} \mid \theta_{iMN} = \begin{cases} 0 \text{ if } \theta_{MN} = 0 \\ 1 \text{ if } \theta_{MN} = 1 \text{ or } 2 \end{cases} \quad \text{for } i = 1, \ldots, n,$$

$$\theta_{i4} \mid \theta_{iMN} = \begin{cases} 0 \text{ if } \theta_{MN} = 0 \text{ or } 1 \\ 1 \text{ if } \theta_{MN} = 2 \end{cases} \quad \text{for } i = 1, \ldots, n,$$

$$\gamma_1 \sim \text{Beta}(21, 6),$$

$$\gamma_{20} \sim \text{Beta}(6, 21),$$

$$\gamma_{21} \sim \text{Beta}(21, 6),$$

$$\gamma_{50} \sim \text{Beta}(6, 21),$$

$$\gamma_{51} \sim \text{Beta}(13.5, 13.5),$$

$$\gamma_{52} \sim \text{Beta}(21, 6),$$

$$\gamma_{MN,0} \sim \text{Dirichlet}(16, 8, 6),$$

$$\gamma_{MN,1} \sim \text{Dirichlet}(12, 10, 8),$$

$$\gamma_{MN,2} \sim \text{Dirichlet}(8, 10, 12),$$

$$\gamma_{MN,3} \sim \text{Dirichlet}(6,8,16),$$

$$\pi_{j1} \sim \text{Beta}(23.5,3.5) \quad \text{for } j=1,\ldots,15,$$

and

$$\pi_{j0} \sim \text{Beta}(3.5,23.5) \quad \text{for } j=1,\ldots,15.$$

14.4.6 WinBUGS

WinBUGS code for the model is given as follows, making use of the pow function, which raises the first argument of the function to the power of the second argument of the function (e.g., `pow(3,2)` corresponds to the formula 3^2).

```
-------------------------------------------------------------------------
#########################################################################
# Model Syntax
#########################################################################
model{

#########################################################################
# Conditional probability of the observables
#########################################################################

for (i in 1:n){
  for(j in 4:4){
    delta[i,j] <- pow(theta[i,1], Q[j,1])*pow(theta[i,2],
Q[j,2])*pow(theta[i,3], Q[j,3])*pow(theta[i,4],
Q[j,4])*pow(theta[i,5], Q[j,5])
    delta_plus_1[i,j] <- delta[i,j] + 1
           x[i,j] ~ dbern(pi_plus_1[delta_plus_1[i,j],j])
      }

  for(j in 6:12){
    delta[i,j] <- pow(theta[i,1], Q[j,1])*pow(theta[i,2],
Q[j,2])*pow(theta[i,3], Q[j,3])*pow(theta[i,4],
Q[j,4])*pow(theta[i,5], Q[j,5])
    delta_plus_1[i,j] <- delta[i,j] + 1
      x[i,j] ~ dbern(pi_plus_1[delta_plus_1[i,j],j])
      }

for(j in 14:20){
    delta[i,j] <- pow(theta[i,1], Q[j,1])*pow(theta[i,2],
Q[j,2])*pow(theta[i,3], Q[j,3])*pow(theta[i,4],
Q[j,4])*pow(theta[i,5], Q[j,5])
    delta_plus_1[i,j] <- delta[i,j] + 1
      x[i,j] ~ dbern(pi_plus_1[delta_plus_1[i,j],j])
      }

}
```

```
#########################################################################
# Prior distribution for the latent variables
#########################################################################
for(i in 1:n){
  theta[i,1] ~ dbern(gamma_1)

  theta_1_plus_1[i] <- theta[i,1] + 1
  theta[i,2] ~ dbern(gamma_2[theta_1_plus_1[i]])

  theta_1_plus_theta_2_plus_1[i] <- theta[i,1] + theta[i,2] + 1
  theta[i,5] ~ dbern(gamma_5[theta_1_plus_theta_2_plus_1[i]])

  theta_1_plus_theta_2_plus_theta_5_plus_1[i] <- theta[i,1] +
theta[i,2] + theta[i,5] + 1
  theta_MN_plus_1[i] ~
dcat(gamma_MN[theta_1_plus_theta_2_plus_theta_5_plus_1[i], ])
  theta_MN[i] <- theta_MN_plus_1[i] - 1

  theta[i,3] <- step(theta_MN[i] - .5)

  theta[i,4] <- step(theta_MN[i] - 1.5)

}

#########################################################################
# Prior distribution for the parameters
# that govern the distribution of the latent variables
#########################################################################
gamma_1 ~ dbeta(21,6)

gamma_2[1] ~ dbeta(6,21)
gamma_2[2] ~ dbeta(21,6)

gamma_5[1] ~ dbeta(6,21)
gamma_5[2] ~ dbeta(13.5,13.5)
gamma_5[3] ~ dbeta(21,6)

gamma_MN[1,1:3] ~ ddirch(alpha_gamma_MN[1, ])
alpha_gamma_MN[1,1] <- 16
alpha_gamma_MN[1,2] <- 8
alpha_gamma_MN[1,3] <- 6

gamma_MN[2,1:3] ~ ddirch(alpha_gamma_MN[2, ])
alpha_gamma_MN[2,1] <- 12
alpha_gamma_MN[2,2] <- 10
alpha_gamma_MN[2,3] <- 8

gamma_MN[3,1:3] ~ ddirch(alpha_gamma_MN[3, ])
alpha_gamma_MN[3,1] <- 8
alpha_gamma_MN[3,2] <- 10
alpha_gamma_MN[3,3] <- 12

gamma_MN[4,1:3] ~ ddirch(alpha_gamma_MN[4, ])
alpha_gamma_MN[4,1] <- 6
```

```
alpha_gamma_MN[4,2] <- 8
alpha_gamma_MN[4,3] <- 16



#########################################################################
# Prior distribution for the measurement model parameters
#########################################################################
for(j in 1:J){
    pi_plus_1[1,j] ~ dbeta(3.5,23.5)
    pi_plus_1[2,j] ~ dbeta(23.5,3.5)
    pi_0[j] <- pi_plus_1[1,j]
    pi_1[j] <- pi_plus_1[2,j]

}

} # closes the model
-------------------------------------------------------------------------
```

For our illustration, we take the publicly available data published by Tatsuoka (2002), which contains dichotomous scores coded as 1/0 corresponding to correct/incorrect responses from 536 middle school students. The model was fit in WinBUGS using three chains from dispersed starting values, and appeared to converge within 100 iterations (see Section 5.7.2). To be conservative, for each chain we discarded the first 1,000 iterations, and ran an additional 9,000 iterations (3,000 per chain), sufficient to get a stable portrait of the posterior, for use in inference. The marginal posterior densities were unimodal and fairly symmetric, with departures from symmetry occurring when the densities for the parameters were located near a boundary of 0 or 1. Table 14.8 contains numerical summaries of the marginal posterior distributions.

Beginning with the measurement model parameters, the conditional probabilities for a correct response given the examinee possesses the requisite skills are all fairly high. For example, for item 4, the probability of a correct response from an examinee that possesses all the requisite skills ($\pi_{4,1}$) has a marginal posterior distribution centered at .89 (and a 95% HPD interval of (.85, .92)). Conversely, the conditional probabilities for a correct response given the examinee does not possess all of the requisite skills are fairly low. For item 4, the probability of a correct response from an examinee that does not possesses all the requisite skills ($\pi_{4,0}$) has a marginal posterior distribution centered at .20 (95% HPD of (.16, .25)). An exception is item 8, which only depends on *Skill 1*. The item requires the examinee to subtract a quantity from itself. Such tasks might be solved correctly by reasoning that anything minus itself is 0, which does not really involve proficiency with respect to fractions per se.

Turning to the parameters for the relationships among the latent variables, we see that the posterior reflects that examinees have higher probabilities of possessing skills if they possess the parent skills on which they depend, an idea that was weakly encoded in the prior distribution, and borne out or at least not strongly contradicted by the data. For example, the posterior indicates that the probability of an examinee possessing *Skill 5* if they possess neither *Skill 1* nor *Skill 2* is fairly low (posterior median for $\gamma_{5,0}$ of .21, 95% HPD interval of (.08, .37)), the probability if they possess exactly one of *Skill 1* or *Skill 2* is higher (posterior median for $\gamma_{5,1}$ of .45, 95% HPD interval of (.27, .62)), and the probability if they possess both *Skill 1* and *Skill 2* is higher still (posterior median for $\gamma_{5,2}$ of .73, 95% HPD interval of (.64, .82)).

TABLE 14.8

Summary of the Posterior Distribution for the Mixed-Number Subtraction Example

Parameter	Median	SD[a]	95% HPD[b] Interval	Parameter	Median	SD[a]	95% HPD[b] Interval
$\pi_{4,1}$	.89	.02	(.85, .92)	γ_1	.80	.02	(.76, .84)
$\pi_{4,0}$	.20	.02	(.16, .25)	$\gamma_{2,0}$	.21	.08	(.08, .38)
$\pi_{6,1}$	.95	.01	(.93, .97)	$\gamma_{2,1}$	.90	.02	(.86, .94)
$\pi_{6,0}$	.13	.04	(.05, .22)	$\gamma_{5,0}$	.21	.08	(.08, .37)
$\pi_{7,1}$	.87	.03	(.82, .92)	$\gamma_{5,1}$	.45	.09	(.27, .62)
$\pi_{7,0}$	.13	.02	(.09, .16)	$\gamma_{5,2}$	.73	.05	(.64, .82)
$\pi_{8,1}$	.83	.02	(.79, .87)	$\gamma_{MN,0,0}$	.53	.09	(.36, .71)
$\pi_{8,0}$	.37	.04	(.29, .46)	$\gamma_{MN,0,1}$	.26	.08	(.12, .43)
$\pi_{9,1}$	.77	.02	(.72, .80)	$\gamma_{MN,0,2}$	.19	.07	(.06, .33)
$\pi_{9,0}$	.30	.04	(.23, .37)	$\gamma_{MN,1,0}$	.40	.08	(.25, .57)
$\pi_{10,1}$	.80	.03	(.75, .85)	$\gamma_{MN,1,1}$	.40	.09	(.24, .57)
$\pi_{10,1}$	.04	.01	(.02, .07)	$\gamma_{MN,1,2}$	.20	.06	(.08, .33)
$\pi_{11,1}$	.92	.02	(.89, .95)	$\gamma_{MN,2,0}$	.14	.04	(.06, .23)
$\pi_{11,0}$	.08	.02	(.05, .11)	$\gamma_{MN,2,1}$	.41	.08	(.26, .55)
$\pi_{12,1}$	.95	.02	(.92, .98)	$\gamma_{MN,2,2}$	.45	.07	(.32, .58)
$\pi_{12,0}$	.12	.03	(.06, .19)	$\gamma_{MN,3,0}$	.05	.02	(.02, .09)
$\pi_{14,1}$	.94	.01	(.91, .96)	$\gamma_{MN,3,1}$	.28	.05	(.19, .37)
$\pi_{14,0}$	.10	.03	(.04, .17)	$\gamma_{MN,3,2}$	.66	.04	(.58, .76)
$\pi_{15,1}$	.92	.02	(.88, .96)				
$\pi_{15,0}$	.17	.02	(.13, .21)				
$\pi_{16,1}$	.90	.02	(.87, .93)				
$\pi_{16,0}$	.11	.03	(.06, .17)				
$\pi_{17,1}$	.87	.02	(.83, .91)				
$\pi_{17,0}$	.05	.01	(.03, .08)				
$\pi_{18,1}$	.85	.02	(.80, .89)				
$\pi_{18,0}$	.14	.02	(.10, .17)				
$\pi_{19,1}$	.81	.03	(.75, .86)				
$\pi_{19,0}$	.03	.01	(.01, .05)				
$\pi_{20,1}$	.83	.02	(.78, .87)				
$\pi_{20,0}$	.02	.01	(.01, .04)				

[a] SD = Standard Deviation.
[b] HPD = Highest Posterior Density.

14.4.7 Inferences for Examinees

The preceding analysis also yields marginal posterior distributions for examinees, and represents a fully Bayesian account. In this section, we illustrate the use of a BN for scoring presuming given values for the measurement model parameters ($\boldsymbol{\pi}$) and the parameters that govern the distribution of the latent variables ($\boldsymbol{\gamma}$). To illustrate the inferences for examinees using the model, we work with a version of the BN that uses the posterior medians of these parameters from fitting the model via MCMC as values for the parameters. This ignores the uncertainty in the measurement model parameters and conditional probabilities in the BN. However, consequences for inferences about the examinees are slight because there is considerable precision in the posterior distribution as indicated by the relatively small posterior standard deviations and narrow credibility intervals in Table 14.8.

A screenshot of the model formulated in Netica (Norsys, 1999–2012), prior to having any values for the observables, is given in Figure 14.8. Each node is depicted with a bar capturing the probability for the node being in that state, which is also represented numerically as a percentage.

We use Examinee 527 as a case study for illustration. Table 14.7 lists the response vector from this examinee. It can be seen that examinee 527 performed well on the items that required combinations of *Skills 1, 2,* and *3,* but struggled with the items that additionally required *Skill 4,* including those that additionally required *Skill 5.* Entering the observed values for this examinee into the BN yields the posterior distribution represented in Figure 14.9. On the basis of the model, we are nearly certain the examinee possesses *Skills 1, 2,* and *3,* and nearly certain the examinee does not possess *Skill 4.* These reflect the broad patterns identified in the examinee's response vector. Note however that our posterior probability for the examinee possessing *Skill 5* is .65, reflecting considerable uncertainty. This can be understood by recognizing that we only have three items (items 7, 15, and 19) that depend on *Skill 5,* and they also depend on *Skill 4,* which the examinee almost certainly does not possess. As a consequence, we have little in the way of evidence about *Skill 5* for this examinee.

14.4.8 Model-Data Fit

The adequacy of the model-data fit may be assessed in a number of ways. Initial checks are first just checking the sensibility of the results, which was basically what the preceding paragraphs were about. Here, we focus on two aspects of model-data fit that are particularly germane to this example, which motivates a subsequent discussion of possible model extensions. In the following examples, we use 3,000 iterations from MCMC (1,000 from each of chain) to conduct PPMC using code written in R.

14.4.8.1 Checking the Hard Prerequisite Relationship

For models that are built with features aimed at reflecting a substantive theory, it is important to critically evaluate how those features of the model are performing. In the current example, theory about student cognition in the domain included that *Skill 3* is a hard prerequisite of *Skill 4.* The model was specified such that if an examinee did not possess *Skill 3,* they necessarily did not—that is, they *could* not—possess *Skill 4.* To the extent that such a specification is indeed commensurate with student cognition and performance on these tasks, we might expect that examinees who perform poorly on tasks that require *Skill 3* will also perform poorly on tasks that require *Skill 4.* Owing to the probabilistic nature of the measurement model reflecting uncertainty of measurement, the model does allow for the possibility that an examinee could perform poorly on a task that requires *Skill 3* while also performing well on a task that requires *Skill 4.* The model allows for these possibilities in two ways. An examinee who possess both skills might still incorrectly complete tasks that require *Skill 3,* the probability of such an occurrence is captured by the value of $1-\pi_{j1}$ for those tasks. Conversely, an examinee who does not possess either skill might still correctly complete tasks that require *Skill 4,* as captured by the value of π_{j0} for those tasks. Nevertheless, we may pursue the adequacy of this feature of the model by examining how often examinees perform poorly on tasks that require *Skill 3* and perform well on tasks that require *Skill 4.*

Let J_3 denote the number of tasks that require *Skill 3* and let y_3 be the sum obtained by adding the scores on these J_3 tasks. Defining J_4 and y_4 analogously, we might then investigate the conditional distributions

$$p(Y_4 = l \mid y_3 = k) \quad k = 0,\ldots,J_3, l = 0,\ldots,J_4. \tag{14.32}$$

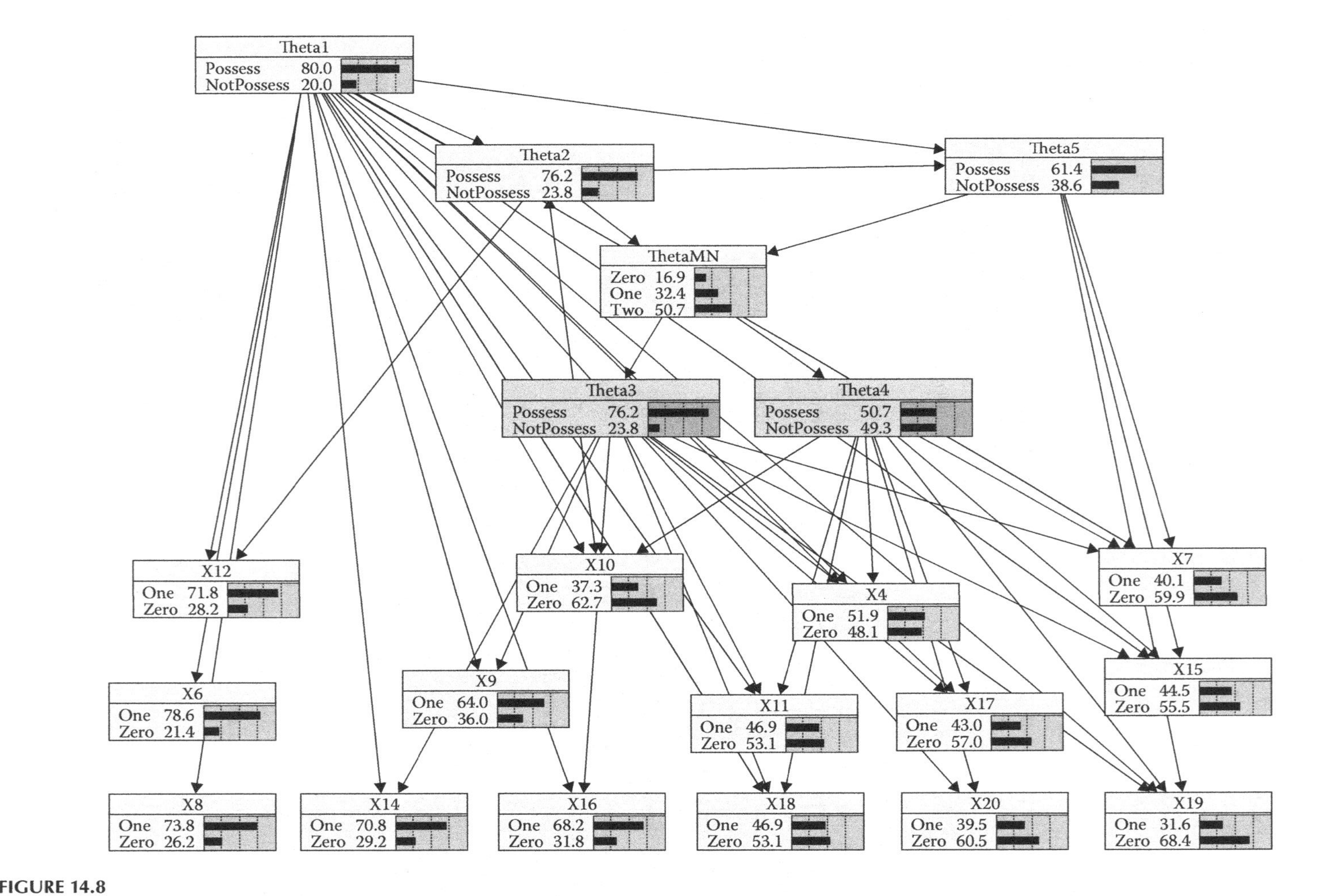

FIGURE 14.8
Netica representation of the mixed-number subtraction example prior to observing any values. The slightly darker shading for the nodes for Theta3 and Theta4 reflect that they are deterministically related to their parent, ThetaMN.

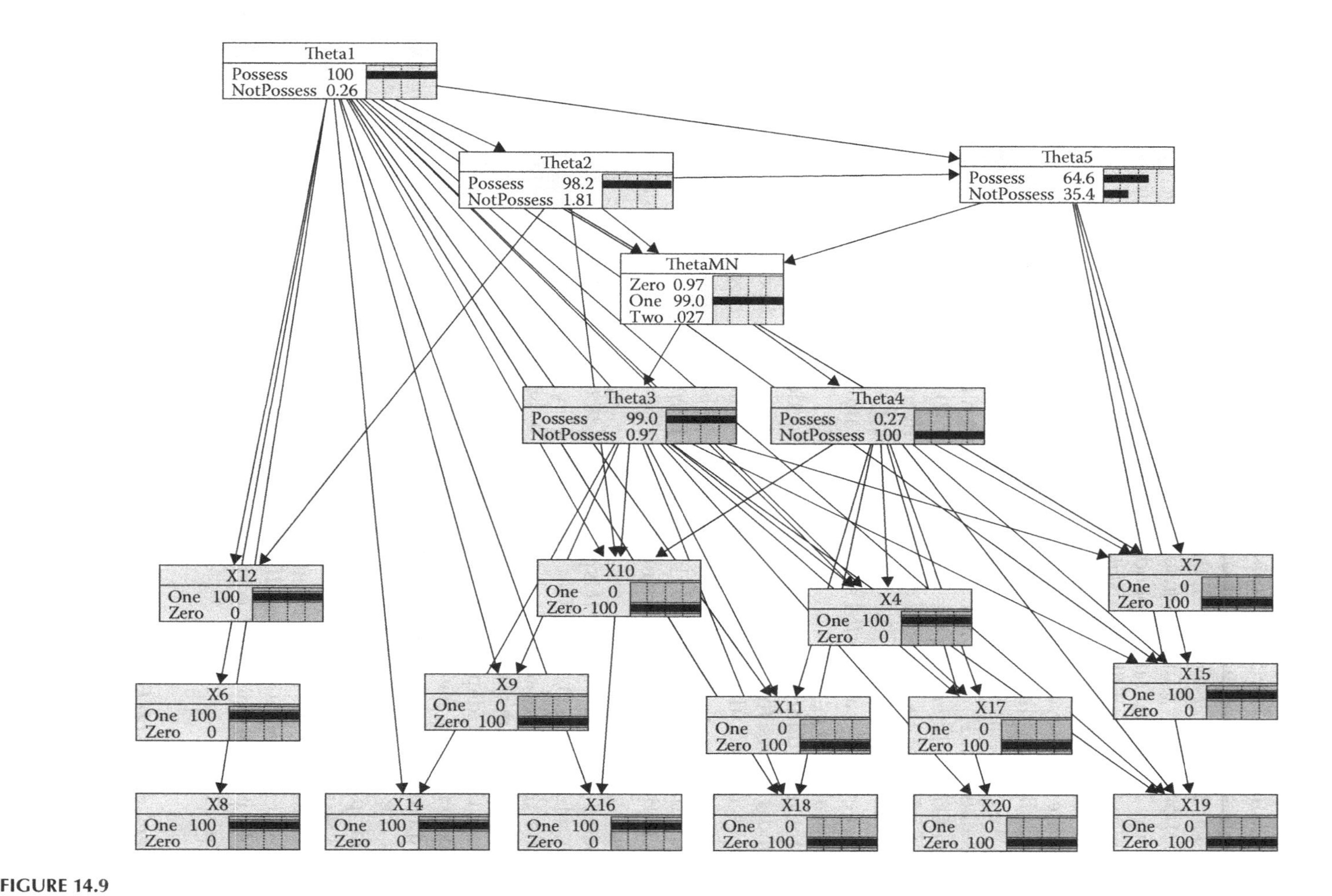

FIGURE 14.9
Netica representation of the posterior distribution for the mixed-number subtraction example for examinee 527. The slightly darker shading for the nodes for Theta3 and Theta4 reflect that they are deterministically related to their parent, ThetaMN. The darker shading for the nodes for the Xs reflect that their values are known with certainty.

If the prerequisite structure holds, we would expect low frequencies of occurrences for l being large when k is small, meaning that it is unlikely that a student would correctly complete the tasks requiring *Skill 4* given that they incorrectly completed the tasks requiring *Skill 3*.

In the current example this approach is complicated by the fact that we do not have tasks that measure *Skill 3* or *Skill 4* in isolation. As the Q-matrix in Table 14.7 expresses, the situation is somewhat messy for our purposes of honing in on performances that depend on *Skill 3* and *Skill 4*. First, all the items require *Skill 1*. Items 9, 14, and 16 are the purest evaluation of *Skill 3* in the sense that they depend on *Skill 3* and *Skill 1* only. We do not have any items that only depend on *Skill 4* in addition to *Skill 1*. Items 4, 11, 17, 18, and 20 require *Skill 4* in addition to *Skill 1* and *Skill 3*. These are the purest evaluation of *Skill 4*, as the other items that depend on *Skill 4* (i.e., items 7, 15, 19, and 10) throw either *Skill 2* or *Skill 5* into the mix. Several options for moving forward with subsets of items are possible. For the following illustration, we take the subsets that are the purest in the sense used above. We use items 9, 14, and 16 for *Skill 3*, even though they additionally depend on *Skill 1*. We use items 4, 11, 17, 18, and 20 for *Skill 4*, even though they additionally depend on *Skill 1* and *Skill 3*. Extending the notation in (14.32), let the sums of the scores from these subsets of items be y_{13} and y_{134}, respectively. Even though these values are clouded reflections of *Skill 3* and *Skill 4*, the underlying logic still holds: the prerequisite relationship between *Skill 3* and *Skill 4* suggests that examinees with low values of y_{13} ought not have high values of y_{134}.

Owing to the uncertainty associated with the dependence of performance on skills, we know the model allows for some deviations from the expected pattern, but how many? And how does the fact that we do not have tasks that only measure *Skill 3* or *Skill 4* in isolation complicate things? To work out the details of the model's implications, we again turn to simulation, and through the use of PPMC we can articulate the model's implications and the correspondence or lack thereof between those implications and the observed data.

We conducted PPMC compiling the posterior predictive frequencies that y_{134} takes on any of its possible values 0,..., 5 given the y_{13} takes on any of its possible values 0,..., 3. The results for the combinations where y_{134} was 3, 4, or 5 (the three highest values) and y_{13} was 0 or 1 (the two lowest values) are depicted in Figure 14.10. For each combination of y_{134} and y_{13}, the plot depicts the histogram of the 3,000 posterior predicted values for the frequencies, and the vertical line is placed at the realized value. For $y_{13} = 0$, the realized frequencies of $y_{134} = 3$, 4, and 5 are all 0, which obviously accords with the substance of the hypothesis, and the model-based expectations expressed in the densities. A similar picture emerges for $y_{13} = 1$. Here, the realized frequencies are 3, 2, and 5 for $y_{134} = 3$, 4, and 5, respectively. The posterior predictive distributions suggest that these are reasonably in accordance with the model-based expectations expressed in the posterior predicted densities. Overall, these results amount to support for this aspect of the model's performance, and correspondingly for the hypothesis that *Skill 3* is a prerequisite for *Skill 4*.

14.4.8.2 Checking Person Fit

We next consider the discrepancy measure of person fit presented in Chapter 13. The formulation here generalizes (13.38) and (13.40) to the case of multiple latent variables (Almond et al., 2015; Yan et al., 2003). As a simple example, the squared Pearson residual for examinee i and observable j is

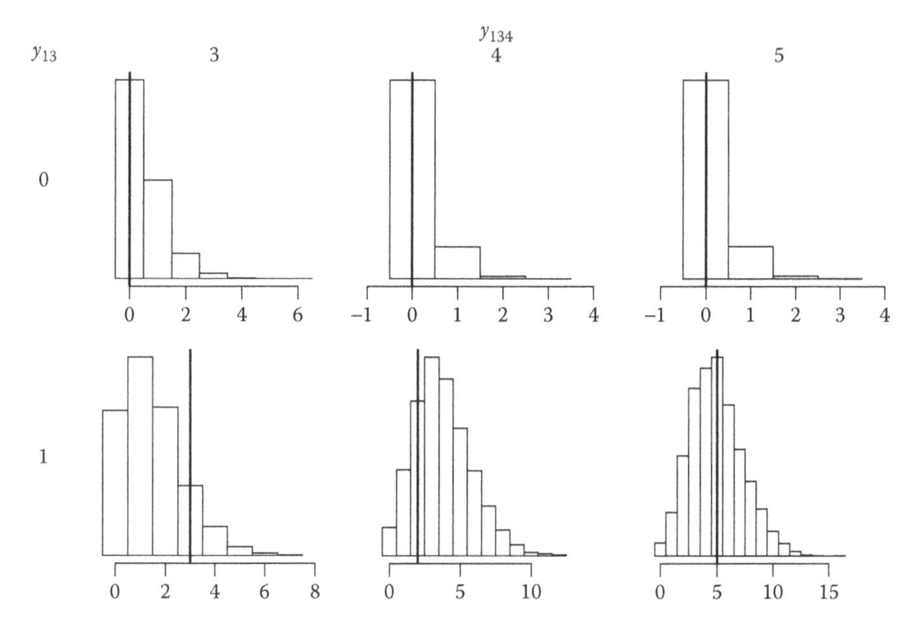

FIGURE 14.10
Posterior predictive model checking results for the conditional distribution of y_{134} given low and high values of y_{13} in the mixed-number subtraction example.

$$V_{ij}(x_{ij},\theta_i,\pi_j)=\frac{(x_{ij}-P_{ij})^2}{P_{ij}(1-P_{ij})}, \tag{14.33}$$

where $P_{ij}=E(x_{ij}\mid\theta_i,\pi_j)$ is the probability of a correct response, which in the current case is π_{j1} if $\delta_{ij}\equiv\prod_{m=1}^{M}\theta_{im}^{q_{jm}}=1$ and π_{j0} if $\delta_{ij}\equiv\prod_{m=1}^{M}\theta_{im}^{q_{jm}}=0$. A person fit discrepancy measure is then given by the root mean square error taken with respect to the values. For examinee i,

$$PF_i(\mathbf{x}_i,\theta,\pi_j)=\left(\frac{1}{J}\sum_{j=1}^{J}V_{ij}(x_{ij},\theta_i,\pi_j)\right)^{\frac{1}{2}}, \tag{14.34}$$

where $\mathbf{x}_i=(x_{i1},\ldots,x_{iJ})$ is the collection of J observed values from examinee i. We conducted PPMC using (14.34), computing 3,000 realized values by employing the observed data for $\mathbf{x}$ along with the 3,000 draws for the parameters, and 3,000 posterior predicted values by using the 3,000 posterior predicted datasets along with the 3,000 draws for the parameters. For each examinee, we computed p_{post} as a numerical summary of the results. We confine our discussion here to the results for two examinees whose response vectors are given in Table 14.7.

Examinee 527 exhibited solid fit ($p_{\text{post}}=.501$), with a response vector that affords an interpretation that coheres with the model, as the examinee performed well on tasks requiring just *Skill 1*, *Skill 2*, or *Skill 3*, but struggled with tasks additionally requiring *Skill 4* and *Skill 5*. In contrast we see poor fit for the model when it comes to examinee 171 ($p_{\text{post}}=.003$), who incorrectly answered the two simplest tasks that only required *Skill 1*, but yet correctly answered some of the more complex tasks that required *Skill 1* as well as additional skills. This pattern does not cohere with the assumed model structure; this discord manifests itself in the lack of person fit.

A lack of person fit may undermine the inferences for the examinee(s) in question, and suggests that the inferential argument expressed in the psychometric model may be dubious, at least for the examinees in question. In the current context, it may be that examinees are using Method A in working through the tasks, which involves a different set of skills. The account provided by the BN for Method B would then be a poor basis for making inferences about the examinees. We might employ a BN developed to represent our reasoning for examinees using Method A (Mislevy, 1995), and then employ the appropriate BN for whichever method the examinee uses. Of course, we may not know which method an examinee is using. To address this, Mislevy (1995) proposed the use of an expanded BN that contains BN fragments for each method, and an additional latent variable indicating which method the examinee is using. The conditional distribution for each observable would then depend on this variable as well as method-specific skills and parameters. If there is evidence that examinees use multiple methods, say, based on perceived features of the task, this could be further extended to include a latent variable capturing which method is being used for each task.

Formally, the expanded model may be seen as a mixture of the method-specific BNs, with the latent variable(s) for which method is being used as defining the components of the mixture. This represents an expanded narrative regarding the situation, and a corresponding expansion to the argument used in reasoning from observed performances to what is of inferential interest (Mislevy et al., 2008).

14.5 BNs in Complex Assessment

Our next example extends the use of BNs to polytomous latent and observable variables and illustrates its use in a complex assessment. We focus on strategies used in specifying the BN psychometric model. The context for this work is Cisco Learning Institute's NetPASS, a simulation-based performance assessment of computer networking proficiencies. A more thorough overview of the assessment, its grounding, and the BN that is of primary interest here is given by Williamson, Almond, Mislevy, and Levy (2006); the details of several of these aspects are described in Williamson et al. (2004), DeMark and Behrens (2004), and Levy and Mislevy (2004).

Briefly, nine tasks were created, three each targeting proficiencies associated with designing, implementing, and troubleshooting computer networks. To complete each task, examinees interacted with a simulator which in turn manipulated hardware. For each task, several work products were produced. Examples include logs of the commands entered when troubleshooting and repairing the network, the final state of the network, and a diagram depicting the network. Examinee work products were automatically evaluated in terms of several features, some of which were used as task level feedback only (e.g., the number of times the examinee used the help system), others of which were evaluated to produce values of observable variables for use in a psychometric model. Examples of the latter included observables capturing whether the final network was in working order, whether the correct diagram was drawn, and how efficiently the examinee arrived at a solution. Each observable could take on three values, corresponding to low, medium, or high levels of performance. These were used to support inferences about examinees in terms of several latent variables, including *Design*, *Implement*, and *Troubleshoot*, which refers to the proficiency in each of these areas, as well as three higher order latent variables: *Networking Disciplinary Knowledge*, *Network Modeling*, and *Network Proficiency*.

In what follows we present a portion of the fuller analysis reported in Levy and Mislevy (2004). Much of the reasoning in developing the full model and analysis was similar to that displayed in the mixed-number fraction subtraction example from Section 14.4, so the goal here is to illustrate some additional issues and how they are handled. In particular, we highlight the use of parametric forms to smooth the elements of a conditional probability table in a BN, while also encoding substantive beliefs about relationships among the variables.

14.5.1 Structure of the Distribution for the Latent Variables

The BN fragment for the latent variables is given in Figure 14.11, which is a just-persons DAG where the replication is assumed over examinees. Each latent variable can take on any of five values: Novice, Semester 1, Semester 2, Semester 3, and Semester 4. These names reflect the desire to characterize the level of the examinee's proficiencies in terms of Cisco's four semester curriculum on computer networking.

The conditional probability table for *Design* is given in Table 14.9. Note that, for each of the five levels of the parent variable *Network Proficiency*, there are five conditional probabilities, four of which need to be specified (the fifth is subject to the constraint that each conditional distribution sum to 1). This is a challenge, be it for subject matter experts defining the values of these conditional probabilities, or for their estimation based on data. To tackle this, Levy and Mislevy (2004) adopted a simplification based on parameterizing these conditional probabilities introduced by Almond et al. (2001) that uses Samejima's graded response model (GRM) from IRT (see Section 11.3) in ways akin to linear logistic specifications of LCA models (Formann, 1985, 1992). Let $\theta_{i,\text{NP}}$ and $\theta_{i,\text{Design}}$ denote the values of the latent variables for *Network Proficiency* and *Design*, respectively, for examinee i that can take on values of 1,..., 5, corresponding to the five levels of each (Novice,..., Semester 4). Letting $\theta^{**}_{i,\text{Design}}$ denote the *effective theta* (defined below) for *Design* for examinee i, the model specifies

$$\begin{aligned} P(\theta_{i,\text{Design}} &= k \mid \theta_{i,\text{NP}}, \mathbf{b}, a) \\ &= P(\theta_{i,\text{Design}} \geq k \mid \theta^{**}_{i,\text{Design}}, b_k, a) - P(\theta_{i,\text{Design}} \geq k+1 \mid \theta^{**}_{i,\text{Design}}, b_{k+1}, a) \end{aligned} \tag{14.35}$$

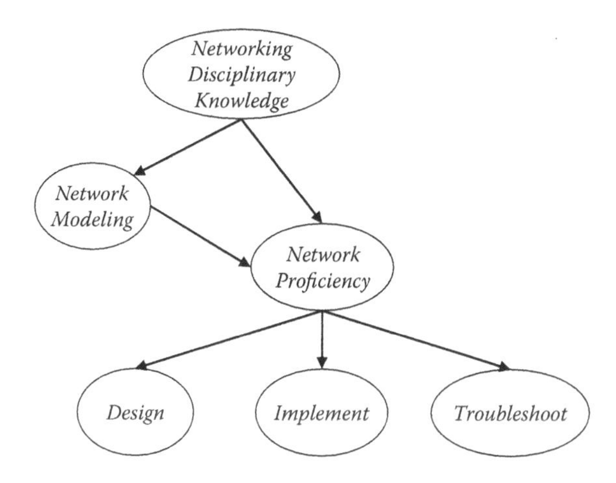

FIGURE 14.11
Directed acyclic graph fragment for the latent variables in the student model for the NetPASS example.

TABLE 14.9

Conditional Probability Table for *Design* in NetPASS, Where θ_{NP} Is the Value of the Latent Variable for *Network Proficiency* and θ^{**}_{Design} Is the Effective Theta Obtained from (14.37) Using Values for the Parameters Based on Subject Matter Expert Beliefs

			p(Design | Network Proficiency)				
Network Proficiency	θ_{NP}	θ^{**}_{Design}	**Novice**	**Semester 1**	**Semester 2**	**Semester 3**	**Semester 4**
Novice	1	−3.8	.690	.253	.049	.007	.001
Semester 1	2	−1.8	.231	.458	.253	.049	.008
Semester 2	3	0.2	.039	.192	.458	.253	.057
Semester 3	4	2.2	.005	.034	.192	.458	.310
Semester 4	5	4.2	.001	.005	.034	.192	.769

Source: Levy, R., & Mislevy, R. J. (2004). Specifying and refining a measurement model for a computer-based interactive assessment. *International Journal of Testing*, *4*, 333–369.

for $k = 1,\ldots, 5$ (for the five possible values of *Design*), where

$$P(\theta_{i,\text{Design}} \geq k \mid \theta^{**}_{i,\text{Design}}, b_k, a) = \text{logit}^{-1}[a(\theta^{**}_{i,\text{Design}} - b_k)] \tag{14.36}$$

for $k = 2,\ldots, 5$. For $k = 5$, the greater than or equal relationship is one of equality. The effective theta method fixes the values for a and the bs in (14.36). This in turn shifts the dependence of *Design* on *Network Proficiency* to the effective theta, defined as

$$\theta^{**}_{i,\text{Design}} = c_{\text{Design,NP}}\theta_{i,\text{NP}} + d_{\text{Design,NP}}, \tag{14.37}$$

where the parameters $c_{\text{Design,NP}}$ and $d_{\text{Design,NP}}$ now govern the dependence of *Design* on *Network Proficiency*. This modeling approach effects a parametric smoothing of the conditional probabilities in the table, yielding a structure that is specified in terms of now just two parameters (here $c_{\text{Design,NP}}$ and $d_{\text{Design,NP}}$) regardless of the number of levels of the parent and child variables. Table 14.9 displays the probabilities resulting from fixing $a = 1$ and $\mathbf{b} = (-3, -1, 1, 3)$ in (14.36) using $c_{\text{Design,NP}} = 2$ and $d_{\text{Design,NP}} = -5.8$. These latter two values were chosen because they best captured the prior beliefs of subject matter experts; likewise for the choices of the parameters of the effective theta model for the remaining conditional probability tables.

WinBUGS code for this portion of the model is given as follows. To begin, the code for the computation of the effective theta for *Design* instantiates (14.37) for the five possible values of *Network Proficiency*:

```
-----------------------------------------------------------------------
#######################################################################
# general form: etDesign<- cDesign*NP+dDesign
#######################################################################

etDesign[1]<- cDesign*1+dDesign
etDesign[2]<- cDesign*2+dDesign
etDesign[3]<- cDesign*3+dDesign
etDesign[4]<- cDesign*4+dDesign
etDesign[5]<- cDesign*5+dDesign
-----------------------------------------------------------------------
```

This takes the five possible values of *Network Proficiency* and yields the associated effective theta. This effective theta is then entered into the GRM. Code for (14.36) is given as follows.

```
---------------------------------------------------------------------------
for (aa in 1:5){                    # index for which effective theta
  for (bb in 1:4){                  # index for which category boundary
    des.p.greater[aa,bb] <- 1/(1+exp((-a)*((etDesign[aa])-(smb[bb]))))
  }
}
---------------------------------------------------------------------------
```

The loop over the index aa refers to the computation for each possible value of the effective theta, and the loop over the index bb refers to the computation for each of the category boundaries. a refers to the a parameter of (14.37). smb refers to the collection of b parameters in (14.37). The first two letters in this name, "sm", indicate that this is the collection of b parameters for the *Student Model* variables, in contrast to the b parameters used in the modeling of the observable variables (discussed in the following section). In the effective theta method, these are fixed. Code for doing so is as follows.

```
---------------------------------------------------------------------------
a<- 1
smb[1]<- -3.0; smb[2]<- -1; smb[3]<- 1; smb[4]<- 3;
---------------------------------------------------------------------------
```

The object des.p.greater contains the probabilities for *Design* exceeding the associated category boundary. To complete the GRM and obtain the probabilities for *Design* being equal to a particular value, the code for the computations in (14.35) is as follows.

```
---------------------------------------------------------------------------
for (aa in 1:5){
  des.p[aa,1] <- 1- des.p.greater [aa,1]
  des.p [aa,2] <- des.p.greater [aa,1]- des.p.greater [aa,2]
  des.p [aa,3] <- des.p.greater [aa,2]- des.p.greater [aa,3]
  des.p [aa,4] <- des.p.greater [aa,3]- des.p.greater [aa,4]
  des.p [aa,5] <- des.p.greater [aa,4]
}
---------------------------------------------------------------------------
```

des.p is the conditional probability table for the *Design* variable, akin to Table 14.9. Finally, the distributional specification for the *Design* variable for examinee i is a categorical distribution with probabilities given by the row of this table corresponding to the examinee's value for *Network Proficiency,* which in the code below is denoted NP.

```
---------------------------------------------------------------------------
for (i in 1:n){
  Design[i] ~ dcat(des.p[NP[i], ])
}
---------------------------------------------------------------------------
```

Structures analogous to that for *Design* were specified for the conditional probabilities for *Implement* and *Troubleshoot.*

A similar structure was specified for the conditional probability table for *Network Modeling* given *Network Disciplinary Knowledge.* However, subject matter experts indicated that a student's level of *Network Modeling* could not be higher than their level for

TABLE 14.10

Conditional Probability Table for *Network Modeling* in NetPASS, with Values Based on Subject Matter Expert Beliefs

Network Disciplinary Knowledge	*p(Network Modeling | Network Disciplinary Knowledge)*				
	Novice	Semester 1	Semester 2	Semester 3	Semester 4
Novice	1	0	0	0	0
Semester 1	.768	.233	0	0	0
Semester 2	.282	.485	.233	0	0
Semester 3	.050	.233	.485	.233	0
Semester 4	.007	.041	.222	.462	.269

Source: Levy, R., & Mislevy, R. J. (2004). Specifying and refining a measurement model for a computer-based interactive assessment. *International Journal of Testing, 4*, 333–369.

Network Disciplinary Knowledge. This constraint was imposed by forcing the resulting probabilities for levels of *Network Modeling* that were higher than *Network Disciplinary Knowledge* to 0 and renormalizing. In effect, *Network Disciplinary Knowledge* acts as a ceiling for *Network Modeling.* Table 14.10 illustrates this structure using $c_{NM,NDK} = 2$ and $d_{NM,NDK} = -8$, where it can be seen that *p(Network Modeling > Network Disciplinary Knowledge)* = 0.

The WinBUGS code for specifying the distribution of *Network Modeling* is similar to that given for *Design,* with the addition of the renormalization. Code mimicking that given above for *Design* is given as follows.

```
-----------------------------------------------------------------------
#########################################################################
# general form: etNM[NDK]<- cNM*NDK+dNM
#########################################################################
etNM[1]<- cNM*1+dNM
etNM[2]<- cNM*2+dNM
etNM[3]<- cNM*3+dNM
etNM[4]<- cNM*4+dNM
etNM[5]<- cNM*5+dNM

for (aa in 1:5){                    # index for which effective theta
  for (bb in 1:4){                  # index for which category boundary
    nm.p.greater[aa,bb] <- 1/(1+exp((-a)*((etNM[aa])-(smb[bb]))));
  }
}
for (aa in 1:5){                    # index for which effective theta
  nm.p[aa,1] <- 1-nmp[aa,1];
  nm.p [aa,2] <- nm.p.greater [aa,1]- nm.p.greater [aa,2]
  nm.p [aa,3] <- nm.p.greater [aa,2]- nm.p.greater [aa,3]
  nm.p [aa,4] <- nm.p.greater [aa,3]- nm.p.greater [aa,4]
  nm.p [aa,5] <- nm.p.greater [aa,4]
}
-----------------------------------------------------------------------
```

At this point, the code enacts the renormalization.

```
-------------------------------------------------------------------------
nm.p.ren[1,1] <- 1
nm.p.ren[1,2] <- 0
nm.p.ren[1,3] <- 0
nm.p.ren[1,4] <- 0
nm.p.ren[1,5] <- 0

nm.p.ren[2,1] <- (nm.p[2,1])/(nm.p[2,1]+nm.p[2,2])
nm.p.ren[2,2] <- (nm.p[2,2])/(nm.p[2,1]+nm.p[2,2])
nm.p.ren[2,3] <- 0
nm.p.ren[2,4] <- 0
nm.p.ren[2,5] <- 0

nm.p.ren[3,1] <- (nm.p[3,1])/(nm.p[3,1]+nm.p[3,2]+nm.p[3,3]);
nm.p.ren[3,2] <- (nm.p[3,2])/(nm.p[3,1]+nm.p[3,2]+nm.p[3,3]);
nm.p.ren[3,3] <- (nm.p[3,3])/(nm.p[3,1]+nm.p[3,2]+nm.p[3,3]);
nm.p.ren[3,4] <- 0
nm.p.ren[3,5] <- 0

nm.p.ren[4,1] <- (nm.p[4,1])/(nm.p[4,1]+nm.p[4,2]+nm.p[4,3]+nm.p[4,4])
nm.p.ren[4,2] <- (nm.p[4,2])/(nm.p[4,1]+nm.p[4,2]+nm.p[4,3]+nm.p[4,4])
nm.p.ren[4,3] <- (nm.p[4,3])/(nm.p[4,1]+nm.p[4,2]+nm.p[4,3]+nm.p[4,4])
nm.p.ren[4,4] <- (nm.p[4,4])/(nm.p[4,1]+nm.p[4,2]+nm.p[4,3]+nm.p[4,4])
nm.p.ren[4,5] <- 0

nm.p.ren[5,1] <- nm.p[5,1]
nm.p.ren[5,2] <- nm.p[5,2]
nm.p.ren[5,3] <- nm.p[5,3]
nm.p.ren[5,4] <- nm.p[5,4]
nm.p.ren[5,5] <- nm.p[5,5]
-------------------------------------------------------------------------
```

`nm.p.ren` contains the renormalized conditional probabilities, as in Table 14.10. The distributional specification for the *Network Model* variable for examinee i is a categorical distribution with probabilities given by the row of this table corresponding to the examinee's value for *Network Disciplinary Knowledge*, which in the code below is denoted `NDK`.

```
-------------------------------------------------------------------------
for (i in 1:n){
  NM[i] ~ dcat(nm.p.ren[NDK[i], ])
}
-------------------------------------------------------------------------
```

A baseline-ceiling relationship was employed for the specification of the conditional distribution of *Network Proficiency* given its two parents: *Network Disciplinary Knowledge* and *Network Modeling*. To develop the model, an *initial* effective theta was defined as a function of *Network Disciplinary Knowledge*. For each examinee this is defined as

$$\theta^{*}_{i,\text{NP}} = c_{\text{NP,baseline}}\theta_{i,\text{NDK}} + d_{\text{NP,baseline}}, \tag{14.38}$$

where $\theta_{i,\text{NDK}}$ is the value of the *Network Disciplinary Knowledge* for examinee *i*. The *final* effective theta is then modifies the initial effective theta, as

$$\theta^{**}_{i,\text{NP}} = \theta^{*}_{i,\text{NP}} + c_{\text{NP,compensatory}}[\theta_{i,\text{NM}} - (\theta_{i,\text{NDK}} - 1)]. \tag{14.39}$$

This allows for the examinee's value of *Network Modeling* ($\theta_{i,\text{NM}}$) to partially compensate for *Network Disciplinary Knowledge*. When *Network Modeling* is one level below *Network Disciplinary Knowledge*, as it was expected to be based on conversations with subject matter experts, the contribution of *Network Modeling* is zero. When *Network Modeling* is equal to *Network Disciplinary Knowledge* the contribution is $c_{\text{NP,compensatory}}$. On the other hand, when *Network Modeling* is two or more levels below *Network Disciplinary Knowledge*, the contribution is negative the number of levels less one multiplied by $c_{\text{NP,compensatory}}$. In principle, there are 25 combinations of the parent variables *Network Disciplinary Knowledge* and *Network Modeling*. However, some of these are rendered impossible by the ceiling relationship (as was the case in Table 14.10). Table 14.11 illustrates this structure for *Network Proficiency* using $c_{\text{NP,baseline}} = 2$, $c_{\text{NP,compensatory}} = 1$, and $d_{\text{NP,baseline}} = -6$.

WinBUGS code for this specification is given below, making use of aa as an index for possible values of *Network Disciplinary Knowledge* and bb as an index for possible values of *Network Proficiency*.

```
--------------------------------------------------------------------
for (aa in 1:5){
 etNP.1[aa]<- c1NP*(aa)+dNP
 for (bb in 1:5){
 etNP.2[aa, bb]<- etNP.1[aa]+c2NP*(bb-(aa-1))
 }
}
--------------------------------------------------------------------
```

The line for etNP.1 corresponds to (14.38) and the line for etNP.2 corresponds to (14.39). etNP.2 is the effective theta that can then be subjected to the GRM and renormalization strategies discussed above.

To complete the model for the latent variables, the $\theta_{i,\text{NDK}}$ are specified as following a categorical distribution in ways discussed previously.

14.5.2 Structure of the Conditional Distribution for the Observable Variables

Figure 14.12 contains a DAG for the BN fragment corresponding to the observables from the first task targeting *Design*. There are three types of variables here: observable variables referred to as *Correctness of Outcome Design 1* and *Quality of Rationale Design 1*; the latent variables *Network Disciplinary Knowledge* and *Design* that are of inferential interest; and intermediary latent variables *DK and Design 1* and *Design Context 1*. As we now discuss,

TABLE 14.11

Conditional Probability Table for *Network Proficiency* in NetPASS, with Values Based on Subject Matter Expert Beliefs

		p(Network Proficiency | *Network Disciplinary Knowledge, Network Modeling)*				
NDK[a]	**NM**[b]	**Novice**	**Semester 1**	**Semester 2**	**Semester 3**	**Semester 4**
Novice	Novice	1	0	0	0	0
Semester 1	Novice	.368	.632	0	0	0
	Semester 1	.238	.762	0	0	0
Semester 2	Novice	.135	.432	.432	0	0
	Semester 1	.065	.303	.632	0	0
	Semester 2	.036	.202	.762	0	0
Semester 3	Novice	.05	.233	.485	.233	0
	Semester 1	.02	.115	.432	.432	0
	Semester 2	.009	.056	.303	.632	0
	Semester 3	.005	.031	.202	.762	0
Semester 4	Novice	.018	.101	.381	.381	.119
	Semester 1	.007	.041	.222	.462	.269
	Semester 2	.003	.016	.101	.381	.500
	Semester 3	.001	.006	.041	.222	.731
	Semester 4	<.001	.002	.016	.101	.881

Source: Levy, R., & Mislevy, R. J. (2004). Specifying and refining a measurement model for a computer-based interactive assessment. *International Journal of Testing, 4*, 333–369.

[a] NDK = Network Disciplinary Knowledge.

[b] NM = Network Modeling.

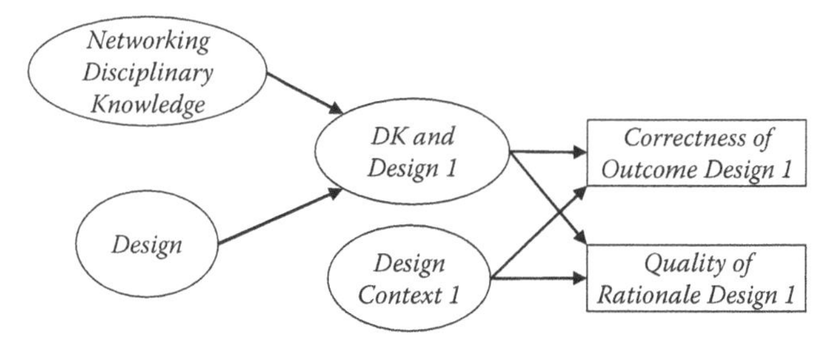

FIGURE 14.12
Directed acyclic graph fragment for the observables from the first task targeting *Design* in the NetPASS example.

these latter variables are not of inferential interest, but are relevant for representing our beliefs about the situation in the psychometric model.

Conversations with subject matter experts indicated that a conjunctive condensation rule was useful for combining *Network Disciplinary Knowledge* and *Design* to specify the intermediary variable *DK and Design 1*, which can take on any of the same five values of its parents (i.e., Novice,…, Semester 4). We expand on the conjunctive models presented in Section 14.4.3 and define a *leaky conjunction* relationship for polytomous variables. For polytomous variables, the conjunction is enacted by taking the minimum value. We therefore begin by defining

$$\theta^{*}_{i,\text{DKandDesign}} = \min(\theta_{i,\text{NDK}}, \theta_{i,\text{Design}}) \tag{14.40}$$

to initially represent the conjunctive relationship. We could use this value as an effective theta, and then zero out possible values of for *DK and Design 1* that exceed this minimum, and renormalize. This would effect a *leaky* conjunction, in which the probabilities are allowed to *leak* below the level that represents the conjunction, but not above. Levy and Mislevy (2004) employed an expanded form of this, allowing for the leak to vary depending on which of the two parent variables was the minimum, specifying the effective theta for *DK and Design 1* as

$$\begin{aligned}\theta^{**}_{i,\text{DKandDesign1}} &= [c_{\text{DKandDesign1},\theta^{*}_{\text{DKandDesign}}}\theta^{*}_{i,\text{DKandDesign}} + d_{\text{DKandDesign1}}] \\ &+ [c_{\text{DKandDesign1,NDK}}(\theta_{i,\text{NDK}} - \theta^{*}_{i,\text{DKandDesign}})] \\ &+ [c_{\text{DKandDesign1,Design}}(\theta_{i,\text{Design}} - \theta^{*}_{i,\text{DKandDesign}})].\end{aligned} \tag{14.41}$$

Table 14.12 illustrates this structure, using $c_{\text{DKandDesign1},\theta^{*}_{\text{DKandDesign}}} = 2$, $d_{\text{DKandDesign1}} = -6$, $c_{\text{DKandDesign1,NDK}} = .2$, and $c_{\text{DKandDesign1,Design}} = .4$.

WinBUGS code for this specification is given below, making use of aa as an index for possible values of *Network Disciplinary Knowledge* and bb as an index for possible values of *Design*.

```
--------------------------------------------------------------------------
for (aa in 1:5){
  for (bb in 1:5){
    etNDKDesign1[aa, bb] <- c1NDKDesign1*(min(aa, bb))
                            + dNDKDesign1
                            + c2NDKDesign1*(aa-min(aa, bb))
                            + c3NDKDesign1*(bb-min(aa, bb))
  }
}
--------------------------------------------------------------------------
```

etNDKDesign1 is the effective theta that can then be subjected to the GRM and renormalization strategies discussed above.

The *Design Context 1* variable is introduced to account for the shared dependencies among the observables derived from performances obtained from the same task (namely, the first *Design* task), akin to parameters in testlet and bi-factor models (Rijmen, 2010). In the current context of a BN, it is defined as having two values, High and Low. As *Design Context 1* has no parents it may be specified as a Bernoulli distribution in ways previously discussed, with a slight modification discussed next.

Formally, the context variable enters the model for the observable via a compensatory specification of the effective theta, akin to a CFA or compensatory MIRT model. For observable j, the effective theta is defined as

$$\theta^{**}_{ij} = c_{j,\text{DKandDesign1}}\theta_{i,\text{DKandDesign1}} + c_{j,\text{DesignContext1}}\theta_{i,\text{DesignContext1}} + d_{j,\text{DKandDesign1}}, \tag{14.42}$$

where $\theta_{i,\text{DesignContext1}}$ takes on a value of 1 or –1. The context variable has the effect of raising or lowering the effective theta by the amount of $c_{j,\text{DesignContext1}}$. The idea here is that there may be a contextual effect that raises or lowers the level of performance as captured by the multiple observables derived from a single task.

TABLE 14.12

Conditional Probability Table for *DK and Design 1* in NetPASS Example, with Values Based on Subject Matter Expert Beliefs

		p(DK and Design 1 | Network Disciplinary Knowledge, Design)				
NDK[a]	**Design**	**Novice**	**Semester 1**	**Semester 2**	**Semester 3**	**Semester 4**
Semester 4	Novice	1	0	0	0	0
	Semester 1	1	0	0	0	0
	Semester 2	1	0	0	0	0
	Semester 3	1	0	0	0	0
	Semester 4	1	0	0	0	0
Semester 1	Novice	1	0	0	0	0
	Semester 1	0.368	0.632	0	0	0
	Semester 2	0.306	0.694	0	0	0
	Semester 3	0.258	0.742	0	0	0
	Semester 4	0.222	0.778	0	0	0
Semester 2	Novice	1	0	0	0	0
	Semester 1	0.335	0.665	0	0	0
	Semester 2	0.065	0.303	0.632	0	0
	Semester 3	0.050	0.256	0.694	0	0
	Semester 4	0.040	0.218	0.742	0	0
Semester 3	Novice	1	0	0	0	0
	Semester 1	0.306	0.694	0	0	0
	Semester 2	0.057	0.279	0.665	0	0
	Semester 3	0.009	0.056	0.303	0.632	0
	Semester 4	0.007	0.043	0.256	0.694	0
Semester 4	Novice	1	0	0	0	0
	Semester 1	0.281	0.719	0	0	0
	Semester 2	0.050	0.256	0.694	0	0
	Semester 3	0.008	0.049	0.279	0.665	0
	Semester 4	0.001	0.006	0.041	0.222	0.731

Source: Levy, R., & Mislevy, R. J. (2004). Specifying and refining a measurement model for a computer-based interactive assessment. *International Journal of Testing, 4,* 333–369.

[a] NDK = Network Disciplinary Knowledge.

This effective theta is then entered into a version of the GRM that specifies the conditional distribution for the observable taking on any of its three possible values. Table 14.13 contains conditional probabilities from setting $a = 1$ and $\mathbf{b} = (-2, 2)$ in (14.36) and $c_{j,\text{DKandDesign1}} = 2$, $c_{j,\text{DesignContext1}} = .4$, and $d_{j,\text{DKandDesign1}} = -5$.

WinBUGS code for this portion of the model begins by specifying the *Design Context 1* variable as a Bernoulli, and then converting it from its usual 0/1 coding to a −1/1 coding:

```
-------------------------------------------------------------------
for (i in 1:n){
  DesignC1[i] ~ dbern(.5)
  DesignContext1[i] <- 2*DesignC1[i]-1
}
-------------------------------------------------------------------
```

TABLE 14.13

Conditional Probability Table for an Observable in the NetPASS, with Values Based on Subject Matter Expert Beliefs

		p(Observable | DK and Design 1, Context)		
DK and Design 1	**Context**	**Low**	**Medium**	**High**
Novice	Low	0.802	0.193	0.004
	High	0.646	0.344	0.010
Semester 1	Low	0.354	0.613	0.032
	High	0.198	0.733	0.069
Semester 2	Low	0.069	0.733	0.198
	High	0.032	0.613	0.354
Semester 3	Low	0.010	0.344	0.646
	High	0.004	0.193	0.802
Semester 4	Low	0.001	0.068	0.931
	High	0.001	0.032	0.968

Source: Levy, R., & Mislevy, R. J. (2004). Specifying and refining a measurement model for a computer-based interactive assessment. *International Journal of Testing, 4*, 333–369.

The code corresponding to (14.42) for an observable j is given by:

```
------------------------------------------------------------------------
for (i in 1:n){
  etDesign1[i, j] <- c1Design1[j]*NDKDesign1[i]
                    + c2Design1[j]*DesignContext1[i]
                    + dDesign1[j]
}
------------------------------------------------------------------------
```

The effective theta is then entered into the GRM, with code as follows, here, using WinBUGS `logit` function to enact (14.37):

```
------------------------------------------------------------------------
for (i in 1:n){
  for (k in 1:K){
    logit(p.greater[i, j, k])<- a*(etDesign1[i, j] - emb[j, k])
  }
}
------------------------------------------------------------------------
```

Here, a refers to the a parameter of (14.37). emb refers to the collection of b parameters in (14.37). The first two letters in this name, "em", indicate that this is the collection of b parameters for the *Evidence Model* variables, in contrast to the b parameters used in the modeling of the Student Model variables (discussed in the preceding section). In the effective theta method, these are fixed. Code for doing so is as follows.

```
------------------------------------------------------------------------
emb[j, 1] <- -2
emb[j, 2] <- 2
------------------------------------------------------------------------
```

The object `p.greater` contains the probabilities for the observable exceeding the associated category boundary. To complete the GRM and obtain the probabilities for the observable being equal to a particular value, the code for the computations in (14.35) is as follows.

```
for (i in 1:n){
  p[i, j, 1] <- 1-p.greater[i, j, 1]
  p[i, j, 2] <- p.greater[i, j, 1] - p.greater[i, j, 2]
  p[i, j, 3] <- p.greater[i, j, 2]
}
```

`p` is the collection of conditional probabilities for examinee i for observable j. Finally, the distributional specification for the observable itself is a categorical distribution with probabilities given by `p` for the examinee and observable:

```
for (i in 1:n){
  x[i, j] ~ dcat(p[i, j, ])
}
```

Analogous specifications were made for the other observables from this and the two other tasks targeting *Design*, which had DAGs akin to that in Figure 14.12, as well as the three tasks targeting *Implement*, and the three tasks targeting *Troubleshooting*. See Levy and Mislevy (2004) for complete details on these specifications, prior distributions based on subject matter experts' expectations, MCMC estimation, and resulting inferences about the tasks and examinees.

14.6 Dynamic BNs

All of the models considered so far have modeled the latent variables as static, in line with an assumption that an examinee's proficiency remains constant during the course of assessment. However, several types of assessment systems are explicitly designed so that learning is at least possible if not desirable during the assessment. Such systems are characterized by the presence of at least a minimum level of feedback to examinees regarding their performance. Examples include remediation triggered by incorrect responses in tutoring systems, or being sent back to the beginning of a level in a game.

Dynamic BNs (DBNs) extend the basic BN to model changes over time and related longitudinal data structures, and so can be gainfully employed in psychometric modeling of change during the course of an assessment. Figure 14.13 lays out the structure over three time slices. The model has two basic components. The first is the *within-time* component. This is the now familiar structure where the observable from that time depends on the latent variable representing the proficiency at that time. The version depicted in Figure 14.13 may be extended to have multiple latent variables and multiple observable

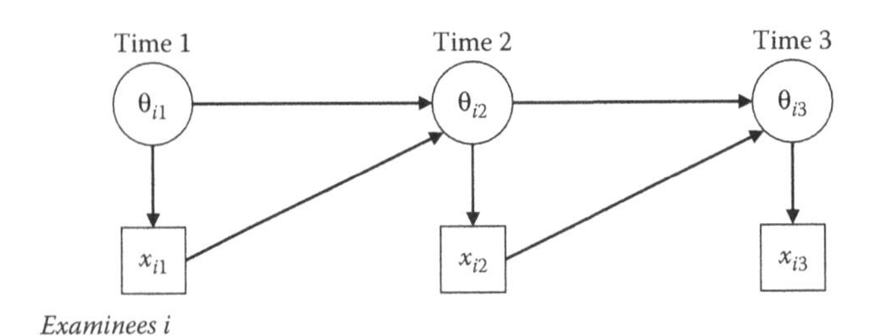

FIGURE 14.13
Directed acyclic graph for a dynamic Bayesian network.

variables within each time slice. The second component is a *between-time* or *transition* component in which the latent variable at a particular time is modeled as dependent on the latent variable and observable at the previous time.

To simplify the situation, suppose we have a model for one dichotomous latent variable that represents possessing a skill and one dichotomous observable variable that corresponds to the correctness of a response to a task. The within-time component is then a latent class model, discussed previously. We focus our attention on the transition component that models the latent variable at any time as dependent on the latent and observable variable at the previous time. Let θ_{it} denote the value of the latent variable for examinee i at time t, taking on a value of 1 or 0 corresponding to examinee i possessing or not possessing the skill at time t. Similarly, let x_{it} denote the observable coded as 1 or 0 for a correct or incorrect response from examinee i at time t. The conditional probability table for the transition component is given in Table 14.14, where γ_{kl} denotes the conditional probability that an examinee with values of $\theta_{it} = k$ and $x_{it} = l$ at time t will possess the skill at time $t + 1$ (i.e., $\theta_{i(t+1)} = 1$).

γ_{11} and γ_{10} represent the conditional probabilities that an examinee will possess the skill at time $t + 1$ given that they possessed the skill at time t, accompanied by a correct and incorrect response at time t, respectively. A common simplifying assumption sets these values at 1, reflecting that if an examinee possesses a skill, they do so at the next time period(s). γ_{01} and γ_{00} represent the transition probabilities from not possessing a skill to possessing a skill following a correct or incorrect response, respectively.

A number of simplifications or extensions of this general structure are possible. For example, Levy (2014) constructed a DBN for *Save Patch*, an educational game targeting rational number addition (Chung et al., 2010) in which

TABLE 14.14

Conditional Probability Table for the Transition Component in a Dynamic Bayesian Network

θ_t	x_t	$x(\theta_{t+1} \mid \theta_t, x_t)$
1	1	γ_{11}
1	0	γ_{10}
0	1	γ_{01}
0	0	γ_{00}

- Players (examinees) complete levels (tasks) by using math skills to navigate from the beginning of the level to the end;
- Feedback occurs in that successfully completing a level leads to being presented a new level, while unsuccessfully completely a level leads to being presented the same level for another attempt;
- The performance of each player on each level of the game was evaluated as representing a correct solution or errors of various kinds, defining a polytomous observable for each attempt at each level;
- Multiple latent variables representing a variety of skills of inferential interest and possible misconceptions were specified;
- The conditional probabilities in the within-time component were smoothed via an IRT parameterization.

See Levy (2014) for complete details, as well as specifications of prior distributions for the unknown parameters of the parametric forms used to enact smoothing, and the results of fitting the model using MCMC. For our current purposes of illustrating DBNs, we present a simplified account of a portion of the model corresponding to performance on Level 19 of the game, making the simplifying assumption that the player does not possess the misconception measured on that level and using point summaries of the posterior distribution obtained from model-fitting.

Table 14.15 presents the conditional probability table for the observable corresponding to performance on Level 19 given the targeted skill, *Adding Unit Fractions*. This structure is assumed to hold, with the same conditional probabilities, over all attempts (time points) at the level.

If a player (examinee) possesses the *Adding Unit Fractions* skill, they will almost certainly complete the level with the Standard Solution, meaning they successfully complete the level in a way that reflects the skill. There is only a small probability that they will provide an Incomplete Solution, or commit an error, possibly one identified as evidencing certain misconceptions (Wrong Numerator Error) or one not associated with any misconceptions (Unknown Error). In contrast, a player who does not possess the skill is much less likely to successfully complete the level and is more likely to make an error.

Table 14.16 presents the conditional probabilities of the possessing the *Adding Unit Fractions* skill at time $t + 1$ given *Adding Unit Fractions* and the observable for performance on the level at time t. The first row reflects the previously mentioned assumption that once a player (examinee) acquires a skill, she will retain it. The second row gives the probabilities of transitioning from a state of not possessing to a state of possessing the skill, which vary based on the performance on the level.

TABLE 14.15

Conditional Probability Table for Performance on Level 19 in *Save Patch*

	p(Observable for Level 19 | Adding Unit Fractions)				
Adding Unit Fractions	**Standard Solution**	**Alternate Solution**	**Incomplete Solution**	**Wrong Numerator Error**	**Unknown Error**
Possess	0.95	0.00	0.01	0.03	0.01
Not Possess	0.58	0.02	0.02	0.25	0.13

TABLE 14.16

Conditional Probabilities for Possessing *Adding Unit Fractions* at Time $t + 1$ on Level 19 in *Save Patch*

	***Observable for Level 19* at Time t**				
Adding Unit Fractions at Time t	**Standard Solution**	**Alternate Solution**	**Incomplete Solution**	**Wrong Numerator Error**	**Unknown Error**
Possess	1	1	1	1	1
Not possess	.38	.17	.19	.20	.09

To complete the specification the current illustration, the probability of possessing *Adding Unit Fractions* at time 1 is set to be .70.

Figure 14.14 contains screenshots from Netica depicting the situation at four different points in time corresponding to four different states of knowledge. In panel (a), no observations have been made. The probability that the player possesses the skill at time 1 is .7. The probabilities for the remaining nodes may be thought of as predictive distributions. For example, the probability of .79 that the player will possess the skill at time 2 is the model-based expectation, obtained by marginalizing over the uncertainty in its parent variables.

Panel (b) represents the state of the network after observing the player commit an Unknown Error on their first attempt at Level 19. The posterior probability that the player possesses the skill at that time is now .15, the drop from .70 due to the evidentiary bearing of having observed this result. Again, the probabilities for the remaining nodes may be thought of as predictive distributions, now posterior to the observation at time 1. The posterior predictive distribution for the player possessing the skill at time 2 is .23, which reflects a synthesis of the two possibilities: they possessed the skill at time 1 and kept it, or they did not possess the skill at time 1 but acquire it (i.e., transition) following their first attempt.

Panel (c) updates the situation based on the observation of an unsuccessful attempt where the player committed a Wrong Numerator Error. This augments the previous data, and as a result we think it is highly unlikely that the player possessed the skill at time 1 (probability = .02) or at time 2 (probability = .03). As before, the model allows for the possibility that the player will possess the skill at time 3, though this is not very likely (probability = .23).

Finally, panel (d) displays the state of the network on observing that the player provides the standard solution on their third attempt. The posterior probability that they possessed the skill at time 3 has increased to .33. The posterior probabilities that they possessed the skill earlier, at timepoints 1 and 2, have increased very slightly. Figure 14.14 illustrates how DBNs are used to represent the evolution of our beliefs about a dynamic latent variable as data arrive.

14.7 Summary and Bibliographic Note

We have described BNs as extensions to LCA models that employ multiple discrete latent variables, illustrating a fully Bayesian approach to BNs in several assessment settings. Almond et al. (2015) provide a textbook length treatment of these and other topics surrounding BNs in assessment.

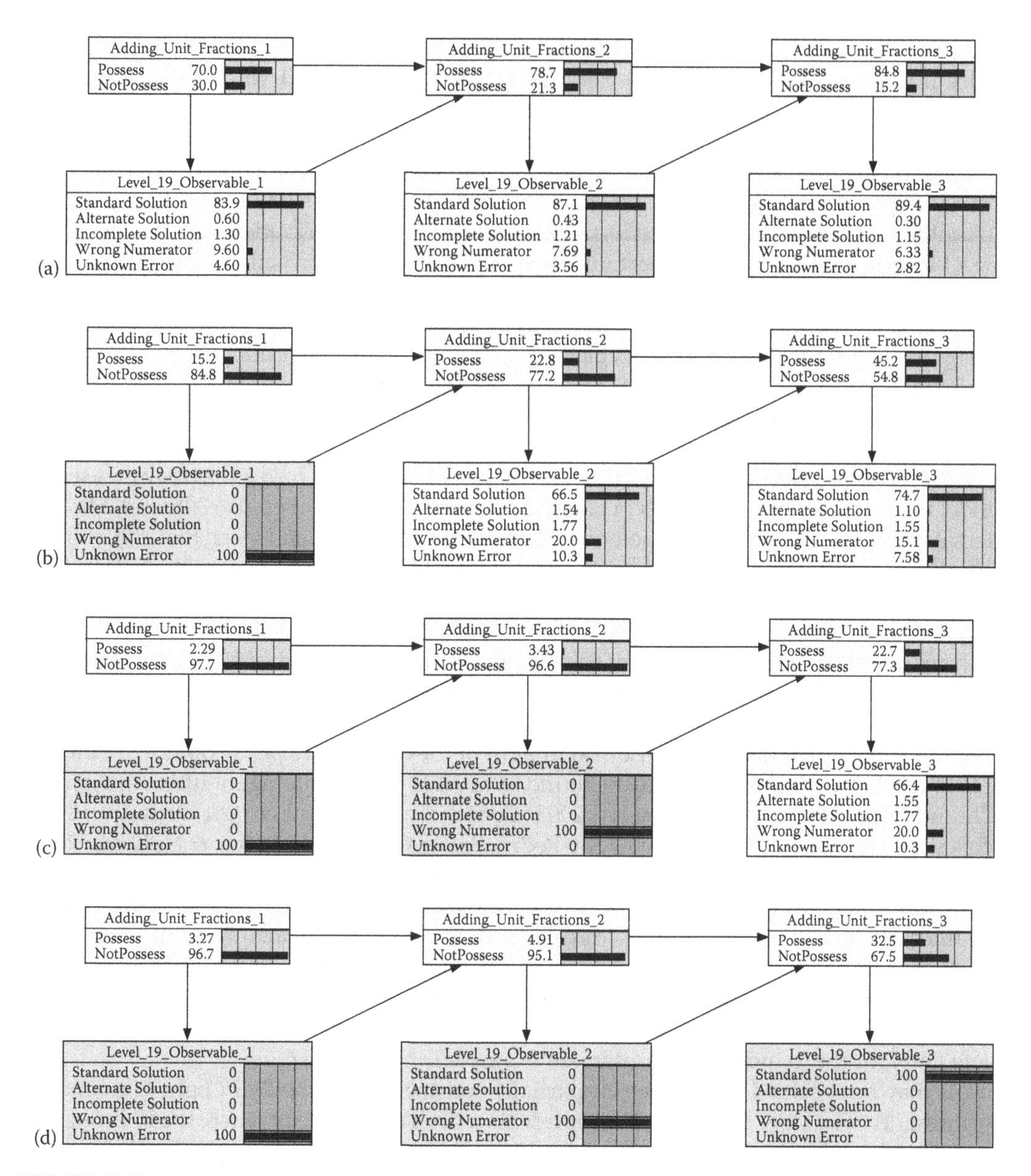

FIGURE 14.14
States of a portion of the dynamic Bayesian network over multiple time points for performance on Level 19 in *Save Patch*: (a) prior to observing any attempts, (b) after observing an unknown error on the first attempt, (c) after observing a wrong numerator error on the second attempt, and (d) after observing a standard solution on the third attempt.

Cast in the current light, DCMs that employ discrete latent and observable variables may be seen as BNs, underscoring the appeal of BNs as a powerful approach to diagnostic assessment (Almond et al., 2007). Bayesian approaches have also been popular for extensions of such models. One line of research incorporates continuous latent variables in several different ways, including as an aspect of proficiency on which observables depend (Bradshaw & Templin, 2014; Hong, Wang, Lim, & Douglas, 2015; Roussos et al., 2007), and as a higher order latent variable to model the dependence among the discrete latent skills (de la Torre & Douglas, 2004). See also Hoijtink, Béland, and Vermeulen (2014) for a Bayesian treatment of DCMs and cognitive diagnostic models of other types.

Another line of research expands DCMs to model differential attractiveness of response options to partially correct examinee understandings (DiBello, Henson, & Stout, 2015) or models slip and guessing parameters as a function of examinees (Huang & Wang, 2014b).

Another extension comes in the form of specifying prior distributions for the entries of the *Q*-matrix, reflecting the uncertainty regarding whether correctly completing a task depends on possessing a particular skill (DeCarlo, 2012). Similarly, a mixture of BNs could be specified to reflect that multiple solution strategies are possible, with different skills being invoked under different strategies, possibly with examinees adopting or switching strategies across tasks (de la Torre & Douglas, 2008; Mislevy, 1995).

A more complete description of construction, MCMC estimation, and use of the BN for the NetPASS example in Section 14.5 is given by Levy and Mislevy (2004). See also Almond et al. (2001) and Mislevy et al. (2002) for descriptions of functional forms that may be used to model the conditional probabilities in BNs in complex assessments. See also Almond, Mulder, Hemat, and Yan (2009), who focused on options for models of dependence that can arise in different ways when multiple observables are derived from complex assessment. This flexibility of BNs has made it an attractive option for complex assessments, such as in the works just cited, VanLehn and Martin (1997), Rupp et al. (2012), and Quellmalz et al. (2012).

The use of discrete variables makes the updating in BNs computationally fast, supporting real-time or near real-time updating as examinees engage with assessments. Accordingly, BNs have become popular for use in adaptive assessment (Almond & Mislevy, 1999) and related systems that adapt what they do in light of the examinee's performance. Examples of BNs used in such capacities may be found in applications of intelligent tutoring systems (Chung, Delacruz, Dionne, & Bewley, 2003; Conati, Gertner, VanLehn, & Druzdzel, 1997; Mislevy & Gitomer, 1996; Reye, 2004; Sao Pedro, Baker, Gobert, Montalvo, & Nakama, 2013; VanLehn, 2008) and games (Iseli, Koenig, Lee, & Wainess, 2010; Rowe & Lester, 2010; Shute, 2011; Shute, Ventura, Bauer, & Zapata-Rivera, 2009). Reye (2004) demonstrated how preceding lines of work on modeling longitudinal patterns assuming learning could be framed as DBNs, paving the way for their applications in tutoring systems (Reye, 2004; VanLehn, 2008) and game-based assessments of learning or change (Iseli et al., 2010; Levy, 2014). See Levy (2014) for a description of the fully Bayesian specification and estimation of the complete DBN for the game-based assessment example in Section 14.6.

BNs may be specified in advance, or partially constructed on the fly in light of examinee performance. The DBN in Section 14.6 represents a simple example of where the BN could be constructed on the fly. Here, the network adds nodes for the next time slice for the current level if the player (examinee) is unsuccessful on the previous attempt, or adds nodes for next level if the player is successful on the previous attempt. Mislevy and Gitomer (1996) described a more complicated scenario in the context of HYDRIVE, a simulation-based assessment of troubleshooting for the F-15 aircraft's hydraulics systems. The open-ended nature of the assessment implies that there are an enormous number of possible states to potentially monitor, which poses challenges to developing psychometric models (Levy, 2013). Rather than model all possibilities, HYDRIVE was designed to recognize when the examinee had worked her or his way into a situation that could be characterized as providing evidence regarding one of the proficiencies of interest. Such situations could occur frequently, sparingly, or not at all depending on the examinee's chosen actions. Once a situation was recognized as such, the BN for the examinee could be augmented to include the appropriate variables.

Exercises

14.1 Reconsider the just-persons DAG for the BN in Figure 14.4

a. Assuming each observable and latent variable is dichotomous and the (conditional) probabilities are unknown parameters, write out the model in terms of the distributions, including conditionally conjugate prior distributions on the unknown parameters.

b. Write out a DAG for the model in (a) that expands Figure 14.4 to include all the entities.

c. For each unknown entity list out the relevant entities that need to be conditioned on in the full conditional distribution. (Hint: This can be done just using the DAG from b.)

d. For each unknown entity, write out the full conditional distribution.

e. Repeat (a)–(d) now assuming each observable and latent variable is polytomous.

14.2 The BN for the mixed-number subtraction example developed in Section 14.4 introduced θ_{MN} to structure the conditional distributions of θ_3 and θ_4 on the remaining latent variables honoring the substantive theory, which states that (a) how many of *Skill 1*, *Skill 2*, and *Skill 5* the examinee possesses is relevant, not which ones; and (b) *Skill 3* is a hard prerequisite for *Skill 4*.

a. Specify a model that honors these substantive beliefs, but (i) expresses the conditional probability of possessing *Skill 3* given *Skill 1*, *Skill 2*, and *Skill 5*, $p(\theta_3 \mid \theta_1, \theta_2, \theta_5)$, in terms of the parameters of the model, and (ii) expresses the conditional probability of possessing *Skill 4* given *Skill 1*, *Skill 2*, *Skill 5*, and *Skill 3*, $p(\theta_4 \mid \theta_1, \theta_2, \theta_5, \theta_3)$, in terms of the parameters of the model.

b. How could the conditional probabilities $p(\theta_3 \mid \theta_1, \theta_2, \theta_5)$ and $p(\theta_4 \mid \theta_1, \theta_2, \theta_5, \theta_3)$ be monitored in WinBUGS?

c. Create an alternative model for the latent variables that honors the substantive theory but does not involve the intermediary variable θ_{MN}. That is, create a model that specifies θ_3 and θ_4 as directly dependent on θ_1, θ_2, and θ_5 (and possibly each other).

14.3 Write out the joint distribution for the entities depicted in Figure 14.13 in terms of recursive conditional distributions given other entities.

14.4 Compute the marginal distributions for the entities in the DBN for *Save Patch* example in Figure 14.14 in the following situations.

a. Using the prior probability of possessing the Adding Unit Fractions skill at Time 1 of .70 and the conditional probability tables in Tables 14.5 and 14.6, compute the marginal distributions for all the entities. Verify your results with Figure 14.14 panel (a).

b. Suppose the examinee exhibits an Unknown Error on their first attempt. Compute the posterior distribution for the Adding Unit Fractions skill at Time 1, and the posterior predictive distribution for the remaining entities. Verify your results with Figure 14.14 panel (b).

c. Suppose the examinee next exhibits a Wrong Numerator Error on their second attempt. Compute the posterior distribution for the Adding Unit Fractions skill at Time 1 and at Time 2, and the posterior predictive distribution for the remaining entities. Verify your results with Figure 14.14 panel (c).

d. Suppose the examinee next completes the level with a Standard Solution on their third attempt. Compute the posterior distribution for the Adding Unit Fractions skill at Time 1, Time 2, and Time 3. Verify your results with Figure 14.14 panel (d).

e. Given the observations of these three attempts, what is the probability that the examinee possesses the Adding Unit Fractions skill at Time 4?

15

Conclusion

We conclude our treatment of Bayesian psychometric modeling in this chapter first by summarizing the main developments in Section 15.1. In Section 15.2, we work through a situation where caution is warranted in applying the principles we have espoused. We conclude the chapter with some parting words in Section 15.3.

15.1 Bayes as a Useful Framework

Our investigation began by seeking a way to support or make claims about examinees in assessment, and we landed on the posterior distribution $p(\theta \mid \mathbf{x})$ as an expression of what we believe about examinees having observed the data. Recognizing that the evidentiary value of the data may vary from task to task in ways we might not precisely know before we collect the data, we extended this out to $p(\theta, \omega \mid \mathbf{x})$ that now expresses what we believe about examinees and the psychometric properties of tasks they interacted with that produced the data. This could be further extended, for example, to include posterior representations for hyperparameters that govern these parameters or data not yet observed, but for our current purposes of summarizing the broad themes, this level of generality is sufficient. Exchangeability, de Finetti's representation theorem, and conditional independence relationships are indispensable tools for setting up probability models in service of arriving at posterior distributions via Bayes' theorem. We have endeavored to articulate how these ideas play out in popular psychometric modeling families by providing introductory treatments of Bayesian approaches to these models, highlighting common choices made in specifying these models. We may summarize the thrust behind adopting a Bayesian approach, which cuts across the modeling families and choices, with two points:

1. $p(\theta, \omega \mid \mathbf{x})$ is what we want. It is the result of first encoding a real-world situation in terms of a probability model that honors and makes explicit our substantive understanding of the situation and purposes, and then reasoning through that model to express what we believe about what we do not know, based on what we do know.
2. $p(\theta, \omega \mid \mathbf{x})$ is attainable. The power of Markov chain Monte Carlo has allowed us to obtain posterior distributions even in very complex models, including situations where other approaches are intractable.

The specifics of these ideas and how their extensions play out in particular psychometric modeling families are instantiations of a general approach described in Chapter 7. We organized our presentation in terms of the psychometric modeling families mainly for didactic reasons. We resist the notion that there are sharp boundaries between these families or that analysts must choose from among them. Instead, we advocate a modular approach to modeling (Rupp, 2002) that views the models we have covered as some of the options in a menu of structures that we can call into service, combine, modify, and extend as needed. These are tools to help us achieve some of inferential goals we have when we employ models, and serve as useful answers to the questions and needs that we have in assessment. Want to characterize examinees in terms of a continuum of proficiency? Use a continuous latent variable (Chapters 8, 9, and 11). Want to also characterize examinees in terms of finite distinct strategies they use? Use a discrete latent variable for that (Chapters 13 and 14). Want to pursue learning during the assessment? Build a transition component for change over time (Chapter 14). Are examinees organized in terms of meaningful clusters, or can tasks be grouped in terms of common design features? Add a multilevel structure with collateral information to render them conditionally exchangeable (Sections 9.7 and 11.7.4). And so on. In this text, we have mentioned some of these situations, with pointers to examples of applications that blur the boundaries between the modeling families as we have organized them.

We advance that Bayesian approaches allow us to better build and reason with statistical models. We have examined how such approaches play out in several different psychometric modeling paradigms, as well as when it comes to statistical issues that transcend psychometric applications such as model-data fit and accommodating missing data. Our view is that a Bayesian perspective provides an attractive *framework* for accomplishing reasoning activities. It allows us to derive things as seemingly disparate as the posterior predictive distribution for missing data on the one hand and Kelley's formula for estimating true scores in CTT on the other hand. It has the technical machinery that allows us to carry our uncertainty from one stage of analysis forward to a later stage, and the technical and conceptual flexibility to build models with different degrees of firmness of constraints to reflect the different firmness of our substantive beliefs or theories. Though Bayes is the not only way to think about the sorts of things that go on in psychometrics, it is a powerful and useful way to approach psychometric challenges, old and new.

Throughout the second part of the book, we have described a number of psychometric activities and applications that lend themselves to, or benefit from, adopting a Bayesian perspective. There are more, many of which may be characterized as capitalizing on one or more of the key advantages of Bayesian inference, such as leveraging prior information, properly accounting for uncertainty, borrowing strength across hierarchical structures, and updating beliefs as information arrives. Examples in assessment include modeling the use of raters (Cao, Stokes, & Zhang, 2010; Johnson, 1996; Patz, Junker, Johnson, & Mariano, 2002), equating (Baldwin, 2011; Karabatsos & Walker, 2009; Liu, Schulz, & Yu, 2008; Mislevy, Sheehan, & Wingerksy, 1993), vertical scaling (Patz & Yao, 2007), determining test length (Hambleton, 1984; Novick & Lewis, 1974), obtaining scores based on subsets of the assessment (de la Torre & Patz, 2005; Edwards & Vevea, 2006), evidence identification with complex work products (Johnson, 1996; Rudner & Liang, 2002), modeling in the absence of predetermined evidence identification rules (Karabatsos & Batchelder, 2003; Oravecz, Anders, & Batchelder, 2013), and instrument development and validation (Gajewski, Price, Coffland, Boyle, & Bott, 2013; Jiang et al., 2014).

Bayesian approaches have also been advanced for examining the viability of a psychometric model and our understandings of the real-world assessment situation, including detecting aberrant responses (Bradlow & Weiss, 2001; Bradlow, Weiss, & Cho, 1998) or response times (Marianti, Fox, Avetisyan, Veldkamp, & Tijmstra, 2014; van der Linden & Guo, 2008), item preknowledge (McLeod, Lewis, & Thissen, 2003; Segall, 2002), cheating (van der Linden & Lewis, 2015), sensitivity to instruction (Naumann, Hochweber, & Hartig, 2014), and measurement noninvariance or differential functioning (Frederickx, Tuerlinckx,

De Boeck, & Magis, 2010; Fukuhara & Kamata, 2011; Sinharay, Dorans, Grant, & Blew, 2009; Soares, Goncalves, & Gamerman, 2009; Verhagen & Fox, 2013; Wang, Bradlow, Wainer, & Muller, 2008; Zwick, Thayer, & Lewis, 1999). The last of these topics is an example of where Bayesian methods have been advanced because they support the formal inclusion of loss or utility functions (Zwick, Thayer, & Lewis, 2000), and because they afford improving our understanding over time as data accrue (Sinharay et al., 2009; Zwick, Ye, Isham, 2012). Bayesian perspectives have also allowed for the resolution of paradoxes and deeper understandings of commonalities that cut across latent variable modeling families (van Rijn & Rijmen, 2015).

We can employ latent variable models to account for the presence of measurement error in larger models that focus on relationships among constructs, sampling, or other aspects of designs. Again, a Bayesian approach has much to recommend it conceptually and in terms of overcoming difficulties associated with conventional approaches. Plausible values methodology is one instance. Others include models that acknowledge measurement error in predictors and outcomes (Fox & Glas, 2003) as in structural equation models (Kaplan & Depaoli, 2012; Lee, 2007; Levy & Choi, 2013; Song & Lee, 2012) and generalized linear latent and mixed models (Hsieh, von Eye, Maier, Hsieh, & Chen, 2013; Segawa et al., 2008), and the integration of latent variable measurement models of the sort described here with generalizability theory (Briggs & Wilson, 2007).

15.2 Some Caution in Mechanically (or Unthinkingly) Using Bayesian Approaches

We should build our model to connect what we want to make an inference about to the data that will serve as evidence in making that inference. Models reflect our beliefs and purposes, which may partially conflict. This necessitates careful thought and underscores the larger point that modeling is often an exercise in making explicit what are the relevant features of the situation to the inference at hand.

To develop this, consider the following situation that could arise in any of the psychometric models considered, but is perhaps easiest to see in the context of Kelley's formula for scoring in CTT. Our discussion here is an abstraction of the types of situations described in Exercise 8.4, and readers may wish to recall their thinking and answers to questions posed there when considering the following. Recall that, under the assumptions articulated in Chapter 8, Kelley's formula may be seen as yielding the posterior mean or EAP estimate for the examinee's true score, expressing how observed scores should be shrunk back to toward the mean of the true scores. Repeating (8.18) here,

$$\mu_{T_i|x_i} = \rho x_i + (1-\rho)\mu_T = \mu_T + \rho(x_i - \mu_T).$$

Suppose we have examinees from two groups that differ in terms of their true score mean on some educational test. The application of Kelley's formula here could then be with respect to the group-specific true score mean (Kelley, 1947). Letting $\mu_{T(g)}$ denote the true score mean for group g and assuming that the reliability ρ is constant across groups and $0 < \rho < 1$, the EAP estimate for examinee i that is a member of group g is

$$\mu_{T_{i(g)}|x_i} = \rho x_i + (1-\rho)\mu_{T(g)} = \mu_{T(g)} + \rho(x_i - \mu_{T(g)}). \tag{15.1}$$

This value may differ for examinees that have the same values of x but belong to different groups. Suppose examinees in group A are on average higher scoring than those in group B; that is, $\mu_{T(A)} > \mu_{T(B)}$. An examinee who is a member of group B will have a lower EAP estimate than a member of a group A, even though their scores on the test are the same. Moreover, the direction that the observed score is shrunk may differ. Let x_0 be a particular test score where $\mu_{T(A)} > x_0 > \mu_{T(B)}$. An examinee in group A with a score of x_0 on the test will have her/his true score estimate shrunk *up*, toward $\mu_{T(A)}$, while an examinee in group B with a score of x_0 on the test will have her/his true score estimate shrunk *down*, toward $\mu_{T(B)}$. Exercise 8.4 illustrated this and contained a more extreme example where an examinee who was a member of group B had a *higher* observed score than an examinee from group A, but had a *lower* estimated true score.

Such an analysis is perfectly in line with Bayesian reasoning. If there are known or believed differences between the groups' true score means, building a model that includes all salient aspects of the real-world situation would yield a multiple-group model with varying $\mu_{T(g)}$ for the different groups. Put another way, if groups differ in ways that matter, we should condition on group membership and incorporate that in our analysis. If an examinee is a member of a lower scoring group, our estimate about their proficiency should be downgraded, relative to an examinee that performed the same but is member of a higher scoring group.

But in many cases, doing so runs afoul of our sense of fairness and would violate some of the spirit if not the letter of the laws and professional standards regarding assessment in education and related settings (AERA, APA, NCME, 2014; Camilli, 2006; Phillips & Camara, 2006). How to reconcile this seeming nefarious implication of adopting a Bayesian approach to scoring? The answer lies in recognizing that modeling is *purpose* oriented and our models should be built to reflect the purposes of inferences as well as our beliefs. We can understand the implications of purposes for psychometric modeling in the current context by distinguishing between using tests as measurements versus tests as contests (Holland, 1994; Wainer & Thissen, 1994). When a test is conceived of as a measurement, accuracy is paramount. If group-specific applications of Kelley's formula yield more accurate estimates, so be it.*

One area where the test-as-measurement metaphor typically works well is medical diagnosis. Two patients with the same symptoms or results from medical tests may have differing posterior probabilities of a disease based on, say, gender, race, or family history of disease. We are perfectly content for doctors to take such group membership into account when forming beliefs about our physical condition. We *want* them to take such things into account on the grounds that it leads to more accurate diagnoses.

What is different about assessment in education and similar settings that would make many of us recoil at the notion that examinees receive different proficiency estimates based on their gender, or race, or family history? The answer, at least partially, is that in education and related settings, assessments are often used as contests with consequences for examinees that are linked to the consequences for the other examinees. Unlike medical diagnosis, where the consequences of one patient's diagnosis are unrelated to those for other patients, in assessments-as-contests, we have notions of someone winning (a student is selected for a reward as in Exercise 8.4, an applicant is selected for admission or hired,

* Interestingly, some have advocated that if we want a more accurate portrait of an examinee's capabilities, we should not only ignore what Kelley's formula implies, but in fact we should shift examinees' scores in the *opposite* direction than what Kelley's formula implies. This has generally not borne out when evaluated for accuracy (Wainer & Brown, 2007).

a student or employee is promoted to the next grade or job) and a corresponding notion of others losing. Accuracy remains important of course, but now principles of fairness take precedence, which dictates that the same performance should yield the same score and assignment of rewards. A seemingly paradigmatic example comes from athletic competitions. At the moment the starter pistol is fired, we might believe that one Olympic marathon runner is faster than another runner based on what country they are from or the past performances of these runners. And even though we might still have these same beliefs about the runners' capabilities broadly construed when the former loses to the latter, we should nevertheless award the medals based on the order they finish the race that day and nothing else.*

On this view, the distasteful nature of the notion that examinees with the same observed performance in an educational test receive different EAP estimates based on their group membership is the result of viewing the test as a contest rather than a measurement. A proper treatment of the test as a contest would preclude the differential EAPs for students with the same observed score on the basis of group membership. The solution here is not to cast aside Kelley's formula or principles of Bayesian modeling on the grounds that using such models that are based on our substantive beliefs leads to unacceptable situations. Rather, we recognize that our models should be built in accordance not only with what we believe about the situation, but our purposes as well. When our beliefs and purposes are in conflict, some of the aspects of the former may need to be excluded from the model to preserve some aspects of the latter. What models we use should reflect our purposes, and we might intentionally use a model that does not fully capture our beliefs about the world if the situation calls for it. In the situation at hand, we would be perfectly content to build a model that does not distinguish between groups, in effect conducting a single group analysis—again using Kelley's formula or Bayesian inference more generally. Importantly, a Bayesian approach to modeling calls for the analyst to make more explicit what aspects of the real-world situation are built into the model.

15.3 Final Words

As the preceding example highlighted, our model is a story, not only of the world as we understand it, but aimed at affecting a certain outcome, facilitating the inferences we would like to make. Regardless of statistical framework we adopt, we strive to get our story straight—aimed at what we desire to reason about, reflecting our beliefs, purposes, and constraints. Bayesian approaches to psychometric modeling serve us in two ways: they enable us to tell better stories, and provide us better ways to tell them.

* Provocatively, some have argued that when the observations of athletic performances are prone to measurement error, the Bayesian approach that yields higher accuracy at the price of fairness as we have presented it is to be preferred, even on the considerations of fairness (Moore, 2009).

Appendix A: Full Conditional Distributions

We develop the full conditional distributions for regression (Chapter 6), CTT (Chapter 8), CFA (Chapter 9), IRT (Chapter 11), and LCA (Chapter 13) models in Sections A.1 through A.5, respectively. For each model, we restate the posterior distribution and then pursue the full conditionals. For completeness and transparency in the full conditionals, we expand beyond what was presented in the chapters to include conditioning on the hyperparameters. When the hyperparameters are specified in advance, they are not random variables in the sense of other model parameters and the data. However, including them serves to highlight that they are involved in the computations and aligns with the use of expanded DAGs that include them, as the full conditional distribution for any unknown is constructed by conditioning on its parents, children, and other parents of its children. Known hyperparameters fall into the first of these categories. In several places, we denote the arguments of the distribution (e.g., mean vector and covariance matrix for a normal distribution) with subscripts denoting that it refers to the full conditional distribution; the subscripts are then just the conditioning notation of the left-hand side.

Many of the specifications involve normal distributions that afford conditional conjugacy relationships through the application of standard Bayesian results for models assuming normality; see Lindley and Smith (1972) and Rowe (2003) for extensive treatments of these results.

A.1 Regression

In regression, the unknowns include the intercept (β_0), the coefficients ($\boldsymbol{\beta} = (\beta_1, \ldots, \beta_J)$) for the predictors ($\mathbf{x}$), and the error variance (σ_ε^2). We seek the posterior distribution for these unknowns given the predictors and the outcomes ($\mathbf{y}$).

We first develop the full conditional distributions for an approach where univariate prior distributions for the parameters are specified. We then develop the full conditional distributions for a more general version of the model, which illustrates certain features of Bayesian inference.

A.1.1 Univariate Priors and Full Conditionals

Under the specifications in Chapter 6, the posterior distribution is given by (6.14), expressed here as

$$p(\beta_0, \boldsymbol{\beta}, \sigma_\varepsilon^2 \mid \mathbf{y}, \mathbf{x}) \propto \prod_{i=1}^{n} p(y_i \mid \beta_0, \boldsymbol{\beta}, \sigma_\varepsilon^2, \mathbf{x}_i) p(\beta_0) \prod_{j=1}^{J} p(\beta_j) p(\sigma_\varepsilon^2), \tag{A.1}$$

where

$$y_i \mid \beta_0, \boldsymbol{\beta}, \sigma_\varepsilon^2, \mathbf{x}_i \sim N(\beta_0 + \beta_1 x_{i1} + \cdots + \beta_J x_{iJ}, \sigma_\varepsilon^2) \text{ for } i=1,\ldots,n, \tag{A.2}$$

$$\beta_0 \sim N(\mu_{\beta_0}, \sigma_{\beta_0}^2), \tag{A.3}$$

$$\beta_j \sim N(\mu_\beta, \sigma_\beta^2) \text{ for } j=1,\ldots,J, \tag{A.4}$$

and

$$\sigma_\varepsilon^2 \sim \text{Inv-Gamma}(\nu_0/2, \nu_0\sigma_{\varepsilon 0}^2/2). \tag{A.5}$$

For the intercept, the full conditional can be expressed as

$$p(\beta_0 \mid \beta_1,\ldots,\beta_J, \sigma_\varepsilon^2, \mu_{\beta_0}, \sigma_{\beta_0}^2, \mathbf{y}, \mathbf{x}) \propto p(\mathbf{y} \mid \beta_0, \beta_1,\ldots,\beta_J, \sigma_\varepsilon^2, \mathbf{x}) p(\beta_0 \mid \mu_{\beta_0}, \sigma_{\beta_0}^2). \tag{A.6}$$

Working with the first term on the right-hand side, we can represent the conditional distribution of the outcomes as

$$\mathbf{y} \mid \beta_0, \boldsymbol{\beta}, \sigma_\varepsilon^2, \mathbf{x} \sim N(\mathbf{1}\beta_0 + \mathbf{x}_1\beta_1 + \cdots + \mathbf{x}_J\beta_J, \sigma_\varepsilon^2 \mathbf{I}), \tag{A.7}$$

where $\mathbf{1}$ is an $(n \times 1)$ vector of 1s, $\mathbf{x}_j$ is the $(n \times 1)$ vector of values for individuals on the jth predictor, and $\mathbf{I}$ is an $(n \times n)$ identity matrix. In the full conditional distribution for β_0, the coefficients $\beta_1,\ldots,\beta_J$ are known. We may therefore rewrite (A.7) as

$$([\mathbf{y} - (\mathbf{x}_1\beta_1 + \cdots + \mathbf{x}_J\beta_J)] \mid \beta_0, \boldsymbol{\beta}, \sigma_\varepsilon^2, \mathbf{x}) \sim N(\mathbf{1}\beta_0, \sigma_\varepsilon^2 \mathbf{I}). \tag{A.8}$$

Using this as the first term on the right-hand side of (A.6) and the prior distribution in (A.3) as the second term, then by standard results of Bayes' theorem it can be shown that the full conditional distribution is (Lindley & Smith, 1972)

$$\beta_0 \mid \beta_1,\ldots,\beta_J, \sigma_\varepsilon^2, \mu_{\beta_0}, \sigma_{\beta_0}^2, \mathbf{y}, \mathbf{x} \sim N(\mu_{\beta_0 \mid \beta_1,\ldots,\beta_J,\sigma_\varepsilon^2,\mu_{\beta_0},\sigma_{\beta_0}^2,\mathbf{y},\mathbf{x}}, \sigma^2_{\beta_0 \mid \beta_1,\ldots,\beta_J,\sigma_\varepsilon^2,\mu_{\beta_0},\sigma_{\beta_0}^2,\mathbf{y},\mathbf{x}}), \tag{A.9}$$

where

$$\mu_{\beta_0 \mid \beta_1,\ldots,\beta_J,\sigma_\varepsilon^2,\mu_{\beta_0},\sigma_{\beta_0}^2,\mathbf{y},\mathbf{x}} = \left(\frac{1}{\sigma_{\beta_0}^2} + \frac{n}{\sigma_\varepsilon^2}\right)^{-1} \left(\frac{\mu_{\beta_0}}{\sigma_{\beta_0}^2} + \frac{\mathbf{1}'[\mathbf{y} - (\mathbf{x}_1\beta_1 + \cdots + \mathbf{x}_J\beta_J)]}{\sigma_\varepsilon^2}\right)$$

and

$$\sigma^2_{\beta_0 \mid \beta_1,\ldots,\beta_J,\sigma_\varepsilon^2,\mu_{\beta_0},\sigma_{\beta_0}^2,\mathbf{y},\mathbf{x}} = \left(\frac{1}{\sigma_{\beta_0}^2} + \frac{n}{\sigma_\varepsilon^2}\right)^{-1}.$$

For the first slope, the full conditional can be expressed as

$$p(\beta_1 \mid \beta_0, \beta_2,\ldots,\beta_J, \sigma_\varepsilon^2, \mu_{\beta_1}, \sigma_{\beta_1}^2, \mathbf{y}, \mathbf{x}) \propto p(\mathbf{y} \mid \beta_0, \beta_1,\ldots,\beta_J, \sigma_\varepsilon^2, \mathbf{x}) p(\beta_1 \mid \mu_{\beta_1}, \sigma_{\beta_1}^2). \tag{A.10}$$

In the full conditional distribution for β_1, the coefficients $\beta_0, \beta_2,\ldots,\beta_J$ are known. We may therefore rewrite (A.7) as

$$([\mathbf{y} - (\mathbf{1}\beta_0 + \mathbf{x}_2\beta_2 + \cdots + \mathbf{x}_J\beta_J)] \mid \beta_0, \boldsymbol{\beta}, \sigma_\varepsilon^2, \mathbf{x}) \sim N(\mathbf{x}_1\beta_1, \sigma_\varepsilon^2 \mathbf{I}). \tag{A.11}$$

Using this as the first term on the right-hand side of (A.10) and the prior distribution in (A.4) as the second term, then by standard results of Bayes' theorem it can be shown that the full conditional distribution is (Lindley & Smith, 1972)

$$\beta_1 \mid \beta_0, \beta_2, \ldots, \beta_J, \sigma_\varepsilon^2, \mu_{\beta_1}, \sigma_{\beta_1}^2, \mathbf{y}, \mathbf{x} \sim N(\mu_{\beta_1 \mid \beta_0, \beta_2, \ldots, \beta_J, \sigma_\varepsilon^2, \mu_{\beta_1}, \sigma_{\beta_1}^2, \mathbf{y}, \mathbf{x}}, \sigma^2_{\beta_1 \mid \beta_0, \beta_2, \ldots, \beta_J, \sigma_\varepsilon^2, \mu_{\beta_1}, \sigma_{\beta_1}^2, \mathbf{y}, \mathbf{x}}), \quad (A.12)$$

where

$$\mu_{\beta_1 \mid \beta_0, \beta_2, \ldots, \beta_J, \sigma_\varepsilon^2, \mu_{\beta_1}, \sigma_{\beta_1}^2, \mathbf{y}, \mathbf{x}} = \left(\frac{1}{\sigma_{\beta_1}^2} + \frac{\mathbf{x}_1'\mathbf{x}_1}{\sigma_\varepsilon^2}\right)^{-1} \left(\frac{\mu_{\beta_1}}{\sigma_{\beta_1}^2} + \frac{\mathbf{x}_1'[\mathbf{y} - (\mathbf{1}\beta_0 + \mathbf{x}_2\beta_2 + \ldots + \mathbf{x}_J\beta_J)]}{\sigma_\varepsilon^2}\right)$$

and

$$\sigma^2_{\beta_1 \mid \beta_0, \beta_2, \ldots, \beta_J, \sigma_\varepsilon^2, \mu_{\beta_1}, \sigma_{\beta_1}^2, \mathbf{y}, \mathbf{x}} = \left(\frac{1}{\sigma_{\beta_1}^2} + \frac{\mathbf{x}_1'\mathbf{x}_1}{\sigma_\varepsilon^2}\right)^{-1}.$$

The same process can be applied for each regression coefficient. To present all of them compactly, we work with a matrix representation of the regression model,

$$\mathbf{y} \mid \boldsymbol{\beta}_A, \sigma_\varepsilon^2, \mathbf{x} \sim N(\mathbf{x}_A \boldsymbol{\beta}_A, \sigma_\varepsilon^2 \mathbf{I}), \quad (A.13)$$

where $\mathbf{x}_A$ is an $(n \times [J+1])$ augmented predictor matrix obtained by combining an $(n \times 1)$ column vector of 1s to the predictor matrix $\mathbf{x}$ and $\boldsymbol{\beta}_A$ is a $([J+1] \times 1)$ augmented vector of coefficients obtained by combining the intercept β_0 with the J coefficients in $\boldsymbol{\beta}$. Let $\boldsymbol{\beta}_{A(-j)}$ denote the $(J \times 1)$ vector obtained by omitting β_j from $\boldsymbol{\beta}_A$. Similarly let $\mathbf{x}_{A(-j)}$ denote the $(n \times J)$ matrix obtained by omitting the column of values of $\mathbf{x}_j$ from $\mathbf{x}_A$. When $j > 0$, $\mathbf{x}_j$ refers to column vector with values for the jth predictor; and when $j = 0$, $\mathbf{x}_j$ refers to the column vector of 1s. The full conditional distribution for β_j, $j = 0, \ldots, J$, is then

$$\beta_j \mid \boldsymbol{\beta}_{A(-j)}, \sigma_\varepsilon^2, \mu_{\beta_j}, \sigma_{\beta_j}^2, \mathbf{y}, \mathbf{x} \sim N(\mu_{\beta_j \mid \boldsymbol{\beta}_{A(-j)}, \sigma_\varepsilon^2, \mu_{\beta_j}, \sigma_{\beta_j}^2, \mathbf{y}, \mathbf{x}}, \sigma^2_{\beta_j \mid \boldsymbol{\beta}_{A(-j)}, \sigma_\varepsilon^2, \mu_{\beta_j}, \sigma_{\beta_j}^2, \mathbf{y}, \mathbf{x}}), \quad (A.14)$$

where

$$\mu_{\beta_j \mid \boldsymbol{\beta}_{A(-j)}, \sigma_\varepsilon^2, \mu_{\beta_j}, \sigma_{\beta_j}^2, \mathbf{y}, \mathbf{x}} = \left(\frac{1}{\sigma_{\beta_j}^2} + \frac{\mathbf{x}_j'\mathbf{x}_j}{\sigma_\varepsilon^2}\right)^{-1} \left(\frac{\mu_{\beta_j}}{\sigma_{\beta_j}^2} + \frac{\mathbf{x}_j'(\mathbf{y} - \mathbf{x}_{A(-j)}\boldsymbol{\beta}_{A(-j)})}{\sigma_\varepsilon^2}\right) \quad (A.15)$$

and

$$\sigma^2_{\beta_j \mid \boldsymbol{\beta}_{A(-j)}, \sigma_\varepsilon^2, \mu_{\beta_j}, \sigma_{\beta_j}^2, \mathbf{y}, \mathbf{x}} = \left(\frac{1}{\sigma_{\beta_j}^2} + \frac{\mathbf{x}_j'\mathbf{x}_j}{\sigma_\varepsilon^2}\right)^{-1}. \quad (A.16)$$

When $j = 0$, μ_{β_j} and $\sigma_{\beta_j}^2$ in (A.15) and (A.16) refer to the prior mean and variance of the intercept, which we had denoted μ_{β_0} and $\sigma_{\beta_0}^2$ in (A.3). When $j > 1$, μ_{β_j} and $\sigma_{\beta_j}^2$ in (A.15) and (A.16) refer to the prior mean and variance of the coefficient for the jth predictor, which under exchangeability was denoted by μ_β and σ_β^2 in (A.4).

Turning to σ_ε^2, the full conditional treats $\boldsymbol{\beta}_A$ as known and the situation is akin to that of the full conditional distribution for the variance in a univariate normal distribution in (5.10) and (5.11). In the current context of regression, the full conditional distribution for σ_ε^2 can be shown to be (Rowe, 2003)

$$\sigma_\varepsilon^2 \mid \beta_A, \nu_0, \sigma_{\varepsilon 0}^2, \mathbf{y}, \mathbf{x} \sim \text{Inv-Gamma}\left(\frac{\nu_0 + n}{2}, \frac{\nu_0\sigma_{\varepsilon 0}^2 + SS(\mathbf{E})}{2}\right), \tag{A.17}$$

where

$$SS(\mathbf{E}) = (\mathbf{y} - \mathbf{x}_A\beta_A)'(\mathbf{y} - \mathbf{x}_A\beta_A).$$

Note the similarity between (A.17) and the full conditional for the variance in a model for a univariate normal distribution in (5.10) and (5.11). In that context, the sums of squares was taken with respect to the model-implied mean of the data. In the current context, each individual has their own model-implied mean, given by the regression model. Hence, the sum of squares is constructed as $(\mathbf{y} - \mathbf{x}_A\beta_A)'(\mathbf{y} - \mathbf{x}_A\beta_A)$.

A.1.2 Multivariate Prior and Full Conditional for Intercept and Regression Coefficients

Here, we develop the full conditional distributions for a more general version of the model in terms of β_A, which aids in illustrating several features of Bayesian inference. Under this specification the posterior distribution is

$$p(\beta_A, \sigma_\varepsilon^2 \mid \mathbf{y}, \mathbf{x}) \propto p(\mathbf{y} \mid \beta_A, \sigma_\varepsilon^2, \mathbf{x})p(\beta_A)p(\sigma_\varepsilon^2). \tag{A.18}$$

The first term on the right-hand side is the conditional distribution of the data given by (A.7).

We specify a multivariate normal prior for β_A,

$$\beta_A \sim N(\mu_{\beta_A}, \Sigma_{\beta_A}), \tag{A.19}$$

where μ_{β_A} is a $(J+1)$ prior mean vector and Σ_{β_A} is a $([J+1] \times [J+1])$ prior covariance matrix. To complete the model, we specify the inverse-gamma prior distribution for σ_ε^2 as in (A.5).

It can be shown that the full conditional distribution for β_A is (Lindley & Smith, 1972)

$$\beta_A \mid \sigma_\varepsilon^2, \mu_{\beta_A}, \Sigma_{\beta_A}, \mathbf{y}, \mathbf{x} \sim N(\mu_{\beta_A|\sigma_\varepsilon^2, \mu_{\beta_A}, \Sigma_{\beta_A}, \mathbf{y}, \mathbf{x}}, \Sigma_{\beta_A|\sigma_\varepsilon^2, \mu_{\beta_A}, \Sigma_{\beta_A}, \mathbf{y}, \mathbf{x}}), \tag{A.20}$$

where

$$\mu_{\beta_A|\sigma_\varepsilon^2, \mu_{\beta_A}, \Sigma_{\beta_A}, \mathbf{y}, \mathbf{x}} = \left(\Sigma_{\beta_A}^{-1} + \frac{1}{\sigma_\varepsilon^2}\mathbf{x}_A'\mathbf{x}_A\right)^{-1}\left(\Sigma_{\beta_A}^{-1}\mu_{\beta_A} + \frac{1}{\sigma_\varepsilon^2}\mathbf{x}_A'\mathbf{y}\right)$$

and

$$\Sigma_{\beta_A|\sigma_\varepsilon^2, \mu_{\beta_A}, \Sigma_{\beta_A}, \mathbf{y}, \mathbf{x}} = \left(\Sigma_{\beta_A}^{-1} + \frac{1}{\sigma_\varepsilon^2}\mathbf{x}_A'\mathbf{x}_A\right)^{-1}.$$

The full conditional for σ_ε^2 remains as in (A.17).

This model allows for a prior covariance matrix specification on the augmented vector of coefficients. The model with the univariate prior distributions may be seen as a special case, enacted by specifying $\mu_{\beta_A} = (\mu_{\beta_0}, \mu_{\beta_1}, \ldots, \mu_{\beta_J})'$and Σ_{β_A} as a diagonal matrix with $(\sigma_{\beta_0}^2, \sigma_{\beta_1}^2, \ldots, \sigma_{\beta_J}^2)$ along the diagonal.

This example highlights several principles of Bayesian modeling. First, note that we can specify the posterior mean for the augmented regression coefficients as

$$\mu_{\beta_A|\sigma_\varepsilon^2,\mu_{\beta_A},\Sigma_{\beta_A},\mathbf{y},\mathbf{x}} = \left(\Sigma_{\beta_A}^{-1} + \frac{1}{\sigma_\varepsilon^2}\mathbf{x}_A'\mathbf{x}_A\right)^{-1}\left(\Sigma_{\beta_A}^{-1}\mu_{\beta_A} + \frac{1}{\sigma_\varepsilon^2}\mathbf{x}_A'\mathbf{x}_A\hat{\beta}_A\right),$$

where

$$\hat{\beta}_A = (\mathbf{x}_A'\mathbf{x}_A)^{-1}\mathbf{x}_A'\mathbf{y}$$

is the point estimate for the augmented regression coefficients from an ML or least-squares solution. The posterior mean $\mu_{\beta_A|\sigma_\varepsilon^2,\mathbf{y},\mathbf{x}}$ is seen as a precision-weighted average of the prior mean μ_{β_A} and the point estimate based on the data alone $\hat{\beta}_A$, where the weights are given by the prior precision $\Sigma_{\beta_A}^{-1}$ (i.e., the inverse of the prior variance), and the precision in the data $\mathbf{x}_A'\mathbf{x}_A/\sigma_\varepsilon^2$ (i.e., the inverse of the sampling variance of $\hat{\beta}_A$).

Second, the model with univariate priors for the intercept and coefficients reflects an assumption of a priori independence between these parameters. In the current representation, this manifests itself as Σ_{β_A} being diagonal. However, inspection of $\Sigma_{\beta_A|\sigma_\varepsilon^2,\mu_{\beta_A},\Sigma_{\beta_A},\mathbf{y},\mathbf{x}}$ reveals that its off-diagonal elements will not necessarily be 0. This reveals that specifying a priori independence does not force the parameters to be independent in the posterior. If the data imply that they are dependent, the posterior distribution will reflect that.

A.2 Classical Test Theory

In CTT, the unknowns include the examinees' continuous true scores (**T**), the mean of the true scores (μ_T), the variance of the true scores (σ_T^2), and the error variance (σ_E^2). We seek the posterior distribution for these unknowns given the continuous observables (**x**). Under the assumptions specified in Chapter 8, the posterior distribution is given by (8.47), re-expressed here as

$$p(\mathbf{T},\mu_T,\sigma_T^2,\sigma_E^2 \mid \mathbf{x}) \propto \prod_{i=1}^{n}\prod_{j=1}^{J} p(x_{ij} \mid T_i,\sigma_E^2)p(T_i \mid \mu_T,\sigma_T^2)p(\mu_T)p(\sigma_T^2)p(\sigma_E^2), \quad \text{(A.21)}$$

where

$$x_{ij} \mid T_i,\sigma_E^2 \sim N(T_i,\sigma_E^2) \quad \text{for } i=1,\ldots,n, j=1,\ldots,J, \quad \text{(A.22)}$$

$$T_i \mid \mu_T,\sigma_T^2 \sim N(\mu_T,\sigma_T^2) \quad \text{for } i=1,\ldots,n, \quad \text{(A.23)}$$

$$\mu_T \sim N(\mu_{\mu_T},\sigma_{\mu_T}^2), \quad \text{(A.24)}$$

$$\sigma_T^2 \sim \text{Inv-Gamma}(\nu_T/2,\nu_T\sigma_{T_0}^2/2), \quad \text{(A.25)}$$

and

$$\sigma_E^2 \sim \text{Inv-Gamma}(\nu_E/2,\nu_E\sigma_{E_0}^2/2). \quad \text{(A.26)}$$

A.2.1 Latent Variables

Given the other parameters and the observed scores for the examinee, the true score for an examinee (T_i) is conditionally independent of the true scores for the other examinees (denoted $\mathbf{T}_{(-i)}$) and the observable values for the other examinees. This allows for the simplification of the full conditional distribution for each examinee's true score,

$$p(T_i \mid \mathbf{T}_{(-i)}, \mu_T, \sigma_T^2, \sigma_E^2, \mathbf{x}) = p(T_i \mid \mu_T, \sigma_T^2, \sigma_E^2, \mathbf{x}_i) \propto p(\mathbf{x}_i \mid T_i, \sigma_E^2) p(T_i \mid \mu_T, \sigma_T^2).$$

In the full conditional distribution μ_T, σ_T^2, and σ_E^2 are treated as known, which was the case discussed in Section 8.2 where we noted that this may be viewed as the situation where a sufficient statistic for the observables is normally distributed. Repeating (8.32),

$$\bar{x}_i \mid T_i, \sigma_E^2 \sim N\left(T_i, \frac{\sigma_E^2}{J}\right).$$

The prior for the unknown mean T_i in is normal with hyperparameters μ_T and σ_T^2 that are treated as known here. This is therefore an instance of a posterior for the unknown mean of a normal distribution treating the variance and hyperparameters as known. The full conditional is

$$T_i \mid \mu_T, \sigma_T^2, \sigma_E^2, \mathbf{x}_i \sim N(\mu_{T_i \mid \mu_T, \sigma_T^2, \sigma_E^2, \mathbf{x}_i}, \sigma^2_{T_i \mid \mu_T, \sigma_T^2, \sigma_E^2, \mathbf{x}_i}), \tag{A.27}$$

where

$$\mu_{T_i \mid \mu_T, \sigma_T^2, \sigma_E^2, \mathbf{x}_i} = \frac{\left(\mu_T / \sigma_T^2\right) + \left(J\bar{x}_i / \sigma_E^2\right)}{\left(1/\sigma_T^2\right) + \left(J/\sigma_E^2\right)}$$

and

$$\sigma^2_{T_i \mid \mu_T, \sigma_T^2, \sigma_E^2, \mathbf{x}_i} = \frac{1}{\left(1/\sigma_T^2\right) + \left(J/\sigma_E^2\right)}.$$

A.2.2 Mean and Variance of the Latent Variables

Given the true scores **T**, the mean of the true scores μ_T is conditionally independent of the observables **x** and the error variance σ_E^2. The full conditional for the mean of the true scores μ_T can therefore be simplified as

$$p(\mu_T \mid \mathbf{T}, \sigma_T^2, \sigma_E^2, \mu_{\mu_T}, \sigma^2_{\mu_T}, \mathbf{x}) = p(\mu_T \mid \mathbf{T}, \sigma_T^2) \propto p(\mathbf{T} \mid \mu_T, \sigma_E^2) p(\mu_T \mid \mu_{\mu_T}, \sigma^2_{\mu_T}),$$

where the prior distribution $p(\mu_T)$ is given in (A.24). Again, this is a situation where variables (here, **T**) are normally distributed with a known variance (σ_T^2) and unknown mean (μ_T), which is normally distributed with known hyperparameters (μ_{μ_T} and $\sigma^2_{\mu_T}$). The full conditional distribution for μ_T is therefore

$$\mu_T \mid \mathbf{T}, \sigma_T^2, \mu_{\mu_T}, \sigma^2_{\mu_T} \sim N(\mu_{\mu_T \mid \mathbf{T}, \sigma_T^2, \mu_{\mu_T}, \sigma^2_{\mu_T}}, \sigma^2_{\mu_T \mid \mathbf{T}, \sigma_T^2, \mu_{\mu_T}, \sigma^2_{\mu_T}}), \tag{A.28}$$

where

$$\mu_{\mu_T \mid \mathbf{T}, \sigma_T^2, \mu_{\mu_T}, \sigma^2_{\mu_T}} = \frac{\left(\mu_{\mu_T} / \sigma^2_{\mu_T}\right) + \left(n\bar{T} / \sigma_T^2\right)}{\left(1/\sigma^2_{\mu_T}\right) + \left(n/\sigma_T^2\right)}$$

and

$$\sigma^2_{\mu_T|\mathbf{T},\sigma_T^2,\mu_{\mu T},\sigma^2_{\mu T}} = \left(\frac{1}{\sigma^2_{\mu T}} + \frac{n}{\sigma_T^2}\right)^{-1}.$$

Given the true scores **T**, the variance of the true scores σ_T^2 is conditionally independent of the observables **x** and the error variance σ_E^2. The full conditional for the variance of the true scores σ_T^2 can therefore be simplified as

$$p(\sigma_T^2 \mid \mathbf{T}, \mu_T, \sigma_E^2, \nu_T, \sigma_{T_0}^2, \mathbf{x}) = p(\sigma_T^2 \mid \mathbf{T}, \mu_T, \nu_T, \sigma_{T_0}^2) \propto p(\mathbf{T} \mid \mu_T, \sigma_T^2) p(\sigma_T^2 \mid \nu_T, \sigma_{T_0}^2),$$

where the prior distribution $p(\sigma_T^2)$ is given in (A.25). This situation is the same as was encountered in Section 4.2, where we have a set of variables (**T**) that are normally distributed with a known mean (μ_T) and unknown variance (σ_T^2), which has an inverse-gamma prior distribution with known hyperparameters (ν_T and $\sigma_{T_0}^2$). The full conditional distribution for σ_T^2 is therefore

$$\sigma_T^2 \mid \mathbf{T}, \mu_T, \nu_T, \sigma_{T_0}^2 \sim \text{Inv-Gamma}\left(\frac{\nu_T + n}{2}, \frac{\nu_T \sigma_{T_0}^2 + SS(\mathbf{T})}{2}\right), \tag{A.29}$$

where

$$SS(\mathbf{T}) = \sum_{i=1}^{n} (T_i - \mu_T)^2.$$

A.2.3 Error Variance

Given the observables **x** and the true scores **T**, the error variance σ_E^2 is conditionally independent of the mean of the true scores μ_T and the variance of the true scores σ_T^2. The full conditional for the error variance σ_E^2 can therefore be simplified as

$$\begin{aligned} p(\sigma_E^2 \mid \mathbf{T}, \mu_T, \sigma_T^2, \nu_E, \sigma_{E_0}^2, \mathbf{x}) &= p(\sigma_E^2 \mid \mathbf{T}, \nu_E, \sigma_{E_0}^2, \mathbf{x}) \\ &\propto p(\mathbf{x} \mid \mathbf{T}, \sigma_E^2) p(\sigma_E^2 \mid \nu_E, \sigma_{E_0}^2) \\ &= \prod_{i=1}^{n} p(\mathbf{x}_i \mid T_i, \sigma_E^2) p(\sigma_E^2 \mid \nu_E, \sigma_{E_0}^2), \end{aligned}$$

where the prior distribution for σ_E^2 is given in (A.26). We have for each examinee i a collection of variables ($\mathbf{x}_i$) that is normally distributed with known mean (T_i) and unknown variance σ_E^2 that is constant across examinees and has an inverse-gamma prior distribution with known hyperparameters (ν_E and $\sigma_{E_0}^2$). This is actually a special case of that derived for the error variance in regression in (A.17). There, the subject-specific mean was given by the regression function. Here, the examinee-specific mean is given by the true score for the examinee, akin to an intercept-only regression model. The full conditional distribution for σ_E^2 is therefore

$$\sigma_E^2 \mid \mathbf{T}, \nu_E, \sigma_{E_0}^2, \mathbf{x} \sim \text{Inv-Gamma}\left(\frac{\nu_E + nJ}{2}, \frac{\nu_E \sigma_{E_0}^2 + SS(\mathbf{E})}{2}\right), \tag{A.30}$$

where

$$SS(\mathbf{E}) = \sum_{i=1}^{n}\sum_{j=1}^{J}(x_{ij} - T_i)^2.$$

A.3 Confirmatory Factor Analysis

In CFA, the unknowns include the examinees' continuous latent variables (Ξ), the means of the latent variables (κ), the covariance matrix for the latent variables (Φ), and the intercepts (τ), loadings (Λ), and error variances (collected in Ψ) for the observables. We seek the posterior distribution for these unknowns given the continuous observables ($\mathbf{x}$).

We first develop the full conditional distributions for an approach where univariate prior distributions for the measurement model parameters are specified. We then develop the full conditional distributions for a more general version of the model that employs multivariate priors for each observable's intercept and loadings.

A.3.1 Univariate Priors and Full Conditionals

Under the assumptions specified in Chapter 9, the posterior distribution is given by (9.31) and can be expressed as

$$p(\Xi,\kappa,\Phi,\tau,\Lambda,\Psi \mid \mathbf{x}) \propto \prod_{i=1}^{n}\prod_{j=1}^{J}\prod_{m=1}^{M} p(x_{ij} \mid \xi_i,\tau_j,\lambda_j,\psi_{jj})p(\xi_i \mid \kappa,\Phi) \times p(\kappa_m)p(\Phi)p(\tau_j)p(\lambda_{jm})p(\psi_{jj}), \tag{A.31}$$

where

$$x_{ij} \mid \xi_i,\tau_j,\lambda_j,\psi_{jj} \sim N(\tau_j + \xi_i\lambda_j',\psi_{jj}) \quad \text{for } i=1,\ldots,n, j=1,\ldots,J, \tag{A.32}$$

$$\xi_i \mid \kappa,\Phi \sim N(\kappa,\Phi) \quad \text{for } i = 1,\ldots,n, \tag{A.33}$$

$$\kappa_m \sim N(\mu_\kappa,\sigma_\kappa^2) \quad \text{for } m = 1,\ldots,M, \tag{A.34}$$

$$\Phi \sim \text{Inv-Wishart}(\Phi_0,d), \tag{A.35}$$

$$\tau_j \sim N(\mu_\tau,\sigma_\tau^2) \quad \text{for } j = 1,\ldots,J, \tag{A.36}$$

$$\lambda_{jm} \sim N(\mu_\lambda,\sigma_\lambda^2) \quad \text{for } j = 1,\ldots,J, m=1,\ldots,M, \tag{A.37}$$

and

$$\psi_{jj} \sim \text{Inv-Gamma}(\nu_\psi/2,\nu_\psi\psi_0/2). \tag{A.38}$$

A.3.1.1 Measurement Model Parameters

Given the latent variables, the measurement model parameters (Λ,τ,Ψ) are conditionally independent of the means of the latent variables and the covariance matrix of the latent variables. Further, given the values for the jth observable $\mathbf{x}_j$ (i.e., the jth column of $\mathbf{x}$), the measurement model parameters for that observable are conditionally independent of the *other* observables and the measurement model parameters for those *other* observables. Suppressing the notation for the hyperparameters for the moment,

$$p(\tau, \Lambda, \Psi \mid \Xi, \kappa, \Phi, \mathbf{x}) = p(\tau, \Lambda, \Psi \mid \Xi, \mathbf{x}) = \prod_{j=1}^{J} p(\tau_j, \lambda_j, \psi_{jj} \mid \Xi, \mathbf{x}_j),$$

and we may proceed with the measurement model parameters one observable at a time. For the jth observable,

$$p(\tau_j, \lambda_j, \psi_{jj} \mid \Xi, \mathbf{x}_j) \propto p(\mathbf{x}_j \mid \Xi, \tau_j, \lambda_j, \psi_{jj}) p(\tau_j) p(\lambda_j) p(\psi_{jj}),$$

where $\lambda_j = (\lambda_{j1}, \ldots, \lambda_{jM})$, $p(\lambda_j) = \prod_{m=1}^{M} p(\lambda_{jm})$, and $p(\tau_j)$, $p(\lambda_{jm})$, and $p(\psi_{jj})$ are the prior distributions specified in (A.36) through (A.38). In these full conditionals the latent variables Ξ are treated as known. Thus the model is the same as the multiple regression model, now with the jth observable $\mathbf{x}_j$ as the outcome and the latent variables Ξ as the (treated-as-known) predictors. We proceed using the strategies adopted there (Section A.1). Let Ξ_A denote the $(n \times [M+1])$ augmented matrix of latent variables obtained by combining an $(n \times 1)$ column vector of 1s to the matrix of latent variables Ξ and let λ_{jA} denote the $([M+1] \times 1)$ augmented vector of loadings obtained by combining the intercept τ_j with the M loadings in λ_j. Let $\lambda_{jA(-m)}$ denote the $(M \times 1)$ vector obtained by omitting λ_{jm} from λ_{jA}. Similarly let $\Xi_{A(-m)}$ denote the $(n \times M)$ matrix obtained by omitting the column of values of Ξ_m from Ξ_A. When $m > 0$, λ_{jm} refers to the loading for observable j on the mth latent variable and Ξ_m refers to column vector with values for the mth latent variable. When $j = 0$, λ_{jm} refers to the intercept for observable j and Ξ_m refers to the column vector of 1s. Reintroducing the hyperparameters into the notation, the full conditional distribution for λ_{jm}, $m = 0,\ldots, M$ is then

$$\lambda_{jm} \mid \Xi, \lambda_{jA(-m)}, \psi_{jj}, \mu_\lambda, \sigma^2_\lambda, \mathbf{x}_j \sim N(\mu_{\lambda_{jm} \mid \Xi, \lambda_{jA(-m)}, \psi_{jj}, \mu_\lambda, \sigma^2_\lambda, \mathbf{x}_j}, \sigma^2_{\lambda_{jm} \mid \Xi, \lambda_{jA(-m)}, \psi_{jj}, \mu_\lambda, \sigma^2_\lambda, \mathbf{x}_j}), \tag{A.39}$$

where

$$\mu_{\lambda_{jm} \mid \Xi, \lambda_{jA(-m)}, \psi_{jj}, \mu_\lambda, \sigma^2_\lambda, \mathbf{x}_j} = \left(\frac{1}{\sigma^2_{\lambda_{jm}}} + \frac{1}{\psi_{jj}} \Xi'_m \Xi_m \right)^{-1} \left(\frac{1}{\sigma^2_{\lambda_{jm}}} \mu_{\lambda_{jm}} + \frac{1}{\psi_{jj}} \Xi'_m (\mathbf{x}_j - \Xi_{A(-m)} \lambda_{jA(-m)}) \right)$$

and

$$\sigma^2_{\lambda_{jm} \mid \Xi, \lambda_{jA(-m)}, \psi_{jj}, \mu_\lambda, \sigma^2_\lambda, \mathbf{x}_j} = \left(\frac{1}{\sigma^2_{\lambda_{jm}}} + \frac{1}{\psi_{jj}} \Xi'_m \Xi_m \right)^{-1}.$$

The full conditional for ψ_{jj} is then

$$\psi_{jj} \mid \Xi, \lambda_{jA}, \nu_\psi, \psi_0, \mathbf{x}_j \sim \text{Inv-Gamma}\left(\frac{\nu_\psi + n}{2}, \frac{\nu_\psi \psi_0 + SS(\mathbf{E}_j)}{2} \right), \tag{A.40}$$

where

$$SS(\mathbf{E}_j) = (\mathbf{x}_j - \Xi_A \lambda_{jA})'(\mathbf{x}_j - \Xi_A \lambda_{jA}).$$

A.3.1.2 Latent Variable Means and Covariance Matrix

Given the latent variables (Ξ), the full conditional distributions for the means of the latent variables are independent of the observables (**x**) and the measurement model parameters (Λ, τ, Ψ),

$$p(\kappa \mid \Xi, \Phi, \Lambda, \tau, \Psi, \mu_\kappa, \Sigma_\kappa, \mathbf{x}) = p(\kappa \mid \Xi, \Phi, \mu_\kappa, \Sigma_\kappa) \propto p(\Xi \mid \kappa, \Phi) p(\kappa \mid \mu_\kappa, \Sigma_\kappa). \tag{A.41}$$

Equation (A.33) implies that the first term on the right-hand side is the multivariate normal conditional distribution of the treated-as-known latent variables given the unknown means and the treated-as-known covariance matrix. The second term is a multivariate normal prior distribution

$$\kappa \sim N(\mu_\kappa, \Sigma_\kappa).$$

This is a generalzation of the prior in (A.34); the specification in (A.34) obtains when Σ_κ is diagonal. It can be shown (Lindley & Smith, 1972) that the full conditional distribution is

$$\kappa \mid \Xi, \Phi, \mu_\kappa, \Sigma_\kappa \sim N(\mu_{\kappa|\Xi,\Phi,\mu_\kappa,\Sigma_\kappa}, \Sigma_{\kappa|\Xi,\Phi,\mu_\kappa,\Sigma_\kappa}). \tag{A.42}$$

where

$$\mu_{\kappa|\Xi,\Phi,\mu_\kappa,\Sigma_\kappa} = (\Sigma_\kappa^{-1} + n\Phi^{-1})^{-1}(\Sigma_\kappa^{-1}\mu_\kappa + n\Phi^{-1}\bar{\xi}),$$

$$\Sigma_{\kappa|\Xi,\Phi,\mu_\kappa,\Sigma_\kappa} = (\Sigma_\kappa^{-1} + n\Phi^{-1})^{-1},$$

and $\bar{\xi}$ is the ($M \times 1$) vector of means of the treated-as-known latent variables over the n examinees.

Given the latent variables, the full conditional distribution for the covariance matrix of the latent variables is independent of the observables and the measurement model parameters,

$$p(\Phi \mid \Xi, \kappa, \Lambda, \tau, \Psi, \Phi_0, d, \mathbf{x}) = p(\Phi \mid \Xi, \kappa, \Phi_0, d) \propto p(\Xi \mid \kappa, \Phi) p(\Phi \mid \Phi_0, d).$$

Equation (A.33) implies that the first term on the right-hand side is the multivariate normal conditional distribution of the treated-as-known latent variables given the treated-as-known means and the unknown covariance matrix, as in (A.41). The second term is the inverse-Wishart prior distribution in (A.35). Treating the latent variables as known, this is a situation where variables (Ξ) are normally distributed with known means (κ) and an unknown covariance. In this case, the inverse-Wishart prior distribution with known hyperparameters (Φ_0 and d) is conjugate and by standard analyses (Rowe, 2003) the full conditional distribution is

$$\Phi \mid \Xi, \kappa, \Phi_0, d \sim \text{Inv-Wishart}(d\Phi_0 + n\mathbf{S}_\xi, d + n), \tag{A.43}$$

where

$$\mathbf{S}_\xi = \frac{1}{n}\sum_i (\xi_i - \kappa)(\xi_i - \kappa)'.$$

A.3.1.3 Latent Variables

The exchangeability and conditional independence assumptions imply that the full conditional distribution for the latent variables may be factored into the product of examinee-specific full conditional distributions

$$p(\Xi \mid \kappa, \Phi, \Lambda, \tau, \Psi, \mathbf{x}) = \prod_{i=1}^{n} p(\xi_i \mid \kappa, \Phi, \Lambda, \tau, \Psi, \mathbf{x}_i) \propto \prod_{i=1}^{n} p(\mathbf{x}_i \mid \xi_i, \Lambda, \tau, \Psi) p(\xi_i \mid \kappa, \Phi),$$

where $\mathbf{x}_i = (x_{i1}, \ldots, x_{iJ})$ is the collection of J observed values from examinee i. The second term on the right-hand side is the prior distribution for the examinee's latent variables, given by (A.33). The first term on the right-hand side is the conditional distribution of the observables for the examinee, given by (9.13) and restated here:

$$\mathbf{x}_i \mid \xi_i, \tau, \Lambda, \Psi \sim N(\tau + \Lambda\xi_i, \Psi).$$

Recognizing that τ is a constant in the full conditional distribution, we may express this as

$$(\mathbf{x}_i - \tau) \mid \xi_i, \Lambda, \Psi \sim N(\Lambda\xi_i, \Psi),$$

which has a form amenable to standard Bayesian analysis (Lindley & Smith, 1972), such that the full conditional distribution is

$$\xi_i \mid \kappa, \Phi, \Lambda, \tau, \Psi, \mathbf{x}_i \sim N(\mu_{\xi_i \mid \kappa, \Phi, \Lambda, \tau, \Psi, \mathbf{x}_i}, \Sigma_{\xi_i \mid \kappa, \Phi, \Lambda, \tau, \Psi, \mathbf{x}_i}), \tag{A.44}$$

where

$$\mu_{\xi_i \mid \kappa, \Phi, \Lambda, \tau, \Psi, \mathbf{x}_i} = (\Phi^{-1} + \Lambda'\Psi^{-1}\Lambda)^{-1}(\Phi^{-1}\kappa + \Lambda'\Psi^{-1}(\mathbf{x}_i - \tau))$$

and

$$\Sigma_{\xi_i \mid \kappa, \Phi, \Lambda, \tau, \Psi, \mathbf{x}_i} = (\Phi^{-1} + \Lambda'\Psi^{-1}\Lambda)^{-1}.$$

A.3.2 Multivariate Priors and Full Conditionals for Each Observable's Intercept and Loadings

As was noted in Section A.1, a multivariate approach to prior specification affords simplicity in the full conditionals for regression models. Recognizing that there is a regression structure for each observable, this section develops the full conditionals for the model using a multivariate prior specification for the intercept and loadings for each observable.

Rather than specify univariate normal priors for the intercept and loadings individually, we employ a multivariate normal prior for the intercept and loadings for each observable j,

$$\lambda_{jA} \sim N(\mu_{\lambda_{jA}}, \Sigma_{\lambda_{jA}}), \tag{A.45}$$

where $\mu_{\lambda_{jA}}$ is an $(M + 1)$ prior mean vector and $\Sigma_{\lambda_{jA}}$ is a $([M + 1] \times [M + 1])$ prior covariance matrix. All other prior distributions remain the same.

The full conditional distribution for λ_{jA} is then

$$\lambda_{jA} \mid \Xi, \psi_{jj}, \mu_{\lambda_A}, \Sigma_{\lambda_A}, \mathbf{x}_j \sim N(\mu_{\lambda_{jA} \mid \Xi, \psi_{jj}, \mu_{\lambda_A}, \Sigma_{\lambda_A}, \mathbf{x}_j}, \Sigma_{\lambda_{jA} \mid \Xi, \psi_{jj}, \mu_{\lambda_A}, \Sigma_{\lambda_A}, \mathbf{x}_j}), \tag{A.46}$$

where

$$\mu_{\lambda_{jA} \mid \Xi, \psi_{jj}, \mu_{\lambda_A}, \Sigma_{\lambda_A}, \mathbf{x}_j} = \left(\Sigma_{\lambda_A}^{-1} + \frac{1}{\psi_{jj}} \Xi_A' \Xi_A\right)^{-1} \left(\Sigma_{\lambda_A}^{-1} \mu_{\lambda_A} + \frac{1}{\psi_{jj}} \Xi_A' \mathbf{x}_j\right),$$

$$\Sigma_{\lambda_j A|\Xi,\psi_{jj},\mu_{\lambda A},\Sigma_{\lambda A},\mathbf{x}_j} = \left(\Sigma_{\lambda A}^{-1} + \frac{1}{\psi_{jj}}\Xi_A'\Xi_A\right)^{-1},$$

and $\mathbf{x}_j = (x_{1j}, \ldots, x_{nj})'$ is the $(n \times 1)$ vector of observed values for observable j. The other full conditional distributions remain the same.

A.4 Item Response Theory

This section develops the full conditionals for the latent response variable formulation of the unidimensional two-parameter normal ogive IRT model described in Section 11.8. Here, the unknowns of inferential interest include the examinees' continuous latent variables (θ), and the location ($\mathbf{d}$) and discrimination ($\mathbf{a}$) parameters. In addition, the continuous latent response variables ($\mathbf{x}^*$) are unknown and part of our posterior, which is formulated given the discrete observables ($\mathbf{x}$). Under the assumptions specified in Section 11.8, the posterior distribution is given by (11.69) and can be expressed as

$$p(\theta,\mathbf{d},\mathbf{a},\mathbf{x}^* \mid \mathbf{x}) \propto \prod_{i=1}^{n}\prod_{j=1}^{J} p(x_{ij} \mid x_{ij}^*)p(x_{ij}^* \mid \theta_i, d_j, a_j)p(\theta_i)p(d_j)p(a_j), \tag{A.47}$$

where

$$x_{ij} \mid x_{ij}^* = \begin{cases} 1 \text{ if } x_{ij}^* \geq 0 \\ 0 \text{ if } x_{ij}^* < 0 \end{cases} \quad \text{for } i = 1,\ldots,n, j=1,\ldots,J, \tag{A.48}$$

$$x_{ij}^* \mid \theta_i, d_j, a_j \sim N(a_j\theta_i + d_j, 1) \quad \text{for } i=1,\ldots,n, j=1,\ldots,J, \tag{A.49}$$

$$\theta_i \sim N(\mu_\theta, \sigma_\theta^2) \quad \text{for } i = 1,\ldots,n, \tag{A.50}$$

$$d_j \sim N(\mu_d, \sigma_d^2) \quad \text{for } j = 1,\ldots,J, \tag{A.51}$$

and

$$a_j \sim N^+(\mu_a, \sigma_a^2) \quad \text{for } j = 1,\ldots,J. \tag{A.52}$$

A.4.1 Measurement Model Parameters

In this formulation of the model, the measurement model parameters characterize the conditional distribution of the latent response variables given the latent variable. Given the latent response variables, the measurement model parameters are conditionally independent of the observables. Further, given the values for the jth latent response variable, denoted $\mathbf{x}_j^*$, the measurement model parameters for that observable, $\omega_j = (d_j, a_j)$, are conditionally independent of the other latent response variables and the other measurement model parameters. This mimics the situation for CFA. Momentarily suppressing the hyperparameters in our notation, we have

$$p(\omega \mid \theta, \mathbf{x}^*, \mathbf{x}) = p(\omega \mid \theta, \mathbf{x}^*) = \prod_{j=1}^{J} p(\omega_j \mid \theta, \mathbf{x}_j^*).$$

The situation is analogous to that of CFA not only in conditional independence relationships, but in distributional forms as well, where the latent response variables $\mathbf{x}_j^*$ play the role of the observables $\mathbf{x}$ in CFA, ω_j play the role of λ_{jA} in CFA, θ play the role of Ξ in CFA, and the error variances in CFA (i.e., the ψ_{jj}s) are equal to 1. The full conditional for ω_j therefore takes the same form as that for λ_{jA} in (A.46), subject to the positivity restriction for each a_j. Reintroducing the hyperparameters into the notation,

$$\omega_j \mid \theta, \mathbf{x}_j^*, \mu_\omega, \Sigma_\omega \sim N(\mu_{\omega_j \mid \theta, \mathbf{x}_j^*, \mu_\omega, \Sigma_\omega}, \Sigma_{\omega_j \mid \theta, \mathbf{x}_j^*, \mu_\omega, \Sigma_\omega})\, I(a_j > 0), \tag{A.53}$$

where

$$\mu_{\omega_j \mid \theta, \mathbf{x}_j^*, \mu_\omega, \Sigma_\omega} = (\Sigma_\omega^{-1} + \theta_A' \theta_A)^{-1} (\Sigma_\omega^{-1} \mu_\omega + \theta_A' \mathbf{x}_j^*),$$

$$\Sigma_{\omega_j \mid \theta, \mathbf{x}_j^*, \mu_\omega, \Sigma_\omega} = (\Sigma_\omega^{-1} + \theta_A' \theta_A)^{-1},$$

θ_A is the $(n \times 2)$ augmented matrix of latent variables obtained by combining an $(n \times 1)$ column vector of 1s to the $(n \times 1)$ column vector of latent variables θ, and I is the indicator function.

A.4.2 Latent Variables

Given the latent response variables, the latent variables are conditionally independent of the observables. Further, the exchangeability and conditional independence assumptions imply that the full conditional distribution for the latent variables may be factored into the product of examinee-specific full conditional distributions

$$\begin{aligned} p(\theta \mid \mathbf{d}, \mathbf{a}, \mathbf{x}^*, \mu_\theta, \sigma_\theta^2, \mathbf{x}) &= p(\theta \mid \mathbf{d}, \mathbf{a}, \mathbf{x}^*, \mu_\theta, \sigma_\theta^2) \\ &= \prod_{i=1}^{n} p(\theta_i \mid \mathbf{d}, \mathbf{a}, \mathbf{x}_i^*, \mu_\theta, \sigma_\theta^2) \\ &\propto \prod_{i=1}^{n} p(\mathbf{x}_i^* \mid \theta_i, \mathbf{d}, \mathbf{a}) p(\theta_i \mid \mu_\theta, \sigma_\theta^2), \end{aligned}$$

where $\mathbf{x}_i$ is the collection of J observed values from examinee i. Equation (A.49) implies that the first term on the right-hand side is the multivariate normal conditional distribution of the latent response variables for the examinee. The second term on the right-hand side is the prior distribution for the examinee's latent variables, given by (A.50).

The similarity of the latent response variable formulation of IRT to CFA manifests itself here as well, and the full conditional distribution for the latent variables here can be recognized as a simpler version of the case of latent variables in CFA. Connecting the notational elements, θ_i here plays the role of ξ_i in CFA, d_j plays the role of τ_j in CFA, and a_j plays the role of λ_j in CFA. What was ψ_{jj} in CFA is now the error variance of the latent response variables, which is set to be 1. As the formulation of the model may be thought of as a CFA model with error variances fixed to 1, and as we are confining our treatment to the unidimensional model with fixed values of the parameters that govern the distribution of the latent variables, the full conditional for each examinee's latent variable is a somewhat simpler version than that developed for CFA in Section A.3:

$$\theta_i \mid \mathbf{d},\mathbf{a},\mathbf{x}_i^*,\mu_\theta,\sigma_\theta^2 \sim N(\mu_{\theta_i|\mathbf{d},\mathbf{a},\mathbf{x}_i^*,\mu_\theta,\sigma_\theta^2},\sigma^2_{\theta_i|\mathbf{d},\mathbf{a},\mathbf{x}_i^*,\mu_\theta,\sigma_\theta^2}), \tag{A.54}$$

where

$$\mu_{\theta_i|\mathbf{d},\mathbf{a},\mathbf{x}_i^*,\mu_\theta,\sigma_\theta^2} = \frac{(\mu_\theta/\sigma_\theta^2) + \sum_j a_j(x_{ij}^* - d_j)}{(1/\sigma_\theta^2) + \sum_j a_j^2}$$

and

$$\sigma^2_{\theta_i|\mathbf{d},\mathbf{a},\mathbf{x}_i^*,\mu_\theta,\sigma_\theta^2} = \left(\frac{1}{\sigma_\theta^2} + \sum_j a_j^2\right)^{-1}.$$

A.4.3 Latent Response Variables

Given the latent variable for examinee i, an observable variable x_{ij}, and the measurement model parameters for the observable $\omega_j = (d_j, a_j)$, the associated latent response variable x_{ij}^* is conditionally independent of all other observables, latent response variables, latent variables, and measurement model parameters:

$$p(\mathbf{x}^* \mid \boldsymbol{\theta},\mathbf{d},\mathbf{a},\mathbf{x}) = \prod_{i=1}^{n}\prod_{j=1}^{J} p(x_{ij}^* \mid \theta_i, d_j, a_j, x_{ij}) \propto \prod_{i=1}^{n}\prod_{j=1}^{J} p(x_{ij} \mid x_{ij}^*) p(x_{ij}^* \mid \theta_i, d_j, a_j).$$

The conditional relationship of x_{ij} given x_{ij}^* is deterministic. Given x_{ij}^*, x_{ij} is known with certainty, as formulated in (A.48). If $x_{ij}^* \geq 0$, $x_{ij} = 1$; if $x_{ij}^* < 0$, $x_{ij} = 0$. Reversing the direction of this relationship, knowing x_{ij} completely determines the sign of x_{ij}^* and nothing else; if $x_{ij} = 1$, $x_{ij}^* \geq 0$; if $x_{ij} = 0$, $x_{ij}^* < 0$. The full conditional for x_{ij}^* therefore follows the form of its prior specification $p(x_{ij}^* \mid \theta_i, \omega_j)$ subject to a truncation dictated by x_{ij}

$$x_{ij}^* \mid \theta_i, d_j, a_j, x_{ij} \sim \begin{cases} N(a_j\theta_i + d_j, 1)\, I(x_{ij}^* \geq 0) & \text{if } x_{ij} = 1 \\ N(a_j\theta_i + d_j, 1)\, I(x_{ij}^* < 0) & \text{if } x_{ij} = 0, \end{cases} \tag{A.55}$$

where I is the indicator function.

A.5 Latent Class Analysis

In LCA, the unknowns include the examinees' discrete latent variables ($\boldsymbol{\theta}$), the proportions of examinees in each class ($\boldsymbol{\gamma}$), and the conditional probabilities for the observables given class membership ($\boldsymbol{\pi}$). We seek the posterior distribution for these unknowns given discrete observables ($\mathbf{x}$).

We first develop the full conditional distributions for the general case of a polytomous latent variable and polytomous observable variables. We then develop the full conditional distributions for the special case of a dichotomous latent variable and dichotomous observable variables.

A.5.1 LCA with Polytomous Latent and Observable Variables

For the general case of polytomous observables and a polytomous latent variable, the posterior distribution is given by (13.13), repeated here as

$$p(\theta,\gamma,\pi \mid \mathbf{x}) \propto \prod_{i=1}^{n}\prod_{j=1}^{J} p(x_{ij} \mid \theta_i,\pi_j)p(\theta_i \mid \gamma)p(\gamma)\prod_{c=1}^{C} p(\pi_{cj}), \tag{A.56}$$

where

$$(x_{ij} \mid \theta_i = c,\pi_j) \sim \text{Categorical}(\pi_{cj}) \quad \text{for } i = 1,\ldots,n, j=1,\ldots,J, \tag{A.57}$$

$$\theta_i \mid \gamma \sim \text{Categorical}(\gamma) \quad \text{for } i = 1,\ldots,n, \tag{A.58}$$

$$\gamma \sim \text{Dirichlet}(\alpha_\gamma), \tag{A.59}$$

and

$$\pi_{cj} \sim \text{Dirichlet}(\alpha_{\pi_c}) \quad \text{for } c = 1,\ldots,C, j=1,\ldots,J. \tag{A.60}$$

A.5.1.1 Measurement Model Parameters

Given the latent variables, the measurement model parameters are conditionally independent of the latent class proportions. Further, given the collection of values for the jth observable $\mathbf{x}_j$, the measurement model parameters for that observable are conditionally independent of the *other* observables and the measurement model parameters for those *other* observables. That is,

$$p(\pi \mid \theta,\gamma,\alpha_\pi,\mathbf{x}) = p(\pi \mid \theta,\alpha_\pi,\mathbf{x}) = \prod_{j=1}^{J} p(\pi_j \mid \theta,\alpha_\pi,\mathbf{x}_j) = \prod_{j=1}^{J}\prod_{c=1}^{C} p(\pi_{cj} \mid \theta,\alpha_{\pi_c},\mathbf{x}_j),$$

where $\alpha_\pi = (\alpha_{\pi_1},\ldots,\alpha_{\pi_C})$ and $\pi_j = (\pi_{1j},\ldots,\pi_{Cj})$ are the conditional probabilities for the jth observable with components that are the class-specific conditional probabilities, each of which is assigned a Dirichlet prior in (A.60). In these full conditionals, the latent variables θ are treated as known, meaning we know which examinees are in each latent class. The implication is that we can construct the joint full conditional for $\boldsymbol{\pi}$ by going observable by observable, and for each observable by going class by class. The full conditional distribution for each $\boldsymbol{\pi}_{cj}$ is

$$p(\pi_{cj} \mid \theta,\alpha_{\pi_c},\mathbf{x}_j) \propto p(\mathbf{x}_j \mid \theta,\pi_{cj})p(\pi_{cj} \mid \alpha_{\pi_c}) = \prod_{i=1}^{r_c} p(x_{ij} \mid \theta_i = c,\pi_{cj})\, p(\pi_{cj} \mid \alpha_{\pi_c}),$$

where r_c is the number of examinees in latent class c. This, along with conditioning on $\theta_i = c$ expresses that we are restricting attention to examinees in class c. Each of the observables is categorically distributed (given membership in latent class c), as expressed in (A.57). Recall that we conducted inference for the unknown parameter of repeated Bernoulli trials by modeling the counts of successes as a binomial. Similarly, we can formulate the model for the current situation in terms of counts. For each observable, for each latent class, we count the number of examinees who respond in each category of the observable. Let r_{cjk} denote the number of examinees in class c who have a value of k for observable j. Let $\mathbf{r}_{cj} = (r_{cj1},\ldots, r_{cjK})$ denote the collection of these counts for class c and observable j, which is sufficient for inference about $\boldsymbol{\pi}_{cj}$. We can re-express the full conditional distribution as

$$p(\pi_{cj} \mid \theta, \alpha_{\pi_c}, \mathbf{x}_j) = p(\pi_{cj} \mid \alpha_{\pi_c}, \mathbf{r}_{cj}) \propto p(\mathbf{r}_{cj} \mid \pi_{cj}) p(\pi_{cj} \mid \alpha_{\pi_c}).$$

The first term on the right-hand side is a multinomial and the second term is the Dirichlet prior. Accordingly, the full conditional distribution is a Dirichlet distribution,

$$\pi_{cj} \mid \theta, \alpha_{\pi_c}, \mathbf{x}_j \sim \text{Dirichlet}(\alpha_{\pi_{c1}} + r_{cj1}, \ldots, \alpha_{\pi_{cK}} + r_{cjK}). \tag{A.61}$$

A.5.1.2 Latent Variables

The exchangeability and conditional independence assumptions imply that the full conditional distribution for the latent variables may be factored into the product of examinee-specific full conditional distributions,

$$p(\theta \mid \gamma, \pi, \mathbf{x}) = \prod_{i=1}^{n} p(\theta_i \mid \gamma, \pi, \mathbf{x}_i) \propto \prod_{i=1}^{n} p(\mathbf{x}_i \mid \theta_i, \pi) p(\theta_i \mid \gamma).$$

The first term on the right-hand side is the conditional distribution of the observables for the examinee

$$p(\mathbf{x}_i \mid \theta_i, \pi) = \prod_{j=1}^{J} p(x_{ij} \mid \theta_i = c, \pi_j). \tag{A.62}$$

where $p(x_{ij} \mid \theta_i = c, \pi_j)$ is given by (A.57). The second term is the prior distribution for the examinee's latent variable, given by (A.58). This is an instance of Bayes' theorem for discrete variables. The posterior distribution is a categorical distribution,

$$\theta_i \mid \gamma, \pi, \mathbf{x}_i \sim \text{Categorical}(\mathbf{s}_i). \tag{A.63}$$

Here, $\mathbf{s}_i = (s_{i1}, \ldots, s_{iC})$ is a vector of probabilities obtained via Bayes' theorem:

$$\begin{aligned} s_{ic} &= \frac{p(\mathbf{x}_i \mid \theta_i = c, \pi) p(\theta_i = c \mid \gamma)}{p(\mathbf{x}_i)} \\ &= \frac{\prod_{j=1}^{J} p(x_{ij} \mid \theta_i = c, \pi_j) p(\theta_i = c \mid \gamma)}{\sum_{g=1}^{C} \prod_{j=1}^{J} p(x_{ij} \mid \theta_i = g, \pi_j) p(\theta_i = g \mid \gamma)} \\ &= \frac{\prod_{j=1}^{J} \prod_{k=1}^{K} (\pi_{cjk})^{I(x_{ij}=k)} \gamma_c}{\sum_{g=1}^{C} \prod_{j=1}^{J} \prod_{k=1}^{K} (\pi_{gjk})^{I(x_{ij}=k)} \gamma_g}, \end{aligned}$$

where I is the indicator function.

A.5.1.3 Latent Class Proportions

Given the latent variables, the full conditional distributions for the latent class proportions are independent of the observables and the measurement model parameters,

$$p(\gamma \mid \theta, \pi, \alpha_\gamma, \mathbf{x}) = p(\gamma \mid \theta, \alpha_\gamma) \propto p(\theta \mid \gamma)p(\gamma \mid \alpha_\gamma) = \prod_{i=1}^{n} p(\theta_i \mid \gamma)p(\gamma \mid \alpha_\gamma),$$

The first term on the right-hand side is the distribution of the treated-as-known latent variables given the unknown latent class proportions, which for each examinee is a categorical distribution. Again, we can work with counts of the latent variables and employ a multinomial distribution. Let r_c denote the number of examinees for whom $\theta = c$ and $\mathbf{r} = (r_1, \ldots, r_C)$ denote the collection of these counts. We can re-express the full conditional distribution as

$$p(\gamma \mid \theta, \alpha_\gamma) = p(\gamma \mid \mathbf{r}, \alpha_\gamma) \propto p(\mathbf{r} \mid \gamma)p(\gamma \mid \alpha_\gamma),$$

where the first term on the right-hand side is a multinomial and the second term is the Dirichlet prior in (A.59). Accordingly, the full conditional distribution is a Dirichlet distribution

$$\gamma \mid \theta, \alpha_\gamma \sim \text{Dirichlet}(\alpha_{\gamma_1} + r_1, \cdots, \alpha_{\gamma_C} + r_C). \tag{A.64}$$

A.5.2 LCA with Dichotomous Latent and Observable Variables

We now consider the special case of dichotomous observables and a dichotomous latent variable (i.e., $C = 2$), formulated as Bernoulli random variables. The posterior distribution is given by (13.26), repeated here as

$$p(\theta, \gamma, \pi \mid \mathbf{x}) \propto \prod_{i=1}^{n} \prod_{j=1}^{J} p(x_{ij} \mid \theta_i, \pi_j)p(\theta_i \mid \gamma)p(\gamma) \prod_{c=1}^{C} p(\pi_{cj}), \tag{A.65}$$

where $\pi = (\pi_1, \ldots, \pi_J)$, $\pi_j = (\pi_{1j}, \pi_{2j})$, π_{cj} is the probability of a value of 1 on observable j given membership in class c, and, adopting the Bernoulli-plus-one construction that codes the latent variable as taking a value of 1 or 2:

$$(x_{ij} \mid \theta_i = c, \pi_j) \sim \text{Bernoulli}(\pi_{cj}) \quad \text{for } i = 1, \ldots, n, j = 1, \ldots, J, \tag{A.66}$$

$$\theta_i \mid \gamma \sim \text{Bernoulli}(\gamma) + 1 \quad \text{for } i = 1, \ldots, n, \tag{A.67}$$

$$\gamma \sim \text{Beta}(\alpha_\gamma, \beta_\gamma), \tag{A.68}$$

and

$$\pi_{cj} \sim \text{Beta}(\alpha_{\pi_c}, \beta_{\pi_c}) \quad \text{for } c = 1, 2, j = 1, \ldots, J. \tag{A.69}$$

On this specification, γ is the proportion of examinees in latent class 2.

A.5.2.1 Measurement Model Parameters

Each π_{cj} is a class-specific conditional probability of a value of 1 on observable j, and is assigned a beta prior distribution. The full conditional distribution for each π_{cj} is given by

$$p(\pi_{cj} \mid \theta, \alpha_{\pi_c}, \beta_{\pi_c}, \mathbf{x}_j) \propto p(\mathbf{x}_j \mid \theta, \pi_{cj})p(\pi_{cj} \mid \alpha_{\pi_c}, \beta_{\pi_c}) = \prod_{i=1}^{r_c} p(x_{ij} \mid \theta_i = c, \pi_{cj})p(\pi_{cj} \mid \alpha_{\pi_c}, \beta_{\pi_c}),$$

where r_c is the number of examinees in latent class c. This, along with conditioning on $\theta_i = c$ expresses that we are restricting attention to examinees in class c. Each of the observables is distributed as a Bernoulli (given membership in latent class c), as expressed in (A.66). Again, we can formulate the model for the current situation in terms of counts. Let r_{cj} denote the number of examinees in class c (out of n_c) who have a value of 1 for observable j, which is sufficient for inference about π_{cj}. We can re-express the full conditional distribution as

$$p(\pi_{cj} \mid \theta, \alpha_{\pi_c}, \beta_{\pi_c}, \mathbf{x}_j) = p(\pi_{cj} \mid \alpha_{\pi_c}, \beta_{\pi_c}, r_{cj}) \propto p(r_{cj} \mid \pi_{cj}) p(\pi_{cj} \mid \alpha_{\pi_c}, \beta_{\pi_c}).$$

The first term on the right-hand side is a binomial and the second term is the beta prior in (A.69). Accordingly, the full conditional distribution is a beta distribution,

$$\pi_{cj} \mid \theta, \alpha_{\pi_c}, \beta_{\pi_c}, \mathbf{x}_j \sim \text{Beta}(\alpha_{\pi_c} + r_{cj}, \beta_{\pi_c} + r_c - r_{cj}). \quad \text{(A.70)}$$

A.5.2.2 Latent Variables

The exchangeability and conditional independence assumptions imply that the full conditional distribution for the latent variables may be factored into the product of examinee-specific full conditional distributions,

$$p(\theta \mid \gamma, \pi, \mathbf{x}) = \prod_{i=1}^{n} p(\theta_i \mid \gamma, \pi, \mathbf{x}_i) \propto \prod_{i=1}^{n} p(x_i \mid \theta_i, \pi) p(\theta_i \mid \gamma).$$

The first term on the right-hand side is the conditional distribution of the observables for the examinee, given by (A.62), where $p(x_{ij} \mid \theta_i = c, \pi_j)$ is given by (A.66). The second term is the prior distribution for the examinee's latent variable, given by (A.67). This is an instance of Bayes' theorem for discrete variables. The posterior distribution is a Bernoulli distribution

$$\theta_i \mid \gamma, \pi, \mathbf{x}_i \sim \text{Bernoulli}(s_i) + 1, \quad \text{(A.71)}$$

where s_i is a probability obtained via Bayes' theorem:

$$\begin{aligned} s_i &= \frac{p(\mathbf{x}_i \mid \theta_i = 2, \pi) p(\theta_i = 2 \mid \gamma)}{p(\mathbf{x}_i)} \\ &= \frac{\prod_{j=1}^{J} p(x_{ij} \mid \theta_i = 2, \pi_j) p(\theta_i = 2 \mid \gamma)}{\sum_{g=1}^{C} \prod_{j=1}^{J} p(x_{ij} \mid \theta_i = g, \pi_j) p(\theta_i = g \mid \gamma)} \\ &= \frac{\prod_{j=1}^{J} (\pi_{2j})^{x_{ij}} (1 - \pi_{2j})^{1 - x_{ij}} \gamma}{\prod_{j=1}^{J} (\pi_{2j})^{x_{ij}} (1 - \pi_{2j})^{1 - x_{ij}} \gamma + \prod_{j=1}^{J} (\pi_{1j})^{x_{ij}} (1 - \pi_{1j})^{1 - x_{ij}} (1 - \gamma)}. \end{aligned}$$

A.5.2.3 Latent Class Proportions

The full conditional distribution for the latent class proportion, γ, is

$$p(\gamma \mid \theta, \pi, \alpha_\gamma, \beta_\gamma, x) = p(\gamma \mid \theta, \alpha_\gamma, \beta_\gamma) \propto p(\theta \mid \gamma) p(\gamma \mid \alpha_\gamma, \beta_\gamma) = \prod_{i=1}^{n} p(\theta_i \mid \gamma) p(\gamma \mid \alpha_\gamma, \beta_\gamma).$$

The first term on the right-hand side is the distribution of the treated-as-known latent variables given the unknown probability of being in latent class 2, which for each examinee is a Bernoulli distribution. Again, we can work with counts of the latent variables and employ a binomial distribution. Let r denote the number of examinees in class 2. We can re-express the full conditional distribution as

$$p(\gamma \mid \theta, \alpha_\gamma, \beta_\gamma) = p(\gamma \mid r, \alpha_\gamma, \beta_\gamma) \propto p(r \mid \gamma) p(\gamma \mid \alpha_\gamma, \beta_\gamma),$$

where the first term on the right-hand side is a binomial and the second term is the beta prior in (A.68). Accordingly, the full conditional distribution is a beta distribution

$$\gamma \mid \theta, \alpha_\gamma, \beta_\gamma \sim \text{Beta}(\alpha_\gamma + r, \beta_\gamma + n - r). \tag{A.72}$$

Appendix B: Probability Distributions

This appendix presents and briefly discusses features of some standard probability distributions, as an aid to understanding their use throughout the book. In each case, we present the notation, probability mass or density function, and summary quantities such as the mean, variance, and mode (where applicable). In addition, we discuss common usage of these distributions. We use the notation x for a random variable and $\mathbf{x}$ for a random vector, except in the cases of the Wishart and inverse-Wishart where we use $\mathbf{W}$ and $\mathbf{W}^{-1}$ to stand for the random matrix, respectively. For deeper discussions of these and other distributions, see Casella and Berger (2008), Gelman et al. (2013), and Jackman (2009). See Lunn et al. (2013) for details on their implementations in WinBUGS and its variants.

B.1 Discrete Variables

B.1.1 Bernoulli

If $x \in \{0,1\}$ has a Bernoulli mass function, denoted $x \sim \text{Bernoulli}(\theta)$, then

$$p(x=1)=\theta$$

and

$$p(x=0)=1-\theta$$

with

- $\theta \in [0,1]$
- $E(x)=\theta$
- $\text{var}(x)=\theta(1-\theta)$.

The Bernoulli may be thought of as a special case of the binomial (which is defined in more detail shortly) where the number of trials $J = 1$. It may also be thought of as a categorical distribution with only two categories, usually with a recoding of the category labels to the 0/1 coding adopted here. The conjugate prior for θ is the beta distribution.

B.1.2 Categorical

The categorical distribution is a generalization of the Bernoulli distribution to the case of $K > 2$ categories, typically assigned values 1, …, K. If x has a categorical distribution with parameter vector θ, denoted by $x \sim \text{Categorical}(\theta)$, it has probability mass function

$$p(x=k)=\theta_k$$

with

- $\theta=(\theta_1,\ldots,\theta_K)$
- $\theta_k \in [0,1] \text{ for } k=1,\ldots,K$
- $\sum_{k=1}^{K}\theta_k=1.$

The categorical distribution may be thought of as a special case of the multinomial (which is described in more detail below) where the number of trials $J = 1$. The conjugate prior for θ is the Dirichlet distribution.

B.1.3 Binomial

A random variable $x \in \{0,1,\ldots,J\}$ has a Binomial distribution with a parameter θ, denoted by $x \sim \text{Binomial}(\theta, J)$, if it has probability mass function

$$p(x) = \binom{J}{x} \theta^x (1-\theta)^{J-x}$$

with

- $\theta \in [0,1]$
- $E(x) = J\theta$
- $\text{var}(x) = J\theta(1-\theta)$.

x is commonly viewed as the sum of J independent Bernoulli variables, each with success probability θ. When $J = 1$, the binomial specializes to the Bernoulli distribution. The conjugate prior for θ is the beta distribution.

B.1.4 Multinomial

The multinomial generalizes the binomial to the situation of more than two categories. If $\mathbf{x} = (x_1, \ldots, x_K)$, where $x_k \in \{0,1,\ldots,J\}$ for $k = 1, \ldots, K$ and $\Sigma_{k=1}^{K} x_k = J$, then $\mathbf{x}$ has a multinomial mass function with a parameter $\boldsymbol{\theta} = (\theta_1,\ldots,\theta_K)$, where $\theta_k \in [0,1]$ for $k = 1, \ldots, K$ and $\Sigma_{k=1}^{K} \theta_k = 1$, denoted by $\mathbf{x} \sim \text{Multinomial}(\theta, J)$, then

$$p(\mathbf{x}) = \frac{J!}{x_1! \cdots x_K!} \theta_1^{x_1} \cdots \theta_K^{x_K}$$

with

- $E(x_k) = J\theta_k$
- $\text{var}(x_k) = J\theta_k(1-\theta_k)$
- $\text{cov}(x_k, x_{k'}) = -J\theta_k\theta_{k'}$.

$\mathbf{x}$ is commonly viewed as the collection of counts of J independent categorical variables that can take on any of K values, where the elements of $\boldsymbol{\theta}$ are the probabilities of any such categorical variable taking on any of the K values. When $J = 1$, the binomial specializes to the categorical distribution. The conjugate prior for $\boldsymbol{\theta}$ is the Dirichlet distribution.

B.1.5 Poisson

A random variable $x \in \{0,1,2,\ldots\}$ has a Poisson distribution with a parameter θ, denoted by $x \sim \text{Poisson}(\theta)$, if it has probability mass function

$$p(x) = \frac{1}{x!} \theta^x \exp(-\theta)$$

with

- $\theta > 0$
- $E(x) = \theta$
- $\text{mode}(x) = \theta$
- $\text{var}(x) = \theta$.

The conjugate prior for θ is the gamma distribution.

B.2 Continuous Variables

B.2.1 Uniform

A random variable x has a uniform distribution, denoted as $x \sim \text{Uniform}(\alpha, \beta)$, if it has a probability density function

$$p(x) = \begin{cases} \dfrac{1}{\beta - \alpha} & x \in [\alpha, \beta] \\ 0 & x \notin [\alpha, \beta] \end{cases}$$

with

- $E(x) = (\alpha + \beta) / 2$
- $\text{var}(x) = (\beta - \alpha)^2 / 12$.

B.2.2 Beta

A random variable $x \in [0,1]$ has a beta density with parameters $\alpha > 0$ and $\beta > 0$, denoted as $x \sim \text{Beta}(\alpha, \beta)$, if it has a probability density function

$$p(x) = \frac{\Gamma(\alpha + \beta)}{\Gamma(\alpha)\Gamma(\beta)} x^{\alpha-1}(1-x)^{\beta-1}$$

where $\Gamma(t)$ is the gamma function, $\int_0^\infty x^{t-1} e^{-x} dx$, with:

- $E(x) = \alpha / (\alpha + \beta)$
- $\text{var}(x) = \alpha\beta / [(\alpha + \beta)^2 (\alpha + \beta + 1)]$
- a unique mode at $\alpha - 1 / (\alpha + \beta - 2)$ when $\alpha, \beta > 1$.

When $\alpha = \beta = 1$, the beta specializes to the uniform distribution, that is, $x \sim \text{Uniform}(0,1)$. The beta distribution is the conjugate prior for the probability parameter in the Bernoulli and binomial distributions.

B.2.3 Dirichlet

A random variable $\mathbf{x} = (x_1, \ldots, x_K)$ has the Dirichlet distribution with parameter vector $\alpha = (\alpha_1, \ldots, \alpha_K)$ if it has a probability density function

$$p(\mathbf{x}) = \frac{\Gamma(\alpha_1 + \cdots + \alpha_K)}{\Gamma(\alpha_1) \cdots \Gamma(\alpha_K)} x_1^{\alpha_1 - 1} \cdots x_K^{\alpha_K - 1}$$

with

- $x_k \geq 0$ for $k = 1, \ldots, K$
- $\alpha = (\alpha_1, \ldots, \alpha_K)$ with $\alpha_k > 0$ for $k = 1, \ldots, K$.

Letting $S = \sum_{k=1}^{K} \alpha_k$,

- The marginal distribution of any x_k is $\text{Beta}(\alpha_k, S - \alpha_k)$
- $E(x_k) = \alpha_k / S$
- $\text{mode}(x_k) = (\alpha_k - 1) / (S - K)$
- $\text{var}(x_k) = \alpha_k (S - \alpha_k) / [S^2 (S + 1)]$
- $\text{cov}(x_k, x_{k'}) = \alpha_k \alpha_{k'} / [S^2 (S + 1)]$.

The Dirichlet distribution may be seen as a multivariate generalization of the beta distribution, or the beta may be seen as a special case of the Dirichlet where $K = 2$. That is, specifying $x \sim \text{Beta}(a, b)$ is equivalent to specifying $(x, 1 - x) \sim \text{Dirichlet}(a, b)$. The Dirichlet distribution is the conjugate prior for the probability parameter in categorical and multinomial distributions.

B.2.4 Normal

A random variable x has a normal distribution, denoted as $x \sim N(\mu, \sigma^2)$, if it has probability density function

$$p(x) = \frac{1}{\sqrt{2\pi}\sigma} \exp\left[\frac{-1}{2\sigma^2}(x - \mu)^2\right]$$

with

- $\sigma^2 > 0$
- $E(x) = \text{mode}(x) = \mu$
- $\text{var}(x) = \sigma^2$.

The normal distribution is the conjugate prior for the mean of a normal distribution.

B.2.5 Lognormal

If $\log x \sim N(\mu, \sigma^2)$, then x has a lognormal distribution, denoted as $x \sim \log N(\mu, \sigma^2)$ with a probability density function

$$p(x) = \frac{1}{\sqrt{2\pi}\sigma x} \exp\left[\frac{-1}{2\sigma^2}(\log(x) - \mu)^2\right]$$

with

- $x > 0$
- $\sigma^2 > 0$
- $E(x) = \exp(\mu + \frac{1}{2}\sigma^2)$
- $\text{var}(x) = \exp(2\mu + \sigma^2)(\exp(\sigma^2) - 1)$
- $\text{mode}(x) = \exp(\mu - \sigma^2)$.

B.2.6 Logit-Normal

If $\text{logit}(x) \sim N(\mu, \sigma^2)$, then x has a logit-normal distribution, denoted as $x \sim \text{logit}N(\mu, \sigma^2)$, with a probability density function

$$p(x) = \frac{1}{\sqrt{2\pi}\sigma x(1-x)} \exp\left[\frac{-1}{2\sigma^2}(\text{logit}(x) - \mu)^2\right]$$

with

- $x \in (0,1)$
- $\sigma^2 > 0$.

B.2.7 Multivariate Normal

If a random variable $\mathbf{x} = (x_1, \ldots, x_K)$ has a multivariate normal distribution with parameters μ and positive-definite Σ, denoted as $\mathbf{x} \sim N(\mu, \Sigma)$, then it has probability density function

$$p(\mathbf{x}) = (2\pi)^{-K/2} |\Sigma|^{-1/2} \exp(-(\mathbf{x} - \mu)'\Sigma^{-1}(\mathbf{x} - \mu)/2)$$

with

- $E(\mathbf{x}) = \mu$
- $\text{var}(\mathbf{x}) = \Sigma$.

If $\mathbf{x}$ is partitioned as $\mathbf{x} = (\mathbf{x}_1, \mathbf{x}_2)'$ such that

$$\mathbf{x} = \begin{bmatrix} \mathbf{x}_1 \\ \mathbf{x}_2 \end{bmatrix} \sim N\left(\begin{bmatrix} \mu_1 \\ \mu_2 \end{bmatrix}, \begin{bmatrix} \Sigma_{11} & \Sigma_{12} \\ \Sigma_{21} & \Sigma_{22} \end{bmatrix}\right)$$

where Σ_{22} is positive definite, then

- $\mathbf{x}_1$ and $\mathbf{x}_2$ are independent if and only if $\Sigma_{21} = 0$
- the marginal density of $\mathbf{x}_2$ is normal: $p(\mathbf{x}_2) = \int p(\mathbf{x}_1, \mathbf{x}_2) d\mathbf{x}_1 = N(\mu_2, \Sigma_{22})$
- the conditional density of $\mathbf{x}_2$ given $\mathbf{x}_1$ is normal: $p(\mathbf{x}_2 \mid \mathbf{x}_1) = N(\mu_{2|1}, \Sigma_{2|1})$, where $\mu_{2|1} = \mu_1 + \Sigma_{12}\Sigma_{22}^{-1}(\mathbf{x}_2 - \mu_2)$ and $\Sigma_{2|1} = \Sigma_{11} - \Sigma_{12}\Sigma_{22}^{-1}\Sigma_{21}$.

The multivariate normal distribution is the conjugate prior for the mean vector of a multivariate normal distribution.

B.2.8 Gamma

A random variable x has a gamma distribution, denoted as $x \sim \text{Gamma}(\alpha, \beta)$, if it has a probability density function

$$p(x) = \frac{\beta^\alpha}{\Gamma(\alpha)} x^{\alpha-1} \exp(-\beta x)$$

with

- $x > 0$
- $\alpha > 0$ and $\beta > 0$
- $E(x) = \alpha / \beta$ if $\alpha > 1$
- $\text{var}(x) = \alpha/\beta^2$
- $\text{mode}(x) = (\alpha - 1)/\beta$ for $\alpha \geq 0$.

If $x \sim \text{Gamma}(\alpha,\beta)$, then $1/x \sim \text{Inv-Gamma}(\alpha,\beta)$. If $x \sim \text{Gamma}(\nu/2,1/2)$, then x follows a χ^2 density with ν degrees of freedom, denoted by $x \sim \chi^2(\nu)$. The gamma is a conjugate prior for the parameter for a Poisson distribution and for the precision (inverse of variance) parameter in a normal distribution, that is, σ^{-2}. In the latter case, a reparameterization where $\alpha = \nu_0/2$ and $\beta = \nu_0\sigma_0^2/2$, that is, $\sigma^{-2} \sim \text{Gamma}(\nu_0/2, \nu_0\sigma_0^2/2)$, yields a convenient interpretation where σ_0^2 may be thought of as a point summary or best estimate of the variance and ν_0 as an associated degrees of freedom or pseudo-sample size associated with that estimate.

B.2.9 Inverse-Gamma

A random variable x is has an inverse-gamma distribution with parameters α and β, denoted as $x \sim \text{Inv-Gamma}(\alpha,\beta)$, if it has probability density function

$$p(x) = \frac{\beta^\alpha}{\Gamma(\alpha)} x^{-\alpha-1} \exp\left(\frac{-\beta}{x}\right)$$

with

- $x > 0$
- $\alpha > 0$ and $\beta > 0$
- $E(x) = \beta/(\alpha-1)$ if $\alpha > 1$
- $\text{var}(x) = \beta^2/[(\alpha-1)^2(\alpha-2)]$ if $\alpha > 2$
- $\text{mode}(x) = \beta/(\alpha+1)$.

If $x \sim \text{Inv-Gamma}(\alpha,\beta)$, then $1/x \sim \text{Gamma}(\alpha,\beta)$. If $x \sim \text{Inv-Gamma}(\nu/2,1/2)$, then x follows an inverse-χ^2 density with ν degrees of freedom, denoted by $x \sim \chi^{-2}(\nu)$. The inverse-gamma is a conjugate prior for the variance parameter in a normal distribution σ^2. In such a case, a reparameterization where $\alpha = \nu_0/2$ and $\beta = \nu_0\sigma_0^2/2$, that is, $\sigma^2 \sim \text{Inv-Gamma}(\nu_0/2, \nu_0\sigma_0^2/2)$, yields a convenient interpretation where σ_0^2 may be thought of as a point summary or best estimate of the variance and ν_0 as an associated degrees of freedom or pseudo-sample size associated with that estimate.

B.2.10 Wishart

A random variable $\mathbf{W}$ has a Wishart distribution, denoted $\mathbf{W} \sim \text{Wishart}(\mathbf{S},\nu)$, if it has a probability density function

$$p(\mathbf{W}) = \frac{|\mathbf{W}|^{\frac{\nu-J-1}{2}}}{2^{\frac{\nu J}{2}} |\mathbf{S}|^{\frac{J}{2}} \Gamma_J(\frac{\nu}{2})} \exp\left(-\frac{\text{tr}(\mathbf{S}^{-1}\mathbf{W})}{2}\right)$$

where

- $\mathbf{W}$ is a $(J \times J)$ symmetric positive definite matrix
- $\mathbf{S}$ is a $(J \times J)$ symmetric positive definite matrix
- $\nu > 0$ is the degrees of freedom
- tr( ) is the trace operator
- Γ_J is the multivariate gamma function $\Gamma_J(\frac{\nu}{2}) = \pi^{\frac{J(J-1)}{2}} \prod_{j=1}^{J} \Gamma\left(\frac{\nu+1-j}{2}\right)$
- $E(\mathbf{W}) = \nu\mathbf{S}$.

A special case obtains when $J = 1$ and $\mathbf{S} = 1$, in which case the Wishart reduces to a χ^2 density with ν degrees of freedom. If $\mathbf{W} \sim \text{Wishart}(\mathbf{S}, \nu)$, then its inverse has an inverse-Wishart distribution, $\mathbf{W}^{-1} \sim \text{Inv-Wishart}(\mathbf{S}, \nu)$. The Wishart generalizes the gamma and χ^2 distributions for precisions to the case of a precision matrix (i.e., the inverse of the covariance matrix), as commonly arise when working with multivariate normal distributions. The Wishart density is the conjugate prior for a precision matrix (inverse of the covariance matrix) of a multivariate normal distribution.

B.2.11 Inverse-Wishart

A random variable $\mathbf{W}$ has an inverse-Wishart distribution, denoted by $\mathbf{W} \sim \text{Inv-Wishart}(\mathbf{S}, \nu)$, if it has a probability density function

$$p(\mathbf{W}) = \frac{|\mathbf{S}|^{\frac{-\nu}{2}} |\mathbf{W}|^{\frac{-(\nu+J+1)}{2}}}{2^{\frac{\nu J}{2}} \Gamma_J(\frac{\nu}{2})} \exp\left(-\frac{\text{tr}(\mathbf{S}^{-1}\mathbf{W}^{-1})}{2}\right)$$

where

- $\mathbf{W}$ is a $(J \times J)$ symmetric positive definite matrix
- $\mathbf{S}$ is a $(J \times J)$ symmetric positive definite matrix
- $\nu > 0$ is the degrees of freedom
- tr( ) is the trace operator
- Γ_J is the multivariate gamma function $\Gamma_J(\frac{\nu}{2}) = \pi^{\frac{J(J-1)}{2}} \prod_{j=1}^{J} \Gamma\left(\frac{\nu+1-j}{2}\right)$
- $E(\mathbf{W}) = (\nu - J - 1)^{-1}\mathbf{S}^{-1}, \nu > J + 1.$

When $J = 1$ and $\mathbf{S} = 1$, the inverse-Wishart reduces to an inverse-χ^2 distribution with ν degrees of freedom. If $\mathbf{W} \sim \text{Inv-Wishart}(\mathbf{S}, \nu)$, then its inverse has a Wishart distribution, $\mathbf{W}^{-1} \sim \text{Wishart}(\mathbf{S}, \nu)$. The inverse-Wishart density generalizes the inverse-gamma and inverse-χ^2 densities for covariances to the case of a covariance matrix, as commonly arise when working with multivariate normal distributions. The inverse-Wishart density is the conjugate prior for a covariance matrix of a multivariate normal distribution.

References

Adams, R. J., Wilson, M., & Wang, W. (1997). The multidimensional random coefficients multinomial logit model. *Applied Psychological Measurement, 21*, 1–23.

AERA, APA, & NCME. (2014). *Standards for educational and psychological testing*. Washington, DC: American Educational Research Association.

Agresti, A., & Coull, B. A. (1998). Approximate is better than "exact" for interval estimation of binomial proportions. *The American Statistician, 52*, 119–126.

Aiken, L. S., West, S. G., & Millsap, R. E. (2008). Doctoral training in statistics, measurement, and methodology in psychology: Replication and extension of Aiken, West, Sechrest, and Reno's (1990) survey of PhD programs in North America. *American Psychologist, 63*, 32–50.

Aitkin, M., & Aitkin, I. (2005). Bayesian inference for factor scores. In A. Maydeu-Olivares & J. J. McArdle (Eds.), *Contemporary psychometrics: A Festschrift to Roderick P. McDonald* (pp. 207–222). Mahwah, NJ: Lawrence Erlbaum Associates.

Albert, J., & Chib, S. (1995). Bayesian residual analysis for binary response regression models. *Biometrika, 82*, 747–759.

Albert, J. H. (1992). Bayesian estimation of normal ogive item response curves using Gibbs sampling. *Journal of Educational and Behavioral Statistics, 17*, 251–269.

Albert, J. H., & Chib, S. (1993). Bayesian analysis of binary and polychotomous response data. *Journal of the American Statistical Association, 88*, 669.

Almond, R. G. (2010). I can name that Bayesian network in two matrixes. *International Journal of Approximate Reasoning, 51*, 167–178.

Almond, R. G. (2013). RNetica: R interface to Netica(R) Bayesian Network Engine (Version 0.3-4). Retrieved from http://ralmond.net/RNetica.

Almond, R. G., DiBello, L. V., Jenkins, F., Senturk, D., Mislevy, R. J., Steinberg, L. S., & Yan, D. (2001). Models for conditional probability tables in educational assessment. In *Artificial Intelligence and Statistics 2001: Proceedings of the 8th International Workshop* (pp. 137–143). San Francisco, CA: Morgan Kaufmann.

Almond, R. G., DiBello, L. V., Moulder, B., & Zapata-Rivera, J.-D. (2007). Modeling diagnostic assessments with Bayesian networks. *Journal of Educational Measurement, 44*, 341–359.

Almond, R. G., & Mislevy, R. J. (1999). Graphical models and computerized adaptive testing. *Applied Psychological Measurement, 23*, 223–237.

Almond, R. G., Mislevy, R. J., Steinberg, L. S., Yan, D., & Williamson, D. M. (2015). *Bayesian networks in educational assessment*. New York: Springer.

Almond, R. G., Mulder, J., Hemat, L. A., & Yan, D. (2009). Bayesian network models for local dependence among observable outcome variables. *Journal of Educational and Behavioral Statistics, 34*, 491–521.

Almond, R. G., Steinberg, L. S., & Mislevy, R. J. (2002). Enhancing the design and delivery of assessment systems: A four-process architecture. *The Journal of Technology, Learning and Assessment, 1*. Retrieved from https://ejournals.bc.edu/ojs/index.php/jtla/article/view/1671.

Andrews, M., & Baguley, T. (2013). Prior approval: The growth of Bayesian methods in psychology. *British Journal of Mathematical and Statistical Psychology, 66*, 1–7.

Andrich, D. (1978). A rating formulation for ordered response categories. *Psychometrika, 43*, 561–573.

Ansari, A., & Jedidi, K. (2000). Bayesian factor analysis for multilevel binary observations. *Psychometrika, 65*, 475–496.

Ansari, A., Jedidi, K., & Dube, L. (2002). Heterogeneous factor analysis models: A Bayesian approach. *Psychometrika, 67*, 49–77.

Arbuckle, J. L. (2007). *AMOS 16.0 user's guide*. Chicago, IL: SPSS Inc.

Arminger, G., & Muthén, B. O. (1998). A Bayesian approach to nonlinear latent variable models using the Gibbs sampler and the Metropolis-Hastings algorithm. *Psychometrika, 63*, 271–300.

Babcock, B. (2011). Estimating a noncompensatory IRT model using Metropolis within Gibbs sampling. *Applied Psychological Measurement, 35*, 317–329.

Bafumi, J., Gelman, A., Park, D. K., & Kaplan, N. (2005). Practical issues in implementing and understanding Bayesian ideal point estimation. *Political Analysis, 13*, 171–187.

Baldwin, P. (2011). A strategy for developing a common metric in item response theory when parameter posterior distributions are known. *Journal of Educational Measurement, 48*, 1–11.

Barnett, V. (1999). *Comparative statistical inference* (3rd ed.). Chichester, UK: Wiley.

Bartholomew, D. (1996). Comment on: Metaphor taken as math: Indeterminancy in the factor model. *Multivariate Behavioral Research, 31*, 551–554.

Bartholomew, D. J. (1981). Posterior analysis of the factor model. *British Journal of Mathematical and Statistical Psychology, 34*, 93–99.

Bartholomew, D. J., Knott, M., & Moustaki, I. (2011). *Latent variable models and factor analysis: A unified approach* (3rd ed.). Chichester, UK: Wiley.

Bayarri, M. J., & Berger, J. O. (2000a). Asymptotic distribution of p values in composite null models: Rejoinder. *Journal of the American Statistical Association, 95*, 1168.

Bayarri, M. J., & Berger, J. O. (2000b). P values for composite null models. *Journal of the American Statistical Association, 95*, 1127.

Beaton, A. E. (1987). *The NAEP 1983-1984 Technical Report*. Princeton, NJ: ETS.

Béguin, A. A., & Glas, C. A. (2001). MCMC estimation and some model-fit analysis of multidimensional IRT models. *Psychometrika, 66*, 541–561.

Behrens, J. T., & DiCerbo, K. E. (2013). *Technological implications for assessment ecosystems: Opportunities for digital technology to advance assessment*. Princeton, NJ: The Gordon Commission on the Future of Assessment. Retrieved from http://researchnetwork.pearson.com/wp-content/uploads/behrens_dicerbo_technlogical_implications_assessments.pdf.

Behrens, J. T., & DiCerbo, K. E. (2014). Harnessing the currents of the digital ocean. In J. A. Larusson & B. White (Eds.), *Learning analytics: From research to practice* (pp. 39–60). New York: Springer.

Behrens, J. T., DiCerbo, K. E., Yel, N., & Levy, R. (2013). Exploratory data analysis. In *Handbook of psychology, Volume 2: Research methods in psychology* (2nd ed., pp. 33–64). Hoboken, NJ: Wiley.

Behrens, J. T., Mislevy, R. J., Dicerbo, K. E., & Levy, R. (2012). Evidence centered design for learning and assessment in the digital world. In M. Mayrath, J. Clarke-Midura, & D. H. Robinson (Eds.), *Technology-based assessments for 21st century skills: Theoretical and practical implications from modern research* (pp. 13–53). Charlotte, NC: Information Age Publishing.

Bem, D. J. (2011). Feeling the future: Experimental evidence for anomalous retroactive influences on cognition and affect. *Journal of Personality and Social Psychology, 100*, 407–425.

Bentler, P. M. (2006). *EQS 6 structural equations program manual*. Encino, CA: Multivariate Software, Inc.

Berger, J. (2006). The case for objective Bayesian analysis. *Bayesian Analysis, 1*, 385–402.

Berger, J. O., & Berry, D. A. (1988). Statistical analysis and the illusion of objectivity. *American Scientist, 76*, 159–165.

Berkhof, J., Van Mechelen, I., & Gelman, A. (2003). A Bayesian approach to the selection and testing of mixture models. *Statistica Sinica, 13*, 423–442.

Bernardo, J. M., & Smith, A. F. M. (2000). *Bayesian theory*. Chichester, UK: Wiley.

Besag, J. (1974). Spatial interaction and the statistical analysis of lattice systems. *Journal of the Royal Statistical Society Series B, 36*, 192–236.

Blackwell, D., & Dubins, L. (1962). Merging of opinions with increasing information. *The Annals of Mathematical Statistics, 33*, 882–886.

Bock, R. D. (1972). Estimating item parameters and latent ability when responses are scored in two or more nominal categories. *Psychometrika, 37*, 29–51.

Bock, R. D., & Aitkin, M. (1981). Marginal maximum likelihood estimation of item parameters: Application of an EM algorithm. *Psychometrika, 46*, 443–459.

Bock, R. D., Gibbons, R., & Muraki, E. (1988). Full-information item factor analysis. *Applied Psychological Measurement, 12*, 261–280.

Bock, R. D., & Lieberman, M. (1970). Fitting a response model for *n* dichotomously scored items. *Psychometrika, 35,* 179–197.
Bock, R. D., & Mislevy, R. J. (1982). Adaptive EAP estimation of ability in a microcomputer environment. *Applied Psychological Measurement, 6,* 431–444.
Bock, R. D., & Moustaki, I. (2007). Item response theory in a general framework. In C. R. Rao & S. Sinharay (Eds.), *Handbook of statistics, Volume 26: Psychometrics* (pp. 469–514). Amsterdam, the Netherlands: North-Holland/Elsevier.
Bolfarine, H., & Bazan, J. L. (2010). Bayesian estimation of the logistic positive exponent IRT model. *Journal of Educational and Behavioral Statistics, 35,* 693–713.
Bollen, K. A. (1989). *Structural equations with latent variables.* New York: Wiley.
Bollen, K. A. (2002). Latent variables in psychology and the social sciences. *Annual Review of Psychology, 53,* 605–634.
Bollen, K. A., & Curran, P. J. (2006). *Latent curve models: A structural equation perspective.* Hoboken, NJ: Wiley.
Bollen, K. A., Ray, S., Zavisca, J., & Harden, J. J. (2012). A comparison of Bayes factor approximation methods including two new methods. *Sociological Methods & Research, 41,* 294–324.
Bolt, D. M., Cohen, A. S., & Wollack, J. A. (2001). A mixture item response model for multiple-choice data. *Journal of Educational and Behavioral Statistics, 26,* 381–409.
Bolt, D. M., Deng, S., & Lee, S. (2014). IRT model misspecification and measurement of growth in vertical scaling. *Journal of Educational Measurement, 51,* 141–162.
Bolt, D. M., & Lall, V. F. (2003). Estimation of compensatory and noncompensatory multidimensional item response models using Markov chain Monte Carlo. *Applied Psychological Measurement, 27,* 395–414.
Bolt, D. M., Wollack, J. A., & Suh, Y. (2012). Application of a multidimensional nested logit model to multiple-choice test items. *Psychometrika, 77,* 339–357.
Box, G. E. P. (1976). Science and statistics. *Journal of the American Statistical Association, 71,* 791–799.
Box, G. E. P. (1979). Some problems of statistics and everyday life. *Journal of the American Statistical Association, 74,* 1–4.
Box, G. E. P. (1980). Sampling and Bayes' inference in scientific modelling and robustness. *Journal of the Royal Statistical Society. Series A (General), 143,* 383.
Box, G. E. P. (1983). An apology for ecumenism in statistics. In G. E. P. Box, T. Leonard, & D. F. J. Wu (Eds.), *Scientific inference, data analysis, and robustness* (pp. 51–84). New York: Academic Press.
Box, G. E. P., & Draper, N. R. (1987). *Empirical model-building and response surfaces.* New York: Wiley.
Box, G. E. P., & Tiao, G. C. (1973). *Bayesian inference in statistical analysis.* Reading, MA: Addison-Wesley.
Bradlow, E. T., & Weiss, R. E. (2001). Outlier measures and norming methods for computerized adaptive tests. *Journal of Educational and Behavioral Statistics, 26,* 85–104.
Bradlow, E. T., Weiss, R. E., & Cho, M. (1998). Bayesian identification of outliers in computerized adaptive tests. *Journal of the American Statistical Association, 93,* 910–919.
Bradshaw, L., & Templin, J. (2014). Combining item response theory and diagnostic classification models: A psychometric model for scaling ability and diagnosing misconceptions. *Psychometrika, 79,* 403–425.
Briggs, D. C., & Wilson, M. (2007). Generalizability in item response modeling. *Journal of Educational Measurement, 44,* 131–155.
Brooks, S., Gelman, A., Jones, G., & Meng, X.-L. (Eds.). (2011). *Handbook of Markov chain Monte Carlo.* Boca Raton, FL: Chapman & Hall/CRC Press.
Brooks, S. P. (1998). Markov chain Monte Carlo method and its application. *The Statistician, 47,* 69–100.
Brooks, S. P., & Gelman, A. (1998). General methods for monitoring convergence of iterative simulations. *Journal of Computational and Graphical Statistics, 7,* 434–455.
Brown, T. A., & Moore, M. T. (2012). Confirmatory factor analysis. In R. H. Hoyle (Ed.), *Handbook of structural equation modeling* (pp. 43–55). New York: Guilford Press.
Burnham, K. P., & Anderson, D. R. (2002). *Model selection and multimodel inference: A practical information-theoretic approach* (2nd ed.). New York: Springer.

Cai, J.-H., & Song, X.-Y. (2010). Bayesian analysis of mixtures in structural equation models with non-ignorable missing data. *British Journal of Mathematical and Statistical Psychology, 63*, 491–508.
Cai, L. (2010). Metropolis-Hastings Robbins-Monro algorithm for confirmatory item factor analysis. *Journal of Educational and Behavioral Statistics, 35*, 307–335.
Camilli, G. (2006). Test fairness. In R. Brennan (Ed.), *Educational measurement* (4th ed., pp. 220–256). Westport, CT: Praeger.
Cao, J., Stokes, S. L., & Zhang, S. (2010). A Bayesian approach to ranking and rater evaluation: An application to grant reviews. *Journal of Educational and Behavioral Statistics, 35*, 194–214.
Carlin, B. P., & Chib, S. (1995). Bayesian model choice via Markov chain Monte Carlo methods. *Journal of the Royal Statistical Society. Series B (Methodological), 57*, 473–484.
Carlin, B. P., & Louis, T. A. (2008). *Bayesian methods for data analysis* (3rd ed.). Boca Raton, FL: Chapman & Hall/CRC Press.
Carroll, J. B. (1945). The effect of difficulty and chance success on correlations between items or between tests. *Psychometrika, 10*, 1–19.
Casella, G., & Berger, R. L. (2008). *Statistical inference*. Pacific Grove, CA: Thomson Press.
Casella, G., & George, E. I. (1992). Explaining the Gibbs sampler. *The American Statistician, 46*, 167–174.
Chang, Y.-W., Tsai, R.-C., & Hsu, N.-J. (2014). A speeded item response model: Leave the harder till later. *Psychometrika, 79*, 255–274.
Chen, W.-H., & Thissen, D. (1997). Local dependence indexes for item pairs using item response theory. *Journal of Educational and Behavioral Statistics, 22*, 265–289.
Cheng, Y. (2009). When cognitive diagnosis meets computerized adaptive testing: CD-CAT. *Psychometrika, 74*, 619–632.
Chib, S. (1995). Marginal likelihood from the Gibbs output. *Journal of the American Statistical Association, 90*, 1313–1321.
Chib, S., & Greenberg, E. (1995). Understanding the Metropolis-Hastings algorithm. *The American Statistician, 49*, 327.
Chib, S., & Jeliazkov, I. (2001). Marginal likelihood from the Metropolis-Hastings output. *Journal of the American Statistical Association, 96*, 270–281.
Cho, S.-J., & Cohen, A. S. (2010). A multilevel mixture IRT model with an application to DIF. *Journal of Educational and Behavioral Statistics, 35*, 336–370.
Cho, S.-J., Cohen, A. S., & Kim, S.-H. (2014). A mixture group bifactor model for binary responses. *Structural Equation Modeling: A Multidisciplinary Journal, 21*, 375–395.
Choi, I.-H., & Wilson, M. (2015). Multidimensional classification of examinees using the mixture random weights linear logistic test model. *Educational and Psychological Measurement, 75*, 78–101.
Choi, J., Levy, R., & Hancock, G. R. (2006). *Markov chain Monte Carlo estimation method with covariance data for structural equation modeling*. Presented at the annual meeting of the American Educational Research Association, San Francisco, CA.
Chow, S. L., O'Leary, R., & Mengersen, K. (2009). Elicitation by design in ecology: Using expert opinion to inform priors for Bayesian statistical models. *Ecology, 90*, 265–277.
Chow, S.-M., Tang, N., Yuan, Y., Song, X., & Zhu, H. (2011). Bayesian estimation of semiparametric nonlinear dynamic factor analysis models using the Dirichlet process prior. *British Journal of Mathematical and Statistical Psychology, 64*, 69–106.
Chung, G., Delacruz, G. C., Dionne, G. B., & Bewley, W. L. (2003). Linking assessment and instruction using ontologies. *Proc. I/ITSEC, 25*, 1811–1822.
Chung, G. K., W. K., Baker, E. L., Vendlinski, T. P., Buschang, R. E., Delacruz, G. C., Michiuye, J. K., & Bittick, S. J. (2010). *Testing instructional design variations in a prototype math game*. Presented at the annual meeting of the American Educational Research Association, Denver, CO.
Chung, H. (2003). *Latent-class modeling with covariates*. State College, PA: Pennsylvania State University.
Chung, H., & Anthony, J. C. (2013). A Bayesian approach to a multiple-group latent class-profile analysis: The timing of drinking onset and subsequent drinking behaviors among U.S. adolescents. *Structural Equation Modeling: A Multidisciplinary Journal, 20*, 658–680.

Chung, H., Flaherty, B. P., & Schafer, J. L. (2006). Latent class logistic regression: Application to marijuana use and attitudes among high school seniors. *Journal of the Royal Statistical Society: Series A (Statistics in Society), 169,* 723–743.

Chung, H., Lanza, S. T., & Loken, E. (2008). Latent transition analysis: Inference and estimation. *Statistics in Medicine, 27,* 1834–1854.

Chung, H., Loken, E., & Schafer, J. L. (2004). Difficulties in drawing inferences with finite-mixture models: A simple example with a simple solution. *The American Statistician, 58,* 152–158.

Chung, H., Walls, T. A., & Park, Y. (2007). A latent transition model with logistic regression. *Psychometrika, 72,* 413–435.

Chung, Y., Rabe-Hesketh, S., Dorie, V., Gelman, A., & Liu, J. (2013). A nondegenerate penalized likelihood estimator for variance parameters in multilevel models. *Psychometrika, 78,* 685–709.

Clark, J. S. (2005). Why environmental scientists are becoming Bayesians: Modelling with Bayes. *Ecology Letters, 8,* 2–14.

Clauser, B. E., Margolis, M. J., & Case, S. M. (2006). Testing for licensure in the professions. In R. Brennan (Ed.), *Educational measurement* (4th ed., pp. 701–731). Westport, CT: Praeger.

Clinton, J., Jackman, S., & Rivers, D. (2004). The statistical analysis of roll call data. *The American Political Science Review, 98,* 355–370.

Clogg, C. C., Rubin, D. B., Schenker, N., Schultz, B., & Weidman, L. (1991). Multiple imputation of industry and occupation codes in census public-use samples using Bayesian logistic regression. *Journal of the American Statistical Association, 86,* 68–78.

Cohen, A. S., & Bolt, D. M. (2005). A mixture model analysis of differential item functioning. *Journal of Educational Measurement, 42,* 133–148.

Cohen, A. S., & Wollack, J. A. (2006). Test administration, security, scoring, and reporting. In R. L. Brennan (Ed.), *Educational measurement* (4th ed., pp. 355–386). Westport, CT: Praeger.

Collins, J. A., Greer, J. E., & Huang, S. X. (1996). Adaptive assessment using granularity hierarchies and Bayesian nets. In C. Frasson, G. Gauthier, & A. Lesgold (Eds.), *Intelligent tutoring systems* (pp. 569–577). Berlin, Germany: Springer.

Collins, L. M., & Lanza, S. T. (2010). *Latent class and latent transition analysis: With applications in the social, behavioral, and health sciences.* Hoboken, NJ: Wiley.

Conati, C., Gertner, A. S., VanLehn, K., & Druzdzel, M. J. (1997). On-line student modeling for coached problem solving using Bayesian networks. In *User Modeling: Proceedings of the 6th International Conference, UM97* (pp. 231–242). Berlin, Germany: Springer.

Congdon, P. (2006). *Bayesian statistical modelling* (2nd ed.). Chichester, UK: Wiley.

Cowles, M. K., & Carlin, B. P. (1996). Markov chain Monte Carlo convergence diagnostics: A comparative review. *Journal of the American Statistical Association, 91,* 883–904.

Crawford, A. V. (2014). *Posterior predictive model checking in Bayesian networks.* Unpublished doctoral dissertation. Arizona State University, Tempe, AZ.

Crocker, L., & Algina, J. (1986). *Introduction to classical and modern test theory.* New York: Cengage Learning.

Croon, M. (1990). Latent class analysis with ordered latent classes. *British Journal of Mathematical and Statistical Psychology, 43,* 171–192.

Curtis, S. M. (2010). BUGS code for item response theory. *Journal of Statistical Software, 36,* 1–34.

Curtis, S. M. (2015). Mcmcplots: Create Plots from MCMC Output (Version 0.4.2). BUGS code.

Dahl, F. A. (2006). On the conservativeness of posterior predictive *p*-values. *Statistics & Probability Letters, 76,* 1170–1174.

Dai, Y. (2013). A mixture Rasch model with a covariate: A simulation study via Bayesian Markov chain Monte Carlo estimation. *Applied Psychological Measurement, 37,* 375–396.

Data Description, Inc. (2011). *Data Desk 6.3.* Ithaca, NY: Data Description.

Davey, T., Oshima, T. C., & Lee, K. (1996). Linking multidimensional item calibrations. *Applied Psychological Measurement, 20,* 405–416.

Dayton, C. M. (1999). *Latent class scaling analysis.* Thousand Oaks, CA: SAGE Publications.

Dayton, C. M., & Macready, G. B. (1976). A probabilistic model for validation of behavioral hierarchies. *Psychometrika, 41,* 189–204.

Dayton, C. M., & Macready, G. B. (2007). Latent class anlaysis in psychometrics. In C. R. Rao & S. Sinharay (Eds.), *Handbook of statistics, Volume 26: Psychometrics* (pp. 421–446). Amsterdam, the Netherlands: North-Holland/Elsevier.
De Ayala, R. J. (2009). *The theory and practice of item response theory*. New York: Guilford Press.
De Boeck, P., & Wilson, M. (Eds.). (2004). *Explanatory item response models: A generalized linear and nonlinear approach*. New York: Springer.
DeCarlo, L. T. (2011). On the analysis of fraction subtraction data: The DINA model, classification, latent class sizes, and the Q-matrix. *Applied Psychological Measurement, 35*, 8–26.
DeCarlo, L. T. (2012). Recognizing uncertainty in the Q-matrix via a Bayesian extension of the DINA model. *Applied Psychological Measurement, 36*, 447–468.
Deely, J. J., & Lindley, D. V. (1981). Bayes empirical Bayes. *Journal of the American Statistical Association, 76*, 833–841.
De Finetti, B. (1931). Funzione caratteristica di un fenomeno aleatorio. *Atti Della R. Academia Nazionale Dei Lincei, Serie 6. Memorie, Classe Di Scienze Fisiche, Mathematice E Naturale, 4*, 251–299.
De Finetti, B. (1937/1964). La prévision: Ses lois logiques, ses sources subjectives. In *Annales de l'Institut Henri Poincaré 7* (pp. 1–68). Translated by Kyburg and Smokler, eds. (1964). *Studies in subjective probability* (pp. 93–158). New York: Wiley.
De Finetti, B. (1974). *Theory of probability, Volume 1*. New York: Wiley.
De la Torre, J. (2009). Improving the quality of ability estimates through multidimensional scoring and incorporation of ancillary variables. *Applied Psychological Measurement, 33*, 465–485.
De la Torre, J., & Douglas, J. A. (2004). Higher-order latent trait models for cognitive diagnosis. *Psychometrika, 69*, 333–353.
De la Torre, J., & Douglas, J. A. (2008). Model evaluation and multiple strategies in cognitive diagnosis: An analysis of fraction subtraction data. *Psychometrika, 73*, 595–624.
De la Torre, J., & Patz, R. J. (2005). Making the most of what we have: A practical application of multidimensional item response theory in test scoring. *Journal of Educational and Behavioral Statistics, 30*, 295–311.
De la Torre, J., & Song, H. (2009). Simultaneous estimation of overall and domain abilities: A higher-order IRT model approach. *Applied Psychological Measurement, 33*, 620–639.
De la Torre, J., Stark, S., & Chernyshenko, O. S. (2006). Markov chain Monte Carlo estimation of item parameters for the generalized graded unfolding model. *Applied Psychological Measurement, 30*, 216–232.
De Leeuw, C., & Klugkist, I. (2012). Augmenting data with published results in Bayesian linear regression. *Multivariate Behavioral Research, 47*, 369–391.
DeMark, S. F., & Behrens, J. T. (2004). Using statistical natural language processing for understanding complex responses to free-response tasks. *International Journal of Testing, 4*, 371–390.
Dempster, A. P., Laird, N. M., & Rubin, D. B. (1977). Maximum likelihood from incomplete data via the em algorithm. *Journal of the Royal Statistical Society. Series B (Methodological), 39*, 1–38.
Depaoli, S. (2012). The ability for posterior predictive checking to identify model misspecification in Bayesian growth mixture modeling. *Structural Equation Modeling: A Multidisciplinary Journal, 19*, 534–560.
Depaoli, S. (2013). Mixture class recovery in GMM under varying degrees of class separation: Frequentist versus Bayesian estimation. *Psychological Methods, 18*, 186–219.
Diaconis, P., & Freedman, D. (1980a). de Finetti's generalizations of exchangeability. In R. C. Jeffrey (Ed.), *Studies in inductive logic and probability* (Vol. 2, pp. 233–249). Berkeley, CA: University of California Press.
Diaconis, P., & Freedman, D. (1980b). Exchangeable sequences. *The Annals of Probability, 8*, 745–764.
Diaconis, P., & Freedman, D. (1986). On the consistency of Bayes estimates. *The Annals of Statistics, 14*, 1–26.
DiBello, L. V., Henson, R. A., & Stout, W. F. (2015). A family of generalized diagnostic classification models for multiple choice option-based scoring. *Applied Psychological Measurement, 39*, 62–79.

DiBello, L. V., Roussos, L., & Stout, W. (2007). Review of cognitively diagnostic assessment and a summary of psychometric models. In C. R. Rao & S. Sinharay (Eds.), *Handbook of statistics, Volume 26: Psychometrics* (pp. 979–1030). Amsterdam, the Netherlands: North-Holland/Elsevier.

Dicerbo, K. E., & Behrens, J. T. (2012). Implications of the digital ocean on current and future assessment. In R. W. Lissitz & H. Jiao (Eds.), *Computers and their impact on state assessment: Recent history and predictions for the future* (pp. 273–306). Charlotte, NC: Information Age Publishing.

Drasgow, F., Luecht, R. M., & Bennett, R. E. (2006). Technology and testing. In R. L. Brennan (Ed.), *Educational Measurement* (4th ed., pp. 471–515). Westport, CT: Praeger.

DuBois, P. H. (1970). *A history of psychological testing*. Needham Heights, MA: Allyn & Bacon.

Dudek, F. J. (1979). The continuing misinterpretation of the standard error of measurement. *Psychological Bulletin, 86*, 335–337.

Duncan, K. A., & MacEachern, S. N. (2008). Nonparametric Bayesian modelling for item response. *Statistical Modelling, 8*, 41–66.

Dunson, D. B. (2000). Bayesian latent variable models for clustered mixed outcomes. *Journal of the Royal Statistical Society: Series B (Statistical Methodology), 62*, 355–366.

Dunson, D. B., Palomo, J., & Bollen, K. (2005). *Bayesian structural equation modeling* (No. 2005-5). Research Triangle Park, NC: Statistical and Applied Mathematical Sciences Institute.

Edwards, M. C. (2010). A Markov chain Monte Carlo approach to confirmatory item factor analysis. *Psychometrika, 75*, 474–497.

Edwards, M. C. (2013). Purple unicorns, true models, and other things I've never seen. *Measurement: Interdisciplinary Research & Perspective, 11*, 107–111.

Edwards, M. C., & Vevea, J. L. (2006). An empirical Bayes approach to subscore augmentation: How much strength can we borrow? *Journal of Educational and Behavioral Statistics, 31*, 241–259.

Edwards, W., Lindman, H., & Savage, L. J. (1963). Bayesian statistical inference for psychological research. *Psychological Review, 70*, 193–242.

Efron, B., & Morris, C. (1977). Stein's paradox in statistics. *Scientific American, 236*, 119–127.

Embretson, S. E. (1997). Multicomponent response models. In W. J. van der Linden & R. K. Hambleton (Eds.), *Handbook of modern item response theory* (pp. 305–321). New York: Springer.

Embretson, S. (1984). A general latent trait model for response processes. *Psychometrika, 49*, 175–186.

Embretson, S. E., & Reise, S. P. (2000). *Item response theory for psychologists*. Mahwah, NJ: Psychology Press.

Emons, W. H. M., Glas, C. A. W., Meijer, R. R., & Sijtsma, K. (2003). Person fit in order-restricted latent class models. *Applied Psychological Measurement, 27*, 459–478.

Enders, C. K. (2010). *Applied missing data analysis*. New York: Guilford Press.

Erosheva, E. A., & Curtis, S. M. (2011). *Dealing with rotational invariance in Bayesian confirmatory factor analysis* (Technical Report no. 589). Seattle, WA: University of Washington. Retrieved from http://citeseerx.ist.psu.edu/viewdoc/download?doi=10.1.1.300.8292&rep=rep1&type=pdf.

Evans, M. (1997). Bayesian inference procedures derived via the concept of relative surprise. *Communications in Statistics—Theory and Methods, 26*, 1125–1143.

Evans, M. (2000). Asymptotic distribution of p values in composite null models: Comment. *Journal of the American Statistical Association, 95*, 1160.

Evans, M. J., Gilula, Z., & Guttman, I. (1989). Latent class analysis of two-way contingency tables by Bayesian methods. *Biometrika, 76*, 557–563.

Fahrmeir, L., & Raach, A. (2007). A Bayesian semiparametric latent variable model for mixed responses. *Psychometrika, 72*, 327–346.

Fan, X., & Sivo, S. A. (2005). Sensitivity of fit indexes to misspecified structural or measurement model components: Rationale of two-index strategy revisited. *Structural Equation Modeling: A Multidisciplinary Journal, 12*, 343–367.

Fan, X., & Sivo, S. A. (2007). Sensitivity of fit indices to model misspecification and model types. *Multivariate Behavioral Research, 42*, 509–529.

Finch, W. H., & French, B. F. (2012). Parameter estimation with mixture item response theory models: A Monte Carlo comparison of maximum likelihood and Bayesian methods. *Journal of Modern Applied Statistical Methods, 11*, 167–178.
Formann, A. K. (1985). Constrained latent class models: Theory and applications. *British Journal of Mathematical and Statistical Psychology, 38*, 87–111.
Formann, A. K. (1992). Linear logistic latent class analysis for polytomous data. *Journal of the American Statistical Association, 87*, 476–486.
Fox, J.-P. (2003). Stochastic EM for estimating the parameters of a multilevel IRT model. *British Journal of Mathematical and Statistical Psychology, 56*, 65–81.
Fox, J.-P. (2005a). Multilevel IRT using dichotomous and polytomous response data. *British Journal of Mathematical and Statistical Psychology, 58*, 145–172.
Fox, J.-P. (2005b). Randomized item response theory models. *Journal of Educational and Behavioral Statistics, 30*, 189–212.
Fox, J.-P. (2010). *Bayesian item response modeling: Theory and applications.* Springer.
Fox, J.-P., Entink Klein, R., & Avetisyan, M. (2014). Compensatory and non-compensatory multidimensional randomized item response models. *British Journal of Mathematical and Statistical Psychology, 67*, 133–152.
Fox, J.-P., & Glas, C. A. (2001). Bayesian estimation of a multilevel IRT model using Gibbs sampling. *Psychometrika, 66*, 271–288.
Fox, J.-P., & Glas, C. A. (2003). Bayesian modeling of measurement error in predictor variables using item response theory. *Psychometrika, 68*, 169–191.
Frederickx, S., Tuerlinckx, F., De Boeck, P., & Magis, D. (2010). RIM: A random item mixture model to detect differential item functioning. *Journal of Educational Measurement, 47*, 432–457.
Freedman, D. A. (1987). As others see us: A case study in path analysis. *Journal of Educational Statistics, 12*, 101–128.
French, B. F., & Oakes, W. (2004). Reliability and validity evidence for the institutional integration scale. *Educational and Psychological Measurement, 64*, 88–98.
Fu, Z.-H., Tao, J., & Shi, N.-Z. (2009). Bayesian estimation in the multidimensional three-parameter logistic model. *Journal of Statistical Computation and Simulation, 79*, 819–835.
Fu, Z.-H., Tao, J., & Shi, N.-Z. (2010). Bayesian estimation of the multidimensional graded response model with nonignorable missing data. *Journal of Statistical Computation and Simulation, 80*, 1237–1252.
Fujimoto, K. A., & Karabatsos, G. (2014). Dependent Dirichlet process rating model. *Applied Psychological Measurement, 38*, 217–228.
Fukuhara, H., & Kamata, A. (2011). A bifactor multidimensional item response theory model for differential item functioning analysis on testlet-based items. *Applied Psychological Measurement, 35*, 604–622.
Gajewski, B. J., Price, L. R., Coffland, V., Boyle, D. K., & Bott, M. J. (2013). Integrated analysis of content and construct validity of psychometric instruments. *Quality & Quantity, 47*, 57–78.
Galindo-Garre, F., Vermunt, J. K., & Bergsma, W. P. (2004). Bayesian posterior estimation of logit parameters with small samples. *Sociological Methods & Research, 33*, 88–117.
Garrett, E. S., Eaton, W. W., & Zeger, S. (2002). Methods for evaluating the performance of diagnostic tests in the absence of a gold standard: A latent class model approach. *Statistics in Medicine, 21*, 1289–1307.
Garrett, E. S., & Zeger, S. L. (2000). Latent class model diagnosis. *Biometrics, 56*, 1055–1067.
Garthwaite, P. H., Kadane, J. B., & O'Hagan, A. (2005). Statistical methods for eliciting probability distributions. *Journal of the American Statistical Association, 100*, 680–701.
Geerlings, H., Glas, C. A., & van der Linden, W. J. (2011). Modeling rule-based item generation. *Psychometrika, 76*, 337–359.
Geisser, S., & Eddy, W. F. (1979). A predictive approach to model selection. *Journal of the American Statistical Association, 74*, 153–160.
Gelfand, A. E. (1996). Model determination using sampling based methods. In W. R. Gilks, S. Richardson, & D. J. Spiegelhalter (Eds.), *Markov chain Monte Carlo in practice* (pp. 145–161). London: Chapman & Hall/CRC Press.

Gelfand, A. E., & Smith, A. F. M. (1990). Sampling-based approaches to calculating marginal densities. *Journal of the American Statistical Association, 85*, 398–409.

Gelfand, A. E., Smith, A. F. M., & Lee, T.-M. (1992). Bayesian analysis of constrained parameter and truncated data problems using Gibbs sampling. *Journal of the American Statistical Association, 87*, 523–532.

Gelman, A. (2003). A Bayesian formulation of exploratory data analysis and goodness-of-fit testing. *International Statistical Review/Revue Internationale de Statistique, 71*, 369–382.

Gelman, A. (2004). Exploratory data analysis for complex models. *Journal of Computational and Graphical Statistics, 13*, 755–779.

Gelman, A. (2007). Comment: Bayesian checking of the second levels of hierarchical models. *Statistical Science, 22*, 349–352.

Gelman, A. (2011). Induction and deduction in Bayesian data analysis. *Rationality, Markets and Morals, 2*, 67–88.

Gelman, A., Carlin, J. B., Stern, H. S., Dunson, D. B., Vehtari, A., & Rubin, D. B. (2013). *Bayesian data analysis* (3rd ed.). Boca Raton, FL: Chapman & Hall/CRC Press.

Gelman, A., & Hill, J. (2007). *Data analysis using regression and multilevel/hierarchical models.* Cambridge: Cambridge University Press.

Gelman, A., Jakulin, A., Pittau, M. G., & Su, Y.-S. (2008). A weakly informative default prior distribution for logistic and other regression models. *The Annals of Applied Statistics, 2*, 1360–1383.

Gelman, A., Meng, X.-L., & Stern, H. (1996). Posterior predictive assessment of model fitness via realized discrepancies. *Statistica Sinica, 6*, 733–760.

Gelman, A., & Rubin, D. B. (1992). Inference from iterative simulation using multiple sequences. *Statistical Science, 7*, 457–511.

Gelman, A., & Shalizi, C. R. (2013). Philosophy and the practice of Bayesian statistics. *British Journal of Mathematical and Statistical Psychology, 66*, 8–38.

Geman, S., & Geman, D. (1984). Stochastic relaxation, Gibbs distributions, and the Bayesian restoration of images. *IEEE Transactions on Pattern Analysis and Machine Intelligence, 6*, 721–741.

Geweke, J. (1992). Evaluating the accuracy of sampling-based approaches to the calculation of posterior moments. In J. M. Bernardo, J. O. Berger, A. P. Dawid, & A. F. M. Smith (Eds.), *Bayesian statistics 4* (pp. 169–193). Oxford, UK: Oxford University Press.

Geweke, J., & Zhou, G. (1996). Measuring the pricing error of the arbitrage pricing theory. *The Review of Financial Studies, 9*, 557–587.

Geyer, C. T., & Thompson, E. A. (1992). Constrained Monte Carlo maximum likelihood for dependent data. *Journal of the Royal Statistical Society, Series B, 54*, 657–699.

Ghosh, J., & Dunson, D. B. (2009). Default prior distributions and efficient posterior computation in Bayesian factor analysis. *Journal of Computational and Graphical Statistics, 18*, 306–320.

Gigerenzer, G. (1993). The superego, the ego, and the id in statistical reasoning. In G. Keren & C. Lewis (Eds.), *A handbook for data analysis in the behavioral sciences: Methodological issues* (pp. 311–339). Hillsdale, NJ: Lawrence Erlbaum.

Gigerenzer, G. (2002). *Calculated risks: How to know when numbers deceive you.* New York: Simon & Schuster.

Gilks, W. R., Richardson, S., & Spiegelhalter, D. J. (1996a). Introducing Markov chain Monte Carlo. In W. R. Gilks, S. Richardson, & D. J. Spiegelhalter (Eds.), *Markov chain Monte Carlo in practice* (pp. 1–19). London: Chapman & Hall/CRC Press.

Gilks, W. R., Richardson, S., & Spiegelhalter, D. (Eds.). (1996b). *Markov chain Monte Carlo in practice.* Boca Raton, FL: Chapman & Hall/CRC Press.

Gilks, W. R., & Wild, P. (1992). Adaptive rejection sampling for Gibbs sampling. *Journal of the Royal Statistical Society. Series C (Applied Statistics), 41*, 337–348.

Gill, J. (2007). *Bayesian methods: A social and behavioral sciences approach* (2nd ed.). Boca Raton, FL: Chapman & Hall/CRC Press.

Gill, J. (2012). BaM: Functions and datasets for books by Jeff Gill (Version 0.99). Retrieved from http://CRAN.R-project.org/package=BaM.

Gilula, Z., & Haberman, S. J. (2001). Analysis of categorical response profiles by informative summaries. *Sociological Methodology, 31*, 129–187.
Glas, C. A. W., & Meijer, R. R. (2003). A Bayesian approach to person fit analysis in item response theory models. *Applied Psychological Measurement, 27*, 217–233.
Goldstein, H., & Browne, W. (2002). Multilevel factor analysis modelling using Markov chain Monte Carlo estimation. In G. A. Marcoulides & I. Moustaki (Eds.), *Latent variable and latent structure models* (pp. 225–243). London: Lawrence Erlbaum Associates.
Goldstein, H., & Browne, W. (2005). Multilevel factor analysis models for continuous and discrete data. In A. Maydeu-Olivares & J. J. McArdle (Eds.), *Contemporary psychometrics: A Festschrift to Roderick P. McDonald.* (pp. 453–475). Mahwah, NJ: Lawrence Erlbaum Associates.
Goldstein, M. (1976). Bayesian analysis of regression problems. *Biometrika, 63*, 51–58.
Good, I. J. (1965). *The estimation of probabilities: An essay on modern Bayesian methods*. Cambridge, MA: MIT Press.
Good, I. J. (1971). 46656 varieties of Bayesians. *American Statistician, 25*, 62–63.
Goodman, L. A. (1974). Exploratory latent structure analysis using both identifiable and unidentifiable models. *Biometrika, 61*, 215–231.
Goodman, S. (2008). A dirty dozen: Twelve p-value misconceptions. *Seminars in Hematology, 45*, 135–140.
Gorsuch, R. L. (1983). *Factor analysis* (2nd ed.). Hillsdale, NJ: Lawrence Erlbaum Associates.
Guilford, J. P. (1936). *Psychometric methods*. New York: McGraw-Hill.
Gulliksen, H. (1961). Measurement of learning and mental abilities. *Psychometrika, 26*, 93–107.
Guttman, I. (1967). The use of the concept of a future observation in goodness-of-fit problems. *Journal of the Royal Statistical Society. Series B (Methodological), 29*, 83–100.
Guttman, L. (1947). On Festinger's evaluation of scale analysis. *Psychological Bulletin, 44*, 451–465.
Hacking, I. (1975). *The emergence of probability*. Cambridge: Cambridge University Press.
Haertel, E. H. (1990). Continuous and discrete latent structure models for item response data. *Psychometrika, 55*, 477–494.
Haertel, E. H. (2006). Reliability. In R. Brennan (Ed.), *Educational Measurement* (4th ed., pp. 65–110). Westport, CT: Praeger.
Hambleton, R. K. (1984). Determining suitable test lengths. In R. Berk (Ed.), *A guide to criterion-referenced test construction* (pp. 144–168). Baltimore, MD: The Johns Hopkins University Press.
Hambleton, R. K., & Han, N. (2005). Assessing the fit of IRT models to educational and psychological test data: A five-step plan and several graphical displays. In W. R. Lenderking & D. A. Revicki (Eds.), *Advancing health outcomes research methods and clinical applications* (pp. 57–77). Washington, DC: Degnon Associates.
Hambleton, R. K., & Jones, R. W. (1994). Item parameter estimation errors and their influence on test information functions. *Applied Measurement in Education, 7*, 171–186.
Hambleton, R. K., Jones, R. W., & Rogers, H. J. (1993). Influence of item parameter estimation errors in test development. *Journal of Educational Measurement, 30*, 143–155.
Hambleton, R. K., & Swaminathan, H. (1985). *Item response theory: Principles and applications*. Hingham, MA: Springer.
Hambleton, R. K., Swaminathan, H., Cook, L. L., Eignor, D. R., & Gifford, J. A. (1978). Developments in latent trait theory: Models, technical issues, and applications. *Review of Educational Research, 48*, 467–510.
Han, C., & Carlin, B. P. (2001). Markov chain Monte Carlo methods for computing Bayes factors: A comparative review. *Journal of the American Statistical Association, 96*, 1122–1132.
Harring, J. R., Weiss, B. A., & Hsu, J.-C. (2012). A comparison of methods for estimating quadratic effects in nonlinear structural equation models. *Psychological Methods, 17*, 193–214.
Hartigan, J. A. (1969). Linear Bayesian methods. *Journal of the Royal Statistical Society. Series B (Methodological), 31*, 446–454.
Harwell, M. R., & Baker, F. B. (1991). The use of prior distributions in marginalized Bayesian item parameter estimation: A didactic. *Applied Psychological Measurement, 15*, 375–389.

Harwell, M. R., Baker, F. B., & Zwarts, M. (1988). Item parameter estimation via marginal maximum likelihood and an EM algorithm: A didactic. *Journal of Educational Statistics, 13,* 243–271.
Hastings, W. K. (1970). Monte Carlo sampling methods using Markov chains and their applications. *Biometrika, 57,* 97–109.
Hayashi, K., & Arav, M. (2006). Bayesian factor analysis when only a sample covariance matrix is available. *Educational and Psychological Measurement, 66,* 272–284.
Hayashi, K., & Yuan, K.-H. (2003). Robust Bayesian factor analysis. *Structural Equation Modeling: A Multidisciplinary Journal, 10,* 525–533.
Heinen, T. (1996). *Latent class and discrete latent trait models: Similarities and differences.* Thousand Oaks, CA: SAGE Publications.
Henson, R. A., Templin, J. L., & Willse, J. T. (2009). Defining a family of cognitive diagnosis models using log-linear models with latent variables. *Psychometrika, 74,* 191–210.
Hewitt, E., & Savage, L. J. (1955). Symmetric measures on Cartesian products. *Transactions of the American Mathematical Society, 80,* 470–501.
Hjort, N. L., Dahl, F. A., & Steinbakk, G. H. (2006). Post-processing posterior predictive p values. *Journal of the American Statistical Association, 101,* 1157–1174.
Ho, M. R., Stark, S., & Chernyshenko, O. S. (2012). Graphical representation of structural equation models using path diagrams. In R. H. Hoyle (Ed.), *Handbook of structural equation modeling* (pp. 43–55). New York: Guilford Press.
Hoeting, J. A., Madigan, D., Raftery, A. E., & Volinsky, C. T. (1999). Bayesian model averaging: A tutorial. *Statistical Science,* 382–401.
Hoijtink, H. (1998). Constrained latent class analysis using the Gibbs sampler and posterior predictive p-values: Applications to educational testing. *Statistica Sinica, 8,* 691–711.
Hoijtink, H., Béland, S., & Vermeulen, J. A. (2014). Cognitive diagnostic assessment via Bayesian evaluation of informative diagnostic hypotheses. *Psychological Methods, 19,* 21–38.
Hojtink, H., & Molenaar, I. W. (1997). A multidimensional item response model: Constrained latent class analysis using the Gibbs sampler and posterior predictive checks. *Psychometrika, 62,* 171–189.
Holland, P. W. (1990). On the sampling theory foundations of item response theory models. *Psychometrika, 55,* 577–601.
Holland, P. W. (1994). Measurements or contests? Comments on Zwick, Bond and Allen/Donoghue. In *Proceedings of the Social Statistics Section of the American Statistical Association* (pp. 27–29). Alexandria, VA: American Statistical Association.
Hong, H., Wang, C., Lim, Y. S., & Douglas, J. (2015). Efficient models for cognitive diagnosis with continuous and mixed-type latent variables. *Applied Psychological Measurement, 39,* 31–43.
Hox, J., van de Schoot, R., & Matthijsse, S. (2012). How few countries will do? Comparative survey analysis from a Bayesian perspective. *Survey Research Methods,* 6, 87–93.
Hsieh, C.-A., von Eye, A., Maier, K., Hsieh, H.-J., & Chen, S.-H. (2013). A unified latent growth curve model. *Structural Equation Modeling: A Multidisciplinary Journal, 20,* 592–615.
Huang, H.-Y., & Wang, W.-C. (2014a). Multilevel higher-order item response theory models. *Educational and Psychological Measurement, 74,* 495–515.
Huang, H.-Y., & Wang, W.-C. (2014b). The random-effect DINA model. *Journal of Educational Measurement, 51,* 75–97.
Hughes, R. I. (1997). Models and representation. *Philosophy of Science, 64,* S325–S336.
Hu, L., & Bentler, P. M. (1999). Cutoff criteria for fit indexes in covariance structure analysis: Conventional criteria versus new alternatives. *Structural Equation Modeling: A Multidisciplinary Journal, 6,* 1–55.
Hulin, C. L., Lissak, R. I., & Drasgow, F. (1982). Recovery of two- and three-parameter logistic item characteristic curves: A Monte Carlo study. *Applied Psychological Measurement, 6,* 249–260.
IPCC. (2014). *Climate change 2013: The physical science basis. Contribution of Working Group I to the Fifth Assessment Report of the Intergovernmental Panel on Climate Change.* Cambridge: Cambridge University Press.

Iseli, M. R., Koenig, A. D., Lee, J. J., & Wainess, R. (2010). *Automated assessment of complex task performance in games and simulations* (No. 775). Los Angeles, CA: University of California, National Center for Research on Evaluation, Standards, Student Testing (CRESST). Retrieved from https://www.cse.ucla.edu/products/reports/R775.pdf.

Jackman, S. (2001). Multidimensional analysis of roll call data via Bayesian simulation: Identification, estimation, inference, and model checking. *Political Analysis, 9*, 227–241.

Jackman, S. (2009). *Bayesian analysis for the social sciences*. Chichester, UK: Wiley.

Jackman, S. (2014). pscl: Classes and Methods for R developed in the Political Science Computational Laboratory, Stanford University (Version 1.4.6). Stanford, CA: Department of Political Science, Stanford University. Retrieved from http://pscl.stanford.edu/.

Jackson, P. H., Novick, M. R., & Thayer, D. T. (1971). Estimating regressions in m groups. *British Journal of Mathematical and Statistical Psychology, 24*, 129–153.

James, W., & Stein, C. (1961). Estimation with quadratic loss. In *Proceedings of the 4th Berkeley Symposium on Mathematical Statistics and Probability* (Vol. 1, pp. 361–379). Berkeley and Los Angeles, CA: University of California Press.

Janssen, R., Tuerlinckx, F., Meulders, M., & De Boeck, P. (2000). A hierarchical IRT model for criterion-referenced measurement. *Journal of Educational and Behavioral Statistics, 25*, 285–306.

Jaynes, E. T. (1988). The relation of Bayesian and maximum entropy methods. In G. J. Erickson & C. R. Smith (Eds.), *Maximum-entropy and Bayesian methods in science and engineering* (Vol. 1, pp. 25–29). Dordrecht, the Netherlands: Kluwer.

Jaynes, E. T. (2003). *Probability theory: The logic of science*. (G. L. Bretthorst, Ed.). Cambridge: Cambridge University Press.

Jeffreys, H. (1961). *Theory of probability* (3rd ed.). Oxford, UK: Clarendon Press.

Jensen, F. V. (1996). *Introduction to Bayesian networks*. New York: Springer.

Jensen, F. V. (2001). *Bayesian networks and decision graphs*. New York: Springer.

Jiang, Y., Boyle, D. K., Bott, M. J., Wick, J. A., Yu, Q., & Gajewski, B. J. (2014). Expediting clinical and translational research via Bayesian instrument development. *Applied Psychological Measurement, 38*, 296–310.

Jiao, H., & Zhang, Y. (2015). Polytomous multilevel testlet models for testlet-based assessments with complex sampling designs. *British Journal of Mathematical and Statistical Psychology, 68*, 65–83.

Jin, K.-Y., & Wang, W.-C. (2014). Generalized IRT models for extreme response style. *Educational and Psychological Measurement, 74*, 116–138.

Johnson, M. S., & Jenkins, F. (2005). *A Bayesian hierarchical model for large-scale educational surveys: An application to the National Assessment of Educational Progress* (Research Report No. RR-04-38). Princeton, NJ: ETS. Retrieved from http://onlinelibrary.wiley.com/doi/10.1002/j.2333-8504.2004.tb01965.x/abstract.

Johnson, M. S., & Junker, B. W. (2003). Using data augmentation and Markov chain Monte Carlo for the estimation of unfolding response models. *Journal of Educational and Behavioral Statistics, 28*, 195–230.

Johnson, M. S., & Sinharay, S. (2005). Calibration of polytomous item families using Bayesian hierarchical modeling. *Applied Psychological Measurement, 29*, 369–400.

Johnson, V. E. (1996). On Bayesian analysis of multirater ordinal data: An application to automated essay grading. *Journal of the American Statistical Association, 91*, 42–51.

Johnson, V. E. (2007). Bayesian model assessment using pivotal quantities. *Bayesian Analysis, 2*, 719–733.

Jones, L. V., & Thissen, D. (2007). A history and overview of psychometrics. In C. R. Rao & S. Sinharay (Eds.), *Handbook of statistics, Volume 26: Psychometrics* (pp. 1–27). Amsterdam, the Netherlands: North-Holland/Elsevier.

Jones, W. P. (2014). Enhancing a short measure of big five personality traits with Bayesian scaling. *Educational and Psychological Measurement, 74*, 1049–1066.

Kadane, J. B. (2011). *Principles of uncertainty*. Boca Raton, FL: Chapman & Hall/CRC Press.

Kadane, J. B., & Wolfson, L. J. (1998). Experiences in elicitation. *The Statistician, 47*, 3–19.

Kamata, A., & Bauer, D. J. (2008). A note on the relation between factor analytic and item response theory models. *Structural Equation Modeling: A Multidisciplinary Journal, 15*, 136–153.
Kang, T., & Cohen, A. S. (2007). IRT model selection methods for dichotomous items. *Applied Psychological Measurement, 31*, 331–358.
Kang, T., Cohen, A. S., & Sung, H.-J. (2009). Model selection indices for polytomous items. *Applied Psychological Measurement, 33*, 499–518.
Kaplan, D. (2014). *Bayesian statistics for the social sciences*. New York: Guilford Press.
Kaplan, D., & Depaoli, S. (2012). Bayesian structural equation modeling. In R. H. Hoyle (Ed.), *Handbook of structural equation modeling* (pp. 650–673). New York: Guilford Press.
Karabatsos, G., & Batchelder, W. H. (2003). Markov chain estimation for test theory without an answer key. *Psychometrika, 68*, 373–389.
Karabatsos, G., & Sheu, C.-F. (2004). Order-constrained Bayes inference for dichotomous models of unidimensional nonparametric IRT. *Applied Psychological Measurement, 28*, 110–125.
Karabatsos, G. (2016). Bayesian nonparametric IRT. In W. J. van der Linden (Ed.), *Handbook of item response theory: Models, statistical tools, and applications, volume 1*. New York: Chapman & Hall/CRC Press.
Karabatsos, G., & Walker, S. G. (2009). A Bayesian nonparametric approach to test equating. *Psychometrika, 74*, 211–232.
Kass, R. E., & Raftery, A. E. (1995). Bayes factors. *Journal of the American Statistical Association, 90*, 773–795.
Kelley, T. L. (1923). *Statistical method*. New York: Macmillan.
Kelley, T. L. (1947). *Fundamentals of statistics*. Cambridge, MA: Harvard University Press.
Kim, S.-H. (2001). An evaluation of a Markov chain Monte Carlo method for the Rasch model. *Applied Psychological Measurement, 25*, 163–176.
Kim, S.-Y., Suh, Y., Kim, J.-S., Albanese, M. A., & Langer, M. M. (2013). Single and multiple ability estimation in the SEM framework: A noninformative Bayesian estimation approach. *Multivariate Behavioral Research, 48*, 563–591.
Klein, M. F., Birnbaum, M., Standiford, S. N., & Tatsuoka, K. K. (1981). *Logical error analysis and construction of tests to diagnose student "bugs" in addition and subtraction of fractions* (Research Report No. 81-6). Urbana, IL: University of Illinois, Computer-Based Education Research Laboratory.
Klein Entink, R. H., Fox, J.-P., & van der Linden, W. J. (2009). A multivariate multilevel approach to the modeling of accuracy and speed of test takers. *Psychometrika, 74*, 21–48.
Kline, R. B. (2010). *Principles and practice of structural equation modeling* (3rd ed.). New York: Guilford Press.
Koopman, R. F. (1978). On Bayesian estimation in unrestricted factor analysis. *Psychometrika, 43*, 109–110.
Kruschke, J. K. (2010). *Doing Bayesian data analysis: A tutorial with R and BUGS*. Burlington, MA: Academic Press.
Kruschke, J. K., Aguinis, H., & Joo, H. (2012). The time has come: Bayesian methods for data analysis in the organizational sciences. *Organizational Research Methods, 15*, 722–752.
Kyburg, H. E., & Smokler, H. E. (Eds.). (1964). *Studies in subjective probability*. New York: Wiley.
Lanza, S. T., Collins, L. M., Schafer, J. L., & Flaherty, B. P. (2005). Using data augmentation to obtain standard errors and conduct hypothesis tests in latent class and latent transition analysis. *Psychological Methods, 10*, 84–100.
Lauritzen, S. L., & Spiegelhalter, D. J. (1988). Local computations with probabilities on graphical structures and their application to expert systems. *Journal of the Royal Statistical Society. Series B (Methodological), 50*, 157–224.
Lazarsfeld, P. F., & Henry, N. W. (1968). *Latent structure analysis*. Boston, MA: Houghton Mifflin.
Lee, S. E., & Press, S. J. (1998). Robustness of Bayesian factor analysis estimates. *Communications in Statistics: Theory and Methods, 27*, 1871–1893.
Lee, S.-Y. (1981). A Bayesian approach to confirmatory factor analysis. *Psychometrika, 46*, 153–160.
Lee, S.-Y. (1992). Bayesian analysis of stochastic constraints in structural equation models. *British Journal of Mathematical and Statistical Psychology, 45*, 93–107.
Lee, S.-Y. (2006). Bayesian analysis of nonlinear structural equation models with nonignorable missing data. *Psychometrika, 71*, 541–564.

Lee, S.-Y. (2007). *Structural equation modeling: A Bayesian approach*. Chichester, UK: Wiley.
Lee, S.-Y., Lu, B., & Song, X.-Y. (2008). Semiparametric Bayesian analysis of structural equation models with fixed covariates. *Statistics in Medicine, 27*, 2341–2360.
Lee, S.-Y., & Song, X.-Y. (2003). Bayesian model selection for mixtures of structural equation models with an unknown number of components. *British Journal of Mathematical and Statistical Psychology, 56*, 145–165.
Lee, S.-Y., & Song, X.-Y. (2004). Evaluation of the Bayesian and maximum likelihood approaches in analyzing structural equation models with small sample sizes. *Multivariate Behavioral Research, 39*, 653–686.
Lee, S.-Y., Song, X.-Y., & Cai, J.-H. (2010). A Bayesian approach for nonlinear structural equation models with dichotomous variables using logit and probit links. *Structural Equation Modeling: A Multidisciplinary Journal, 17*, 280–302.
Lee, S.-Y., Song, X.-Y., & Tang, N.-S. (2007). Bayesian methods for analyzing structural equation models with covariates, interaction, and quadratic latent variables. *Structural Equation Modeling: A Multidisciplinary Journal, 14*, 404–434.
Lee, S.-Y., & Xia, Y.-M. (2008). A robust Bayesian approach for structural equation models with missing data. *Psychometrika, 73*, 343–364.
Lee, S.-Y., & Zhu, H.-T. (2000). Statistical analysis of nonlinear structural equation models with continuous and polytomous data. *British Journal of Mathematical and Statistical Psychology, 53*, 209–232.
Lee, S.-Y., & Zhu, H.-T. (2002). Maximum likelihood estimation of nonlinear structural equation models. *Psychometrika, 67*, 189–210.
Lei, P.-W., & Wu, Q. (2012). Estimation in structural equation modeling. In R. H. Hoyle (Ed.), *Handbook of structural equation modeling* (pp. 164–180). New York: Guilford Press.
Levy, R. (2006). *Posterior predictive model checking for multidimensionality in item response theory and Bayesian networks*. Unpublished doctoral dissertation. University of Maryland, College Park, MD.
Levy, R. (2009). The rise of Markov chain Monte Carlo estimation for psychometric modeling. *Journal of Probability and Statistics, 2009*, 1–18.
Levy, R. (2011). Bayesian data-model fit assessment for structural equation modeling. *Structural Equation Modeling: A Multidisciplinary Journal, 18*, 663–685.
Levy, R. (2013). Psychometric and evidentiary advances, opportunities, and challenges for simulation-based assessment. *Educational Assessment, 18*, 182–207.
Levy, R. (2014). *Dynamic Bayesian network modeling of game based diagnostic assessments* (No. 837). Los Angeles, CA: University of California, National Center for Research on Evaluation, Standards, Student Testing (CRESST). Retrieved from http://www.cse.ucla.edu/products/reports/R837.pdf.
Levy, R., Behrens, J. T., & Mislevy, R. J. (2006). Variations in adaptive testing and their on-line leverage points. In D. D. Williams, S. L. Howell, & M. Hricko (Eds.), *Online assessment, measurement and evaluation: Emerging practices* (pp. 180–202). Hershey, PA: Information Science Publishing.
Levy, R., & Choi, J. (2013). Bayesian structural equation modeling. In G.R. Hancock & R.O. Mueller (Eds.), *Structural equation modeling: A second course* (2nd ed., pp. 563–623). Charlotte, NC: Information Age Publishing.
Levy, R., & Crawford, A. V. (2009). Incorporating substantive knowledge into regression via a Bayesian approach to modeling. *Multiple Linear Regression Viewpoints, 35*, 4–9.
Levy, R., & Hancock, G. R. (2007). A framework of statistical tests for comparing mean and covariance structure models. *Multivariate Behavioral Research, 42*, 33–66.
Levy, R., & Hancock, G. R. (2011). An extended model comparison framework for covariance and mean structure models, accommodating multiple groups and latent mixtures. *Sociological Methods & Research, 40*, 256–278.
Levy, R., & Mislevy, R. J. (2004). Specifying and refining a measurement model for a computer-based interactive assessment. *International Journal of Testing, 4*, 333–369.

Levy, R., Mislevy, R. J., & Behrens, J. T. (2011). MCMC in educational research. In S. Brooks, A. Gelman, G. L. Jones, & X.-L. Meng (Eds.), *Handbook of Markov chain Monte Carlo: Methods and applications* (pp. 531–545). London: Chapman & Hall/CRC Press.

Levy, R., Mislevy, R. J., & Sinharay, S. (2009). Posterior predictive model checking for multidimensionality in item response theory. *Applied Psychological Measurement, 33*, 519–537.

Levy, R., & Svetina, D. (2011). A generalized dimensionality discrepancy measure for dimensionality assessment in multidimensional item response theory. *British Journal of Mathematical and Statistical Psychology, 64*, 208–232.

Levy, R., Xu, Y., Yel, N., & Svetina, D. (2015). A standardized generalized dimensionality discrepancy measure and a standardized model-based covariance for dimensionality assessment for multidimensional models. *Journal of Educational Measurement, 52*, 144–158.

Lewis, C. (1986). Test theory and Psychometrika: The past twenty-five years. *Psychometrika, 51*, 11–22.

Lewis, C. (2007). Selected topics in classical test theory. In C. R. Rao & S. Sinharay (Eds.), *Handbook of statistics, Volume 26: Psychometrics* (pp. 29–43). Amsterdam, the Netherlands: North-Holland/Elsevier.

Lewis, C., & Sheehan, K. (1990). Using Bayesian decision theory to design a computerized mastery test. *Applied Psychological Measurement, 14*, 367–386.

Li, F., Cohen, A. S., Kim, S.-H., & Cho, S.-J. (2009). Model selection methods for mixture dichotomous IRT models. *Applied Psychological Measurement, 33*, 353–373.

Li, Y., Bolt, D. M., & Fu, J. (2006). A comparison of alternative models for testlets. *Applied Psychological Measurement, 30*, 3–21.

Lindley, D. V. (1970). *A Bayesian solution for some educational prediction problems* (No. RB-70-33). Princeton, NJ: ETS.

Lindley, D. V. (1971). The estimation of many parameters. In V. P. Godambe & D. A. Sprott (Eds.), *Foundations of statistical inference* (pp. 435–455). Toronto, Ontario, Canada: Holt, Rinehart & Winston.

Lindley, D. V., & Novick, Melvin R. (1981). The role of exchangeability in inference. *The Annals of Statistics, 9*, 45–58.

Lindley, D. V., & Phillips, L. D. (1976). Inference for a Bernoulli process (a Bayesian view). *The American Statistician, 30*, 112–119.

Lindley, D. V., & Smith, A. F. M. (1972). Bayes estimates for the linear model. *Journal of the Royal Statistical Society. Series B, 34*, 1–41.

Lindsay, B., Clogg, C. C., & Grego, J. (1991). Semiparametric estimation in the Rasch model and related exponential response models, including a simple latent class model for item analysis. *Journal of the American Statistical Association, 86*, 96–107.

Linzer, D. A., & Lewis, J. B. (2011). poLCA: An R package for polytomous variable latent class analysis. *Journal of Statistical Software, 42*, 1–29.

Liu, Y., Schulz, E. M., & Yu, L. (2008). Standard error estimation of 3PL IRT true score equating with an MCMC method. *Journal of Educational and Behavioral Statistics, 33*, 257–278.

Little, R. J. A., & Rubin, D. B. (2002). *Statistical analysis with missing data* (2nd ed.). New York: Wiley.

Loken, E. (2004). Multimodality in mixture models and factor analysis. In *Applied Bayesian modeling and causal inference from incomplete-data perspectives: An essential journey with Donald Rubin's statistical family* (pp. 203–213). Chichester, UK: Wiley.

Loken, E. (2005). Identification constraints and inference in factor models. *Structural Equation Modeling: A Multidisciplinary Journal, 12*, 232–244.

Loken, E., & Rulison, K. L. (2010). Estimation of a four-parameter item response theory model. *British Journal of Mathematical and Statistical Psychology, 63*, 509–525.

Longford, N. T. (1995). *Model-Based Methods for Analysis of Data from 1990 NAEP Trial State Assessment* (No. 95-696). Washington, DC: National Center for Education Statistics. Retrieved from http://nces.ed.gov/pubsearch/pubsinfo.asp?pubid=95696.

Lopes, H. F., & West, M. (2004). Bayesian model assessment in factor analysis. *Statistica Sinica, 14*, 41–68.

Lord, F. M. (1974). Estimation of latent ability and item parameters when there are omitted responses. *Psychometrika, 39,* 247–264.
Lord, F. M. (1980). *Applications of item response theory to practical testing problems.* Hillsdale, N.J: Lawrence Erlbaum Associates.
Lord, F. M. (1986). Maximum likelihood and Bayesian parameter estimation in item response theory. *Journal of Educational Measurement, 23,* 157–162.
Lord, F. M., & Novick, M. R. (1968). *Statistical theories of mental test scores.* Reading, MA: Addison-Wesley.
Lucke, J. F. (2005). The α and the ω of congeneric test theory: An extension of reliability and internal consistency to heterogeneous tests. *Applied Psychological Measurement, 29,* 65–81.
Lunn, D., Jackson, C., Best, N., Thomas, A., & Spiegelhalter, D. (2013). *The BUGS book: A practical introduction to Bayesian analysis.* Boca Raton, FL: Chapman & Hall/CRC Press.
Lunn, D., Spiegelhalter, D., Thomas, A., & Best, N. (2009). The BUGS project: Evolution, critique and future directions. *Statistics in Medicine, 28,* 3049–3067.
Lynch, S. M. (2007). *Introduction to applied Bayesian statistics and estimation for social scientists.* New York: Springer.
MacCallum, R. C. (2003). Working with imperfect models. *Multivariate Behavioral Research, 38,* 113–139.
Macready, G. B., & Dayton, C. M. (1992). The application of latent class models in adaptive testing. *Psychometrika, 57,* 71–88.
Maier, K. S. (2002). Modeling incomplete scaled questionnaire data with a partial credit hierarchical measurement model. *Journal of Educational and Behavioral Statistics, 27,* 271–289.
Maraun, M. D. (1996). Metaphor taken as math: Indeterminacy in the factor analysis model. *Multivariate Behavioral Research, 31,* 517–538.
Marianti, S., Fox, J.-P., Avetisyan, M., Veldkamp, B. P., & Tijmstra, J. (2014). Testing for aberrant behavior in response time modeling. *Journal of Educational and Behavioral Statistics, 39,* 426–451.
Marin, J.-M., & Robert, C. (2007). *Bayesian core: A practical approach to computational Bayesian statistics.* New York: Springer.
Maris, E. (1999). Estimating multiple classification latent class models. *Psychometrika, 64,* 187–212.
Maris, G., & Maris, E. (2002). A MCMC-method for models with continuous latent responses. *Psychometrika, 67,* 335–350.
Markman, A. B. (1999). *Knowledge representation.* Mahwah, NJ: Psychology Press.
Marshall, S. P. (1981). Sequential item selection: Optimal and heuristic policies. *Journal of Mathematical Psychology, 23,* 134–152.
Marsh, H. W., Hau, K.-T., & Wen, Z. (2004). In search of golden rules: Comment on hypothesis-testing approaches to setting cutoff values for fit indexes and dangers in overgeneralizing Hu and Bentler's (1999) findings. *Structural Equation Modeling: A Multidisciplinary Journal, 11,* 320–341.
Martin, A. D., Quinn, K. M., & Park, J. H. (2011). MCMCpack: Markov chain Monte Carlo in R. *Journal of Statistical Software, 42,* 1–21.
Martin, J. K., & McDonald, R. P. (1975). Bayesian estimation in unrestricted factor analysis: A treatment for Heywood cases. *Psychometrika, 40,* 505–517.
Masters, G. N. (1982). A Rasch model for partial credit scoring. *Psychometrika, 47,* 149–174.
Matson, J. (2011, September 26). Faster-than-light neutrinos? Physics luminaries voice doubts. Retrieved from http://www.scientificamerican.com/article/ftl-neutrinos/.
Mavridis, D., & Ntzoufras, I. (2014). Stochastic search item selection for factor analytic models. *British Journal of Mathematical and Statistical Psychology, 67,* 284–303.
Maydeu-Olivares, A. (2013). Goodness-of-fit assessment of item response theory models. *Measurement: Interdisciplinary Research & Perspective, 11,* 71–101.
Maydeu-Olivares, A., Drasgow, F., & Mead, A. D. (1994). Distinguishing among parametric item response models for polychotomous ordered data. *Applied Psychological Measurement, 18,* 245–256.
Mazzeo, J., Lazer, S., & Zieky, M. J. (2006). Monitoring educational progress with group-score assessments. In R. L. Brennan (Ed.), *Educational Measurement* (4th ed., pp. 681–699). Westport, CT: Praeger.

McCutcheon, A. L. (1987). *Latent class analysis.* Newbury Park, CA: SAGE Publications.
McDonald, R. P. (1999). *Test theory: A unified treatment.* Mahwah, NJ: Lawrence Erlbaum Associates.
McDonald, R. P. (2010). Structural models and the art of approximation. *Perspectives on Psychological Science, 5,* 675–686.
McGrayne, S. B. (2011). *The theory that would not die: How Bayes' rule cracked the enigma code, hunted down Russian submarines, and emerged triumphant from two centuries of controversy.* New Haven, CT: Yale University Press.
McLeod, L., Lewis, C., & Thissen, D. (2003). A Bayesian method for the detection of item preknowledge in computerized adaptive testing. *Applied Psychological Measurement, 27,* 121–137.
McManus, I. C. (2012). The misinterpretation of the standard error of measurement in medical education: A primer on the problems, pitfalls and peculiarities of the three different standard errors of measurement. *Medical Teacher, 34,* 569–576.
Meng, X.-L. (1994a). Multiple-imputation inferences with uncongenial sources of input. *Statistical Science, 9,* 538–558.
Meng, X.-L. (1994b). Posterior predictive *p*-values. *The Annals of Statistics, 22,* 1142–1160.
Messick, S. (1994). The interplay of evidence and consequences in the validation of performance assessments. *Educational Researcher, 23,* 13.
Metropolis, N., Rosenbluth, A. W., Rosenbluth, M. N., Teller, A. H., & Teller, E. (1953). Equation of state calculations by fast computing machines. *The Journal of Chemical Physics, 21,* 1087–1092.
Meyer, J. P. (2010). A mixture Rasch model with item response time components. *Applied Psychological Measurement, 34,* 521–538.
Millsap, R. E. (2011). *Statistical approaches to measurement invariance.* New York: Routledge.
Mislevy, R. J. (1984). Estimating latent distributions. *Psychometrika, 49,* 359–381.
Mislevy, R. J. (1986). Bayes modal estimation in item response models. *Psychometrika, 51,* 177–195.
Mislevy, R. J. (1991). Randomization-based inference about latent variables from complex samples. *Psychometrika, 56,* 177–196.
Mislevy, R. J. (1994). Evidence and inference in educational assessment. *Psychometrika, 59,* 439–483.
Mislevy, R. J. (1995). Probability-based inference in cognitive diagnosis. In P. Nichols & R. Brennan (Eds.), *Cognitively diagnostic assessment* (pp. 43–71). Hillsdale, NJ: Lawrence Erlbaum Associates.
Mislevy, R. J. (2006). Cognitive psychology and educational assessment. In R. Brennan (Ed.), *Educational measurement* (4th ed., pp. 257–305). Westport, CT: Praeger.
Mislevy, R. J. (2008). How cognitive science challenges the educational measurement tradition. *Measurement: Interdisciplinary Research & Perspective, 6,* 124.
Mislevy, R. J. (2010). Some implications of expertise research for educational assessment. *Research Papers in Education, 25,* 253–270.
Mislevy, R. J. (2013). Evidence-centered design for simulation-based assessment. *Military Medicine, 105,* 107–114.
Mislevy, R. J. (2016). Missing responses in item response theory. In W. J. van der Linden (Ed.), *Handbook of item response theory: Models, statistical tools, and applications, volume 2* (pp. 171–194). New York: Chapman & Hall/CRC Press.
Mislevy, R. J., Almond, R. G., & Lukas, J. F. (2004). *A brief introduction to evidence-centered design* (No. CSE Report 632). Los Angeles, CA: National Center for Research on Evaluation, Standards, and Student Testing (CRESST) Center for the Study of Evaluation (CSE). Retrieved from http://files.eric.ed.gov/fulltext/ED483399.pdf.
Mislevy, R., Almond, R., Yan, D., & Steinberg, L. S. (1999). Bayes nets in educational assessment: Where do the numbers come from? *Appears in Proceedings of the 15th Conference on Uncertainty in Artificial Intelligence,* 437–446.
Mislevy, R. J., Beaton, A. E., Kaplan, B., & Sheehan, K. M. (1992). Estimating population characteristics from sparse matrix samples of item responses. *Journal of Educational Measurement, 29,* 133–161.
Mislevy, R. J., Behrens, J. T., Bennett, R. E., Demark, S. F., Frezzo, D. C., Levy, R., ... Winters, F. I. (2010). On the roles of external knowledge representations in assessment design. *The Journal of Technology, Learning and Assessment, 8.* Retrieved from http://napoleon.bc.edu/ojs/index.php/jtla/article/view/1621.

Mislevy, R. J., Behrens, J. T., Dicerbo, K. E., & Levy, R. (2012). Design and discovery in educational assessment: Evidence-centered design, psychometrics, and educational data mining. *JEDM-Journal of Educational Data Mining, 4,* 11–48.

Mislevy, R. J., & Gitomer, D. H. (1996). The role of probability-based inference in an intelligent tutoring system. *User Modeling and User-Adapted Instruction, 5,* 253–282.

Mislevy, R. J., Johnson, E. G., & Muraki, E. (1992). Scaling procedures in NAEP. *Journal of Educational Statistics, 17,* 131–154.

Mislevy, R. J., Levy, R., Kroopnick, M., & Rutstein, D. (2008). Evidentiary foundations of mixture item response theory models. In G. R. Hancock & K. M. Samuelsen (Eds.), *Advances in latent variable mixture models* (pp. 149–175). Charlotte, NC: Information Age Publishing.

Mislevy, R. J., & Riconscente, M. M. (2006). Evidence-centered assessment design. In S. Downing & T. Haladyna (Eds.), *Handbook of test development* (pp. 61–90). Mahwah, NJ: Lawrence Erlbaum Associates.

Mislevy, R. J., Senturk, D., Almond, R. G., Dibello, L. V., Jenkins, F., Steinberg, L. S., & Yan, D. (2002). *Modeling conditional probabilities in complex educational assessments* (No. CSE Technical Report 580). Los Angeles, CA: University of California, National Center for Research on Evaluation, Standards, Student Testing (CRESST). Retrieved from http://citeseerx.ist.psu.edu/viewdoc/download?doi=10.1.1.322.4516&rep=rep1&type=pdf.

Mislevy, R. J., Sheehan, K. M., & Wingersky, M. (1993). How to equate tests with little or no data. *Journal of Educational Measurement, 30,* 55–78.

Mislevy, R. J., Steinberg, L. S., & Almond, R. G. (2003). On the structure of educational assessments. *Measurement: Interdisciplinary Research and Perspectives, 1,* 3–62.

Mitchell, L. (2009). *Examining the structural properties and competing models for the institutional integration scale.* Unpublished thesis. Arizona State University, Tempe, AZ.

Molenaar, I. W., & Hoijtink, H. (1990). The many null distributions of person fit indices. *Psychometrika, 55,* 75–106.

Moore, C. (2009, February 7). Bayesian umpires. Retrieved from http://baseballanalysts.com/archives/2009/12/bayesian_umpire.php.

Morey, R. D., Romeijn, J.-W., & Rouder, J. N. (2013). The humble Bayesian: Model checking from a fully Bayesian perspective. *British Journal of Mathematical and Statistical Psychology, 66,* 68–75.

Mosteller, F., & Tukey, J. W. (1977). *Data analysis and regression: A second course in statistics.* Reading, MA: Pearson.

Mosteller, F., & Wallace, D. L. (1964). *Inference and disputed authorship: The Federalist.* Reading, MA: Addison-Wesley.

Moustaki, I., & Knott, M. (2000). Weighting for item non-response in attitude scales by using latent variable models with covariates. *Journal of the Royal Statistical Society: Series A (Statistics in Society), 163,* 445–459.

Mulaik, S. A. (2009). *Linear causal modeling with structural equations.* Boca Raton, FL: Chapman & Hall/CRC Press.

Muraki, E. (1992). A generalized partial credit model: Application of an EM algorithm. *Applied Psychological Measurement, 16,* 159–176.

Muthén, B., & Asparouhov, T. (2012). Bayesian structural equation modeling: A more flexible representation of substantive theory. *Psychological Methods, 17,* 313–335.

Muthén, B. O., & Muthén, L. K. (1998). *Mplus user's guide* (7th ed.). Los Angeles, CA: Muthén & Muthén.

Nadarajah, S., & Kotz, S. (2006). R programs for truncated distributions. *Journal of Statistical Software, 16.* Retrieved from http://www.jstatsoft.org/v16/c02.

Natesan, P., Limbers, C., & Varni, J. W. (2010). Bayesian estimation of graded response multilevel models using Gibbs sampling: Formulation and illustration. *Educational and Psychological Measurement, 70,* 420–439.

Naumann, A., Hochweber, J., & Hartig, J. (2014). Modeling instructional sensitivity using a longitudinal multilevel differential item functioning approach. *Journal of Educational Measurement, 51,* 381–399.

Neapolitan, R. E. (2004). *Learning Bayesian networks.* Upper Saddle River, NJ: Prentice Hall.

Norsys Software Corporation. (1999). *Netica manual.* Vancouver, BC: Author.

Novick, M. R. (1964). *On Bayesian logical probability* (No. RB-64-22). Retrieved from http://onlinelibrary.wiley.com/doi/10.1002/j.2333-8504.1964.tb00330.x/abstract.

Novick, M. R. (1969). Multiparameter Bayesian indifference procedures. *Journal of the Royal Statistical Society. Series B, 31,* 29–64.

Novick, M. R., & Jackson, P. (1974). *Statistical methods for educational and psychological research.* New York: McGraw-Hill.

Novick, M. R., Jackson, P. H., & Thayer, D. T. (1971). Bayesian inference and the classical test theory model: Reliability and true scores. *Psychometrika, 36,* 261–288.

Novick, M. R., Jackson, P. H., Thayer, D. T., & Cole, N. S. (1972). Estimating multiple regressions in *m* groups: A cross-validation study. *British Journal of Mathematical and Statistical Psychology, 25,* 33–50.

Novick, M. R., & Lewis, C. (1974). Prescribing test length for criterion-referenced measurement. In C. W. Harris, M. C. Alkin, & W. J. Popham (Eds.), *Problems in criterion-referenced measurement* (pp. 139–158). Los Angeles, CA: Center for the Study of Evaluation, University of California, Los Angeles.

Novick, M. R., Lewis, C., & Jackson, P. H. (1973). The estimation of proportions in *m* groups. *Psychometrika, 38,* 19–46.

Nye, C. D., & Drasgow, F. (2011). Assessing goodness of fit: Simple rules of thumb simply do not work. *Organizational Research Methods, 14,* 548–570.

O'Hagan, A. (1998). Eliciting expert beliefs in substantial practical applications. *Journal of the Royal Statistical Society: Series D (The Statistician), 47,* 21–35.

O'Hagan, A., Buck, C. E., Daneshkhah, A., Eiser, J. R., Garthwaite, P. H., Jenkinson, D. J., Oakley, J. E., & Rakow, T. (2006). *Uncertain judgements: Eliciting experts' probabilities.* London: Wiley.

O'Neill, B. (2009). Exchangeability, correlation, and Bayes' effect. *International Statistical Review, 77,* 241–250.

Oravecz, Z., Anders, R., & Batchelder, W. H. (2013). Hierarchical Bayesian modeling for test theory without an answer key. *Psychometrika,* 1–24.

Owen, R. J. (1969). *Tailored testing* (No. 69–92). Princeton, NJ: ETS.

Owen, R. J. (1975). A Bayesian sequential procedure for quantal response in the context of adaptive mental testing. *Journal of the American Statistical Association, 70,* 351–356.

Pan, J.-C., & Huang, G.-H. (2014). Bayesian inferences of latent class models with an unknown number of classes. *Psychometrika, 79,* 621–646.

Pascarella, E. T., & Terenzini, P. T. (1980). Predicting freshman persistence and voluntary dropout decisions from a theoretical model. *The Journal of Higher Education, 51,* 60–75.

Pastor, D. A., & Gagné, P. (2013). Mean and covariance structure mixture models. In G. R. Hancock & R. O. Mueller (Eds.), *Structural equation modeling: A second course* (2nd ed., pp. 343–393). Greenwich, CT: Information Age Publishing.

Patton, J. M., Cheng, Y., Yuan, K.-H., & Diao, Q. (2013). The influence of item calibration error on variable-length computerized adaptive testing. *Applied Psychological Measurement, 37,* 24–40.

Patz, R. J., & Junker, B. W. (1999a). Applications and extensions of MCMC in IRT: Multiple item types, missing data, and rated responses. *Journal of Educational and Behavioral Statistics, 24,* 342–366.

Patz, R. J., & Junker, B. W. (1999b). A straightforward approach to Markov chain Monte Carlo methods for item response models. *Journal of Educational and Behavioral Statistics, 24,* 146–178.

Patz, R. J., Junker, B. W., Johnson, M. S., & Mariano, L. T. (2002). The hierarchical rater model for rated test items and its application to large-scale educational assessment data. *Journal of Educational and Behavioral Statistics, 27,* 341–384.

Patz, R. J., & Yao, L. (2007). Methods and models for vertical scaling. In N. J. Dorans, M. Pommerich, & P. W. Holland (Eds.), *Linking and aligning scores and scales* (pp. 253–272). New York: Springer.

Pearl, J. (1988). *Probabilistic reasoning in intelligent systems: Networks of plausible inference.* San Francisco, CA: Morgan Kaufmann.

Pearl, J. (2009). *Causality: Models, reasoning and inference* (2nd ed.). Cambridge: Cambridge University Press.

Phillips, S. E., & Camara, W. J. (2006). Legal and ethical issues. In R. Brennan (Ed.), *Educational measurement* (4th ed., pp. 734–755). Westport, CT: Praeger.
Plummer, M. (2003). JAGS: A program for analysis of Bayesian graphical models using Gibbs sampling. In *Proceedings of the 3rd International Workshop on Distributed Statistical Computing* (pp. 20–22). Vienna, Austria: Technische Universität Wien.
Plummer, M. (2008). Penalized loss functions for Bayesian model comparison. *Biostatistics, 9,* 523–539.
Plummer, M. (2010). *JAGS version 2.0.0 user manual.* Lyon, France. Retrieved from http://www-fis.iarc.fr/~martyn/software/jags/.
Plummer, M., Best, N., Cowles, K., & Vines, K. (2006). CODA: Convergence diagnosis and output analysis for MCMC. *R News, 6,* 7–11.
Preacher, K. J. (2006). Quantifying parsimony in structural equation modeling. *Multivariate Behavioral Research, 41,* 227–259.
Press, S. J. (1989). *Bayesian statistics: Principles, models, and applications.* New York: Wiley.
Press, S. J., & Shigemasu, K. (1997). *Bayesian inference in factor analysis (revised)* (No. 243). Riverside, CA: University of California, Riverside. Retrieved from http://citeseerx.ist.psu.edu/viewdoc/download?doi=10.1.1.14.6968&rep=rep1&type=pdf.
Proctor, C. H. (1970). A probabilistic formulation and statistical analysis of Guttman scaling. *Psychometrika, 35,* 73–78.
Quellmalz, E., Timms, M., Buckley, B., Levy, R., Davenport, J., Loveland, M., & Silberglitt, M. (2012). 21st century dynamic assessment. In J. Clarke-Midura, M. Mayrath, & D. H. Robinson (Eds.), *Technology-based assessments for 21st century skills: Theoretical and practical implications from modern research* (pp. 55–90). Charlotte, NC: Information Age Publishing.
Rabe-Hesketh, S., Skrondal, A., & Pickles, A. (2004). GLLAMM Manual (Second Edition). U.C. Berkeley Division of Biostatistics Working Paper Series. Berkeley, CA: University of California, Berkeley.
Raftery, A. E. (1993). Bayesian model selection in structural equation models. In K. A. Bollen & J. S. Long (Eds.), *Testing structural equation models* (pp. 163–180). Newbury Park, CA: SAGE Publications.
Ramsay, J. O. (1991). Kernel smoothing approaches to nonparametric item characteristic curve estimation. *Psychometrika, 56,* 611–630.
Rasch, G. (1960). *Probabilistic models for some intelligence and attainment tests.* Copenhagen, Denmark: Danish Institute for Educational Research.
Raudenbush, S. W. (1988). Educational applications of hierarchical linear models: A review. *Journal of Educational Statistics, 13,* 85–116.
Raudenbush, S. W., Fotiu, R. P., & Cheong, Y. F. (1999). Synthesizing results from the trial state assessment. *Journal of Educational and Behavioral Statistics, 24,* 413–438.
R Core Team. (2014). *R: A language and environment for statistical computing.* Vienna, Austria: R Foundation for Statistical Computing. Retrieved from http://www.R-project.org/.
Reckase, M. D. (2009). *Multidimensional item response theory.* New York: Springer.
Reese, W. J. (2013). *Testing wars in the public schools: A forgotten history.* Cambridge, MA: Harvard University Press.
Reye, J. (2004). Student modelling based on belief networks. *International Journal of Artificial Intelligence in Education, 14,* 63–96.
Richardson, S., & Green, P. J. (1997). On Bayesian analysis of mixtures with an unknown number of components. *Journal of the Royal Statistical Society. Series B (Methodological), 59,* 731–792.
Rijmen, F. (2010). Formal relations and an empirical comparison among the bi-factor, the testlet, and a second-order multidimensional IRT model. *Journal of Educational Measurement, 47,* 361–372.
Rijmen, F., & De Boeck, P. (2002). The random weights linear logistic test model. *Applied Psychological Measurement, 26,* 271–285.
Robert, C., & Casella, G. (2011). A short history of Markov chain Monte Carlo: Subjective recollections from incomplete data. *Statistical Science, 26,* 102–115.
Roberts, G. O. (1996). Markov chain concepts related to sampling algorithms. In W. R. Gilks, S. Richardson, & D. J. Spiegelhalter (Eds.), *Markov chain Monte Carlo in practice* (pp. 45–57). London: Chapman & Hall/CRC Press.

Roberts, J. S., & Thompson, V. M. (2011). Marginal maximum a posteriori item parameter estimation for the generalized graded unfolding model. *Applied Psychological Measurement, 35*, 259–279.
Robins, J. M., van der Vaart, A., & Ventura, V. (2000). Asymptotic distribution of p values in composite null models. *Journal of the American Statistical Association, 95*, 1143–1156.
Rodgers, J. L. (2010). The epistemology of mathematical and statistical modeling: A quiet methodological revolution. *American Psychologist, 65*, 1–12.
Rodríguez, C. E., & Walker, S. G. (2014). Label switching in Bayesian mixture models: Deterministic relabeling strategies. *Journal of Computational and Graphical Statistics, 23*, 25–45.
Roussos, L. A., DiBello, L. V., Stout, W., Hartz, S. M., Henson, R. A., & Templin, J. L. (2007). The fusion model skills diagnosis system. In J. P. Leighton & M. J. Gierl (Eds.), *Cognitive diagnostic assessment for education: Theory and applications* (pp. 275–318). New York: Cambridge University Press.
Rowe, D. B. (2003). *Multivariate Bayesian statistics: Models for source separation and signal unmixing*. Boca Raton, FL: Chapman & Hall/CRC Press.
Rowe, J. P., & Lester, J. C. (2010). Modeling user knowledge with dynamic Bayesian networks in interactive narrative environments. In G. M. Youngblood & V. Bulitko (Eds.), *AIIDE*. Retrieved from http://aaai.org/ocs/index.php/AIIDE/AIIDE10/paper/view/2149.
Rubin, D. B. (1976). Inference and missing data. *Biometrika, 63*, 581–592.
Rubin, D. B. (1977). Formalizing subjective notions about the effect of nonrespondents in sample surveys. *Journal of the American Statistical Association, 72*, 538–543.
Rubin, D. B. (1978). Multiple imputations in sample surveys-a phenomenological Bayesian approach to nonresponse. In *Proceedings of the survey research methods section of the American Statistical Association, 1*, 20–34.
Rubin, D. B. (1984). Bayesianly justifiable and relevant frequency calculations for the applied statistician. *The Annals of Statistics, 12*, 1151–1172.
Rubin, D. B. (1987). *Multiple imputation for nonresponse in surveys*. New York: Wiley.
Rubin, D. B. (1988). Using the SIR algorithm to simulate posterior distributions. In J. M. Bernardo, M. H. DeGroot, D. V. Lindley, & A. F. M. Smith (Eds.), *Bayesian Statistics 3* (pp. 395–402). Oxford, UK: Oxford University Press.
Rubin, D. B. (1996a). Comment: On posterior predictive p-values. *Statistica Sinica, 6*, 787–792.
Rubin, D. B. (1996b). Multiple imputation after 18+ years. *Journal of the American Statistical Association, 91*, 473–489.
Rubin, D. B., & Stern, H. S. (1994). Testing in latent class models using a posterior predictive check distribution. In A. von Eye & C. C. Clogg (Eds.), *Latent variable analysis: Applications for developmental research* (pp. 420–438). Thousand Oaks, CA: SAGE Publications.
Rudner, L. M. (2009). Scoring and classifying examinees using measurement decision theory. *Practical Assessment, Research & Evaluation, 14*. Retrieved from http://pareonline.net/getvn.asp?v=14&n=8.
Rudner, L. M., & Liang, T. (2002). Automated essay scoring using Bayes' theorem. *The Journal of Technology, Learning and Assessment, 1*. Retrieved from http://napoleon.bc.edu/ojs/index.php/jtla/article/view/1668.
Rupp, A. A. (2002). Feature selection for choosing and assembling measurement models: A building-block-based organization. *International Journal of Testing, 2*, 311–360.
Rupp, A. A., Dey, D. K., & Zumbo, B. D. (2004). To Bayes or not to Bayes, from whether to when: Applications of Bayesian methodology to modeling. *Structural Equation Modeling: A Multidisciplinary Journal, 11*, 424–451.
Rupp, A. A., Levy, R., DiCerbo, K. E., Sweet, S., Crawford, A. V., Calico, T., … Behrens, J. T. (2012). Putting ECD into practice: The interplay of theory and data in evidence models within a digital learning environment. *Journal of Educational Data Mining, 4*, 49–110.
Rupp, A. A., & Mislevy, R. J. (2007). Cognitive foundations of structured item response models. In J. P. Leighton & M. J. Gierl (Eds.), *Cognitive diagnostic assessment for education: Theory and applications* (pp. 205–241). New York: Cambridge University Press.
Rupp, A. A., Templin J., & Henson, R. A. (2010). *Diagnostic measurement: Theory, methods, and applications*. New York: Guilford Press.

Sahu, S. K. (2002). Bayesian estimation and model choice in item response models. *Journal of Statistical Computation and Simulation, 72*, 217–232.

Samejima, F. (1983). Some methods and approaches of estimating the operation characteristics of discrete item responses. In H. Wainer & S. Messick (Eds.), *Principals of modern psychological measurement: A Festschrift for Frederic M. Lord* (pp. 154–182). Hillsdale, NJ: Lawrence Erlbaum Associates.

Samejima, F. (1997). Graded response model. In W. J. van der Linden & R. K. Hambleton (Eds.), *Handbook of modern item response theory* (pp. 85–100). New York: Springer-Verlag.

Samejima, F. (1969). Estimating of latent ability using a response pattern of graded scores. *Psychometrika Monograph Supplement, No. 17.* Richmond, VA: The William Byrd Press. Retrieved from https://www.psychometricsociety.org/sites/default/files/pdf/MN17.pdf

Samejima, F. (1973). A comment on Birnbaum's three-parameter logistic model in the latent trait theory. *Psychometrika, 38*, 221–233.

Samuelsen, K. M. (2008). Examining differential item function from a latent class perspective. In G. R. Hancock & K. M. Samuelsen (Eds.), *Advances in latent variable mixture models* (pp. 177–197). Charlotte, NC: Information Age Publishing.

Sao Pedro, M. A., Baker, R. S. J. de, Gobert, J. D., Montalvo, O., & Nakama, A. (2011). Leveraging machine-learned detectors of systematic inquiry behavior to estimate and predict transfer of inquiry skill. *User Modeling and User-Adapted Interaction, 23(1)*, 1–39.

Savage, L. J. (1971). Elicitation of personal probabilities and expectations. *Journal of the American Statistical Association, 66*, 783–801.

Schafer, J. L. (2003). Multiple imputation in multivariate problems when the imputation and analysis models differ. *Statistica Neerlandica, 57*, 19–35.

Schafer, J. L., & Graham, J. W. (2002). Missing data: Our view of the state of the art. *Psychological Methods, 7*, 147–177.

Scheines, R., Hoijtink, H., & Boomsma, A. (1999). Bayesian estimation and testing of structural equation models. *Psychometrika, 64*, 37–52.

Schum, D. A. (1987). *Evidence and inference for the intelligence analyst* (2nd ed.). Lanham, MD: University Press of America.

Schum, D. A. (1994). *The evidential foundations of probabilistic reasoning.* New York: Wiley.

Schwarz, G. (1978). Estimating the dimension of a model. *The Annals of Statistics, 6*, 461–464.

Scott, S. L., & Ip, E. H. (2002). Empirical Bayes and item-clustering effects in a latent variable hierarchical model: A case study from the National Assessment of Educational Progress. *Journal of the American Statistical Association, 97*, 409–419.

Sedransk, J., Monahan, J., & Chiu, H. Y. (1985). Bayesian estimation of finite population parameters in categorical data models incorporating order restrictions. *Journal of the Royal Statistical Society. Series B (Methodological), 47*, 519–527.

Segall, D. O. (1996). Multidimensional adaptive testing. *Psychometrika, 61*, 331–354.

Segall, D. O. (2002). An item response model for characterizing test compromise. *Journal of Educational and Behavioral Statistics, 27*, 163–179.

Segawa, E. (2005). A growth model for multilevel ordinal data. *Journal of Educational and Behavioral Statistics, 30*, 369–396.

Segawa, E., Emery, S., & Curry, S. J. (2008). Extended generalized linear latent and mixed model. *Journal of Educational and Behavioral Statistics, 33*, 464–484.

Senn, S. (2011). You may believe you are a Bayesian but you are probably wrong. *Rationality, Markets and Morals, 2*, 48–66.

Sheng, Y., & Wikle, C. K. (2007). Comparing multiunidimensional and unidimensional item response theory models. *Educational and Psychological Measurement, 67*, 899–919.

Sheng, Y., & Wikle, C. K. (2008). Bayesian multidimensional IRT models with a hierarchical structure. *Educational and Psychological Measurement, 68*, 413–430.

Sheriffs, A. C., & Boomer, D. S. (1954). Who is penalized by the penalty for guessing? *Journal of Educational Psychology, 45*, 81–90.

Shute, V. J. (2011). Stealth assessment in computer-based games to support learning. In S. Tobias & J. D. Fletcher (Eds.), *Computer games and instruction* (pp. 503–524). Charlotte, NC: Information Age Publishing.

Shute, V. J., Ventura, M., Bauer, M., & Zapata-Rivera, D. (2009). Melding the power of serious games and embedded assessment to monitor and foster learning. In U. Ritterfeld, M. J. Cody, & P. Vorderer (Eds.), *Serious games: Mechanisms and effects* (Vol. 2, pp. 295–321). Philadelphia, PA: Routledge/LEA.

Sijtsma, K., & Molenaar, I. W. (2002). *Introduction to nonparametric item response theory.* Thousand Oaks, CA: SAGE Publications.

Sinharay, S. (2004). Experiences with Markov chain Monte Carlo convergence assessment in two psychometric examples. *Journal of Educational and Behavioral Statistics, 29*, 461–488.

Sinharay, S. (2005). Assessing fit of unidimensional item response theory models using a Bayesian approach. *Journal of Educational Measurement, 42*, 375–394.

Sinharay, S. (2006a). Bayesian item fit analysis for unidimensional item response theory models. *British Journal of Mathematical and Statistical Psychology, 59*, 429–449.

Sinharay, S. (2006b). Model diagnostics for Bayesian networks. *Journal of Educational and Behavioral Statistics, 31*, 1–33.

Sinharay, S., & Almond, R. G. (2007). Assessing fit of cognitive diagnostic models: A case study. *Educational and Psychological Measurement, 67*, 239–257.

Sinharay, S., Dorans, N. J., Grant, M. C., & Blew, E. O. (2009). Using past data to enhance small sample DIF estimation: A Bayesian approach. *Journal of Educational and Behavioral Statistics, 34*, 74–96.

Sinharay, S., Johnson, M. S., & Stern, H. S. (2006). Posterior predictive assessment of item response theory models. *Applied Psychological Measurement, 30*, 298–321.

Sinharay, S., Johnson, M. S., & Williamson, D. M. (2003). Calibrating item families and summarizing the results using family expected response functions. *Journal of Educational and Behavioral Statistics, 28*, 295–313.

Sinharay, S., Stern, H. S., & Russell, D. (2001). The use of multiple imputation for the analysis of missing data. *Psychological Methods, 6*, 317–329.

Skrondal, A., & Rabe-Hesketh, S. (2004). *Generalized latent variable modeling: Multilevel, longitudinal, and structural equation models.* Boca Raton, FL: Chapman & Hall/CRC Press.

Smith, A. F. M. (1973). A general Bayesian linear model. *Journal of the Royal Statistical Society. Series B (Methodological), 35*, 67–75.

Smith, A. F. M., & Roberts, G. O. (1993). Bayesian computation via the Gibbs sampler and related Markov chain Monte Carlo methods. *Journal of the Royal Statistical Society. Series B, 55*, 3–23.

Snijders, T. A. (2001). Asymptotic null distribution of person fit statistics with estimated person parameter. *Psychometrika, 66*, 331–342.

Soares, T. M., Goncalves, F. B., & Gamerman, D. (2009). An integrated Bayesian model for DIF analysis. *Journal of Educational and Behavioral Statistics, 34*, 348–377.

Song, H., & Ferrer, E. (2012). Bayesian estimation of random coefficient dynamic factor models. *Multivariate Behavioral Research, 47*, 26–60.

Song, X.-Y., & Lee, S.-Y. (2001). Bayesian estimation and test for factor analysis model with continuous and polytomous data in several populations. *British Journal of Mathematical and Statistical Psychology, 54*, 237–263.

Song, X.-Y., & Lee, S.-Y. (2002a). Analysis of structural equation model with ignorable missing continuous and polytomous data. *Psychometrika, 67*, 261–288.

Song, X.-Y., & Lee, S.-Y. (2002b). A Bayesian approach for multigroup nonlinear factor analysis. *Structural Equation Modeling: A Multidisciplinary Journal, 9*, 523–553.

Song, X.-Y., & Lee, S.-Y. (2004). Bayesian analysis of two-level nonlinear structural equation models with continuous and polytomous data. *British Journal of Mathematical and Statistical Psychology, 57*, 29–52.

Song, X.-Y., & Lee, S.-Y. (2012). *Basic and advanced Bayesian structural equation modeling: With applications in the medical and behavioral sciences.* Chichester, UK: Wiley.

Song, X.-Y., Lee, S.-Y., & Hser, Y.-I. (2009). Bayesian analysis of multivariate latent curve models with nonlinear longitudinal latent effects. *Structural Equation Modeling: A Multidisciplinary Journal, 16*, 245–266.

Song, X.-Y., Lee, S.-Y., & Zhu, H.-T. (2001). Model selection in structural equation models with continuous and polytomous data. *Structural Equation Modeling: A Multidisciplinary Journal, 8*, 378–396.

Song, X.-Y., Lu, Z.-H., Cai, J.-H., & Ip, E. H.-S. (2013). A Bayesian modeling approach for generalized semiparametric structural equation models. *Psychometrika, 78*, 624–647.

Song, X.-Y., Xia, Y.-M., Pan, J.-H., & Lee, S.-Y. (2011). Model comparison of Bayesian semiparametric and parametric structural equation models. *Structural Equation Modeling: A Multidisciplinary Journal, 18*, 55–72.

Spearman, C. (1904). "General Intelligence," objectively determined and measured. *The American Journal of Psychology, 15*, 201–292.

Spiegelhalter, D. J., Best, N. G., Carlin, B. P., & Van Der Linde, A. (2002). Bayesian measures of model complexity and fit. *Journal of the Royal Statistical Society: Series B (Statistical Methodology), 64*, 583–639.

Spiegelhalter, D. J., & Lauritzen, S. L. (1990). Sequential updating of conditional probabilities on directed graphical structures. *Networks, 20*, 579–605.

Spiegelhalter, D. J., Thomas, A., Best, A. G., & Lunn, D. (2007). *WinBUGS user manual: Version 1.4.3.* Cambridge: MRC Biostatistics Unit.

Stein, C. (1956). Inadmissibility of the usual estimator for the mean of a multivariate normal distribution. In *Proceedings of the 3rd Berkeley Symposium on Mathematical Statistics and Probability, Volume 1: Contributions to the Theory of Statistics Proceedings of the 3rd Berkeley Symposium on Mathematical Statistics and Probability* (pp. 197–206). Berkeley, CA: University of California Press.

Stein, C. M. (1962). Confidence sets for the mean of a multivariate normal distribution. *Journal of the Royal Statistical Society. Series B, 24*, 265–296.

Stephens, M. (2000). Dealing with label switching in mixture models. *Journal of the Royal Statistical Society: Series B (Statistical Methodology), 62*, 795–809.

Stern, H. S. (2000). Comment. *Journal of the American Statistical Association, 95*, 1157–1159.

Stone, C. A., & Hansen, M. A. (2000). The effect of errors in estimating ability on goodness-of-fit tests for IRT models. *Educational and Psychological Measurement, 60*, 974–991.

Stone, L. D., Keller, C. M., Kratzke, T. M., & Strumpfer, J. P. (2014). Search for the wreckage of Air France Flight Af 447. *Statistical Science, 29*, 69–80.

Stout, W., Habing, B., Douglas, J., Kim, H. R., Roussos, L., & Zhang, J. (1996). Conditional covariance-based nonparametric multidimensionality assessment. *Applied Psychological Measurement, 20*, 331–354.

Sturtz, S., Ligges, U., & Gelman, A. E. (2005). R2WinBUGS: A package for running WinBUGS from R. *Journal of Statistical Software, 12*, 1–16.

Süli, E., & Mayers, D. F. (2003). *An introduction to numerical analysis.* Cambridge: Cambridge University Press.

Swaminathan, H., & Gifford, J. A. (1982). Bayesian estimation in the Rasch model. *Journal of Educational Statistics, 7*, 175–191.

Swaminathan, H., & Gifford, J. A. (1985). Bayesian estimation in the two-parameter logistic model. *Psychometrika, 50*, 349–364.

Swaminathan, H., & Gifford, J. A. (1986). Bayesian estimation in the three-parameter logistic model. *Psychometrika, 51*, 589–601.

Swaminathan, H., Hambleton, R. K., & Rogers, H. J. (2007). Assessing the fit of item response models. In C. R. Rao & S. Sinharay (Eds.), *Handbook of statistics, Volume 26: Psychometrics* (pp. 683–718). Amsterdam, the Netherlands: North-Holland/Elsevier.

Takane, Y., & De Leeuw, J. (1987). On the relationship between item response theory and factor analysis of discretized variables. *Psychometrika, 52*, 393–408.

Tanner, M. A., & Wong, W. H. (1987). The calculation of posterior distributions by data augmentation. *Journal of the American Statistical Association, 82*, 528–540.

Tatsuoka, C. (2002). Data analytic methods for latent partially ordered classification models. *Journal of the Royal Statistical Society: Series C (Applied Statistics), 51*, 337–350.
Tatsuoka, K. K. (1984). *Analysis of errors in fraction addition and subtraction problems* (NIE Final Rep. for Grant No. NIE-G-81-002). Urbana, IL: University of Illinois, Computer-Based Education Research Laboratory. Retrieved from http://eric.ed.gov/?id=ED257665.
Tatsuoka, K. K. (1987). Validation of cognitive sensitivity for item response curves. *Journal of Educational Measurement, 24*, 233–245.
Tatsuoka, K. K. (1990). Toward an integration of item response theory and cognitive error diagnosis. In N. Frederiksen, R. Glaser, A. Lesgold, & M. G. Shafto (Eds.), *Diagnostic monitoring of skill and knowledge acquisition* (pp. 453–488). Hillsdale, NJ: Lawrence Erlbaum Associates.
Tatsuoka, K. K. (2009). *Cognitive assessment: An introduction to the rule space method.* New York: Routledge.
Thissen, D. (2001). Psychometric engineering as art. *Psychometrika, 66*, 473–485.
Thompson, M. S., & Green, S. B. (2013). Evaluating between-group differences in latent variable means. In G. R. Hancock & R. O. Mueller (Eds.), *Structural equation modeling: A second course* (2nd ed., pp. 163–218). Greenwich, CT: Information Age Publishing.
Thurstone, L. L. (1947). *Multiple-factor analysis: A development & expansion of the vectors of mind.* The University of Chicago Press.
Tierney, L. (1994). Markov chains for exploring posterior distributions. *The Annals of Statistics, 22*, 1701–1728.
Tierney, L., & Kadane, J. B. (1986). Accurate approximations for posterior moments and marginal densities. *Journal of the American Statistical Association, 81*, 82–86.
Toulmin, S. E. (1958). *The uses of argument.* Cambridge, UK: Cambridge University Press.
Tsutakawa, R. K. (1992). Prior distribution for item response curves. *British Journal of Mathematical and Statistical Psychology, 45*, 51–74.
Tsutakawa, R. K., & Johnson, J. C. (1990). The effect of uncertainty of item parameter estimation on ability estimates. *Psychometrika, 55*, 371–390.
Tsutakawa, R. K., & Lin, H. Y. (1986). Bayesian estimation of item response curves. *Psychometrika, 51*, 251–267.
Tsutakawa, R. K., & Soltys, M. J. (1988). Approximation for Bayesian ability estimation. *Journal of Educational Statistics, 13*, 117–130.
Tukey, J. W. (1962). The future of data analysis. *The Annals of Mathematical Statistics, 33*, 1–67.
Tukey, J. W. (1969). Analyzing data: Sanctification or detective work? *American Psychologist, 24*, 83–91.
Tukey, J. W. (1977). *Exploratory data analysis.* Reading, MA: Pearson.
Tukey, J. W. (1986). Exploratory data analysis as part of a larger whole. In *The collected works of John W. Tukey: Vol. IV. Philosophy and principles of data analysis: 1965–1986* (pp. 793–803). Pacific Grove, CA: Wadsworth.
Van der Linden, W. J. (1998). Bayesian item selection criteria for adaptive testing. *Psychometrika, 63*, 201–216.
Van der Linden, W. J. (2007). A hierarchical framework for modeling speed and accuracy on test items. *Psychometrika, 72*, 287–308.
Van der Linden, W. J., & Glas, C. A. (2000). Capitalization on item calibration error in adaptive testing. *Applied Measurement in Education, 13*, 35–53.
Van der Linden, W. J., & Glas, C. A. W. (Eds.). (2010). *Elements of adaptive testing.* New York: Springer.
Van der Linden, W. J., & Guo, F. (2008). Bayesian procedures for identifying aberrant response-time patterns in adaptive testing. *Psychometrika, 73*, 365–384.
Van der Linden, W. J., & Hambleton, R. K. (Eds.). (1997). *Handbook of modern item response theory.* New York: Springer.
Van der Linden, W. J., & Lewis, C. (2015). Bayesian checks on cheating on tests. *Psychometrika, 80*, 689–706.
Van der Linden, W. J., & Pashley, Peter J. (2010). Item selection and ability estimation in adaptive testing. In W. J. van der Linden & C. A. W. Glas (Eds.), *Elements of adaptive testing* (pp. 3–30). New York: Springer.

Van der Linden, W. J., & Ren, H. (2014). Optimal Bayesian adaptive design for test-item calibration. *Psychometrika, 80*, 1–26.
VanLehn, K. (2008). Intelligent tutoring systems for continuous, embedded assessment. In C. Dwyer (Ed.), *The future of assessment: Shaping teaching and learning* (pp. 113–138). Mahwah, NJ: Lawrence Erlbaum.
VanLehn, K., & Martin, J. (1997). Evaluation of an assessment system based on Bayesian student modeling. *International Journal of Artificial Intelligence in Education, 8*, 179–221.
Van Onna, M. J. H. (2002). Bayesian estimation and model selection in ordered latent class models for polytomous items. *Psychometrika, 67*, 519–538.
Van Rijn, P. W., & Rijmen, F. (2012). *A note on explaining away and paradoxical results in multidimensional item response theory* (No. ETS RR-12-13). Princeton, NJ: ETS. Retrieved from http://onlinelibrary.wiley.com/doi/10.1002/j.2333-8504.2012.tb02295.x/abstract.
Van Rijn, P., & Rijmen, F. (2015). On the explaining-away phenomenon in multivariate latent variable models. *British Journal of Mathematical and Statistical Psychology, 68*, 1–22.
Verhagen, A. J., & Fox, J. P. (2013). Bayesian tests of measurement invariance. *British Journal of Mathematical and Statistical Psychology, 66*, 383–401.
Von Davier, M. (2008). A general diagnostic model applied to language testing data. *British Journal of Mathematical & Statistical Psychology, 61*, 287–307.
Von Davier, M., Sinharay, S., Oranje, A., & Beaton, A. (2007). The statistical procedures used in National Assessment of Educational Progress: Recent developments and future directions. In C. R. Rao & S. Sinharay (Eds.), *Handbook of statistics, Volume 26: Psychometrics* (pp. 1039–1055). Amsterdam, the Netherlands: North-Holland/Elsevier.
Vos, H. J. (1999). Applications of Bayesian decision theory to sequential mastery testing. *Journal of Educational and Behavioral Statistics, 24*, 271–292.
Wagenmakers, E.-J. (2007). A practical solution to the pervasive problems of *p* values. *Psychonomic Bulletin & Review, 14*, 779–804.
Wagenmakers, E., Wetzels, R., Borsboom, D., & van der Maas, H. L. J. (2011). Why psychologists must change the way they analyze their data: The case of psi: Comment on Bem (2011). *Journal of Personality and Social Psychology, 100*, 426–432.
Wainer, H. (2000). Introduction and history. In H. Wainer, N. J. Dorans, R. Flaugher, B. F. Green, R. J. Mislevy, L. Steinberg & D. Thissen (Eds.), *Computerized adaptive testing: A primer* (2nd ed.). Mahwah, NJ: Routledge.
Wainer, H., Bradlow, E. T., & Wang, X. (2007). *Testlet response theory and its applications*. New York: Cambridge University Press.
Wainer, H., & Brown, L. M. (2007). Three statistical paradoxes in the interpretation of group differences: Illustrated with medical school admission and licensing data. In *Handbook of Statistics* (Vol. 26, pp. 893–918). Amsterdam, the Netherlands: Elsevier.
Wainer, H., Dorans, N. J., Eignor, D., Flaugher, R., Green, B. F., Mislevy, R. J., Steinberg, L., & Thissen, D. (2000). *Computerized adaptive testing: A primer* (2nd ed.). Mahwah, NJ: Routledge.
Wainer, H., & Thissen, D. (1994). On examinee choice in educational testing. *Review of Educational Research, 64*, 159–195.
Wall, M. M. (2009). Maximum likelihood and Bayesian estimation for nonlinear structural equation models. In R. E. Millsap & A. Maydeu-Olivares (Eds.), *The SAGE Handbook of Quantitative Methods in Psychology* (pp. 540–567). London: SAGE Publications.
Wang, C., Fan, Z., Chang, H.-H., & Douglas, J. A. (2013). A semiparametric model for jointly analyzing response times and accuracy in computerized testing. *Journal of Educational and Behavioral Statistics, 38*, 381–417.
Wang, W.-C., Liu, C.-W., & Wu, S.-L. (2013). The random-threshold generalized unfolding model and its application of computerized adaptive testing. *Applied Psychological Measurement, 37*, 179–200.
Wang, X., Bradlow, E. T., Wainer, H., & Muller, E. S. (2008). A Bayesian method for studying DIF: A cautionary tale filled with surprises and delights. *Journal of Educational and Behavioral Statistics, 33*, 363–384.

Warm, T. A. (1989). Weighted likelihood estimation of ability in item response theory. *Psychometrika, 54*, 427–450.

Warner, H. R., Toronto, A. F., Veasey, L. G., & Stephenson, R. (1961). A mathematical approach to medical diagnosis: application to congenital heart disease. *Journal of the American Medical Association, 177*, 177–183.

Warner, S. L. (1965). Randomized response: A survey technique for eliminating evasive answer bias. *Journal of the American Statistical Association, 60*, 63–69.

Weber, J. E. (1973). *Historical aspects of the Bayesian controversy: With comprehensive bibliography*. Tucson, AZ: Division of Economic and Business Research, University of Arizona.

Welch, R. E., & Frick, T. W. (1993). Computerized adaptive testing in instructional settings. *Educational Technology Research and Development, 41*, 47–62.

West, S. G., Taylor, A. B., & Wu, W. (2012). Model fit and model selection in structural equation modeling. In R. H. Hoyle (Ed.), *Handbook of structural equation modeling* (pp. 209–231). New York: Guilford Press.

Whitely, S. E. (1980). Multicomponent latent trait models for ability tests. *Psychometrika, 45*, 479–494.

Williamson, D. M., Almond, R. G., Mislevy, R. J., & Levy, R. (2006). An application of Bayesian networks in automated scoring of computerized simulation tasks. In D. M. Williamson, R. J. Mislevy, & I. I. Bejar (Eds.), *Automated scoring of complex tasks in computer-based testing* (pp. 201–257). Mahwah, NJ: Lawrence Erlbaum Associates.

Williamson, D. M., Bauer, M., Steinberg, L. S., Mislevy, R. J., Behrens, J. T., & DeMark, S. F. (2004). Design rationale for a complex performance assessment. *International Journal of Testing, 4*, 303–332.

Williamson, J. (2010). *In defence of objective Bayesianism*. Oxford: Oxford University Press.

Winkler, R. L. (1972). *An introduction to Bayesian inference and decision*. New York: Holt McDougal.

Wirth, R. J., & Edwards, M. C. (2007). Item factor analysis: Current approaches and future directions. *Psychological Methods, 12*, 58–79.

Wise, S. L., Plake, B. S., Johnson, P. L., & Roos, L. L. (1992). A comparison of self-adapted and computerized adaptive tests. *Journal of Educational Measurement, 29*, 329–339.

Wollack, J. A., Bolt, D. M., Cohen, A. S., & Lee, Y.-S. (2002). Recovery of item parameters in the nominal response model: A comparison of marginal maximum likelihood estimation and Markov chain Monte Carlo estimation. *Applied Psychological Measurement, 26*, 339–352.

Woodward, B. (2011, May 12). Death of Osama bin Laden: Phone call pointed U.S. to compound—and to "the pacer." Retrieved January 20, 2015, from http://www.washingtonpost.com/world/national-security/death-of-osama-bin-laden-phone-call-pointed-us-to-compound--and-to-the-pacer/2011/05/06/AFnSVaCG_story.html.

Wright, S. (1934). The method of path coefficients. *The Annals of Mathematical Statistics, 5*, 161–215.

Yanai, H., & Ichikawa, M. (2007). Factor analysis. In C. R. Rao & S. Sinharay (Eds.), *Handbook of statistics, Volume 26: Psychometrics* (pp. 257–296). Amsterdam, the Netherlands: North-Holland/Elsevier.

Yan, D., Almond, R., & Mislevy, R. (2004). *A comparison of two models for cognitive diagnosis* (No. RR-04-02). Princeton, NJ: ETS. Retrieved from http://onlinelibrary.wiley.com/doi/10.1002/j.2333-8504.2004.tb01929.x/abstract.

Yan, D., Mislevy, R. J., & Almond, R. G. (2003). *Design and analysis in a cognitive assessment* (No. RR-03-32). Princeton, NJ: ETS.

Yang, M., Dunson, D. B., & Baird, D. (2010). Semiparametric Bayes hierarchical models with mean and variance constraints. *Computational Statistics & Data Analysis, 54*, 2172–2186.

Yao, L., & Boughton, K. A. (2007). A multidimensional item response modeling approach for improving subscale proficiency estimation and classification. *Applied Psychological Measurement, 31*, 83–105.

Yao, L., & Schwarz, R. D. (2006). A multidimensional partial credit model with associated item and test statistics: An application to mixed-format tests. *Applied Psychological Measurement, 30*, 469–492.

Yen, W. M., Burket, G. R., & Sykes, R. C. (1991). Nonunique solutions to the likelihood equation for the three-parameter logistic model. *Psychometrika, 56*, 39–54.

Zellner, A. (1971). *An introduction to Bayesian inference in econometrics.* New York: Wiley.

Zhang, Z., Lai, K., Lu, Z., & Tong, X. (2013). Bayesian inference and application of robust growth curve models using Student's t distribution. *Structural Equation Modeling: A Multidisciplinary Journal, 20*, 47–78.

Zhu, H.-T., & Lee, S.-Y. (2001). A Bayesian analysis of finite mixtures in the LISREL model. *Psychometrika, 66*, 133–152.

Zhu, X., & Stone, C. A. (2011). Assessing fit of unidimensional graded response models using Bayesian methods. *Journal of Educational Measurement, 48*, 81–97.

Zhu, X., & Stone, C. A. (2012). Bayesian comparison of alternative graded response models for performance assessment applications. *Educational and Psychological Measurement, 72*, 774–799.

Zwick, R. (2006). Higher education admissions testing. In R. Brennan (Ed.), *Educational measurement* (4th ed., pp. 647–679). Westport, CT: Praeger.

Zwick, R., Thayer, D. T., & Lewis, C. (1999). An empirical Bayes approach to Mantel-Haenszel DIF analysis. *Journal of Educational Measurement, 36*, 1–28.

Zwick, R., Thayer, D. T., & Lewis, C. (2000). Using loss functions for DIF detection: An empirical Bayes approach. *Journal of Educational and Behavioral Statistics, 25*, 225–247.

Zwick, R., Ye, L., & Isham, S. (2012). Improving Mantel-Haenszel DIF estimation through Bayesian updating. *Journal of Educational and Behavioral Statistics, 37*, 601–629.

Index

Note: Page numbers followed by f and t refer to figures and tables, respectively

C

D

E

I

J

K

L

M

Printed in the United States
by Baker & Taylor Publisher Services